CONVERSION FACTORS
English to Metric

English Measure	*Multiply by*	*Metric Measure*
LENGTH		
Inches (in)	2.54	Centimeters (cm)
Feet (ft)	0.3048	Meters (m)
Yards (yd)	0.9144	Meters (m)
Miles (mi)	1.6093	Kilometers (km)
AREA		
Square inches (in^2)	6.452	Square centimeters (cm^2)
Square feet (ft^2)	0.0929	Square meters (m^2)
Square yards (yd^2)	0.836	Square meters (m^2)
Square miles (mi^2)	2.59	Square kilometers (km^2)
Acres	0.4	Hectares (ha)
VOLUME		
Cubic inches (in^3)	16.387	Cubic centimeters (cm^3)
Cubic feet (ft^3)	0.028	Cubic meters (m^3)
Cubic yards (yd^3)	0.7646	Cubic meters (m^3)
Fluid ounces (fl oz)	29.57	Milliliters (ml)
Pints (pt)	0.47	Liters (l)
Quarts (qt)	0.946	Liters (l)
Gallons (gal)	3.785	Liters (l)
MASS (weight)		
Ounces (oz)	28.35	Grams (g)
Pounds (lb)	0.4536	Kilograms (kg)
Tons (2000 lb)	907.18	Kilograms (kg)
Tons (2000 lb)	0.90718	Tonnes (t)

Alpine pool, dawn glow, Mosquito Range, Colorado
David Muench Photography

Editor-in-chief: **James M. Smith**
Executive Editor: **Jeffrey L. Hahn**
Assistant Editor: **Susan Gilday**
Project Manager: **Karen Edwards**
Production Editor: **Richard Barber**
Designer: **Elizabeth Fett**

Illustrations: **Foerster Illustration, Inc.**
JAK Graphics
Cartography: **Maryland Cartographics, Inc.**
Steven Simpson

Printed in the United States of America

Mosby–Year Book, Inc.
11830 Westline Industrial Drive
St. Louis, Missouri 63146

Library of Congress Cataloging in Publication Data

Bradshaw, Michael.
Physical geography : an introduction to earth environments / Michael Bradshaw, Ruth Weaver,—1st ed.
p. cm.
Includes bibliographical references and index.
ISBN 0-8016-0298-X
1. Physical geography. I. Title.
GB54.5.B73 1992
910'.02—dc20 92-33499
CIP

93 94 95 96 97 GW/CD/VH 9 8 7 6 5 4 3 2 1

Preface

This text is designed for introductory physical geography courses: it aims to provide an understanding of Earth's physical environments. It is innovative while maintaining many features that will be familiar. Its particular contributions can be summarized under three heads—themes, organization, and presentation.

THEMES

Five main themes are developed throughout the text, although some chapters place more emphasis on specific themes than others.

Earth Environments. The concept of "environment" is the basic framework through which the study of the physical nature of our planet is approached. An environment is the sum total of conditions in which an organism lives. The varied conditions of, for example, weather, soils, and other organisms in the environment affect the way in which an organism acts. In physical geography, human beings are the central organisms within Earth physical environments. Moreover, although the term "environment" has been misused, it is irrevocably linked to environmental issues that provide a motivation for students' interest in physical geography. An environment-based approach also provides an introduction to the fuller use of a systems approach in more advanced courses.

Oneness of Earth's Environments. In any environmental study, the collective influence of conditions is a vital consideration. Earth's surface physical environment—the home of human beings—must be studied as a whole by physical geographers because of the many interactions among its separate parts. This oneness is emphasized throughout the book by the grouping of specific environmental conditions into four major environments and through a series of highlighted linkages that emphasize the interactions among all four environments. Comparisons are also made between Earth and the inner planets of the solar system.

The four major environments are based on the basic sources of energy and materials that determine the ways in which each environment functions.

The *atmosphere-ocean environment* is powered by energy from the sun. Movements within this environment produce short-term weather and longer-term climatic conditions. The interactions between ocean waters and atmospheric gases are seen as increasingly important in influencing the characteristics of this major environment.

The *solid-earth environment* derives its energy from within the planet, driving internal processes that cause continents to move and mountains to rise.

The *surface-relief environment* is a narrow zone on the continental surfaces where the first two major environments interact. The detailed molding of the mountains and continents produced by solid-earth environment processes is carried out by running water, moving ice, and other agents linked to processes in the atmosphere-ocean environment.

The *living-organism environment* also gains its energy supply from the sun, but interacts with the other three major environments in cycling chemical elements. This provides the basic conceptual framework for its study—the ecosystem.

Changing Earth Environments. Change over time is a feature of all Earth environments. Weather is subject to daily changes; climate changes over longer periods; earthquakes and volcanoes signal change within Earth; the continents move around on Earth's surface; mountains rise and are torn down; rivers, glaciers, and wind etch patterns on the surface; plants and animals migrate and ecosystems develop. Human beings must live in the knowledge that their environment will change in the future.

The Impacts of Human Interventions. Human activities affect the operation of environmental processes, and the results often rebound on humans. Human intervention is an increasing concern of physical geography. The human factor is an important consideration in the workings of each of the four major environments. The role of human impacts in climatic change, landform changes, and ecosystem changes are particularly emphasized in this text. The nature of the physical environment as resource and hazard—two concepts based on the human dimension—are considered.

The Physical Environment is Studied by Scientific Methodology. Observation, generalization, application, and prediction, by individuals and project groups, are the basis of scientific investigation. Physical geography is an investigative science, and much knowledge is still required so that better predictions and decisions can be made.

ORGANIZATION

Five Parts. The text is organized in five main parts, of which Part 1 provides an introduction to Physical Geography, its ideas and subject matter (Chapter 1), and its methodology and tools (Chapter 2). The remaining four parts each cover one of the major environments. Each part consists of chapters that emphasize the structure, chapters that focus on processes, and a final chapter that unites these conditions.

Part 2, The Atmosphere-Ocean Environment, Chapters 3 to 9, focuses on the study of meteorology and climatology. Chapter 3 is about the materials and structure of the atmosphere and oceans. Chapters 4 to 6 cover the weather that arises from the processes at work—heating, winds and currents, and the circulation of water. Chapter 7 combines these in a study of weather systems. Chapter 8 is a further synthesizing chapter related to the long-term atmosphere-ocean conditions, and leads in to Chapter 9 on climatic change.

Part 3, The Solid-Earth Environment, Chapters 10 to 12, is a study of the ways in which internal earth processes produce the continents, ocean basins, and mountain ranges. Chapter 10 is about the materials and structure of the layered solid earth. Chapter 11 describes the processes at work in forming major relief features through plate tectonics. Chapter 12 combines the study of materials and processes in a consideration of the origins of mountain systems and continents.

Part 4, The Surface-Relief Environment, Chapters 13 to 18, completes the study of landforms by considering those that are formed by atmospheric and marine processes acting on the continental surfaces. Each chapter is based on a set of processes: Chapter 13 on weathering and mass movement; Chapter 14 on river action; Chapter 15 on ice action; Chapter 16 on wind action; and Chapter 17 on action by the sea along coastlines. Within each chapter the theme of environment is developed in relation to landforms produced, rock and climate factors are examined, and the role of human activities is assessed. Chapter 18 combines the work of the different agents in a study of both detailed and also larger landform origins.

Part 5, The Living-Organism Environment, Chapters 19 to 21, concludes the study of physical geography by a consideration of ecosystems and soils. Chapter 19 considers the nature of materials, structures, and processes in this environment. Chapter 20 is about the development and use of soils within an ecosystem framework. Chapter 21 combines the previous chapters and parts in a study of world-scale ecosystems, the biomes.

Chapter Contents. Each chapter provides a rich variety of teaching and learning resources. The main text is organized around a clear and logical development of the material.

Within each chapter there are two to three *boxes* in which themes of human interaction and methodology are developed, including environmental issues, natural hazards and resources, the roles of particular scientists, and some aspects of studying physical geographic phenomena.

In some of the chapters there are studies reporting the role of current *investigations* in the sciences that contribute knowledge to physical geography. These emphasize the dynamic nature of the subject.

At the end of each chapter there are several resources:

- A section headed "*Frontiers,*" which lists and discusses some of the developments and issues currently under study in this section of physical geography.
- A *summary* of the main themes of the chapter.
- A list of *key terms* introduced, which are also defined in the Glossary.
- *Questions for review and exploration.*
- A list of *further reading* that provides examples of recent studies in the area and details of one or two other texts that can be used to obtain further information. Virtually all these references are from the mid 1980s to early 1990s.

At the end of the main text, there is a reference section that includes a *Glossary of Key Terms, Weather and Topographic Map Symbol Keys,* and a summary of the *Canadian Soil Classification.*

PRESENTATION

The *diagrams* are designed to make straightforward points, and the color and three-dimensional representations are used to help students understand the concepts being studied.

The *images*—photographs and satellite images—have been selected for their high quality. A special feature of the text is the inclusion of results from the latest work at NASA, such as global images providing information about clouds, temperatures, and ocean currents. The photos taken by Space Shuttle astronauts are another feature of this text. Other federal agencies, such as USGS, NOAA, and USDA have provided photographs for the book.

The chapters are also illustrated by relevant *excerpts* from a variety of literature sources, including novels, travel accounts, and early scientific writings.

Additional Instructor Materials. Additional materials are available for instructors that will enrich the teaching and learning that arises from use of this book. They include:

Instructor's Resource Manual with Test Bank
Transparency Acetates
Computest III Computerized Test Bank
Student Study Guide
"Our Environment" videodisc by Optilearn, Inc.
Video Tapes
Slide Set

ACKNOWLEDGMENTS

This new text has been reviewed extensively at the first and second draft stages. Each chapter of the second draft was subjected to expert reviewers. The text and its content have gained much from these processes. The following are thanked for their contributions to these reviews.

Alabama

Lary M. Dilsaver
University of S. Alabama

David Icenogle
Auburn University

Arizona

Anthony Brazel
Arizona State University

Randall S. Cerveny
Arizona State University

Mel Marcus
Arizona State University

California

Ned Greenwood
San Diego State University

Guy King
Cal State—Chico

Thomas S. Krabacher
Cal State—Sacramento

Julie Laity
Cal State—Northridge

Joel Michaelson
UC—Santa Barbara

Amalie Jo Orme
Cal State—Northridge

James R. Powers
Pasadena City College

John Wolcott
USC

Colorado

Susan Beatty
University of Colorado

Glen D. Weaver
Colorado State University

Florida

Joann Mossa
University of Florida

Idaho

Eric C. Ewert
University of Idaho

Illinois

Scott Isard
University of Illinois

Bruce Rhoads
University of Illinois

Colin Thorn
University of Illinois

Indiana

Katherine Price
DePauw University

Kevin M. Turcotte
Ball State University

Iowa

David W. May
University of Northern Iowa

Kansas

Karen M. Trifonoff
University of Kansas

Kentucky

Conrad Moore
Western Kentucky University

Maryland

Brent R. Skeeter
Salisbury State University

Massachusetts

Vernon Domingo
Bridgewater State University

Laurence Lewis
Clark University

William D. McCoy
University of Massachusetts

Michigan

Jay Harman
Michigan State University

Randy Schaetzl
Michigan State University

Julie Winkler
Michigan State University

Minnesota

Charles G. Parson
Bemidji State University

Richard Scaggs
University of Minnesota—TC

Rod Squires
Univ. of Minnesota—
Twin Cities

Graham Tobin
University of Minnesota—
Duluth

Nevada

Christopher Exline
University of NV—Reno

Anne Wyman
UNLV

New Jersey

David A. Robinson
Rutgers University—
New Brunswick

New York

David de Laubenfels
Syracuse University

Ohio

Jerry Green
Miami University

Kenneth Hinkel
University of Cincinnati

Jeffrey C. Rodgers
The Ohio State University

Oregon

Patricia F. McDowell
University of Oregon

Pennsylvania

Richard Crooker
Kutztown University

Tennessee

Carol P. Harden
University of Tennessee

Texas

Jeffrey A. Lee
Texas Tech. University

Vatche P. Tchakerian
Texas A and M

Vermont

Harold Meeks
University of Vermont

West Virginia

Robert Hanham
West Virginia University

Wisconsin

Barbra Borowiecki
University of Wisconsin—
Milwaukee

Waltraud Brinkman
University of Wisconsin—
Madison

Canada

O.W. Archibold
University of Saskatchewan

A.M. Davis
University of Toronto

Peter Herrem
University of Calgary

Thomas Merideth
McGill University

Kenneth Tinkler
Brock University

In the early stages of the project, a questionnaire was mailed to all physical geography instructors in American and Canadian colleges. This also provided a basis for decisions about the book, and all who responded are thanked.

The text also owes much to others whom we would like to thank. A number have made special contributions to the illustrations. Gene Feldman of NASA read the first draft and put us in touch with several of his colleagues. Michael Helfert, also of NASA, has been very helpful in suggesting suitable photos from the Space Shuttle missions. Susan Russell-Robinson of the U.S. Geological Survey Information Office, Tim McCabe of the Soil Conservation Service, and Linda Kremkau and Carla Wallace of NOAA also helped greatly in finding appropriate photographs from their extensive collections.

MICHAEL BRADSHAW
College of St. Mark and St. John
Plymouth, England
RUTH WEAVER
University of Plymouth, England

Brief Contents

Contents

Part 1 Foundations of Physical Geography

Part 2 The Atmosphere-Ocean Environment

Part 4 The Surface-Relief Environment

Appendices

FOUNDATIONS OF PHYSICAL GEOGRAPHY

In a very short time, human activity has become so varied and complex that it is having effects not only at local and national levels, but on the whole world itself. Having discovered only the other day that our world is round, we are suddenly finding it uncomfortably small and fragile.

Mankind has always been capable of great good and great evil. This is certainly true of our role as custodians of our planet. We have a moral duty to look after our planet and to hand it on in good order to future generations.

This Common Inheritance, 1990

Over the time scale of a human lifetime, some changes are manifest even to individuals throughout the world. Memories of a succession of particularly hard winters or of hot, dry summers provide an awareness of climate changes on a scale of tens of years. Drought on this time scale can bring tragedy to marginal life zones such as the African Sahel. Substantial alterations in ecosystems, such as the loss of much of the primary Eastern woodlands and Midwestern grasslands of North America, and the disappearance and reappearance of important fish stocks, are also in the public memory.

Over longer time scales, historical records reveal century-long excursions of climate (e.g., the Little Ice Age of the 15th through the 18th centuries in Europe, and the drying of the climate in the North American Southwest in an earlier period).

U.S. Global Change Research Program, 1991

In June 1990 Mount Pinatubo, the Philippines, shot millions of tons of pulverized rock and gases straight up into the air. Much of the debris fell back to Earth as snowy ash that buried nearby towns and Clark Air Force Base. But some 20 to 30 million tons of particles kept on racing through the troposphere—the layer of air that begins at Earth's surface, extends 10 miles up, and contains nearly all the clouds and winds that we know as weather.

The immense gassy plume lodged in the next higher layer, the stratosphere. By the end of September the stratosphere's winds, which began blowing north-south in late summer, had scattered the volcano's plume across the entire planet.

Discover, January 1992

CHAPTER 1

The Study of Earth Environments

When the 20th Century began, the world was home to about 1.6 billion of the human species. Although pollution and environmental degradation were common—some cities lived under a pall of smoke, soot, and ash—the problems were local. The world as a whole seemed vast, with huge regions virtually untouched by its human inhabitants.

By midcentury, airplanes and radio broadcasts had begun to shrink distance and bring the peoples of the world, now 2.5 billion, into greater contact. Industrial growth had multiplied per capita consumption of natural resources—and per capita pollution—in many countries.

As the 1990s begin, world population has more than doubled since 1950, to 5.2 billion people, and world economic activity has nearly quadrupled. To local concerns about environmental degradation have been added new, global worries.

World Resources 1990-91.

A large-scale program to restore damaged and polluted wetlands, rivers, streams, and lakes throughout the United States should be put into action promptly to prevent permanent ecological damage, concluded a National Research Council committee.

Wetlands—such as bogs, marshes, mud flats, riverbanks, and coastal areas—serve an important role by helping to purify polluted waters, absorbing floodwaters, and providing habitat for a diverse array of fishes, birds, and other wildlife. Because the diversity of America's wetlands makes them difficult to describe with a single definition, scientific and legal views of what constitutes a wetland often are conflicting and controversial.

While successful projects are underway in some areas, "the practice of wetland restoration needs to move from a trial-and-error process to a predictive science," said the committee.

National Research Council news release, December 1991

High in the Trans-Antarctic Mountains, with the sun glaring at midnight through a hole in the sky, I stood in the unbelievable coldness and talked with a scientist in the late fall of 1988 about the tunnel he was digging through time. Slipping his parka back to reveal a badly burned face that was cracked and peeling, he pointed to the annual layers of ice in a core sample dug from the glacier on which we were standing. He moved his finger back in time to the ice two decades ago. "Here's where the U.S. Congress passed the Clean Air Act," he said. At the bottom of the world, two continents away from Washington, D.C., even a small reduction in one country's emissions had changed the amount of pollution found in the remotest and least accessible place on earth.

Sen. Al Gore, Earth in the Balance

To understand global change and the increasing demands of human activity, it is essential that we document and comprehend how the Earth works as a system. The international scientific community is organizing research efforts to advance our knowledge of both natural and human-induced global change.

The Earth Observing System (EOS) of satellites will integrate the measurements now being taken by short-term research missions. It will provide the first coordinated simultaneous measurements of the interactions of the atmosphere, oceans, solid earth, and hydrologic and biogeochemical cycles.

EOS: a Mission to Planet Earth, NASA, 1990

The excerpts from government documents and other publications on the opening pages reflect the growing importance of environmental issues. From a position of marginal concern, these issues have moved to take a central place in government policy and scientific research. Several significant points emerge.

First, many environmental issues are related to events and processes at the global scale. Subjects such as global warming, the ozone hole, and acid rain transcend the local, regional, and national scales and require international attention and agreement. Other issues, such as the cleanup of toxic wastes at disposal sites, have impacts on a smaller scale. When discussing aspects of the physical environment that influence humanity, establishing the appropriate *geographic scale* is important.

Second, *environmental issues are complex.* Many processes affecting the inhabited world involve interactions among atmosphere, oceans, land, and living organisms. It is too simple, for instance, to say that if the carbon dioxide in the atmosphere increases by so much, the global temperatures will increase by so much and the sea level will rise by so much. In nature, unexpected factors complicate relationships among the different parts of the global environment. The fact that the understanding of how the environment works is still incomplete and is likely to remain so also makes simple conclusions suspect.

Third, scientists realize today as never before that the natural environment is not static and unchanging, but is subject to shifts. Climatic change and changes caused by internal Earth forces produce a *dynamic "stage"* on which human activities take place. The time scales involved in environmental change range from a few seconds in an earthquake to hundreds of millions of years in the formation of mountains. People are most conscious of short-term changes in their surroundings occurring within their lifetimes, but these changes are often part of longer-term shifts and cycles.

Fourth, *human activities* are affected by the natural environment, but increasingly humans are changing the environment and producing new environments. In some places human activities enhance the natural environment, but in many, pollution and the alteration of natural processes degrade Earth's ability to renew itself and sustain life.

Fifth, humans have a *moral responsibility* for the stewardship of their environment and are obligated to manage it to ensure a long-term future for the species. Although it is seldom mentioned in a textbook, this stewardship is just as much a part of being human as is the constant effort to thrive and prosper during one's own lifetime.

Sixth, although humans have learned much about the natural environment by scientific study, our knowledge is still deficient. Scientists and policymakers are realizing the importance of regular *environmental monitoring.* Measuring variables such as temperature and rainfall to provide a frequent and uniform world coverage, for example, helps researchers to follow the trends in climatic change and in the impacts of human intervention. The realization that humans have much to learn about their planet helps to generate technology that will deliver more complete information.

Seventh, it is not enough to learn more about how the natural environment functions. It is necessary to *apply the findings of scientific research.* This involves choosing among the research programs that compete for budgetary resources, and a willingness to make hard decisions on the basis of the partial understanding that is available. Government and industry leaders must stay informed about the findings of scientists if governments, businesses, and international bodies are to acquire the political will to take the necessary—and often costly—steps to maintain and improve the quality of Earth's environment. Voters who elect politicians and consumers who buy goods and services should also understand aspects of the physical environment that affect them on global, national, regional, and local scales.

Figure 1-1 summarizes these points. It places environmental issues within a web of factors including geographic scale, complexity of interacting factors, changes over time, and human intervention. These factors are linked to human responsibility in terms of capability to recognize the challenge of declining environmental quality and of government and corporate ability to organize responses to that challenge.

PHYSICAL GEOGRAPHY AND ENVIRONMENTAL ISSUES

Physical geography is the study of Earth's natural environment, with a particular emphasis on *spatial* characteristics, which pertain to location and extent of coverage, and *temporal* characteristics, which involve change over time. The study of spatial characteristics includes the consideration of *where* a phenomenon exists, *how much* of Earth's surface it covers, and the *causes* and *significance* of such a distribution. For instance, the black and dark brown soils found over large areas of the Midwest are

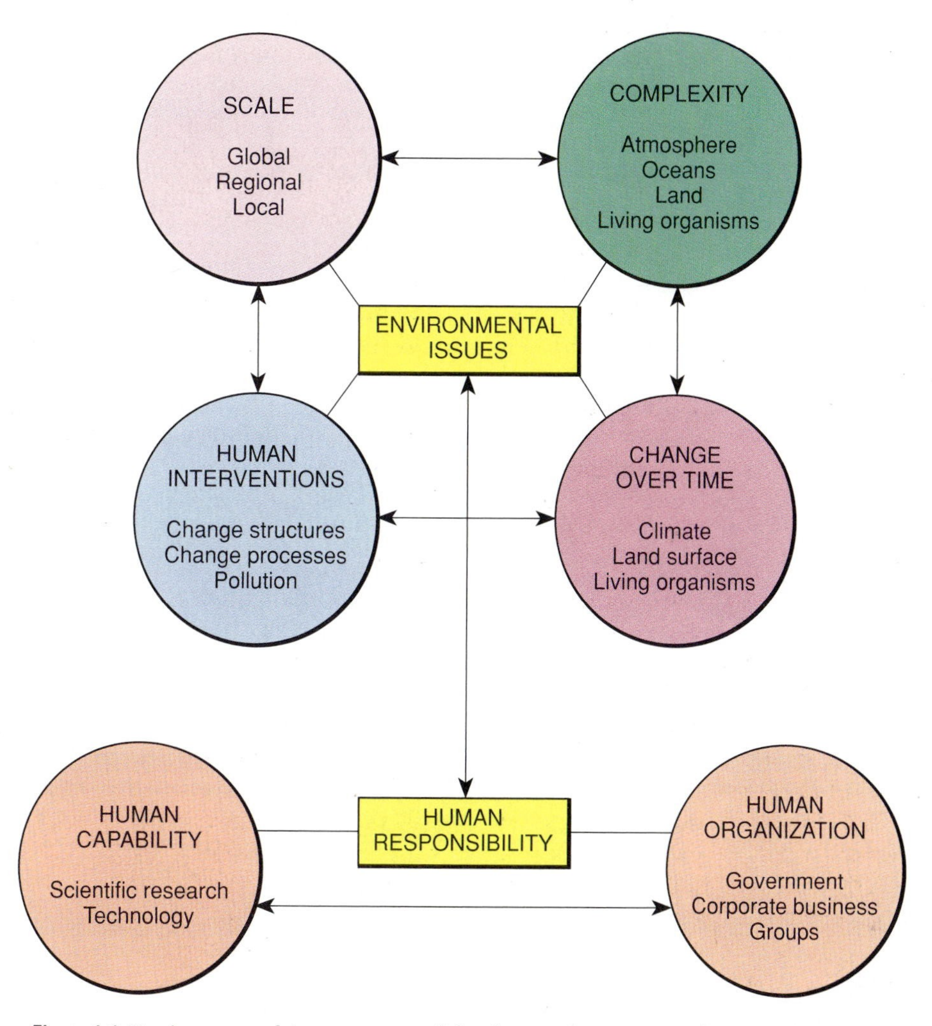

Figure 1-1 Environmental issues: some of the factors that are significant in their study.

among the most fertile in the world; similar varieties occur in Ukraine and Argentina. In each case they are linked to midlatitude climates with hot, wet summers and cold winters, and to natural grassland vegetation that was for a long time somewhat modified by human burning. These areas have become some of the most productive farmlands of the world.

The study of temporal characteristics includes the frequency of particular natural events and the rate at which changes take place—whether an area is subject to more or less change within a given time. Tornado occurrence provides a good example of frequency, since they happen often in the interior plains of the United States but hardly at all west of the Rockies. Differences in rates of change can be illustrated by comparing the northeast and southeast United States. The former was partly covered by an ice sheet 15,000 years ago, and its melting over the next 10,000 years led to shifts from glaciers to rivers, from bare earth and tundra to forest cover, and to rising sea levels. These changes affected the southeast, but in less dramatic ways, since rivers continued to flow and most of the region continued to be forested.

Physical geographers study the variety, distribution, duration, and significance of weather and climatic characteristics, ocean circulation, relief forms (hills, mountains, valleys), soils, vegetation, and animals. Such studies are not merely descriptive but also investigate processes, origins, and changes; they attempt to explain the patterns that are discovered and the role of human activities.

Physical geographers contribute to the wider study of *geography*, which emphasizes the spatial aspects of how humans use Earth's surface. They are inevitably drawn into the study of interactions among different parts of the natural environment, and of interactions of people with the environment.

BOX 1-1

FOUR-MILE RUN WATERSHED

Human activities have impacts on natural processes that may produce problems. An understanding of the potential impacts can lead to better management programs that overcome the problems. Proper planning and assessment of environmental impacts are aided by a physical geographer's knowledge.

In the case of Four-Mile Run, an understanding of local drainage patterns was needed to repair the effects of suburban expansion on streams that formerly flowed through farmland. Four-Mile Run is a stream in which such impacts led to better management. Four-Mile Run and its tributaries drain part of the northern Virginia suburbs of Washington, D.C., one of the most heavily urbanized sectors of the D.C. metropolitan area.

The streams that flow into Four-Mile Run drain an area of about 21 square km (10 square miles), which was rapidly developed from fields and woodland after 1945. By 1970, 140,000 people lived there, 85 percent of the area had been developed, and 35 percent was covered by roads and buildings. Underground storm sewers had replaced much of the former network of streams in the higher part of the drained area. The recently constructed buildings, concrete, and asphalt prevented rainwater from getting into the soil, while the storm sewers hurried it off the surface. Rapid stream runoff after storms increased. The consequent flooding in the lower part of the valley destroyed homes and delayed air traffic since the waters overflowed the edge of Washington National Airport.

In March 1974, the U.S. Congress authorized a $49 million U.S. Army Corps of Engineers flood control project to enlarge the stream channel just above the outlet into the Potomac River. This is where the tributaries come together and concentrate the stream flow and where the storm sewers empty into the river. The combination of converging streams and sewer pipes produced the flooding. The project was designed to prevent flooding, but would be useless if building development continued in the area drained by Four-Mile Run and brought increased storm runoff to the outlet. Accordingly, the authorization required four local jurisdictions (Fairfax and Arlington counties, and the cities of Alexandria and Falls Church) to develop a plan for managing the runoff of water within the basin.

Enough is known about water movement between the time when rain reaches the ground and runoff reaches the stream that scientists can predict the effects of different land use types on the flow in a particular area. This knowledge was used in a computer program developed to assess the impact of new development proposals brought to the four councils. If a proposed development would increase storm runoff and so overrun the capacity of the Corps of Engineers' new channel, the developers were expected to construct runoff control facilities, such as small dams, or they might be required to reduce the size of the development so that less of the absorbent soil was paved over. A more expensive alternative would be to build multilevel parking instead of a larger asphalt-covered area at ground level. The total impact of all new developments in the basin is monitored to ensure that flood capacity is not exceeded.

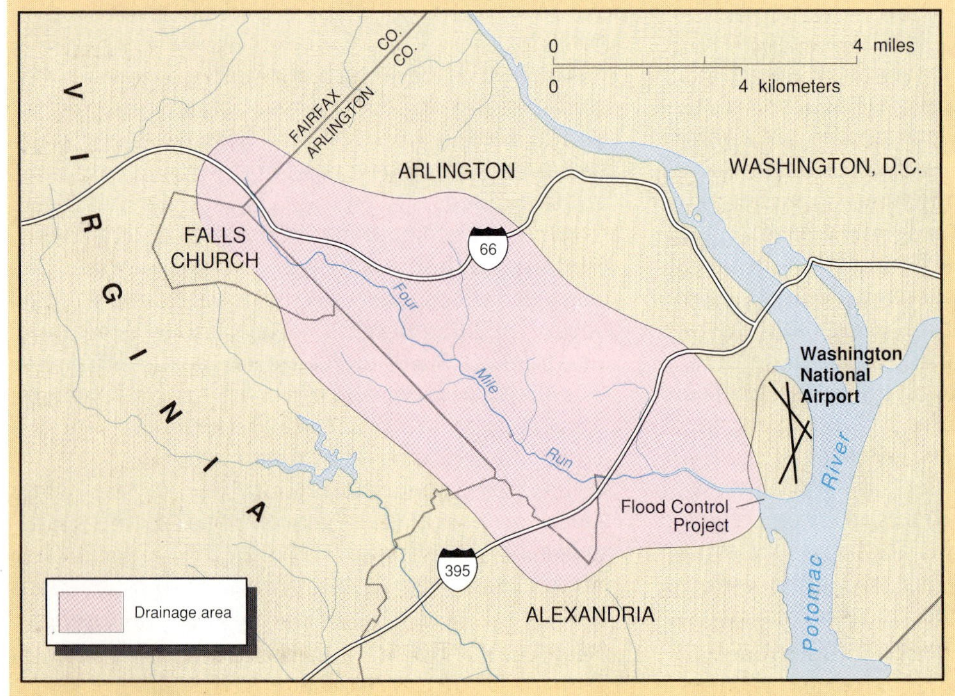

Map of area. The Four-mile Run watershed in the southern suburbs of Washington, D.C.

Such studies may have significance for environmental concerns or processes of economic development. Although their work is often based on the results of other scientists, such as chemists, physicists, biologists, geologists, soil scientists, ecologists, meteorologists, and economists, they may bring together findings from several other sciences or focus more closely on the global distributions of a phenomenon than other scientists. Physical geographers also carry out investigations that delve deeper into specific aspects of natural environments or emphasize the interactions between humans and natural environments.

Physical geographers are thus particularly concerned with the use of environmental data that have a spatial context—that is, information about a phenomenon that varies from place to place. Temperature, wind, rain, land height and shape, soil, and plants are all examples of phenomena that vary from place to place. This information can be mapped and is often important in planning and environmental impact discussions. Data that can be mapped can also be compiled in computerized Geographic Information Systems (see Chapter 2). With such knowledge and information resources, physical geographers provide a link between the workings of the physical environment and the human geographer's understanding of economics and culture. For instance, soil erosion may be explained in terms of the flow of rainwater or wind over bare, loose soil (physical factors), but it is often linked to human practices such as the plowing of unsuitable land.

Physical geographers have an important part to play in late twentieth-century environmental concerns. They are trained to understand a variety of environmental phenomena and to discern links between them at particular geographic (or "spatial") scales. They recognize the significance of the natural changes in the environment and the ways in which human actions can transform the environment (see Box 1-1: Four-Mile Run). Hence, many physical geographers have experience as consultants to governments and corporations in matters regarding environmental policy. Environmental impact statements assessing the influence of large developments (e.g., ports, pipelines, power stations) on the local environment have received valuable input from physical geographers.

As this book will show, physical geography belongs at the center of the current public concern and political focus on the environment. The rest of this chapter demonstrates the oneness—and possible uniqueness—of the complex natural environments that interact with human activities on Earth's surface to provide humans with homes, food, energy, and other needs. It also introduces the approach of the rest of the book.

THE ONENESS OF EARTH'S ENVIRONMENTS

The concept of "environment" provides the main organizing framework of this book on physical geography. The **environment** of a place, object, person, or event is the set of surrounding conditions that act on it and give it a particular character. A body of air near the ground is affected by such conditions as heating from the sun, winds, and the nature of the ground surface; the flow of a river reflects the supply of water from rain, the steepness of the land, and the plant cover in the area being drained; an animal's activities are influenced by seasonal temperature changes, the presence of predators and competitors, and the availability of food and water. People relate to natural environments (the subject of this book), and to the built environments of the towns and cities in which they live.

Environments occur at a variety of scales: the **global scale** includes the whole Earth; the **regional scale** involves large divisons of Earth's surface covering several thousands of square kilometers; and the **local scale** refers to smaller divisions of Earth's surface from a few tens to a few hundreds of square kilometers—the area individuals commonly experience in their daily movements.

Major Earth Environments

Four major natural environments are recognized at the global scale (Figure 1-2): the atmosphere-ocean environment, the solid-earth environment, the surface-relief environment, and the living-organism environment. Because physical geography is concerned with all aspects of Earth's natural environment, it provides a basis for studying how these major environments interact with each other.

Each major environment has its special conditions. The atmosphere and oceans—all considered together as one world ocean—compose the **atmosphere-ocean environment** (Figure 1-3). They are combined because of their important exchanges of heat, gases, water, and small particles of solid matter. The **solid-earth environment** consists of the rocky part of the planet (Figure 1-4). It is composed of very different materials than the atmosphere and ocean, and it has its own internal source of energy. The first two major environments interact with each other in the **surface-relief environment,** the narrow zone where landforms such as

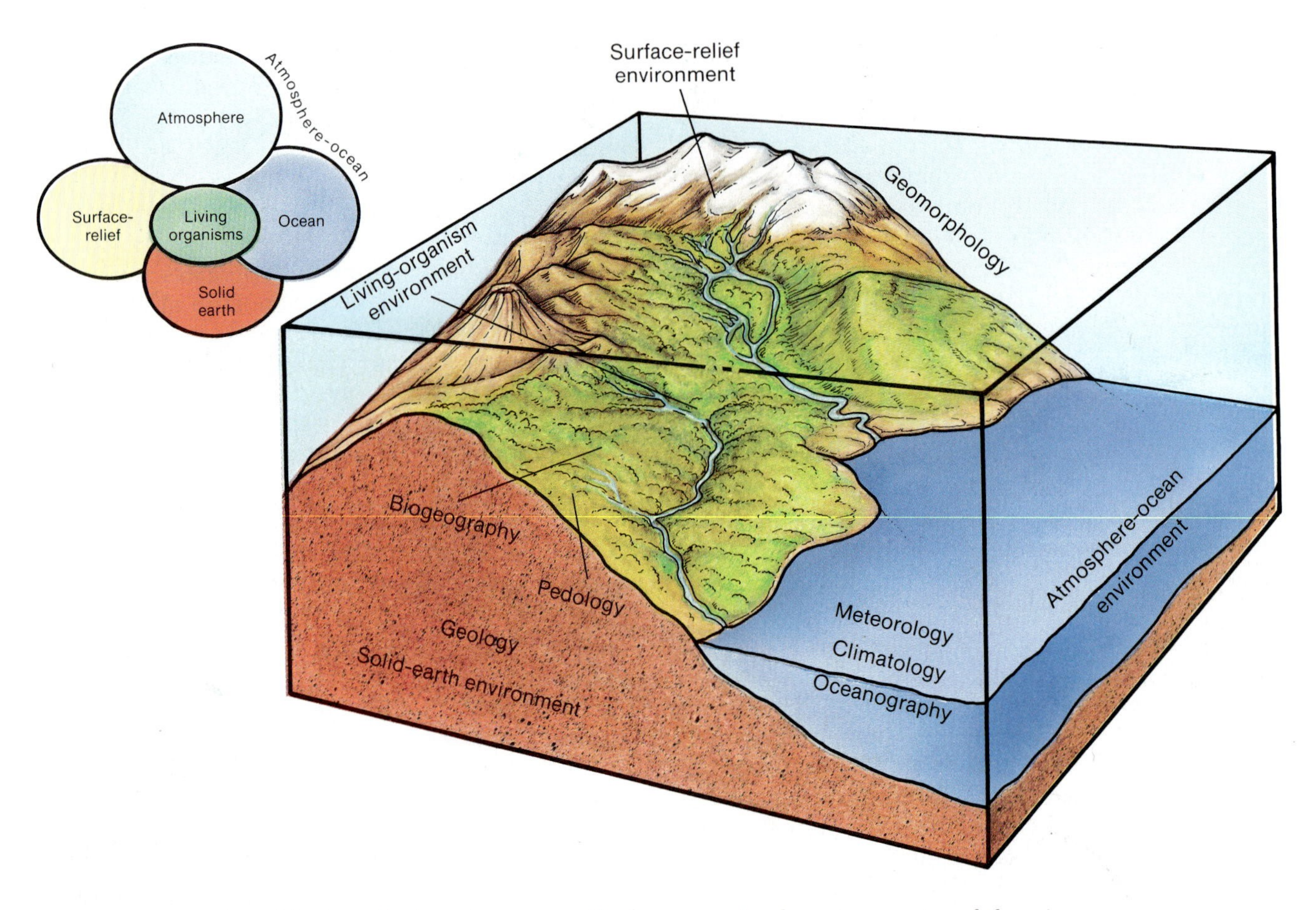

Figure 1-2 Major Earth environments. The four major Earth environments and the sciences that study them. (The circles diagram emphasizes the linkages.)

Figure 1-3 The atmosphere-ocean environment. The ocean surface and clouds in the atmosphere. The water that forms the clouds comes from evaporation of the ocean water.

FRONTIERS IN KNOWLEDGE
Physical Geography

Physical geography, and the sciences on which it is based, have not achieved a complete understanding of Earth. There is much that scientists do not fully comprehend, and in some areas understanding is only beginning. New discoveries are made and new theories proposed all the time, and there is ample work for generations of physical geographers to come. At the end of each chapter in this book is a short section that highlights the particular needs for further understanding of the chapter's subject as perceived by scientists at the time of writing. These sections also emphasize the fact that physical geographers need to be aware of developments in related sciences.

Figure 1-4 The solid-earth environment. The eruption of lava—molten rock—on Hawaii demonstrates the heat of Earth's interior.

hills and valleys are produced (Figure 1-5). Portions of all the other three major environments interact to form an arena for the existence of living organisms—the **living-organism environment** (Figure 1-6). Plants and animals depend on energy from the sun, the circulation of water, the materials present in the rocks and soils, and the form of the land.

Each environment consists of a structure of various parts and processes that link the parts and are responsible for changes. The **structure** of each major environment is its constituent elements and the forms each takes—the chemical composition and physical properties of the atmosphere, oceans, continents, and ecosystems; the minerals, rocks, landforms, plants, and animals that are parts of the larger environments; and their shapes and geographic distributions. The **processes** that produce these forms and link them together in environments include energy pathways and the circulation of materials. Energy from the sun, or from Earth's interior, passes through the environments where it produces physical and chemical transformations. The sun's rays, for example, are absorbed by plant leaves, and the energy they contain is transformed into chemical energy that provides food for animals that eat the plant. The continuous circulation of water, rock materials, and plant nutrients results in environmental change and development. The processes acting in each environment ensure that it does not stay the same for very long.

The four major earth environments can be used as vehicles for studying Earth as a whole, or at the regional or local levels. At the global scale, processes in the atmosphere-ocean environment are fueled by a daily replenishment of energy from the sun.

Figure 1-5 The surface-relief environment. The Alps near Grindelwald, Switzerland. Glaciers have carved the high-level features, water is at work on the lower valleys.

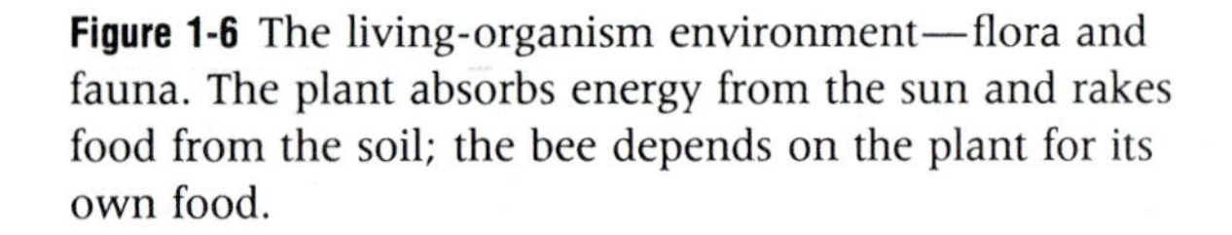

Figure 1-6 The living-organism environment—flora and fauna. The plant absorbs energy from the sun and rakes food from the soil; the bee depends on the plant for its own food.

All the energy in the solid-earth environment arises from within Earth. The other major environments result from interactions between these two, since solar and internal Earth sources of energy produce a variety of outcomes. A landform such as a mountain, for example, often results from a combination of uplift as a result of internal movements and surface sculpting by rain and wind.

The substances forming the major environments are recycled repeatedly to provide continuing sources of materials. For instance, volcanic eruptions may release particles in the air that are brought to the ground by rain and then washed into the soil. After chemical changes, the particles erupted by the volcano become plant foods and are taken up into a plant. If the plant is eaten by an animal, the substance is returned to the soil when the animal dies or is taken by rivers to the sea and incorporated in a deposit on the sea floor. This deposit becomes a rock, is lifted above sea level by internal earth movements, and the cycle begins again.

Within each major earth environment there are sets of distinctive processes that bring about the subdivisions listed in Table 1-1. The atmosphere-ocean environment is characterized by a set of *climatic environments,* each of which has its own distinctive patterns of temperature and rainfall. The solid-earth environment has the *surface environments* of ocean basin and continent and the *interior environments* of mantle and core. The surface-relief environment has groups of *landforms,* such as valleys and plains, which are determined by interactions between climatic and geologic conditions. The living-organism environment is organized in ecosystems at various scales from the local pond to global-scale assemblages of plants and animals.

The concept of the four environments also operates at the regional and local levels. For instance, some weather phenomena, such as wind circulation, affect the whole globe; some, such as hurricanes, affect broad regions; and some, such as thunderstorms, have a local influence. The processes in the solid-earth environment that cause continents to move around Earth's surface are global in nature, but volcanic eruptions have a regional effect, and earthquakes can be local events.

Table 1-1 Major Earth Environments: Subdivisions and Change Over Time

MAJOR ENVIRONMENTS	ATMOSPHERE-OCEAN	SOLID-EARTH	SURFACE-RELIEF	LIVING-ORGANISM
Processes	Solar heating driving winds and water cycle, plus ocean water movements	Internal heat source driving movements of continents and forming mountain systems	Interactions of atmosphere, ocean, and solid rock: uplift, breakdown by weathering, water movement, ice, wind, and waves	Feeding, growth, reproduction, movement, dispersal; based on supply of solar energy, water, and nutrients
Major Environment Subdivisions	Climate regions: polar; midlatitude—*west coast, interior, east coast;* tropical—*arid, seasonal, trade wind, equatorial*	Surface: *ocean basins, continents* Interior: *crust, mantle, core*	Climate-landform regions: *glacial, periglacial, humid midlatitude, arid, seasonal tropical, equatorial*	Global ecosystems: forest—*tropical, midlatitude;* grassland—*tropical, midlatitude;* tundra; desert; oceans
Human Intervention	Increasing	Slight	Major influence in some subdivisions	Major influence in all ecosystems
Type of Change Over Time	Climatic change in cycles of varied length	Volcanoes, earthquakes; plate tectonics; supercontinent cycle	Interactions of climatic change and solid-earth movements; human actions where high densities of population	Evolution related to genetic and environmental factors; increasing role of human actions

Changing Earth Environments

Physical geographers focus on the fact that Earth's environments are subject to change over time. Until the middle of the twentieth century, however, their descriptions of how landscapes evolved were often incomplete, because they only considered processes and structures occurring under a single set of climatic conditions. For instance, landforms produced by running water were described as if streams had always been the only processes operating. With the new emphasis on interacting environments, the complexity of the forces leading to changes over time must be considered.

Scientists and the general public are particularly conscious of climatic change at present. Studies have shown that such change is not merely a matter of fluctuating amounts of energy coming from the sun. It also involves the way that the atmosphere-ocean environment receives the energy and interacts with the solid-earth and living-organism environments. Climate change is an ongoing process. Many midlatitude areas that are temperate today were buried beneath great masses of ice 20,000 years ago. The salt flats of Lake Bonneville in Utah testify to the higher rainfall that once fell there: it produced huge lakes that dried up when rainfall declined, leaving behind the salt that had been dissolved in the water. The distribution of vegetation types has fluctuated over the last few thousand years as forests shift and deserts contract and expand.

Changes affecting the different environments take place over varied time scales, which influence the nature of the resulting environmental conditions. For instance, change in climate can occur over a few hundred years, as happened during the "Little Ice Age," a cold phase between 1400 AD and 1800 AD, or the warming that followed it from the start of the nineteenth century. Or climate change can take tens of thousands of years, as happened in the advances and retreats of great masses of ice across North America and northern Europe during the last 2 million years. There are also climatic fluctuations of shorter duration—periods of 6 to 10 wetter or drier years. The physical geographer studies climatic changes at different time scales by combining evidence from a variety of sources. Weather records based on instrument readings extend back only 100 years for much of the world. Older changes have to be studied from a combination of historical and environmental evidence, including written accounts, tree rings, ice layers, and ocean-floor sediments. Such studies require the sort of integration of data and working with scientists in a variety of fields that is part of the physical geographer's training and experience.

Human Activities and Earth Environments

The influence *human beings* have on the natural environment has become a special concern of physical geographers. Even in their earliest days on Earth, people modified the natural environment by bringing other species to the point of extinction and using fire to extend grassland. The capacity for changing the natural environment increased as the basis of the human economy passed from hunting to agriculture and industry. Human impacts continue to intensify as the world population grows and developing countries industrialize. Although in the past physical geographers paid scant attention to the influence of human factors, that influence is now impossible to ignore. At one stage, physical geography was studied to explain the impacts of natural environments on human activities; that concern continues, but today human activities are having an increasing impact on the natural environment. Whereas climatic conditions have been used to explain the different levels of economic development in midlatitude and tropical areas, for instance, today's focus is on human influences such as the impact of the burning of tropical forest on climatic change, which may in turn affect economic geography. In this way, recognition of the growing role of the human factor is a further integrating element in the study of Earth environments.

The effects of human interventions are least obvious in the atmosphere-ocean and solid-earth environments and most obvious in those environments where atmosphere, ocean, and land interact (Figure 1-7). Even in the surface-relief and living-organism environments, human impacts are variable.

Studying Earth Environments

Another important element of physical geography concerns how Earth environment investigations are carried out. The methods used are discussed further in Chapter 2, and the Investigations articles throughout the text focus on particular implications of these methods. The natural sciences, which provide the basis for understanding Earth environments, emphasize observation, measurement, and description as a framework for explaining cause-effect relationships and drawing conclusions. These conclusions may be based on the recognition of patterns, or on the division of forms and processes into their constituent parts for study. Generalizations that can be framed to cover a wide range of observations can be tested against further observations, and may be used to predict outcomes. Weather forecasts, for example, are based on ob-

Figure 1-7 The coast of Mozambique, Africa, as seen from the Space Shuttle. The smoke, blown by winds from the southeast, is rising from burning of grass and woodland. The clouding of the ocean by sediment off the river mouths is a result of soil erosion stemming from poor cultivation methods. The Zambezi River is in the center and the coastline shown is about 300 km (200 miles) long.

servations of temperature, cloud cover, humidity, and rainfall at the surface and at several levels in the atmosphere. These observations can be combined on maps and related to general descriptions of weather systems. Knowledge of the history of how these weather systems work makes it possible to forecast several days ahead.

One aspect of applying all these approaches to natural environments is that the linkages and interactions are so complex that the understanding of events is always incomplete. Simple cause-effect relationships are rare. For instance, the accuracy of weather forecasting is limited; meteorologists cannot always tell how much it will rain, when it will rain, or even if it will rain. But forecasts have improved greatly in the last 30 years as more and better reporting stations are built and satellite images have become available. Yet despite the wide range of data available and the use of the most powerful computers, the forecasts can be improved even more. Other environmental scientists face the same situation: geologists, oceanographers, geomorphologists (who study landforms and their origins), and ecologists all admit to a level of uncertainty in their predictions, and to the need for more research into the workings of Earth's environments.

Although much has been written in this chapter about physical geographers having a special role in bringing together a range of results from natural scientists and applying them to geographic interpretation and people-environment interactions in a holistic way, it is unusual for any one physical geographer to be knowledgeable in all aspects of the subject. Each physical geographer tends to focus research on one area, such as climate, the study of landforms (geomorphology), the study of soils (pedology), or the study of distributions of living organisms (biogeography), or even on a more limited specialty. However, an emphasis is often placed on the wider environmental linkages involved in such specialized studies, or on the relevance of such studies to human activities. For instance, a study of water flow in an arid environment needs to consider climate and ecosystems and can be helpful in selecting sites for urban expansion.

Box 1-2, about Mount St. Helens, illustrates a number of the points made here about physical geography, and in particular the interaction of different Earth environmental processes that occur in one place at the same time, the impacts of such processes on people, and the limited state of human knowledge about many physical environmental processes.

BOX 1-2

MOUNT ST. HELENS

The recent history of Mount St. Helens in Washington State illustrates what physical geography is about. A catastrophe that affected the lives of thousands of people, the eruption also had a powerful influence on other aspects of the surrounding environments. The mountain was remolded by a volcanic eruption, and the associated earthquakes, ash clouds, avalanches, and mudflows had widespread effects.

The main eruption of May 18, 1980, was not an isolated or unexpected event. Mount St. Helens is one of several volcanoes along the Cascade Mountains just inland from the Pacific coast as shown on the map. These volcanoes form part of the "ring of fire" around the Pacific Ocean. Mount St. Helens itself rose over the last 2500 years on the remains of an older volcano, as layers of molten lava and fragments of rock accumulated. Quiet periods with little volcanic activity have been interrupted by great explosions. Until 1400 AD, the quiet periods lasted up to 500 years, but since that date they have lasted only 200 years at most. The previous large eruption, in about 1800 AD, was witnessed by local Native Americans and a few early settlers. Smaller eruptions continued until 1857. Increasing knowledge about the history of Mount St. Helens led geologists in 1975 to predict another large eruption before the end of the century.

From the middle of 1979 to early 1980, a bulge 150 meters (470 feet) high formed on the northern flank. The first signs of the new eruption occurred on March 20, 1980, when a moderate earthquake began a sequence of stronger and more frequent events. A small eruption occurred on March 27, when a gush of steam and ash rose 2100 meters (7000 feet) and left behind a summit crater. People living within 25 km (16 miles) of the mountain were advised to leave. Further eruptions opened up a larger central crater 450 meters (1500 feet) deep, and the constant shaking of the ground alerted geologists that molten rock material was moving underneath. Scientists, reporters, and sightseers gathered, and access to the immediate area of the mountain had to be controlled.

On May 18, 1980, the main explosion occurred. Following a strong earthquake, 3 cubic km (1.2 cubic miles) of rock on the bulged northern flank of the mountain collapsed. An enormous column of finer material shot 20 km (12 miles) into the atmosphere. The collapse of the northern flank was accompanied by a powerful blast of gas and ash fragments that moved at 160 km per hour (100 miles per hour) and reached temperatures of over 800°C (1400°F). The air blast devastated an arc-shaped area extending 16 to 25 km (10 to 16 miles) to the north, felling and charring trees and destroying

Map of the area around Mount St. Helens showing other volcanoes in the area.

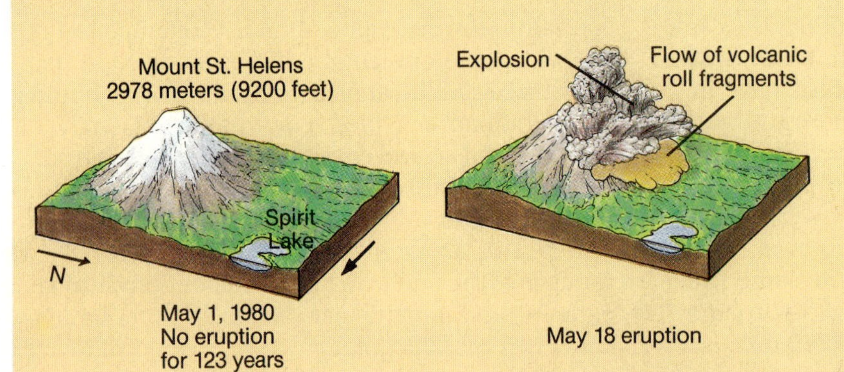

The eruption of May 18, 1980.

trucks, bridges, and homes. Melted snow and water spilled from lakes, mixed with the ash, and cascaded down the valleys in a mudflow that swept away logging equipment and killed fish in the Toutle River. The navigation channel in the Columbia River was blocked by the mud and ash dropped in a swath over 1000 km (600 miles) to the northeast. The ash that carried eastward in the upper atmosphere was blamed for cool, wet June and July weather in Europe.

A further eruption occurred on May 25, and others continued through the following years. A small, dome-shaped mound of lava built up in the great hollow formed in the central part of the mountain. Since then, Mount St. Helens has continued to be a focus of geological study, and minor activity has continued from time to time.

People's reactions to the eruption varied. Some were overawed by the experience; some resisted the official warnings and advice before moving out at the last moment, while others never left; some worried about the harm the eruption might do to the regional economy; and some were concerned that the eruption would continue for years. In fact, many of the worries about the long-term effects were not borne out in practice, since tourists came in greater numbers, and the volcanic activity soon died down.

The Mount St. Helens eruption illustrates one reason why earth scientists need to understand more about how Earth environments function—to make more informed decisions about possible impacts on human affairs. Humans ignore the impact of such events at their peril. Mount St. Helens had not erupted for 123 years, and habitation had crept closer and closer to the peak. There was no specific emergency procedure devised to deal with the catastrophe. Although geologists knew that an eruption was very likely to take place within the next few years and had posted more urgent warnings closer to the event, they could not be sure of the timing and size of the eruption. Certainly, there was no chance that the explosion could be modified or controlled in any way. The Mount St. Helens eruption took place in a sparsely inhabited part of the United States, yet 58 people died. Other natural catastrophes, such as those caused by tropical cyclones in Bangladesh and by earthquakes in China and Armenia, caused greater loss of life.

Physical geographers are particularly concerned with the human consequences of such extreme natural events. In a case such as Mount St. Helens, they examine the impact of the catastrophe on the whole physical environment—the local and regional weather, ecosystems, rivers, and human activities and infrastructure. In cases such as Bangladesh, China, and Armenia, there are also significant economic, social, and political implications of the events.

Mount St. Helens, before (top) and after the eruption of May 1980.

COMMENTS ABOUT THE ERUPTION OF MOUNT ST. HELENS

"It was like one of those Biblical epics. You felt overwhelmed. You felt like falling down on your knees and covering your face." (observer from 12 miles)

"The United States Geological Survey issued an updated hazard warning to state and local officials and to the U.S. Forest Service on 30 April 1980." (USGS press release)

"We tried to let people know where the hazards were, but nobody listened." (county sheriff)

"We told the National Guard to take off, that we could take care of ourselves. But when they said that it wasn't the mudflows they were worried about but poison gas—we got aboard the rescue helicopter." (logger)

"If the mountain goes, I'm gonna stay right here." (83-year-old who refused to be evacuated, and died)

"I hate to think what would happen if the eruptions kept up for years." (federal official)

"If this thing looks like its going to carry on for 10 more years, there's no way we're going to stick around. It's not worth it." (inhabitant of Toutle)

"Overnight we've gone from one of the most liveable cities in the U.S. to one of the least liveable." (Portland, Oregon, editor)

"We have a public relations problem. Portland suddenly has a reputation as a place to avoid." (manager of a major hotel)

"We simply don't know too much about what is going on in that mountain or what it will do in the future." (geologist)

"We're entirely at the mercy of the mountain." (U.S. Forest Service official)

"I've known the mountain all my life. I've climbed part of it, I've fished its waters. It's always been there to look at, all snowy and so grand looking. I loved Mount St. Helens, but look what she's done. I hate her." (80-year-old woman)

EARTH'S PLACE IN THE SOLAR SYSTEM

To appreciate the oneness of Earth's environment, consider Earth as one of the planets in the solar system. This perspective provides a set of comparisons for Earth and its characteristics and shows this planet is unique among its neighbors (Figure 1-8).

It has been possible to make detailed comparisons among the moons and planets only since the 1960s. Before unmanned and manned spacecraft provided a series of remarkable pictures, knowledge of other planets was limited and often inaccurate. In the 1980s, as the Voyager craft passed Jupiter, Saturn, and Neptune, views of these outer planets led to new discoveries.

Earth's Moon and Mercury

As the nearest body to Earth, the moon has received much attention, including manned landings that have brought back rocks from the surface. The moon has no atmosphere or ocean, probably because it is too small to generate the gravitational force necessary to retain an envelope of water and gases. Its surface includes a combination of craters and flat areas long thought to be seas but now known to be formed by flows of molten volcanic rock. Studies of the relationships between craters and smooth areas (called maria, or "seas"), together with the dating of rock samples brought back to Earth, have made it possible to work out a general history of the moon's landscape.

The moon was bombarded heavily by meteorites some 4 billion years ago. This churned up the surface rocks, and remnants of the moonscapes produced are now found in the heavily cratered "highlands." Between 3.8 and 3.0 billion years ago, huge basins were gouged out by a further series of meteor impacts; the heat generated caused interior material to melt and flood the basins to form the maria. Little has happened to alter the surface features of the moon over the last 3 billion years. There have been few meteorite impacts and no volcanic eruptions. The lack of an atmosphere means that there is no rain or chemical activity at the surface. The moon has a "fossil" landscape that has been relatively unchanged for billions of years—a stark contrast to Earth's constantly changing surface. The moon has no atmosphere-ocean or living-organism environment.

Mercury is about the same size as Earth's moon and resembles it in many ways: it has no atmo-

Figure 1-8 The solar system planets as they might be viewed from Earth's moon. This emphasizes the barren nature of the moon, the distinctiveness of Earth, and the features of cloudy Venus *(bottom left)*, Jupiter *(top left)*, and Mars.

sphere, and its surface is dominated by meteor impact craters formed during a major period of cratering around 4 billion years ago. The similarities with the moon suggest that the early cratering phase was followed by little surface change for long periods of time. Mercury is the nearest planet to the sun, and its surface temperatures reach several hundred degrees Celsius.

Venus

Venus is almost the same size as Earth but is closer to the sun. It has a dense atmosphere and may have had water on its surface at one time. However, its atmosphere has become dominated by clouds as the result of a "runaway greenhouse effect" in which the high levels of solar heating, coupled with high concentrations of heat-trapping carbon dioxide, led to a rise in atmospheric temperatures (Chapter 3). Any surface water "boiled off" to form the clouds, which totally obscure the surface and contain sulfuric and hydrofluoric acids. The surface temperature of Venus is now around 500°C (950°F). Its high surface temperatures are due more to the atmosphere trapping heat than to its proximity to the sun.

The surface features of Venus have been studied by cloud-penetrating radar from orbiting spacecraft and appear to be more subdued than landforms on other planets. Many volcanic features but few impact craters have been identified. The dense atmosphere causes meteorites to burn up on entry. The Magellan spacecraft in 1990 sent pictures back to Earth that showed much surface evidence of internal movements in the planet, such as fractures and volcanic features. But the features were often found to be very different from those on Earth.

The Venusian environment has active atmosphere-ocean, solid-earth, and probably surface-relief environments, but no living-organism environment exists. Living tissue could not survive at the surface of the planet.

Earth

Earth also has a dense atmosphere, but one that is largely transparent, as photographs from space show. At any moment, clouds occupy less than half the globe's surface.

Earth is unique among solar system planets in many ways. Some two thirds of its surface is covered by liquid water. The relief of its continents and ocean basins is marked by linear features—mountain ranges and undersea ridges—rising above their surroundings. Only a few impact craters have been identified, but volcanic activity and frequent earthquakes testify to continuing internal activity. Running water, moving ice, gravitational processes, the sea, and the wind mold the surface. A wide range of vegetation covers most of the land surface, and many creatures populate the continents and the oceans. Humans, another unique feature of Earth, dominate the other creatures and change the ways in which natural processes operate.

Like the other planets, Earth is a sphere—almost. It is slightly flattened at the poles, making a shape known as an **oblate spheroid.** Its equatorial diameter is 12,757 km (7927 miles), and its diameter from pole to pole is 12,714 km (7900 miles). That makes it a spheroid, but the North Pole is 45 meters (140 feet) farther from an equatorial cross section than the South Pole. In practical terms in physical geography, Earth can be treated as a sphere.

The spherical nature of Earth and its rotation around its axis every 24 hours have significant consequences concerning Earth's heating by the sun (Chapter 4). These circumstances result in variations in the receipt of solar energy between day and night and from place to place. Virtually all of Earth's surface sees the sun for a part of every 24 hours, however. Other planets have different lengths of day, from 5832 hours on Venus down to 9.8 hours on Jupiter. The significance of these differences is not fully understood, but the 24-hour rotation of Earth means that surface temperatures do not have time to fall very low during the night.

Mars

Mars's volume is about one-fourth that of Earth's, and it has a thin atmosphere. The Viking spacecraft landings in 1976 greatly increased scientists' understanding of this planet's environment, with its ice caps and dust storms. The ice caps are formed mainly of water ice, but in the coldest season, carbon dioxide freezes on the surface. Martian surface temperatures range from −50°C (−58°F) at the equator in summer to below −120°C (−184°F) at high latitudes in winter. They are much lower than those on Earth, because Mars is farther from the sun and has a thinner atmosphere. The daily range of temperature is also greater on Mars, despite the fact that the planet's day length is similar to that of Earth. The greater temperature range is caused by the thinner Martian atmosphere.

Mars has a wider variety of landforms than Earth's moon or Mercury, but less variety than Earth (Figure 1-9). Mars has extensive areas of cratering rather like the moon, but the Martian craters are shallower, and their rims are not as sharply defined. This is because of the presence of an atmosphere and consequent weathering processes. There are also volcanic lava plains.

The largest features of the Martian landscape are huge volcanic mountains such as Olympus Mons, which is 600 kilometers (400 miles) across and thousands of meters high. This mountain, larger than any volcano on Earth, is built of many individual lava flows that poured out at the same place over millions of years. The ice caps are surrounded by plateaus thought to be made of layers of dust and ice deposited by water melted from the ice caps. An immense canyon stretches along the equator of Mars for 4000 kilometers (2500 miles); in parts it is 200 km (120 miles) wide and 7 km (4 miles) deep. Such huge features suggest that there have been movements in the planet's crust, followed by erosion.

The present Martian environment thus differs from that of Earth in many ways, including a lack of flowing surface water, mountain ranges, or living creatures. Its atmosphere-ocean environment is composed of a thin atmosphere without oceans; its solid-earth environment appears to be inactive; its surface-relief environment is subject to weak forces of change; and there is no evidence of a living-organism environment.

Outer Planets

The planets beyond Mars include Jupiter and Saturn, which are huge, Uranus and Neptune, which are both larger than Earth, and tiny Pluto. The nearest of these planets is more than five times farther from the sun than Earth is. Solar heating makes little impression at that distance. The planets are composed largely of frozen gaseous materials surrounding relatively small rocky cores. Each has a number of moons, which are sometimes rocky in composition, but the environments of these planets and their moons have little in common with Earth, and comparisons provide little information for studies of this planet.

Figure 1-9 The surface of Mars showing the equatorial canyonlands and volcanic peaks.

The Origin of the Solar System

The latest theories about the origin of the solar system build on an understanding of the planets and their features. In one account, the origin of the solar system goes back over 10 billion years to the time when a spinning mass of dust and gas toward the edges of the Milky Way galaxy began to change. The central parts warmed up around the developing sun, and rotation concentrated the dust and gas into clumps of material. The clumps near the center contained more high-density materials such as iron, nickel, and manganese, which consolidated into the cores of the inner rocky planets from Mercury out to Mars. Moons often began as similar clumps that gravity drew into orbit around the larger masses.

The phase of major cratering that left its imprint on Earth's moon, Mercury, and Mars was experienced by all the inner planets as they absorbed large amounts of the loose material still unattached in the solar system some 4 billion years ago. Each future planet accumulated a distinctive mix and amount of materials. Some planets, such as Earth and Venus, grew large enough that internal energy sources remained sufficient to produce upheavals of mountains and volcanic activity; others were too small to support such activity for long before surface rocks and deeper rocks cooled and solidified. Earth, Mars, and Venus experienced surface upheavals that involved volcanic eruptions and the uplift of mountains. The larger rocky planets attracted and retained the lighter gaseous elements to form atmospheres, which received additional gases from volcanic eruptions.

Present knowledge suggests that only on Earth were the temperatures in the right range to produce precipitation of water on the surface to fill the ocean basins. On Earth alone, the conditions were right for living organisms to exist.

The Evolution of Earth and Its Major Environments

Earth's history goes back over 4 billion years, although there is little evidence for what happened in the early stages. As time progressed, however, the amount of evidence increased, and it is clear that Earth has experienced a series of changing environments. Many of the changes can be seen as evolutionary, leading to the combination of surface environments studied by physical geographers today.

Following the early accumulation of iron-rich compounds to form Earth's core, and the bombardment of this core by somewhat less dense rocky materials around 4 billion years ago, the lightest materials were also attracted by gravity to produce an envelope of atmosphere around Earth. **Vola-**

tiles are substances that evaporate at relatively low temperatures and so normally exist in gaseous form. They may have been added by meteorite impacts that also transferred water, rare gases, and even organic chemicals—the basis for living organisms—from the outer parts of the solar system.

The *atmosphere-ocean environment* developed out of interactions among the early atmosphere, surface rocks, and living organisms. Debate continues as to whether the early atmosphere is still here or whether Earth lost its first atmosphere in a gigantic blast of solar wind. If the latter is true, the present atmosphere probably was formed by the release of gases resulting from the meteorite impacts and volcanic eruptions around 4 billion years ago. After this "outgassing" (if it occurred), Earth's atmosphere was formed largely of water vapor, carbon dioxide, and nitrogen. Soon after, the oceans were formed by rains as the Earth's environment cooled (Figure 1-10). The oxygen content of the atmosphere probably built up slowly, first following the breakup of water molecules reacting with solar radiation and later as the result of photosynthesis (see Chapter 19) by the increasing number of plants in the oceans and on land. The first evidence of marine life appeared over 3 billion years ago, but land life began to appear only 400 million years ago. Land organisms did not appear until quite late in the history of living organisms on Earth because earlier living organisms needed protection against ultraviolet solar rays in order to evolve, and not until 400 million years ago had plants provided sufficient atmospheric oxygen to form a shield against it. Only then could plants and animals live out of water on the continental surfaces.

The *solid-earth environment* has continued to be active. Evidence in rocks shows that the processes of mountain building and volcanic activity, which are linked to the movement of continental masses and the changing shapes of ocean basins, were active 3 billion years ago. These processes have renewed Earth's surface features so that most evidence of early meteorite craters has disappeared. Ocean floor rocks are renewed every 200 million years, and the rocks of continents are worn down and changed over periods of several hundred million years. The continents contain the oldest Earth rocks, which have been dated as nearly 4 billion years old. Major changes in the positions and sizes of continents and in the formation of mountain ranges take place over hundreds of millions of years.

The *surface-relief environment* has been shaped by the interaction of the air and sea with the surface rocks and mountains. Most of its landforms have been formed over periods of a few thousand to a few million years, although some are older.

The *living-organism environment* has developed from interactions among plants, animals, and the other major environments. The oldest signs of life are fossils of microscopic aquatic plants in rocks dated at around 3 billion years old. Living organisms have developed continuously ever since. Although fluctuations have occurred, there was a general increase in numbers and diversity until the advent of humans.

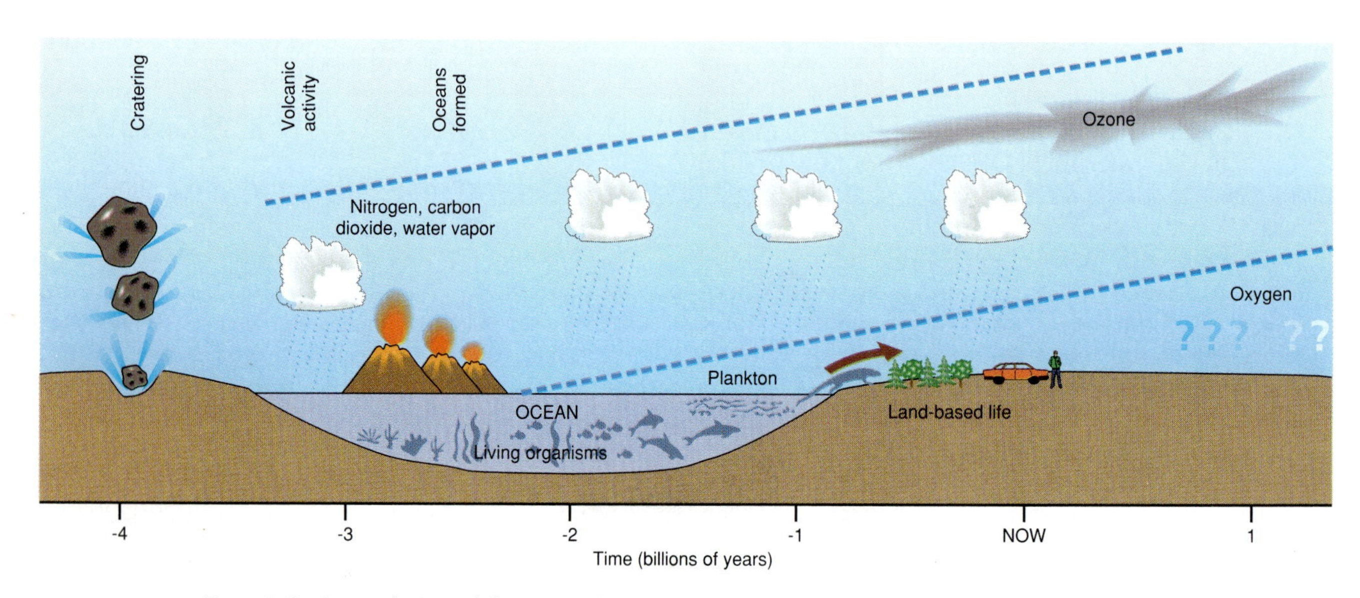

Figure 1-10 The evolution of the atmosphere-ocean environment. The diagram reads from the left (4 billion years ago) to the right. The ozone layer at the right-hand side keeps out some of the dangerous fractions of solar radiation.

A Unique Planet?

The study of Earth as part of the solar system indicates its uniqueness. It is possible that this is the only planet in the universe on which life exists. The existence of life anywhere else has not been proven, and it is unlikely to be found in the foreseeable future, although many scientists say that there is a good probability that it exists somewhere in the vastness of the universe. Significant "coincidences" have brought about Earth's special characteristics. For instance, the surface temperature ranges are related to the planet's distance from the sun, as well as the composition of the atmosphere, oceans, and continents. Earth developed a protective, yet transparent atmosphere and ocean cover in which the complex chemical changes required to produce living organisms could occur. Such interactions demonstrate the delicate balance between the development of this hospitable planet and the origins of the harsh conditions in the rest of the solar system.

For the practical purposes of humanity, Earth is unique. There is no escape if Earth is ruined. The millennia of evolution have produced a complex and fragile environment that is now rapidly being changed by humans. Whether one believes that life on Earth occurred by chance or as the result of supernatural creation, each person has a responsibility to ensure that the environment here is not degraded or destroyed.

A CHALLENGE TO PHYSICAL GEOGRAPHERS

The need to conserve the environments that sustain life on Earth in the face of natural and human-induced changes in these environments has presented a number of challenges to all environmental scientists, including physical geographers. A better understanding of the natural environment must involve research scientists, officials in government and business who influence general policy, and ordinary citizens who make everyday decisions concerning the use of the environment. Scientists need better systems for monitoring the environment and its changes on local, regional, and global scales. They can then provide improved information on which others can base their decisions.

More information from improved monitoring facilities will make it possible to improve the models that summarize knowledge of the workings of Earth's environments, and so make it possible to make more accurate predictions about the near future. Such information, modeling, and predictions by scientists will be of little use unless accompanied by a greater understanding on the part of politicians, corporate leaders, and citizens. The will and the ability to make changes depend on them.

Hence, a course in physical geography provides a basis for the greater awareness of Earth environments and, it is hoped, a foundation for greater participation in reaching the judgments and decisions that will ensure the continuing existence of "the living planet."

SUMMARY

1. Environmental issues are characterized by:
   a. varied geographic scales of impact: some issues are globally significant, but others affect only local or regional areas;
   b. complexity of interaction within and among Earth environments;
   c. dynamic change;
   d. the impacts of human intervention;
   e. the moral responsibility of humans to conserve for future generations;
   f. incomplete understanding of environmental processes and the need for continuing scientific study;
   g. the need for political and corporate action and private choices.

2. Physical geography is the study of Earth environments with a particular emphasis on spatial characteristics and change over time. It involves a study of interactions between different aspects of the natural environments, and interactions between human activities and the environments.

3. Earth's environments are linked. The four major environments are the atmosphere-ocean environment, the solid-earth environment, the surface-relief environment, and the living-organism environment.

4. Each environment has a structure and a related set of processes; the processes interact with and change the structure. Each environment is subject to change over time and to the intervention of human activities.

5. Earth and its environments exist in the wider setting of the solar system and its planets. Studies of these planets provide a context for assessing the physical geography of Earth.

6. Earth has a long history of some 4 billion years. The present character of Earth environments has resulted from continuous development during that period.

KEY TERMS

physical geography, p. 4
environment, p. 7
global scale, p. 7
regional scale, p. 7
local scale, p. 7
atmosphere-ocean environment, p. 7
solid-earth environment, p. 7
surface-relief environment, p. 7
living-organism environment, p. 10
structure, p. 10
process, p. 10
oblate spheroid, p. 18
volatiles, p. 19

QUESTIONS FOR REVIEW AND EXPLORATION

1. Collect examples of local, regional, national, and global environmental issues. How do they relate to the set of seven points made about such issues at the beginning of this chapter?
2. Write your own definitions of "geography" and "physical geography."
3. Summarize the main features of each of the four major Earth environments.
4. How are "structure" and "process" linked in major Earth environments?
5. Compare planetary environments in the solar system. Account for the lack of evidence of living organisms on any planet other than Earth.
6. Discuss the question of human responsibility in relation to the conservation of Earth environments. On what principles is it based? How can it be worked out in practice?
7. Define specific roles for physical geographers in relation to environmental issues such as those collected for question 1.

FURTHER READING

Chipman EG et al: *A meeting with the universe: science discoveries from the space program*, Washington, D.C., 1981, NASA. A summary of much of the knowledge about the solar system gleaned from NASA activities. Profusely illustrated.

Cloud P: *Cosmos, Earth, and man*, New Haven, Conn., 1978, Yale University Press.

Committee on Earth Sciences: *Our changing planet: FY1991 Research Plan*, Washington, D.C., 1990, U.S. Government Printing Office. The priorities set for funding projects within this program.

Cooke RU, Doornkamp JC: *Geomorphology in environmental management*, Oxford (U.K.), 1990, Oxford University Press. Contains many examples of specialized studies in aspects of physical geography that involve understanding of interactions between elements of the natural environment, and examines applications in planning and environmental management.

Department of the Environment: *This common inheritance: Britain's environmental strategy*, London, 1990, Her Majesty's Stationery Office. A survey of a wide range of areas requiring environmental policies.

Dunn G, Crump DJ: *Exploring your world—the adventure of geography*, Washington, D.C., 1989, National Geographic Society.

Gore A: *Earth in the balance*, New York, 1992, Houghton Mifflin. A readable survey, recounting Senator Gore's wide travels and study of environmental issues.

NASA: *Earth system science*, 1988. Materials that relate the integrated study of Earth environments to global changes and to environmental monitoring from satellites can be obtained from: Office for Interdisciplinary Earth Studies, University Corporation for Atmospheric Research, PO Box 3000, Boulder, CO 80307, USA.

NASA: *EOS: A mission to planet earth*, Washington, D.C., 1990, U.S. Government Printing Office. The Earth Observing System and plans for increased monitoring of Earth environments from satellites during the 1990s.

Piel J, editor: *Exploring space: special issue*, 1990, Scientific American. A number of recent articles covering the study of the solar system and beyond.

Tilling RI: *Eruptions of Mount St Helens: past, present and future*, U.S. Geological Survey, 1987, Washington, D.C. A good illustrated account of the eruptions and their impacts.

Weiner J: *Planet Earth*, New York, 1986, Bantam Books. An illustrated book covering many of the topics discussed in this textbook. Prepared in conjunction with a television series aired by the Public Broadcasting System in 1986.

FOUNDATIONS OF PHYSICAL GEOGRAPHY

Antarctica map reading: Olympus Range, Dry Valleys, Victoria Land Camera Colin Monteath/Mountain

The Physical Geographer at Work

2

CHAPTER

When I had mapped the pond by the scale of ten rods to an inch, and put down the soundings, more than a hundred in all, I observed this remarkable coincidence. Having noticed that the number indicating the greatest depth was apparently in the centre of the map, I laid a rule on the map lengthwise, and then breadthwise, and found, to my surprise, that the line of greatest length intersected the line of greatest breadth exactly at the point of greatest depth, notwithstanding that the middle is so nearly level, the outline of the pond far from regular, and the extreme length and breadth were got by measuring into the coves; and I said to myself, Who knows but this hint would conduct to the deepest part of the ocean as well as of a pond or puddle? Is not this the rule also for the height of mountains, regarded as the opposite of valleys?

If we knew all the laws of Nature, we should need only one fact, or the description of one actual phenomenon, to infer all the particular results at that point. Now we know only a few laws, and our result is vitiated, not, of course, by any confusion or irregularity in Nature, but by our ignorance of essential elements in the calculation. Our notions of law and harmony are commonly confined to those instances which we detect; but the harmony which results from a far greater number of seemingly conflicting, but really concurring laws, which we have not detected, is still more wonderful.

Henry David Thoreau, "Walden, or Life in the Woods"

Physical geographers examine the structure and the processes of the major Earth environments at different spatial scales and how structure and process combine to create the natural environment at one place and time. They try to answer the general question, "Why is it like it is here?" This question may take many specific forms. Some examples are: "Why does the Rio Grande still run in dry periods? Why is it so cloudy in the tropics? Why do glaciers form? How do cacti and other plants survive in the desert? Where is the coldest place on Earth?"

The essential purpose of physical geography is to gain a better understanding of how the total Earth environment works, in order to understand how it limits human activity and to judge how it is influenced by human activities. With this knowledge, humans could keep the planet fit for future generations. Many scientific disciplines, including ecology, geology, and meteorology, contribute to physical geography. This "borrowing" is physical geography's real strength; the subject incorporates the results of other disciplines and undertakes research at the boundaries between subjects that might otherwise be ignored. The result is a unique understanding of how the complex and dynamic Earth environments interact to form a living planet. Physical geography is important in an everyday sense, too; the type of natural environment in which people live affects the way they live, and in some extreme cases determines whether they live at all. Figure 2-1 shows some very different physical environments around the world; you might try working out how different your life might be if you lived in one of these. The understanding of the natural environment provided by physical geography will help human beings to live in these environments in 20, 200, and 2000 years' time.

Physical geography has three main approaches or pathways to understanding. The first is *comparison* and *classification*. The geographer compares and contrasts features in the Earth environments, describing how some are similar to each other and how others are different. How, for example, does the shape of a mountain range in America differ from one in Europe and another in Asia? This type of comparison requires classification; classifying mountain ranges is the geographic equivalent to recognizing and naming different species of butterflies. Classification is outwardly mundane but is the basis of all science (see Investigation in Physical Geography, Chapter 8). By grouping together similar features of Earth environments geographers can understand the common processes by which they are formed and how they will develop in the future.

The second approach to physical geography concentrates on examining how geographic features and the natural landscape have *changed over time.* Physical geographers want to know how long it takes to build a mountain range and to wear it down again, and how the climate of a place has changed since the last glacial phase. On a shorter time scale, geographers study how long it takes for a forest to develop on a bare soil, and how rapidly rainfall runs off the land into a stream.

The third approach examines how the *features of the environment are arranged* next to each other in space and why this arrangement occurs. How, for example, does the shape of a stream channel change from the stream's source to its mouth at a lake or the sea, and why are kangaroos found in Australia but nowhere else? The importance of spatial location is the strongest axis in geographic work. Physical geographers who specialize in the workings of one Earth environment, such as a biogeographer, or a geomorphologist (who studies the structure

Figure 2-1 Physical environments around the world.

A tropical island

The arctic

and processes of the surface-relief environment) draw on the spatial aspects of other environmental sciences such as biology or geology.

Physical geography can be thought of as a cube defined by the three pathways described above. Each axis contributes to every part of the subject: physical geography examines Earth environments in all three dimensions. This chapter examines how physical geographers go about studying and understanding Earth environments, what data they use, how these data are analyzed and interpreted, and how the results are evaluated. The first stage in any research is to outline the nature of the problem and to design an experiment around it. The next stage is the collection of geographic data, either directly through field or laboratory work, or from maps, remotely sensed images, and other types of published data. Finally the data must be analyzed, presented, and the results and interpretations circulated to other scientists for criticism and review.

THE PROCESS OF DISCOVERY IN PHYSICAL GEOGRAPHY

As noted in Chapter 1, studying Earth environments and their interaction is an immense and complex undertaking. A single geographer can concentrate on only a particular section of the problem, but together many geographers can build up a core of shared knowledge. Individual researchers need some kind of framework to guide their study, so that their experiments complement those of other scientists and their results are comparable. A common framework also helps researchers get the best returns, in terms of understanding of Earth environments, for the effort expended in gathering and analyzing data.

Observation, Explanation, and Prediction

Geographic studies typically begin with an observation about the natural environment. This can be a direct observation of some phenomenon in the landscape, such as the time lag between heavy rain and peak river flow, the increased melting rate of a glacier, the remarkably constant angle at which loose rocks come to rest on a slope, the difference in vegetation between north-facing and south-facing slopes at the same elevation, or the fact that similar soils are found under pine trees despite widely different underlying rocks. In other cases, the "observation" is a combination of direct observation in the field or laboratory, personal experience, and studying the published work of other scientists. However the observation arises, it poses the question, "Why is it like it is here?" The geographer finds some feature of the structure or process of the natural environment that cannot be explained by current knowledge. The unexpected and the inexplicable are the most interesting and valuable types of observation since they lead to further questions and eventually to new knowledge. The basis of modern geographic investigation lies in describing what does not fit existing knowledge and finding an explanation for it.

The early physical geographers were literally explorers; their job was to visit uncharted territory and to observe and describe what they found there. In our shrinking world, however, there is very little true exploration left to do; human beings have established themselves on, or at least visited, virtually the entire land surface on Earth. Even inhospitable areas such as Antarctica and the sea floor have been mapped to some extent. The emphasis of modern geography has shifted from exploration, observa-

A desert

A coast

A high mountain

DATA COLLECTION

The basic information for all investigations in physical geography is found in Earth environments. Geographers may collect **primary data** directly from such environments through fieldwork; they may bring back samples for more detailed measurement in the laboratory. They may also use **secondary data,** one step removed from the original field environment, from published sources such as meteorologic records or maps, or take measurements from *remotely sensed* data such as aerial photographs or satellite images. They may also *simulate* real world conditions in the laboratory or on a computer. Each of these sources of data has a place in geographic investigations, and most experiments include data from a number of sources. If the experiment is designed well, it should be clear what data are needed and how they should be analyzed.

Figure 2-4 An environmental cabinet being prepared for use. Conditions of temperature and humidity in the cabinet can be adjusted to simulate very different climates.

Sources of Primary Data

To many physical geographers, "real" physical geography involves getting out into the field, scaling mountains, fording streams, and generally experiencing the excitement of studying interactions between Earth environments firsthand. Fieldwork in physical geography ranges from excavating recent glacial deposits in high mountains, to monitoring the buildup of sediment on a beach, or conducting a detailed survey of the natural vegetation in an area and its response to increased grazing. Fieldwork is certainly one of the great pleasures of the subject and leads both to a healthy respect for Earth environments and to a better understanding of the complex interactions summarized by diagrams like Figure 2-3. However, not all geographic investigations can be carried out fully in the field, and geographers use both laboratory techniques and secondary data sources to fill in the gaps.

The abstract laboratory experiments conducted in physics, chemistry, and biology are rare in physical geography. Here laboratory work has two roles. First, it can provide a controlled imitation of the natural environment in which to simulate and measure natural processes. An example of this type of use is a *flume,* in which water is directed across a "bed" of sediment, such as stones and sand, mimicking a stream. Individual variables, such as the volume of water input, the type of sediment, and the tilt of the stream channel, can be controlled one at a time, and their effects on features such as the speed and pattern of flow in the water can be measured. This type of experiment simulates the working of a stream system in easy stages. New variables, such as mixes of sediment sizes and the effects of tributary streams, can be added one by one, until conditions in the flume are close to the complex situation in the field.

Another example of this type of simulation exercise is the use of an *environmental cabinet* (Figure 2-4). This piece of equipment can run through a year's fluctuations in climate in a matter of hours, and can conveniently move from a desert environment to a snowy mountain peak at the flick of a switch. The cabinet is used to accelerate the natural processes of weathering (Chapter 13), so that rocks disintegrate before the scientist's eyes. The effects of varying heat and moisture regimes can be compared for different rock types.

The common feature of these experiments is that the complex natural environment is reduced to a small set of variables, operating in a small space at a time scale that can be grasped by the human observer. The greatest value of these experiments is that they let the scientist observe and manipulate "natural" systems in a way that is impossible in the field. There is, however, the danger that the complexities of the real system will be underestimated, or that an important variable will be missed.

Figure 2-5 Particle size analysis of glacial deposits. The equipment is a series of sieves with, from the top down, progressively finer meshes. Particles of different sizes are separated into different sieves. Each size class can be expressed as a proportion of the total amount of sediment.

The second type of laboratory work is more closely tied to field investigation. Typically it is a more detailed examination of *samples collected in the field*, but requires time periods, power supplies, or complex equipment not readily available at the field site. An example of this type of data collection is the particle size analysis of sediments, which may come from glacial deposits, stream beds, or soil (Figure 2-5). The proportion of large, medium, and small particles in a sample of sediment can tell the researcher about its origin and how it was deposited. Other examples of laboratory work include measuring the amount of moisture, plant and animal tissues, and nutrients in different types of soils, or at different depths in the same soil, measuring the acidity of rainwater and lake water, and describing the mineral composition of rocks or broken deposits through powerful microscopes. The results of laboratory analysis of field samples are also primary data. Here the fieldwork and laboratory work go together; one is virtually useless without the other, as generations of geographers with poorly labeled field samples have found to their frustration.

Computers allow further extrapolation from both field and laboratory data through *mathematical models*. Here the primary data—usually in the form of numerical values—are used to describe the initial state of a geographic feature. One example might be that a certain amount of beach sand, moved by offshore currents at a particular speed in a particular direction, produces a sand bank of a particular size, shape, and position at the shore. The researcher can then manipulate the inputs of sand and water movement to predict the evolution of the sand bank if, for example, sand is dredged from the bay, or if the sea level rises.

Sources of Secondary Data

The nature and quality of primary data from field and laboratory experiments are directly controlled by the experimenter. However, the geographer can work with only a tiny portion, or sample, of the data that could be collected, and the sample is usually limited in both space and time. If the study is to be more widely applicable, the geographer must use secondary data sources. Examples might include historically recorded data of rainfall and river flow to extend the study through time or contemporary or historical maps, air photographs, or satellite images to extend the study through space.

MAPS AND MAPMAKING

Maps are a basic tool to the geographer. They are a particularly important source of secondary data, and are also used in all stages of geographic research, from the first observations to the presentation, interpretation, and communication of results. Maps are also important in everyday life. Everybody carries around a mental map of his or her neighborhood that, consciously or unconsciously, shows features of all the Earth environments. For example, college students might think of the campus as downhill and across the river from their apartment; they may also have mental pictures of all the intervening landscapes between the college and their hometowns. Whether this mental map, to the mind's eye, is like a road map, snapshots of street corners, patches of vegetation and other landmarks, the information it holds on spatial relationships is used to organize our everyday lives. To the physical geographer maps are more than this; the geographer can look at a two-dimensional map of Earth's surface and picture the full three-dimensional landscape, with as much detail as the map allows. From this picture the geographer can gather information about the structure and processes of the environment at that place, often as accurately and with as much geographic understanding as someone who has visited the area.

The geographer needs at least some elementary information about mapmaking, or **cartography,** to "read" the landscape depicted on the map. Maps are complicated mathematical constructs; they are a flat, two-dimensional representation of a three-dimensional landscape that is also wrapped around a nearly spherical globe. A map is also a partial and imperfect representation of Earth's complex environment. The mapmaker starts with a blank sheet of paper and copies onto it the important features of the scene; another observer might have different priorities and record a different set of features.

Latitude and Longitude

Any feature on a map has a location, and its location must be described so that anyone can find it. The international reference system for locating any place on Earth's surface is latitude and longitude.

Latitude describes how far north or south of the equator a place is. Latitude is a measure of angle rather than surface distance and is therefore quoted in degrees rather than kilometers. A place at 45 degrees of latitude lies where a line from the surface to the center of Earth intersects the plane of the equator at 45 degrees (Figure 2-6). Many lines can be drawn that meet this definition; if they are connected up on Earth's surface, they form a circle around the globe that is parallel to the equator. One such circle can be drawn to the north of the equator, and one to the south; these are designated 45 degrees north (abbreviated to 45°N) and 45 degrees south (45°S), respectively, and are called lines, or *parallels,* of latitude. All parallels of latitude are parallel to each other and to the equator.

The equator is at 0° latitude, and the poles, where the axis of Earth's rotation intersects the surface of the planet, are at 90°N and 90°S, respectively. Given that the distance along Earth's surface from either pole to the equator is 20,000 km (12,500 miles), some simple arithmetic shows that two parallels of latitude 1 degree apart are about 112 km (70 miles) apart on Earth's surface. This is a rather imprecise system of location, so a degree of latitude is broken down into smaller units. There are 60 minutes (written 60′) in a degree; parallels of latitude 1′ apart are about 1.9 km (1.2 miles) apart at Earth's surface. There are 60 seconds (60″) in one minute; parallels of latitude 1″ are about 30 meters (80 feet) apart at Earth's surface.

The following pairs of cities are at nearly the same latitude: Philadelphia, Pennsylvania (39°57′N) and Beijing, China (39°55′N); Phoenix, Arizona (33°27′N) and Baghdad, Iraq (33°20′N); Saskatoon, Canada (52°07′N) and Warsaw, Poland (52°15′N); Washington, D.C. (38°55′N) and Athens, Greece (38°00′N); Buenos Aires, Argentina (34°40′S) and Sydney, Australia (33°55′S). If necessary, points at Earth's surface can be located very precisely. For example, a marker at Meades Ranch, Kansas, the control point for a survey of the United States, is located at latitude 39°13′ 26.686″ N. Note that it is important to state whether a latitude is north or south of the equator.

The *Tropic of Cancer* is the parallel of latitude at 23°30′N and the *Tropic of Capricorn* is the parallel of latitude at 23°30′S. Within these two parallels the sun will be directly overhead at some time of the year. The tropical area of the world, usually abbreviated as the *tropics,* lies between the Tropics of Cancer and Capricorn. The Arctic circle is the parallel of latitude at 66°32′N, and the Antarctic circle is the parallel of latitude at 66°32′S. At latitudes above both these parallels the sun never sets on one day of the year, the summer *solstice.* The summer solstice is June 21st in the northern hemisphere and December 22nd in the southern hemisphere. Similarly, the sun never rises along these parallels on the winter solstice (December 22nd in the northern hemisphere and June 21st in the southern hemisphere). Poleward of the Arctic and Antarctic circles, the number of completely light and completely dark days increases. The midlati-

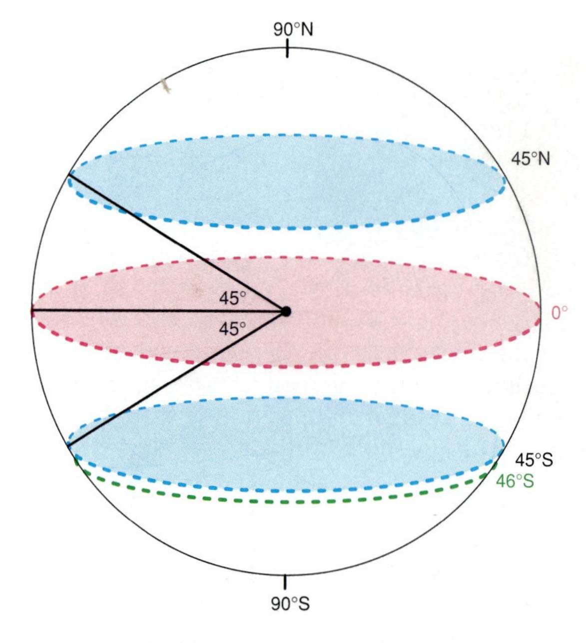

Figure 2-6 Measurement of latitude.

tudes are the areas of the world that lie between the tropics and the Arctic and Antarctic circles.

Lines of latitude give a precise north-south location to any point on Earth's surface. Lines of **longitude** are called *meridians* and complete the global referencing system by giving a precise east-west position. There are no natural starting points for the lines of longitude, as the poles are for latitude. The line of 0° longitude, or the *prime meridian*, is therefore an arbitrary choice. It was agreed upon internationally in 1884 and is sometimes called the Greenwich meridian, because it passes through the old Royal Observatory at Greenwich, near London, England.

The prime meridian is half of a line that goes right around the world, passing through both poles. That circular line is the edge of a plane that passes through the center of the globe and therefore divides it in half. All lines around the globe with this property are called *great circles;* all meridians are half great circles. The principle of measuring degrees of longitude is shown in Figure 2-7. The angular measurement is that between the plane of the prime meridian and the plane of the great circle, passing through the poles, on which the location lies. Longitude is therefore measured in degrees east (°E) or degrees west (°W) of the prime meridian to a maximum of 180° in either direction. In contrast to parallels of latitude, meridians are not parallel to each other. At any particular latitude they are evenly spaced around the globe, but they are closer together nearer the poles than they are at the equator (Figure 2-7). The full latitude and longitude coordinates for some of the cities listed above are as follows:

Saskatoon, Canada: 52°07′N 106°38′W
Warsaw, Poland: 52°15′N 21°00′E
Washington, D.C.: 38°55′N 77°00′W
Athens, Greece: 38°00′N 23°44′E
Buenos Aires, Argentina: 34°40′S 59°25′W
Sydney, Australia: 33°55′S 151°10′E
Meades Ranch, Kansas is at 39°13′26.686″ N, 98°32′30.506″ W.

Many countries have devised a more manageable grid system to locate small features more precisely. Examples are the Universal Transverse Mercator coordinates used by the U.S. Geological Survey in North America and the National Grid used by the Ordnance Survey in Britain. Both these grid systems can be set in the context of latitude and longitude to give a worldwide location.

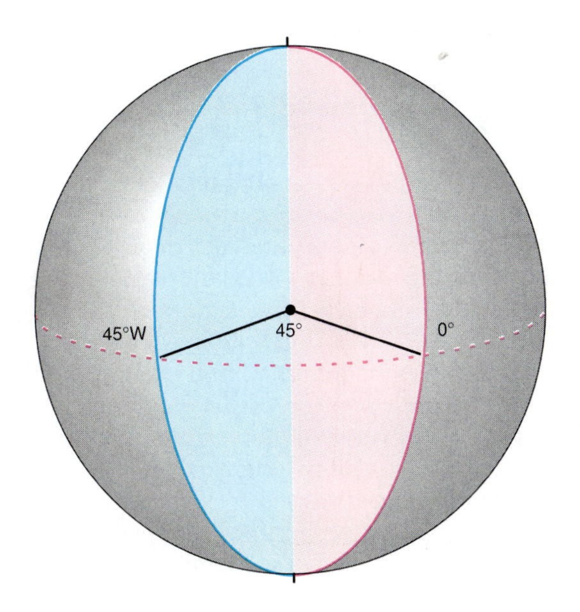

Figure 2-7 Measurement of longitude.

Time Zones

Before the era of railroads, telegraphs, and transcontinental travel, timekeeping at a place was fixed by local solar noon—the time when the sun was at its highest in the sky. This time was taken to be 12 noon and the rest of the day and night fitted around it. Because Earth spins on its axis, solar noon occurs at different times on different meridians. In fact, Earth rotates about 15° of longitude in one hour, so the solar noon at two places 15° of longitude apart differs by one hour. This system of local time worked well enough until improving physical and electronic communications meant that some kind of worldwide standard was needed. In 1884 an international conference in Washington, D.C., divided the world into 24 time zones, each one about 15° of longitude wide. The first time zone is centered on the prime meridian, and others are centered every 15th meridian east and west from there. Within each time zone there is no local variation in time, and clocks are set to the same time.

The time in the 15° centered on the prime meridian is variously called universal time or Greenwich mean time. All other time zones are counted as hours forward or hours behind this. Places to the west of Greenwich are behind Greenwich mean time; places to the east are ahead. In practice, the time zones are adjusted by national and physical boundaries so that they do not completely follow the meridians. The conterminous United States uses four time zones (Figure 2-8). Countries at high latitudes have more time zones because the meridians converge toward the poles. Russia, for example, has 11 time zones.

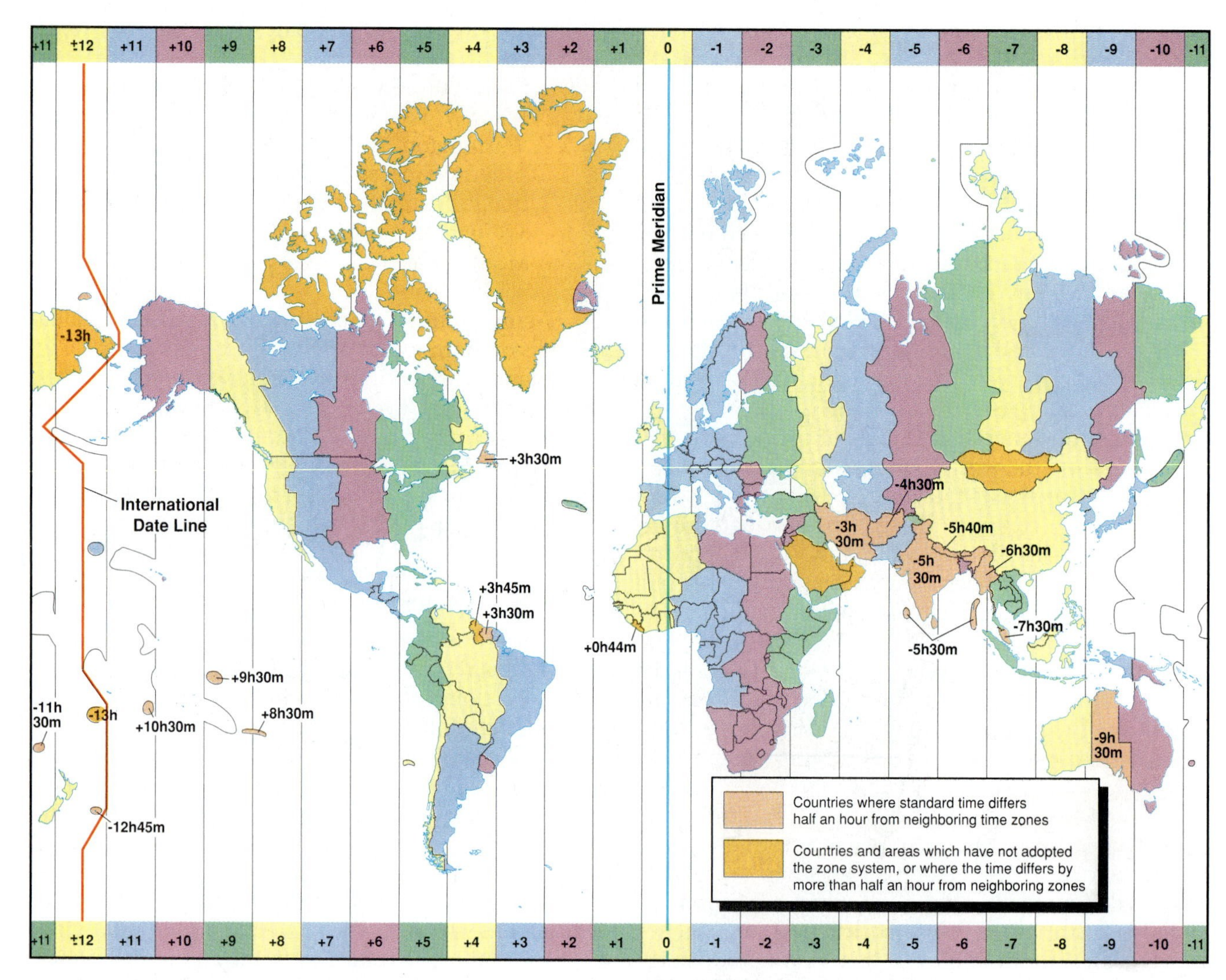

Figure 2-8 The world time zones and International Date Line.

The time differences on either side of the prime meridian accumulate, so that on the other side of the globe the areas either side of 180° of longitude are a whole day apart. The 180° meridian therefore marks the *International Date Line*. (Like the time zones, the International Date Line is shifted a little from the 180° meridian to encompass national boundaries.) Those crossing it from east to west skip a day, going, for example, from 11:00 PM on Saturday night to 11:00 PM on Sunday night in a split second. Those crossing the International Date Line from west to east at the same time have a longer weekend, going from Sunday night back to Saturday. Traveling through time zones, particularly on long east-to-west or west-to-east airplane flights, is largely responsible for the phenomenon of jet lag.

Map Projections

A map should show the spatial relationships of features at Earth's surface as accurately as possible in terms of distance, area, and direction. These requirements can be met when the cartographer is dealing with a relatively small area, such as a county or a small state in the United States. Over this area Earth's curvature is minimal, and the surfaces can be mapped onto a piece of flat paper with little distortion. However, at a global scale the problems are acute; the skin of a sphere will simply not lie flat without being distorted in some way. No flat map can portray shape, distance, area, and direction over the spherical globe accurately. The different portraits of Earth shown in Figure 2-9 are different **map projections,** each of which attempts to min-

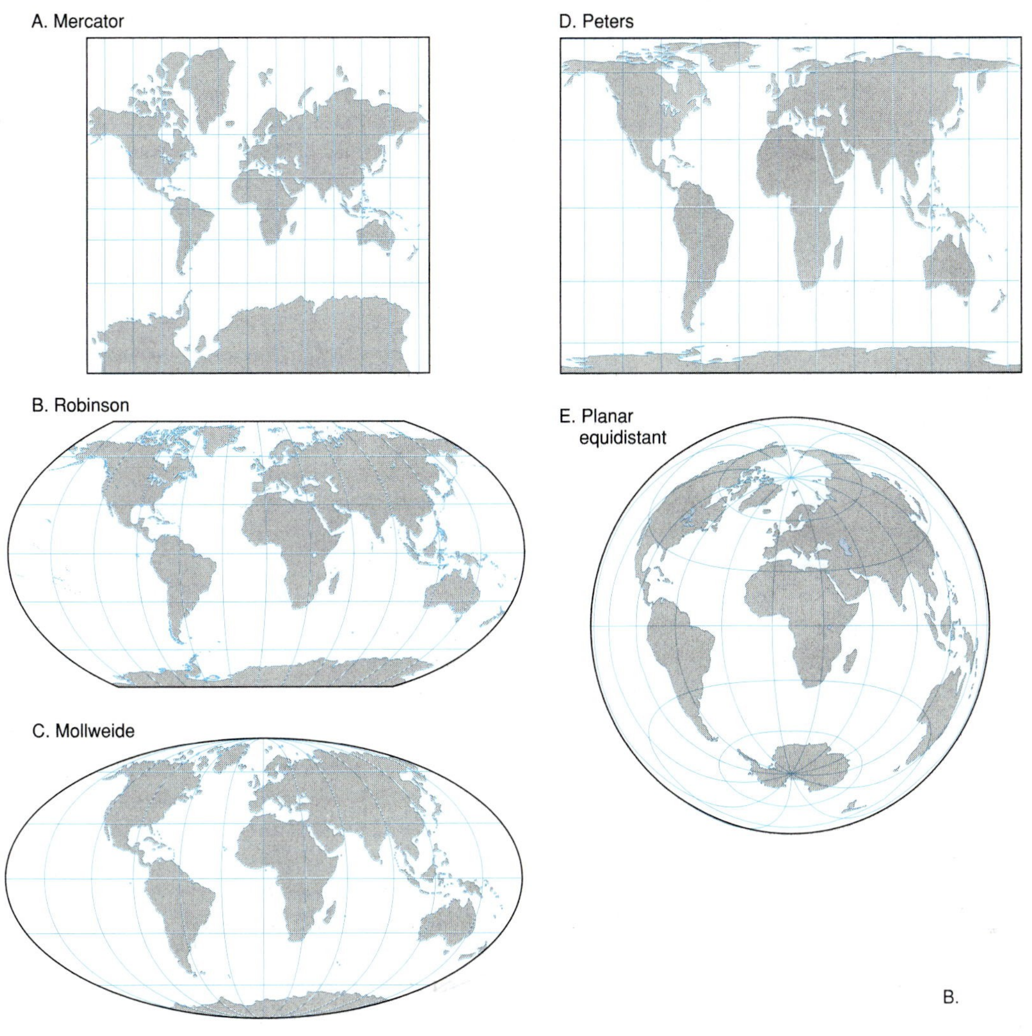

Figure 2-9 Examples of map projections in common use.

imize overall distortion, or to maximize the accuracy of one measure of space. There are no right or wrong map projections, because they all contain distortion of one sort or another. In particular, in virtually all projections, linear scale is not constant across the map, meaning that measurement of long distances will be inaccurate.

The simplest way to imagine constructing a map projection is to think of a transparent globe with a light bulb inside it and a sheet of paper touching the globe in one of the three ways shown in Figure 2-10. The outline of the continents and the lines of latitude and longitude are silhouetted on the paper to form the map. In fact, most projections are not constructed as simply as this; they are calculated mathematically and may be a compromise among

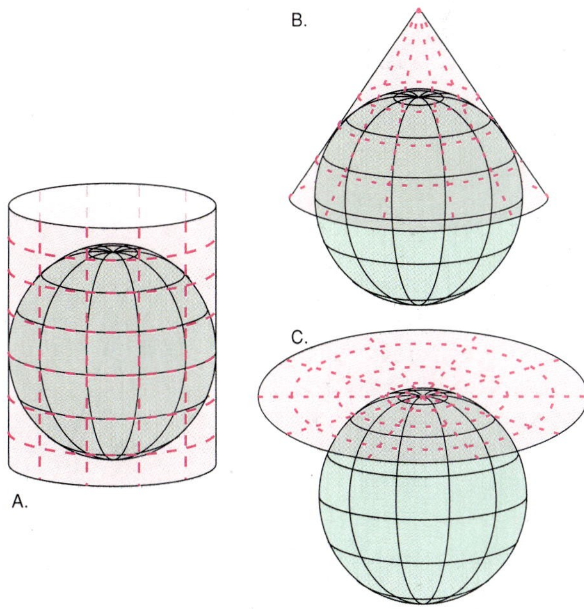

Figure 2-10 Methods of constructing map projections. **A,** Cylindrical. **B,** Conical. **C,** Planar. The cylindrical and conical projections are "unwrapped" from the globe to give a flat map.

a number of methods, designed to preserve the best features of each. There are three main desirable properties in a map, and different projections attempt to preserve each one.

Equal-area projections, as the name suggests, maintain the areas of the continents in their correct proportions across the globe. They are used to depict distributions such as population, or zones of climate, soils, or vegetation. The Robinson projection used for world maps in this book is an equal-area projection. The scale distortion in *conformal maps* is equal in the two main directions from the projection's origin but increases away from the origin. This means that conformal projections maintain the correct shape over small areas, but the outline of continents and oceans mapped over a larger area is distorted. *Azimuthal maps* are constructed around a point or *focus*. Lines of constant bearing, or compass direction, radiating from the focus are straight lines on an azimuthal map. Distortion of shape and area is symmetrical around the central point and increases away from it. The azimuthal projection is used most often in geography to portray the polar regions, which are distorted in projections designed for lower latitudes. Azimuthal maps can also be equal-area or conformal.

It is mathematically impossible to combine the properties of equal area and reasonably correct shape on one map. The commonly used Mercator projection is a conformal map and demonstrates graphically how poor this projection is for showing geographic distributions. The Mercator projection was developed in 1569 by a Flemish cartographer, Gerardus Mercator, to help explorers and navigators. It has the important property, for navigation, that a straight line drawn anywhere on the map, in any direction, gives the true compass bearing between these two points. However, the size of land masses in the middle and high latitudes (toward the poles) is grossly distorted in this projection. Alaska, for example, appears to be the same size as Brazil, although Brazil is actually five times as large. Greenland appears similar in size to South America. This distortion is partly because the meridians, which actually come together toward the poles, are shown with uniform spacing throughout the Mercator map. Despite its unsuitability for most geographic purposes, the Mercator is still very widely used.

Figure 2-11 Extract from a 1:25,000 map of the Boston area, showing the representative fraction and bar scale.

Map Scale and Symbolism

Maps use relatively small pieces of paper to portray the complexity of Earth environments. This has two implications; first, it means that a single map can show only a selection of the features at the ground. Second, the map must use a **scale** to relate a measured distance on the ground to its representation on the map. Scale is most often quoted as a *ratio* or a **representative fraction,** such as 1:10,000 or 1/10,000, respectively. This means that 1 unit on the map represents 10,000 of the same units on the ground. For example, on this map 1 cm on the map represents 10,000 cm, or 100 m, on the ground. Most maps also have a *bar scale* to show this graphically (Figure 2-11). Maps with a scale of up to 1:100,000 are called *large-scale* maps. Those with scales between 1:1 million and 1:100 million are called *small-scale* maps. These terms can be rather confusing. It helps to think of large-scale maps needing a large piece of paper to represent a given ground area, whereas small-scale maps would fit the same area on a smaller piece of paper.

Vertical scale—measurement of elevation showing hills and valleys—is shown on maps by **contour lines,** which are lines connecting points of equal height above a fixed base, usually sea level. Figure 2-12 shows how the surface relief of an area is mapped using contours. The *contour interval,* or vertical distance between neighboring contours, is chosen to fit the horizontal scale of the map. With some practice the physical geographer can use the pattern of contours to "read" the physical form of the landscape and gain some idea of the physical processes that shaped it. Contours that are close together, for example, indicate steep slopes.

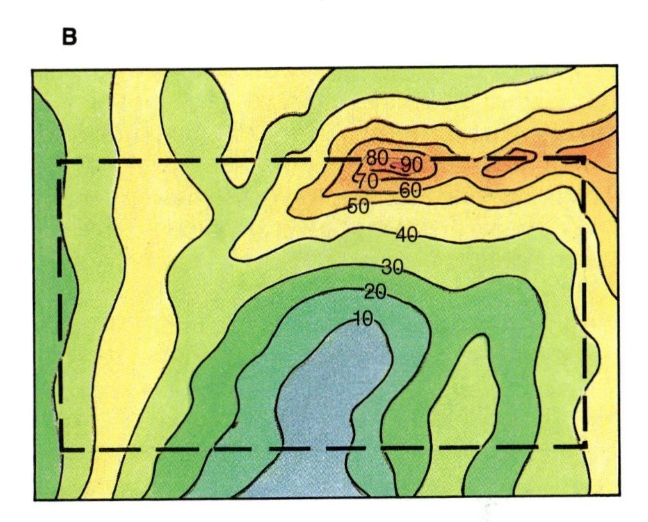

Figure 2-12 Representing height with contours. **A,** The pecked lines on the landscape join points of equal height above sea level. **B,** The landscape is represented by the contour lines in this map; note how the contours are closest together on the steepest slopes.

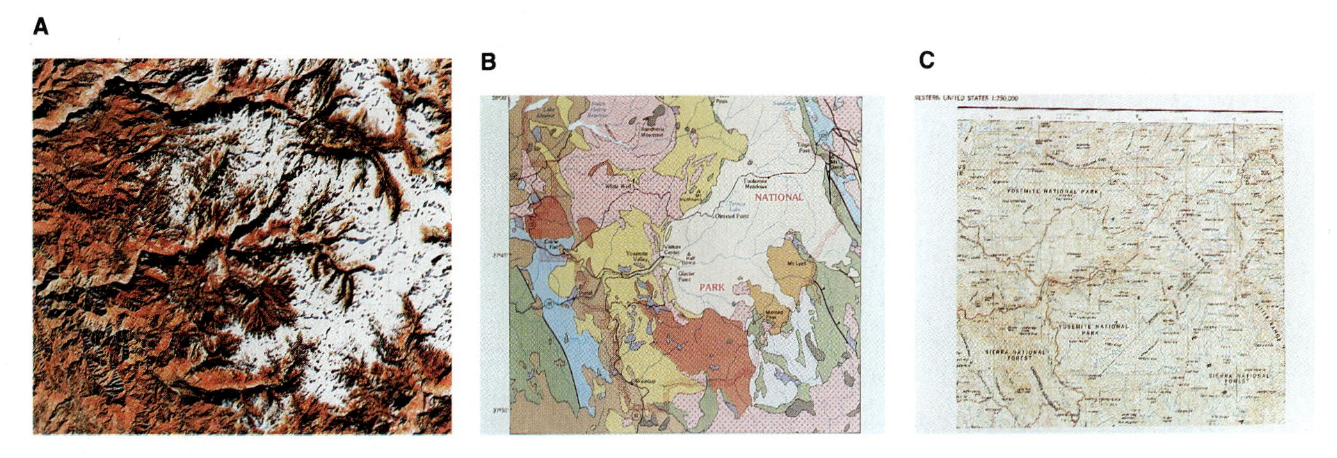

Figure 2-13 A, Satellite image of Yosemite National Park and two examples of the specialized maps and charts produced by the USGS of the same area. **B,** Land-use map. **C,** Topographic map.

Information depicted on the map is a simplification of conditions on the ground. The cartographer selects the features considered important and shows them on the map. Since different things are important to different map users, many maps, each showing different features of the human and physical landscape, can be made for one place. Most maps are multipurpose: they show features of both the physical and human environment that are more or less static, or change only over very long periods of time. These features include relief (mountains, hills, plains, and valleys), rivers, coastlines, major settlements, and transportation routes. Other geographic phenomena are best represented by *thematic maps*, which portray only a single feature and are more easily updated; examples include population, natural vegetation, and the distribution of crops and soils (Figure 2-13). More specialized maps are produced for specific purposes, such as the charts used by navigators in the air and at sea.

Not everything can be drawn on a map at exactly the right scale or exactly how it looks on the ground. For example, to show a transportation network clearly, the cartographer will exaggerate the width of the roads on the map and use different colors or sorts of lines to differentiate between different types of roads. Similarly, urban landmarks, such as schools, may be represented by a symbol, and take up more room on the map than they really do on the ground. Every map should have a key or a *legend*, a list that explains what the map's symbols mean.

All maps are "snapshots" of conditions in the Earth environment at one point in time, and most of them are at least slightly out of date by the time they are committed to paper. This perennial problem of updating has been overcome to some degree by the increasing use of computers for collecting and storing cartographic data. Maps can be updated continually, so that the most up-to-date version is always available in the data bank.

The USGS Map Series

National programs of relief mapping and the production of nautical and aeronautical charts in the United States are the responsibility of the Federal government. The U.S. Geological Survey is the lead agency in the National Mapping Program, producing a family of general purpose maps that show the key features of the human and the physical environments, as well as more specialized products. The symbols used in the general purpose maps are set out in the Topographic Map Symbols section. The variety of human and physical environments in the United States requires maps at a comprehensive range of scales; the widely used scales are listed in Table 2-1. The area represented on each map is a four-sided block called a quadrangle, located by latitude and longitude. Initially the maps were produced entirely from laborious ground survey; these days aerial photographs and some satellite imagery are also used.

Geographic Information Systems

A **Geographic Information System** (GIS) is a recent development from computer-assisted cartography. A GIS is a computer database in which every piece of information is tied to a geographic location. It is used to collect, store, retrieve, analyze, and display spatial data. The data may be input or output as maps, numbers, graphs, and diagrams, or in any other form that makes sense to the user. The importance of GIS, compared to computer cartography, is that GIS moves away from a static representation in map form to the dynamic analysis of many kinds of spatial data.

The data in a GIS can be thought of as a series of layers, all relating to a base grid of location. Figure 2-14 shows some of the types of geographic information that might be held in a GIS designed to monitor a river system that feeds a reservoir. The outputs from this system could include a graph of average river flow compared to rainfall, or a map

Table 2-1 Scales of Topographic Maps Produced by U.S. Geological Survey

SCALE	USGS SERIES	QUADRANGLE SIZE (LONGITUDE × LATITUDE)	AREA COVERED IN QUADRANGLE (SQUARE MILES)
1:24,000	7.5 minute	7.5′ × 7.5′	49 to 71
1:50,000	Intermediate	—	County
1:62,500	15 minute	15′ × 15′	197 to 282
1:100,000	Intermediate	30′ × 60′	1145 to 2167
1:250,000	United States	1° × 2° or 3°	4580 to 8669
1:1,000,000	United States	4° × 6°	73,734 to 102,759

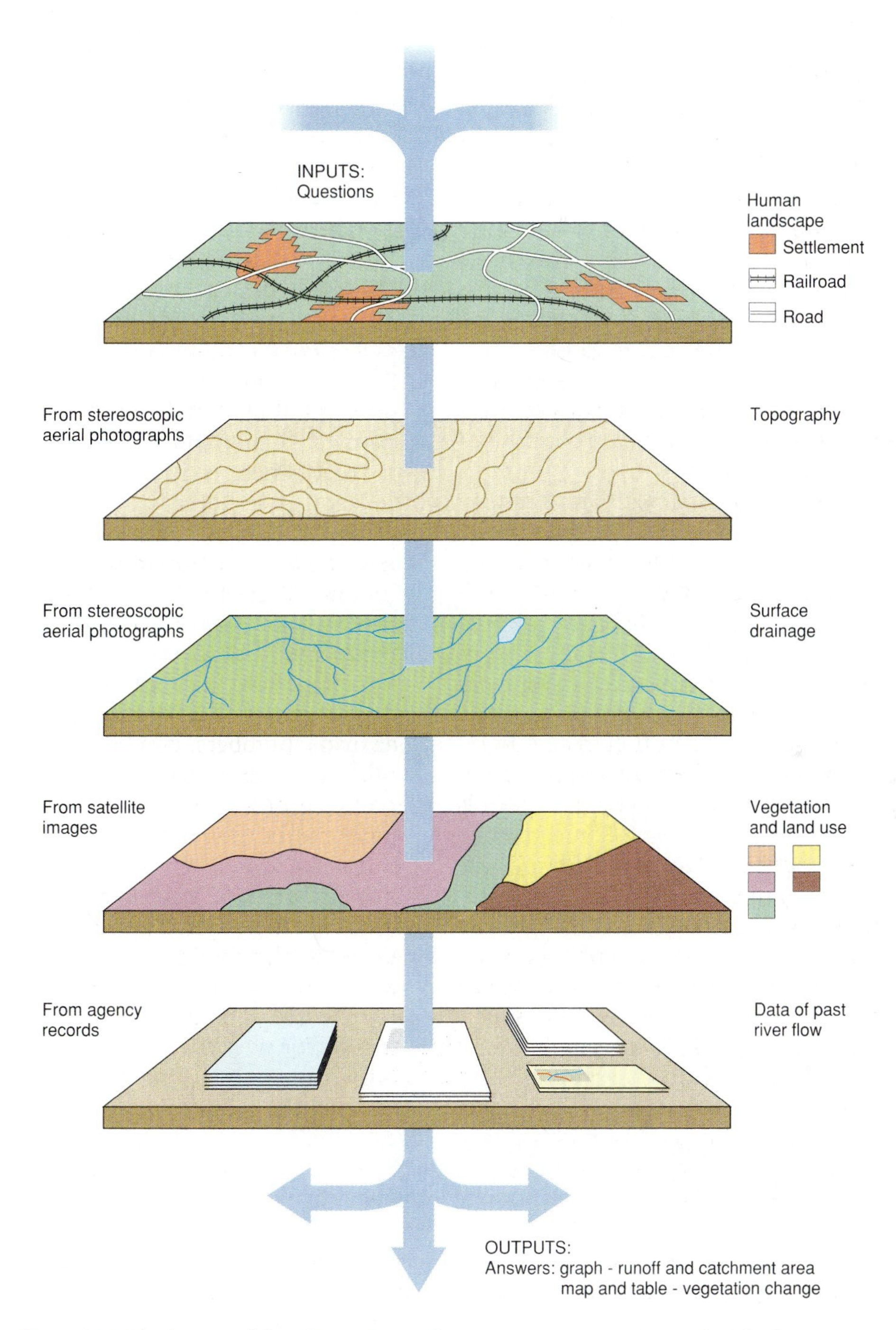

Figure 2-14 The layers of data in a GIS used to monitor a river system that feeds a reservoir. The output of this system could include graphs of seasonal streamflow, divided by drainage basin and/or dominant vegetation type, and changing land use in the drainage basin.

and table of yearly vegetation change. Much of the current research effort in GIS is in integrating disparate data sets about Earth environments, collected to different standards and at different scales. This is particularly important if remotely sensed data, from aircraft and satellites, are to be fully integrated with other forms of geographic data.

REMOTE SENSING

Remote sensing is the use of aerial photographs or satellite images to gather information about Earth environments. In contrast to maps, remotely sensed images show virtually all surface features in the landscape and can, in most cases, be rapidly updated. Remote sensing extends the researcher's access to Earth environments over time and particularly over space. It is a cost-effective way to survey large or inaccessible areas.

Principles of Remote Sensing

Remote sensors gather information about an object without actually touching it; eyes and ears are remote sensors that humans use constantly to find out about their surroundings. The remote sensors in aircraft and satellites are electromechanical copies of these living sensors, with some important differences.

Most remote sensors in aircraft and satellites collect and record the amount of solar energy reflected toward them from Earth's surface. Some remote sensing systems are sensitive to the radiation given out by Earth itself. Others, called active systems (usually radars), have an internal source of energy; these systems send short pulses of energy to the ground and record how much of it bounces back. Different objects at Earth's surface or in the atmosphere reflect or emit all these kinds of energy in different ways. Geographers can therefore use the measurements of energy received by the sensor to differentiate and to recognize geographic features in Earth environments.

If the geographers were in the plane or the satellite themselves, with some powerful binoculars, they would be able to differentiate and recognize some features of the Earth environment, but many of their observations would be incomplete. This is because human eyes are sensitive to only a small part of the energy coming from the sun and Earth, called light, or, more technically, the visible part of the electromagnetic spectrum (Chapter 4). If humans could see other types of energy, in other parts of the spectrum, the world would look very different, and the airborne geographers would be able to distinguish many more features of their environment. Luckily, artificial remote sensing systems can receive, and distinguish among, many types of electromagnetic energy. The images they produce literally give humans a new view of the world. In particular, remote sensing systems that are sensitive to near infrared energy show good differentiation of vegetation. Typically a remote sensing system records energy in several distinct parts of the spectrum, called *wavebands.* Sensors are usually designed to differentiate particular surfaces, such as different types of vegetation. The wavebands recorded by each sensor are the ones in which the objects to be recognized are most clearly separated.

Remote sensors can be mounted on more or less any kind of platform, from the back of a truck to a manned or unmanned balloon or a satellite deep in space. There is also a choice of ways to collect the data, either directly on film as a photograph, or as a series of electronic signals from a digital scanner. Physical geographers most often use cameras mounted in aircraft to produce **aerial photographs,** and digital scanners mounted in spacecraft to generate **digital satellite images.**

There are two distinctive types of orbit for remote sensing spacecraft (Figure 2-15). Satellites in **geostationary orbit** are about 36,000 km (22,500 miles) away from Earth; their speed is adjusted so that it matches Earth's rotation, and they therefore appear to be stationary over a particular point of Earth's surface. There is an international network of geostationary satellites around the equator, which gives complete coverage of the low and middle latitudes. Typically these sensors send down very frequent images, as many as one every half hour, of about half the globe. They are used mainly to monitor weather systems.

In contrast, satellites in **polar orbit** are much closer to the ground, at heights of around 900 km (500 miles), and move in relation to Earth's surface. A polar-orbiting satellite therefore takes fewer images of any particular place. However, because polar orbiters are close to Earth, they show much greater spatial detail and are more useful for local or regional scale studies in physical geography. The wavebands to which remote sensing systems are sensitive, the amount of spatial detail they can discern, and the frequency with which images are collected are all important characteristics of remote sensing data for the physical geographer.

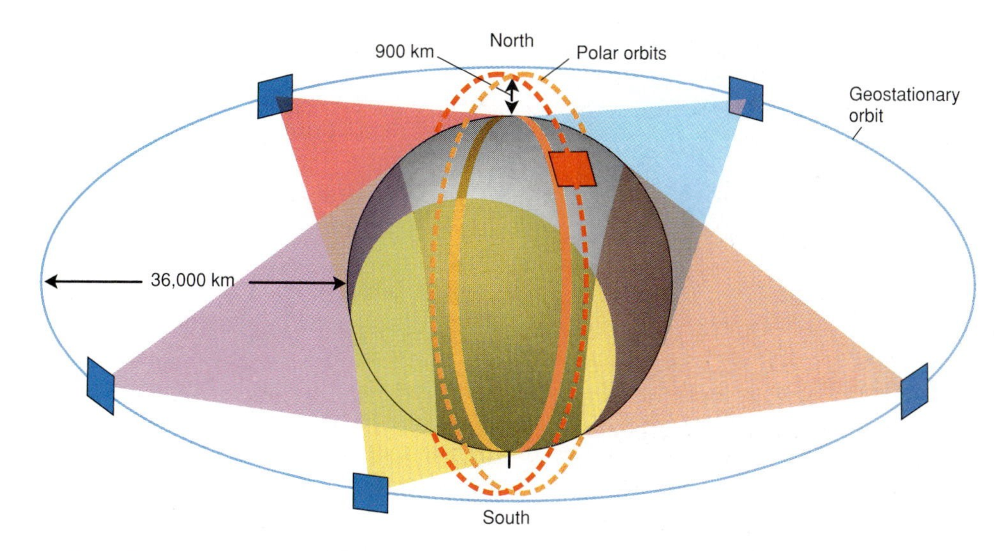

Figure 2-15 Geostationary and polar orbits for meteorologic and earth observation satellites. Geostationary satellites view a large area of the globe at a time, but in poor detail. Polar orbiting satellites record a more detailed image but take a longer time, up to 16 days, to cover the whole of Earth's surface.

Aerial Photographs

Photographs are made by letting the sun's energy fall onto a film that undergoes chemical changes where light (or some form of infrared energy) strikes it. A combination of filters and different types of film provides several types of aerial photograph for research purposes. Of these the standard black-and-white photograph is cheapest, most widely available, and most widely used by physical geographers. The spatial detail that can be seen in the image is determined by the flying height of the plane and the composition of the film, which is chosen on the basis of available light conditions and the speed of flight. Where no ground reference is available, the scale of the photograph can be worked out from the focal length of the camera and the flying height, both of which are normally recorded at the edge of the print.

Most physical geographers practice the art of **photointerpretation** with only limited training, using clues such as shading, texture, size, and shape to identify different features of Earth environments. **Photogrammetry,** on the other hand, is a highly technical skill that requires professional training. Photogrammetrists take precise measurements of horizontal and particularly vertical distances from photographs, typically to produce maps. Measurement of relief from pairs of two-dimensional photographs is possible because of the phenomenon of **parallax.** This is the apparent displacement of an object when viewed from different angles; you can try this yourself by seeing how an object in your room "moves" when you look at it with just your right eye, and then with just your left. By taking two overlapping photographs that show the same scene from different angles, scientists can take advantage of parallax. Using overlapping pairs of photos and an instrument called a *stereoscope,* the scientists can effectively lift themselves to the flying height of the plane and construct a three-dimensional version of the combined photos. The French SPOT satellite is the first to provide stereoscopic images from space that can be recreated on the computer screen.

Digital Satellite Data

Collecting satellite images is a more complex process than taking aerial photographs for a number of reasons. The satellite platform is a long way from Earth and must continue working in the depths of space without maintenance. The data recorded in each waveband, plus information on the satellite's position and general health, are collected and transmitted to the ground as a series of electronic pulses. These signals are decoded and rearranged by complex computer programs into usable data. Further complicating the process, the satellite has to "see" through the distortions of Earth's atmosphere to get useful information about the surface. Not surprisingly, satellites and receiving stations take a long time to design and build and cost billions of dollars.

Figure 2-16 Part of a satellite image of Plymouth, England, taken from the Thematic Mapper sensor on Landsat 4. The image is made up of a grid of picture elements, or pixels, each one representing an area 30 meters by 30 meters on the ground.

The precision of the engineering involved throughout the system is staggering, and it is a tribute to its excellence that scientists can receive such high-quality data about Earth's environment from so far away.

A close look at a satellite image shows that it is made up of a grid of little squares called **pixels** (short for "picture elements"). This is a clue to the way in which the data are collected (Figure 2-16). The ground area represented by a pixel determines the level of spatial detail in the image. Currently the most detailed spatial information, a single pixel showing an area on the ground that is 10 meters by 10 meters, is available from the French SPOT satellite.

The energy reaching the satellite from Earth is recorded by a series of tiny detectors, each of which is sensitive to energy in a particular waveband. The detector "sees" the area of a single pixel at a time; it cannot pick out any variation within the pixel (Figure 2-17). A polar-orbiting satellite builds up a complete and continuous picture of the area beneath it as it moves through its orbit. The data are held as a grid of electronic signals; each pixel has a series of signals or numbers assigned to it, which represent the amount of energy recorded for that

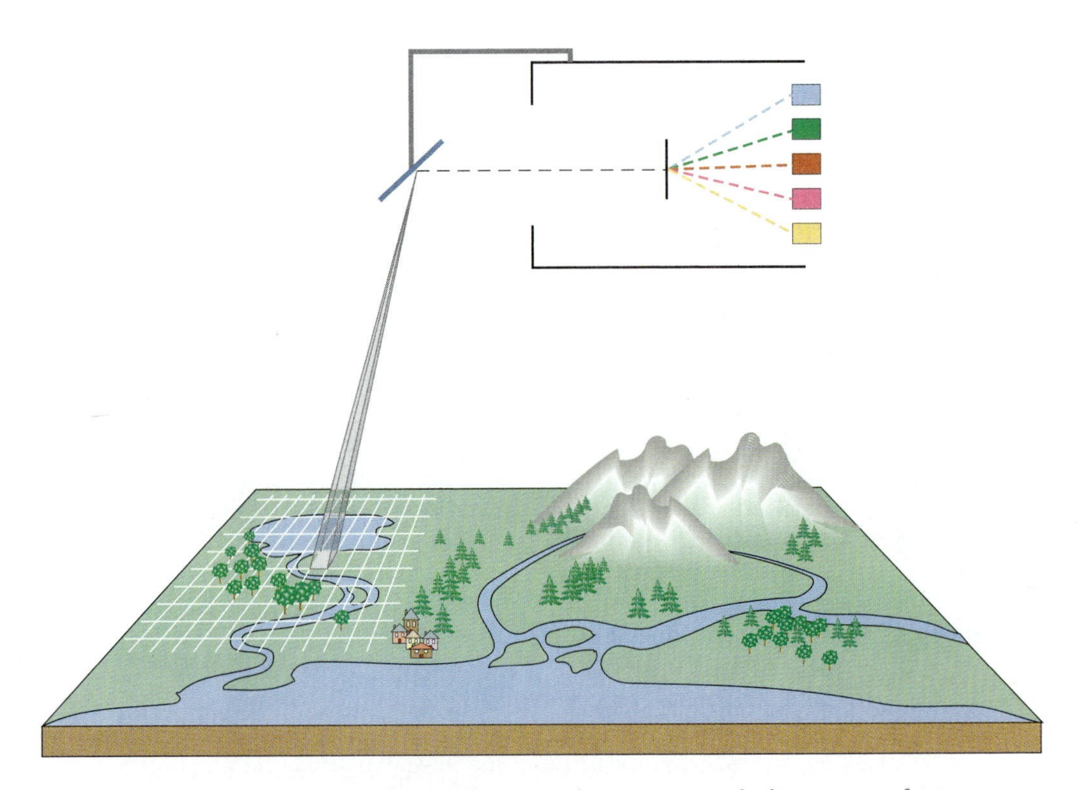

Figure 2-17 Data collection by a digital sensor. The sensor records the amount of energy reflected or emitted from each pixel in turn. Each waveband is recorded separately.

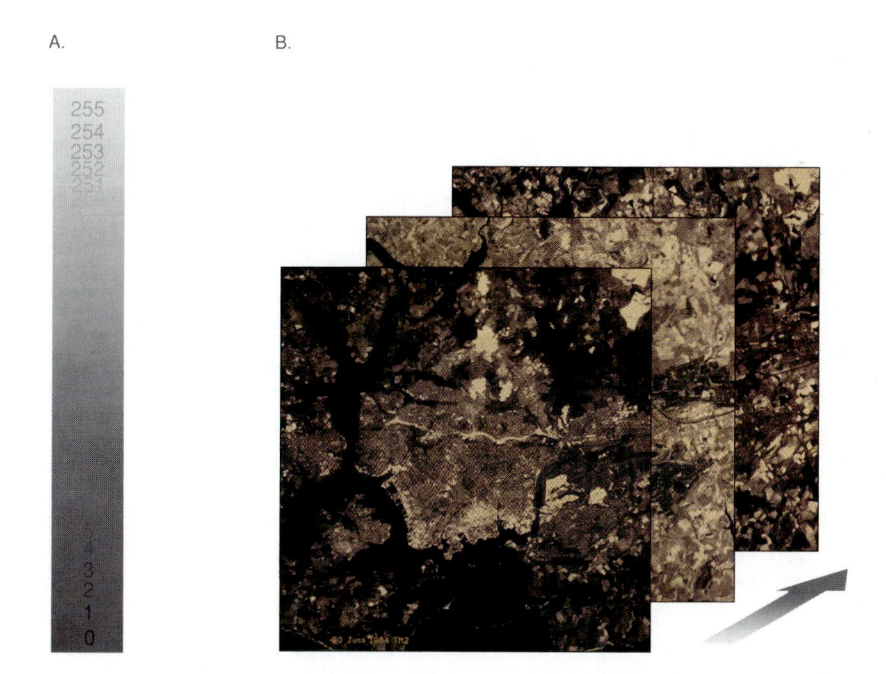

Figure 2-18 The way in which digital data form an image. **A,** Low levels of recorded energy appear dark on the image and pixels with a high recorded energy appear light. **B,** Three single waveband images of Plymouth, England, show different information about land use in the region.

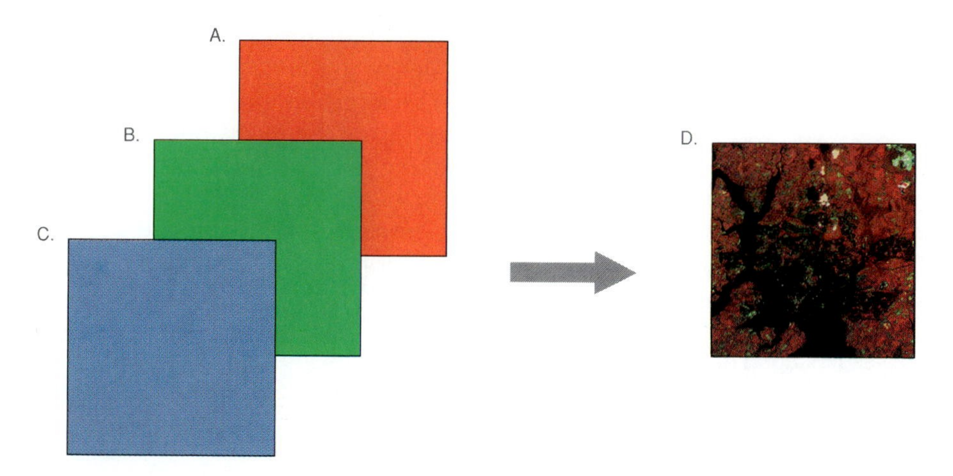

Figure 2-19 Color composite images. The human eye is more sensitive to variation in color than it is to shades of gray and most satellite images are used in color form. One waveband image is shown in shades of red *(A)*, a second in shades of green *(B)*, and the third in shades of blue *(C)*. The three images are overlaid in a computer to produce a color composite image *(D)*. The wavebands can be assigned to blue, green, and red in different combinations to give different color images.

pixel in each waveband. These data are transmitted down to a ground receiving station, and after some processing, are ready to be put back together as an image.

The way this is done is shown in Figure 2-18, *A*. The numbers recorded in one waveband usually fall between 0 and 255. They can be translated into an image by showing any pixel recorded as zero in black on a specialized computer screen, and any pixel recorded as 255 in white. All other numbers end up as a shade of gray between these two extremes. Figure 2-18, *B*, shows the images created in this way for three wavebands from the Landsat Thematic Mapper sensor. It is clear that each of these wavebands contains different information about Earth environments; to use them, each could be interpreted in turn and compared to the others. However, human eyes are not particularly sensitive to differences in shades of gray; they are much better at seeing subtle differences in color. Figure 2-19 shows how three of the original waveband images can be blended together to form a *color composite* image. This is the form in which satellite data are most often used by physical geographers.

The analysis and manipulation of the data need not stop here. Because the data are still held on the computer as a series of numbers, geographers can do all sorts of mathematical operations on them and display the results as images. Images of the same region at different dates can also be digitally matched to one another to examine changes over time. This is an important advantage of digital data. They have much greater flexibility in analysis than photographs. Geographers are likely to be able to extract more useful information from these data with a computer than by standard photointerpretation of the original image.

DATA ANALYSIS AND EVALUATION OF RESULTS

Fieldwork, laboratory experiments, maps, GIS, and remote sensing all provide data about Earth environments. To turn these data into useful geographic information, some kind of synthesis, summary, or analysis of the data is needed. To turn them into an explanation or understanding requires a critical interpretation and evaluation of the analysis and the results. Much of the data gathered from the sources discussed above is in numerical form, and a central part of many research programs lies in hunting for meaningful numerical patterns in large data sets, a process much helped by the increasing use of sophisticated computers.

An elementary understanding of numerical analysis is important to the physical geographer, to interpret and evaluate the published work of other researchers. A data set can be usefully summarized by **descriptive statistics,** which convey the most important features of the data set in a compact and manageable form, such as a mean or average. Most data collected in geography are a sample of measurements from a much larger population of all possible measurements. For example, rainfall is usually measured at only a few points, and river flow may be measured every hour rather than continuously. In most cases the geographer is trying to draw conclusions about the whole population from the sample; **inferential statistics** help in this. Inferential statistics work by calculating the probability that a pattern established in the sample by a series of measurements, such as a relationship between rainfall and the volume of river runoff, also exists in the population of all rainfall and river flow figures for the same valley. If this probability is acceptable, the geographer can be confident about generalizing from the initial results. Larger samples generally are more likely to match the characteristics of the population.

The results of the experiment show whether the researcher can answer "no" or "maybe" to the hypothesis. For this contribution to knowledge to be useful the physical geographer must constantly evaluate and criticize the experiment. "Have I got the hypothesis right? Did I overlook any variables? Am I using the most appropriate data sources? Am I taking samples in the right places? Is this the best form of analysis?" The acid test comes when the experiment is written up, published, and circulated to other physical geographers for critical comment. Geographers' understanding of Earth environments advances when this building block of knowledge is accepted and forms the basis for another cycle of observation, experimentation, and explanation.

CASE STUDY OF INVESTIGATIONS IN PHYSICAL GEOGRAPHY

This case study shows how geographers employ, in active investigation, the tools described in this chapter. The "Investigations in Physical Geography" boxes look at other specific problems in geography and some of the new methods of collecting and analyzing geographic data for these problems.

Surging Glaciers in Svalbard, Norway

Physical geographers interested in the effects of moving ice have observed that particular forms of the surface-relief environment are the result of particular processes of ice movement. In this study, the researchers (a group from Europe and the United States) had noted that certain accumulations of debris at the edges of glaciers, called push moraines or composite ridges, seem to be associated with a particularly rapid type of glacier movement called surging. The aim of this study was to investigate the relationship between the landform—piles of rock debris—and the behavior of the glacier. The study results have two important implications. First, in environments that are glaciated today, they will explain in practical terms the link between processes in the atmosphere-ocean environment (the formation and movement of ice) and the features at Earth's surface. Second, the results help to explain the nature of landscapes that were once glaciated but now enjoy a warmer climate. The hypothesis to be tested was that push moraines are formed by surging glaciers rather than by a more gradual movement. The data collected concerned conditions around about 30 glaciers in Svalbard, Norway.

The data for this survey came from a wide variety of sources. Primary observations of moraines in the field were made at some sites, and these were backed up by laboratory work that identified the origin of the debris forming the moraines. The study was extended to other areas through secondary data in the form of aerial photographs. Satellite images were used as well, but these proved to have too coarse a spatial resolution to be useful. Data on the behavior of individual glaciers were also extracted from the published records of the World Glacier Monitoring Service. The results confirmed the researchers' original observation, that under certain conditions of valley slope and sediment types, composite ridges are most likely to result from surging glaciers.

FRONTIERS IN KNOWLEDGE
Methods of Geographic Research

One of the biggest issues in physical geography is how to piece together the results of our experiments to give a coherent picture of Earth's environment at all scales from the local to the global. Most geographic experiments are at the local scale; building them up to the regional and global scale is an uncertain business. A second and related problem is the analysis and interpretation of data that come from a great variety of sources and in a great variety of forms. This is particularly important in examining geographic questions at the global scale. If we are to compare results from Brazil, Australia, Russia, France, and the United States, the data must be collected to the same degree of precision and in a similar form. This means that there will, ideally, be a growing level of international collaboration and standardization in data collection.

INVESTIGATIONS in Physical Geography

Earth Systems Science

Chapter 1 of this book highlights the growing significance of environmental issues in public debate. In the 1990s, researchers, including physical geographers, are investigating new and urgent questions about Earth environments. The key words in this new research are *global, dynamic, interaction,* and *human impact.*

Satellite remote sensing, backed by detailed ground measurement and sophisticated computer analysis, has given scientists literally a new view of Earth. The global view shows that Earth is a dynamic and interacting system, that structures in the atmosphere, such as eddies and fronts, are mirrored in the ocean, and that flows in both the atmosphere and the ocean are affected by the distribution of land masses. Scientists cannot explain many of the patterns shown in satellite images, and international collaboration is needed to investigate them further.

The sense of urgency in global research has increased since the late 1980s as scientists realize the extent of human influence in global change. Until recently, scientists considered human impact on Earth environments to be minimal and localized. However, the global view suggests that the cumulative effect of human impact on Earth environments is far greater than had been thought. As a whole, individual cases of air and water pollution, deforestation, and agricultural expansion show that humans are altering the composition of the atmosphere and oceans and redistributing Earth's vegetation at an unprecedented rate. The global importance of human action was noted as early as the 1960s and 1970s in the phenomenon of acid rain. Sulfur output from factories and power plants combines with water droplets to fall as a weak acid that can damage lakes and vegetation hundreds of miles from the smokestack. But the most urgent research in this decade involves global warming (Chapters 3, 4, and 9) and the decrease in the ozone layer (Chapter 3). Both problems are partly the result of human activity; humans contribute to global warming through the emission of carbon dioxide and other gases to the atmosphere; in addition, they contribute to the ozone hole through the release of chemical compounds that attack ozone. Both problems threaten the future of humans on Earth.

To determine the extent of human impact, scientists must first understand the processes of the natural system, so that, for example, natural short-term fluctuations are not interpreted as long-term directional changes caused by humans. Scientists' concern about human impact stems in part from their poor understanding of the natural global system. The challenge to scientists in the 1990s and beyond is to describe, explain, and predict the interactions of Earth environments on a global scale.

In response to this challenge, a new form of research, called Earth Systems Science, is emerging. The National Aeronautics and Space Administration (NASA), in addition to other national and international organizations, has investigated the key issues of Earth Systems Science and identified directions and methodologies for research. According to NASA, the goal of Earth Systems Science is:

> *To obtain a scientific understanding of the entire Earth system on a global scale by describing how its component parts and their reactions have evolved, how they function, and how they may be expected to continue to evolve on all timescales.*
>
> —Earth Systems Science Committee, NASA Advisory Council

The panel also defines a challenge to the researchers who practice Earth Systems Science:

> *To develop the capability to predict those changes that will occur in the next decade to century, both naturally and in response to human activity.*

The central philosophy of Earth Systems Science is that research into global systems must cross the traditional boundaries between scientific disciplines.

For example, a scientist studying the effect of volcanic activity must follow the transfer of materials and chemical reactions from the solid-earth through the atmosphere-ocean to the living-organism environment. In practice, since no one person has enough knowledge for detailed study of the many processes involved, the work is done by a team. This kind of interdisciplinary work seems logical to geographers, who have filled this role in science for a long time. In its simplest form, Earth Systems Science is geography at a global scale. Physical geographers therefore have an important role to play in undertaking and coordinating Earth Systems Science research.

The programs of research in Earth Systems Science share three characteristics. First, at least to some extent, they are cross-disciplinary. In most cases a large interdisciplinary program is split, by subject specialty, to manageable units. The team organizing the project will have the important task of drawing together findings from all the units and communicating them to the wider scientific community. Second, the data collection program is organized on a continental or oceanic scale and usually involves international collaboration. Third, the funding for these programs is

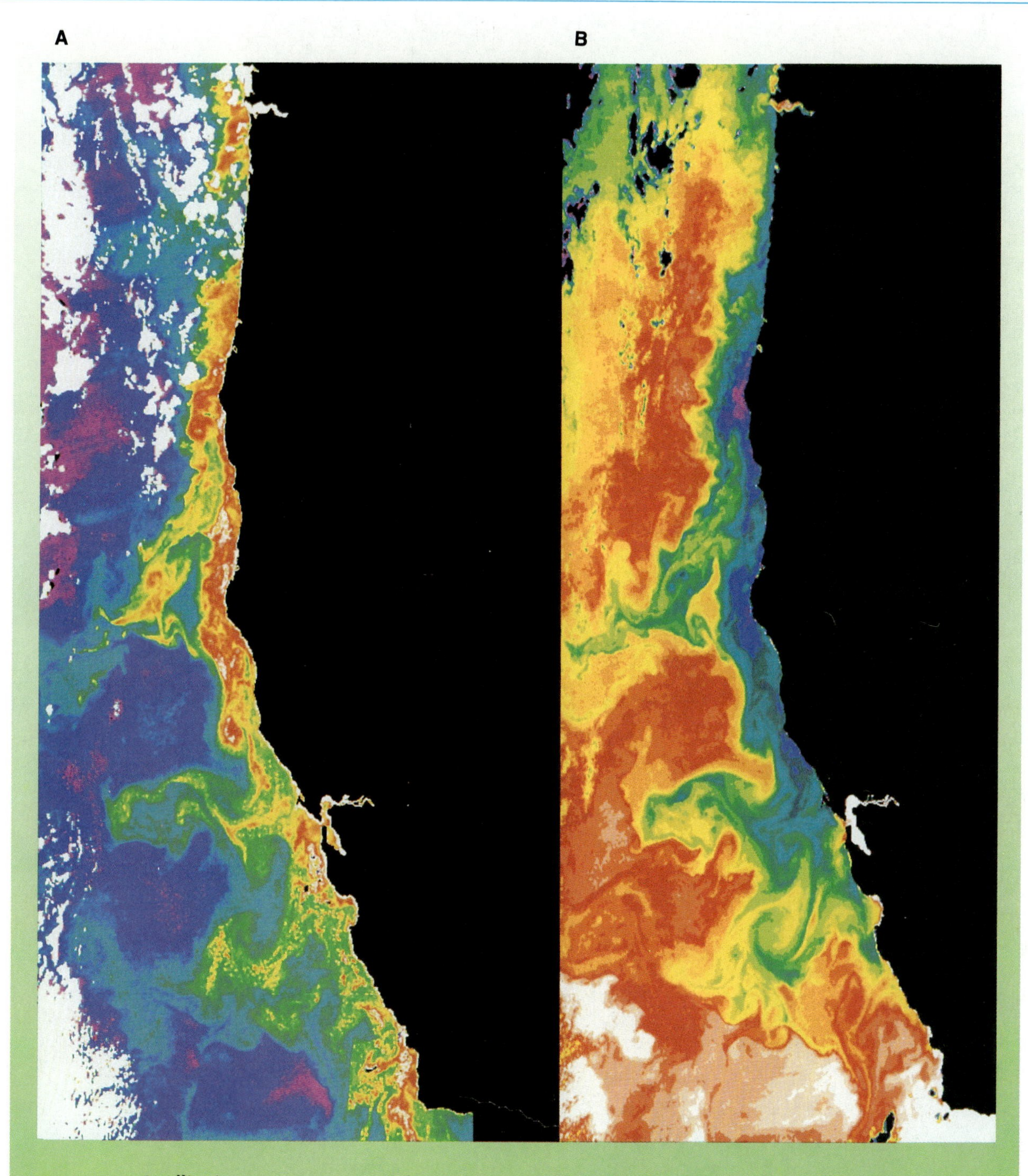

Satellite images: two images taken 8 hours apart on July 8, 1981, show sea surface temperature (right, *A*) and concentration of phytoplankton (left, *B*) in the Pacific Ocean off the west coast of the United States. **A,** In the temperature image, the California Current (yellow and red) meanders south, mixing with the cooler coastal waters. **B,** The coastal waters are rich in nutrients and provide a base for high concentrations of phytoplankton, shown in reds and yellows. These satellite images have altered scientists' perceptions about the flow of the California Current and the patterns of phytoplankton distribution, demonstrating that both are more complex and dynamic than previously thought.

planned for decades. Virtually all science funding comes from federal budgets on which there are many competing demands. Long-term financial commitment to Earth Systems Science programs marks a major breakthrough in science funding and demonstrates the importance with which governments and their electors perceive the issue of global change.

Most of the Earth Systems Science programs are actually part of larger scientific plans, organized by international bodies such as the International Council for Scientific Unions, the World Meteorological Organization, and the United Nations Environmental Program. Many programs are complementary. For example, the World Ocean Circulation Experiment (WOCE) is related to the Joint Global Ocean Flux Study (described in Chapter 4). The circulation experiment is a worldwide exercise to develop computer models of ocean circulation. Particular topics of interest are the transfers of heat among the Pacific, Indian, and Atlantic Oceans, and the interaction of the atmosphere and oceans in this process. The results from this experiment will help scientists working on the flux program, which concentrates on the exchange of chemical elements, particularly carbon, between the atmosphere and ocean and the factors affecting this exchange.

The task facing Earth Systems Science is enormous and, compared with other scientific programs, the costs are high and the results long-term and uncertain. Most scientists and policymakers agree, however, that the costs of *not* meeting the challenge of Earth Systems Science are higher still. For many scientists, including physical geographers, changing the focus of their work to the global interactions of Earth environments is a leap into the unknown. NASA's Committee on Earth Systems Science encourages them with this quotation:

Whatever you do, or dream you can, begin it.
Boldness has genius, power and magic in it.
Begin it now.

—Goethe

Human settlements, such as the urban area of Los Angeles shown here, can have a dramatic effect on Earth environments. Emissions from industry and from car exhausts alter the structure of the atmosphere, causing local pollution and contributing to global changes in climate.

SUMMARY

1. Physical geographers examine the structure and the processes of Earth environments and their interaction.
2. Physical geography examines the variability of the natural environment, its change over time, and particularly over space.
3. In studying the environment, the physical geographer typically moves from the observation of an unexplained feature, through its description, to its explanation, and possibly its prediction and management.
4. The physical geographer may use primary data from the field or laboratory, and secondary data from remotely sensed images, maps, or other recorded data.
5. A map is a partial representation of Earth's surface. Features on the map are located by latitude and longitude and often by another local referencing system. Distances on a map are related to those on the ground by a scale.
6. All maps are representations of a curved Earth surface on a flat piece of a paper and are therefore distorted. Different map projections have different distortions.
7. A Geographic Information System is a computer database in which every piece of information can be tied to a geographic location. The principal advantage of GIS over conventional cartography is that a GIS allows more dynamic analysis of the data.
8. Remote sensing data such as aerial photographs or satellite images are an important source of information in physical geography. Remote sensors record the amount of electromagnetic energy reflected or emitted from a surface and use this to differentiate and recognize different surfaces.
9. An investigation in physical geography is not complete until the results have been fully criticized and evaluated.

KEY TERMS

hypothesis, p. 28
primary data, p. 30
secondary data, p. 30
cartography, p. 32
latitude, p. 32
longitude, p. 33
map projections, p. 34
scale, p. 37
representative fraction, p. 37
contour lines, p. 37
Geographic Information System, p. 38
remote sensing, p. 40
aerial photographs, p. 40
digital satellite images, p. 40
geostationary orbit, p. 40
polar orbit, p. 40
photointerpretation, p. 41
photogrammetry, p. 41
parallax, p. 41
pixels, p. 42
descriptive statistics, p. 44
inferential statistics, p. 44

QUESTIONS FOR REVIEW AND EXPLORATION

1. How is physical geography distinct from other scientific disciplines? Give some examples, similar to those presented in the chapter, of the type of questions that physical geographers might ask about Earth's environment.
2. Draw a diagram, similar to Figure 2-3, of an experiment to study one of your questions.
3. What is the relative importance of primary and of secondary data in physical geography?
4. Compare two maps of different scales for the area around your home. What are the differences in information content between them? Which do you think gives the better impression of the environment? Find the representative fraction and bar scale on each map. For each map: what does 1 cm on the map represent at the ground? What distance on the map represents 50 km at the ground? Is it a large-scale or a small-scale map?
5. Find out how the spatial referencing system of your local map works, and how it is related to latitude and longitude.
6. For the geographic problem outlined in your answer to Question 2, consider how a GIS might help you to explain this feature of the environment. What data would you need? In what form would you input the data? What type of output might you use?
7. Compare the types of information you could gather from aerial photographs and from satellite images. In what form of study might one form of remote sensing have an advantage over the other?
8. Explain how digital data from a satellite sensor are reconstructed as an image.
9. Outline the importance of publishing and criticizing the methodology and results of an experiment in physical geography.

FURTHER READING

Dent B: *Cartography,* ed 2, Dubuque, Iowa, 1990, WC Brown.

Gould P: *The geographer at work,* New York, 1985, Routledge & Kegan Paul. An account of recent developments in the study of geography and the ways in which geographers approach specific investigations.

Greenhood D: *Mapping,* Chicago, 1964, The University of Chicago Press.

Hall S: *Mapping the next millennium,* New York, 1991, Random House. An illustrated survey of innovations in mapping and imaging, including applications to medicine and astronomy as well as geography.

Makower J, editor: *The map catalog: every kind of map and chart on Earth and even some above it,* ed 2, New York, 1986, Vintage Books.

Pike J, Thelin GP: Building a better map, *Earth,* 44-51, January 1992. A description of digital shaded relief maps presenting "Landforms of the Conterminous United States," a beautiful map prepared for the U.S. Geological Survey in 1991.

Robinson A et al: *Elements of cartography,* ed 2, New York, 1985, John Wiley & Sons.

Thompson MM: *Maps for America,* ed 3, Washington, D.C., 1988, U.S. Department of Interior. This book contains not only details of the range of maps produced to cover the United States, but many sections on new mapmaking technology, a history of U.S. mapmaking, map projections, and satellite imagery.

The Atmosphere-Ocean Environment

PART 2

The focus of Part 2 is weather and climate. Weather is the result of day-to-day changes in the atmosphere; climate is the longer-term character of the atmosphere-ocean environment. Both can be studied for their local, regional, and global variations, and so form an important part of physical geography. Mark Twain realized that weather is an important part of human experience but felt it was difficult to do justice to its influence in his fiction because of its complexity:

> No weather will be found in this book. This is an attempt to pull a book through without weather. Nothing breaks up an author's progress like having to stop every few pages to fuss-up the weather. Of course weather is necessary to a narrative of human experience. That is conceded. But it ought to be the ablest weather that can be had, not ignorant, poor-quality, amateur weather. Weather is a literary specialty, and no untrained hand can turn out a good article of it. The present author can do only a few trifling ordinary kinds of weather, and he cannot do those very good. So it has seemed wisest to borrow such weather as is necessary for the book from qualified and recognized experts—giving credit, of course.
>
> "The Weather in this Book," *by Mark Twain*

Part 2 of this book is written to ensure that physical geography students do not "do weather" in an amateur way.

The atmosphere-ocean environment is characterized by the materials that compose it, the ways in which they are arranged (the structure), and the processes that give rise to action and variability within the environment. The *compositions* and *structures* of the mixture of gases in the atmosphere and the waters filling the ocean basins are studied in Chapter 3. The processes acting between atmosphere and oceans are so complex that weather elements—measurable aspects of weather such as temperature, wind, cloud cover, and precipitation—are first studied separately. Chapter 4 investigates the causes of *temperature* differences around the world. Chapter 5 explains what makes atmospheric gases move in *winds* and ocean waters move in *currents*. Chapter 6 is about *moisture*—how it gets into the atmosphere, forms clouds and fog, and is precipitated back to Earth's surface.

The atmosphere-ocean environment also functions as a whole, and this is how rocks at Earth's surface,

Indian Ocean Rainbows
Frans Lanting/Minden Pictures

humans, and other living organisms experience it. The weather elements act collectively to produce their effects. On one day it may be sunny, hot, cloudless, and almost windless; on another it may be cloudy with rain and gales. *Weather systems* (Chapter 7) are distinct, short-term, and repetitive patterns of weather elements. Understanding such weather systems makes it possible to forecast the weather up to a few days ahead. Climate, the longer-term outcome of the workings of the atmosphere-ocean environment and its interactions with land surfaces, is studied within two time frameworks. An understanding of *present climatic environments* (Chapter 8) and how *climates change over time* (Chapter 9) is basic to today's increasing interest in environmental change.

Although humans think of themselves as living *on* Earth, in a very important way they and other animals and the plants live *in* the planet. Living organisms exist within a fluid medium of air and water whose properties are constantly changing, and these changes can affect people profoundly. The composition and structure of the atmosphere and oceans are the framework within which the processes of heating, movement, and water circulation take place to cause weather. This chapter provides a basis for understanding the weather-forming processes studied in Chapters 4 through 6. It also highlights some of the interactions among the atmosphere-ocean environment, other Earth environments, and human activities.

COMPOSITION AND STRUCTURE OF THE ATMOSPHERE

Composition of the Atmosphere

Earth's **atmosphere** forms a relatively thin envelope of air, dust, and water droplets around the planet. It fades into space several hundred kilometers above the surface, but 99 percent of the atmosphere's mass is in the lowest 100 km layer (a mere 0.25 percent of Earth's diameter). Relatively, it is such a shallow layer that its thickness has been compared to that of the fuzz on a peach (Figure 3-1). Despite this shallowness, the atmosphere is vital to the existence of living organisms and plays a major part in molding surface landforms.

The atmosphere is composed of a variety of gases, suspended dust particles, and condensed moisture droplets (Figure 3-2). The atmospheric composition is different in the lower part compared with the upper reaches. In the 80 km (50 miles) nearest the ground, the major atmospheric gases are nitrogen (78.03 percent by volume of pure, dry air), oxygen (20.99 percent), and argon (0.94 percent), as shown in Figure 3-3. The proportions of these three major gases in this layer do not vary, and so it is often known as the *homogeneous atmosphere*. The gases near the surface exist as **molecules,** combinations of atoms held together by electric bonds. An **atom** is the smallest particle of an element that can exist alone or combine with atoms of other elements. In a molecule the combined atoms may be of the same element, as in oxygen (O_2), or of different elements, as in water (H_2O), a **compound.**

In the outermost atmosphere, the bonds holding the atoms together are split apart by the sun's energy to form **ions,** atoms or molecules that have a positive or negative electric charge. These ions of gases exist where there is an extremely low concentration of matter because gravity draws the main mass down nearer the ground. The composition of the highest atmospheric layers includes some of the lightest gases, such as **hydrogen** and **helium.**

Figure 3-1 Earth's atmosphere: a section through as viewed from the Space Shuttle. Several layers are distinguished by color. In the lowest layer, tall black clouds are visible over the Amazon basin.

The present balance of gases making up the lowest 80 km of the atmosphere is maintained by continuous exchanges of molecules and atoms among the atmosphere, the oceans, living organisms, soil, and minerals. These exchanges maintain the composition of the atmosphere over periods of hundreds of thousands of years. Although the atmosphere's composition changed dramatically over longer periods in the early stages of Earth's evolution, the present balance has been established over more recent time. The rates and roles of such exchanges reflect the chemical activity of the individual atmospheric gases.

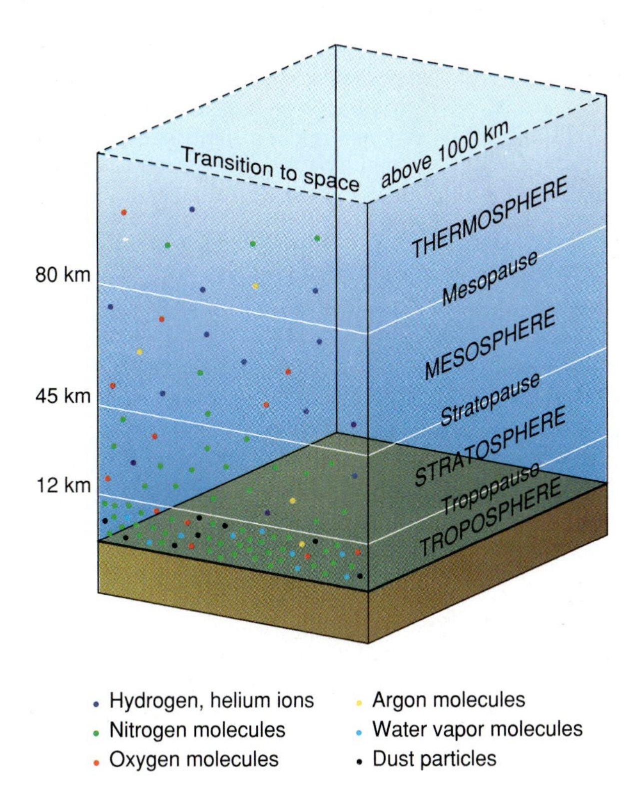

Figure 3-2 The composition of the atmosphere. Seventy-five percent of the total weight of atmospheric gases is concentrated in the troposphere, plus all the water vapor and dust.

The Major Gases of the Atmosphere. Atmospheric **nitrogen** by volume makes up over 75 percent of the lower homogeneous atmosphere. It is not very chemically active, and only a small proportion is removed from the atmosphere by land-based microorganisms (14 billionths of the total atmospheric nitrogen), by lightning, and by industrial processes such as fertilizer manufacture. In such cases, nitrogen leaves the atmosphere and is incorporated or "fixed" in chemical compounds. Before the growth of the fertilizer industry, bacteria balanced the amounts removed from the atmosphere by releasing nitrogen from these compounds back into the atmosphere. On average, a nitrogen molecule spent about 42 million years in the atmosphere during one cycle; the time spent fixed in chemical compounds varied. It is believed that the chemical industry removes more nitrogen from the atmosphere than is returned to it. At present rates it would take 250 million years to remove half the atmospheric nitrogen, so the total impact is still relatively small. However, the excess of fixed nitrogen builds up in streams and lakes, where nitrogen provides extra food for microorganisms and causes them to go on "binges" that foul the waters (see Chapter 19 for a fuller discussion of the links between atmospheric gases and living organisms).

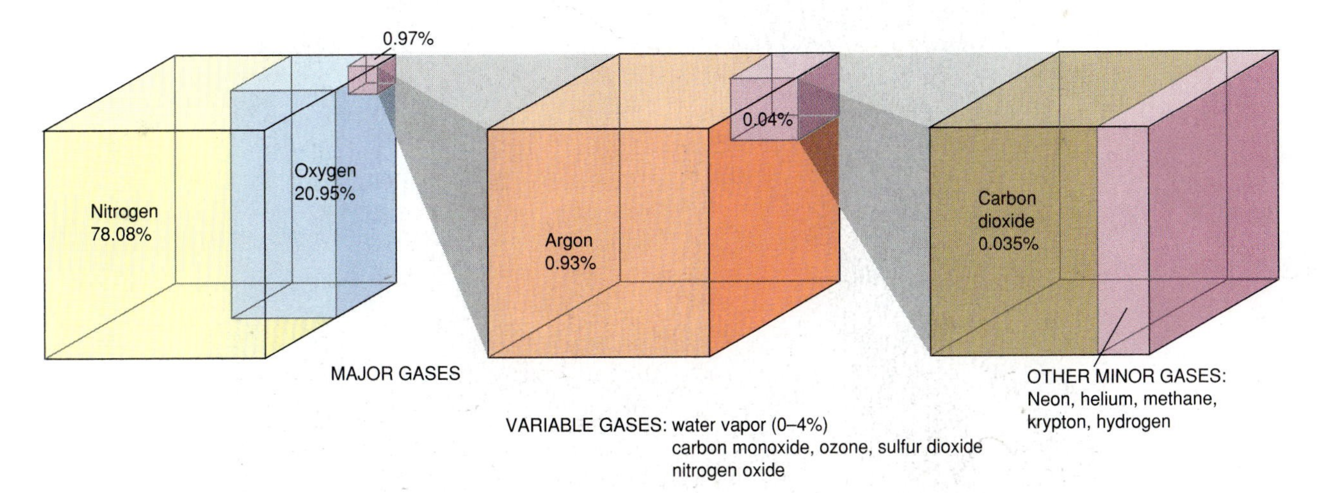

Figure 3-3 The proportions of gases in the lowest 80 km of the atmosphere. The major gases dominate the atmosphere by volume, but some of the minor gases play important roles in weather.

BOX 3-1

THE OZONE ISSUE

Ozone (O_3), a toxic and destructive gas, is a near neighbor of the life-giving oxygen found in the atmosphere-ocean environment. Oxygen (O_2) is vital to life on Earth; its presence in the atmosphere allows organisms to breathe and to survive.

At close quarters, ozone is toxic. It is a dangerous constituent of smog, which is formed near ground level by the interaction of sunlight with pollutants, particularly nitrogen oxides. It can severely damage and even kill city plants and is an irritant and health hazard to humans. At a distance, however, ozone is a lifesaver. Its concentration in the upper stratosphere forms a shield around Earth, preventing most of the incoming ultraviolet solar radiation from reaching the troposphere and Earth's surface. The most destructive forms of ultraviolet energy can break apart important molecules in living tissue; milder forms (longer wavelengths) are associated with skin cancer, cataracts, and breakdown of the immune system. Sunscreen products are designed to stop the small amounts of these harmful rays that reach the troposphere from damaging the skin. The ozone layer in the stratosphere acts as a sunscreen for the whole planet.

In 1985, scientists taking regular measurements of ozone concentration over Antarctica announced that the amount of ozone in the springtime atmosphere had fallen by 40 percent in the period 1977-1984. Other sources, including satellite data, confirmed these findings. In the Antarctic spring, a hole the size of the United States and about 10 km (6 miles) thick appears in the ozone layer, as shown in the satellite images.

Scientists were taken by surprise by these figures, which have considerable implications for the future of life on Earth. Three crucial and interlinked research questions are:

- Is the Antarctic ozone hole the first sign of a general and rapid erosion of the ozone layer over Earth, or is it confined to the unique stratospheric conditions of the Antarctic?
- What has caused the ozone hole?
- What can be done to prevent further destruction of stratospheric ozone?

No place on Earth is experiencing such a depletion in ozone levels as that over the Antarctic. The Antarctic atmosphere is the coldest part of Earth's atmosphere and is effectively isolated from the rest of the globe by the strong westerly winds surrounding it in the midlatitudes of the southern hemisphere. This isolation allows materials from outside to enter in small quantities and to accumulate there but makes it difficult for them to get out. These factors result in a concentration of the effects of ozone depletion there but do not stop it from occurring on a smaller scale elsewhere.

Detection of the phenomenon of the ozone hole over Antarctica led to further research to assess the situation over the Arctic and midlatitudes. Flights are now taking place over the Arctic to gather further information. In the Arctic the losses are not so great as in the Antarctic because its weather has more exchanges with the surrounding areas and is invaded by warmer midlatitude air before the sun reappears in spring. Wintertime ozone levels above the midlatitudes in the northern hemisphere declined by 3 to 6 percent between 1970 and 1990.

Sunbathing will become more dangerous if the ozone layer in the stratosphere continues to be depleted.

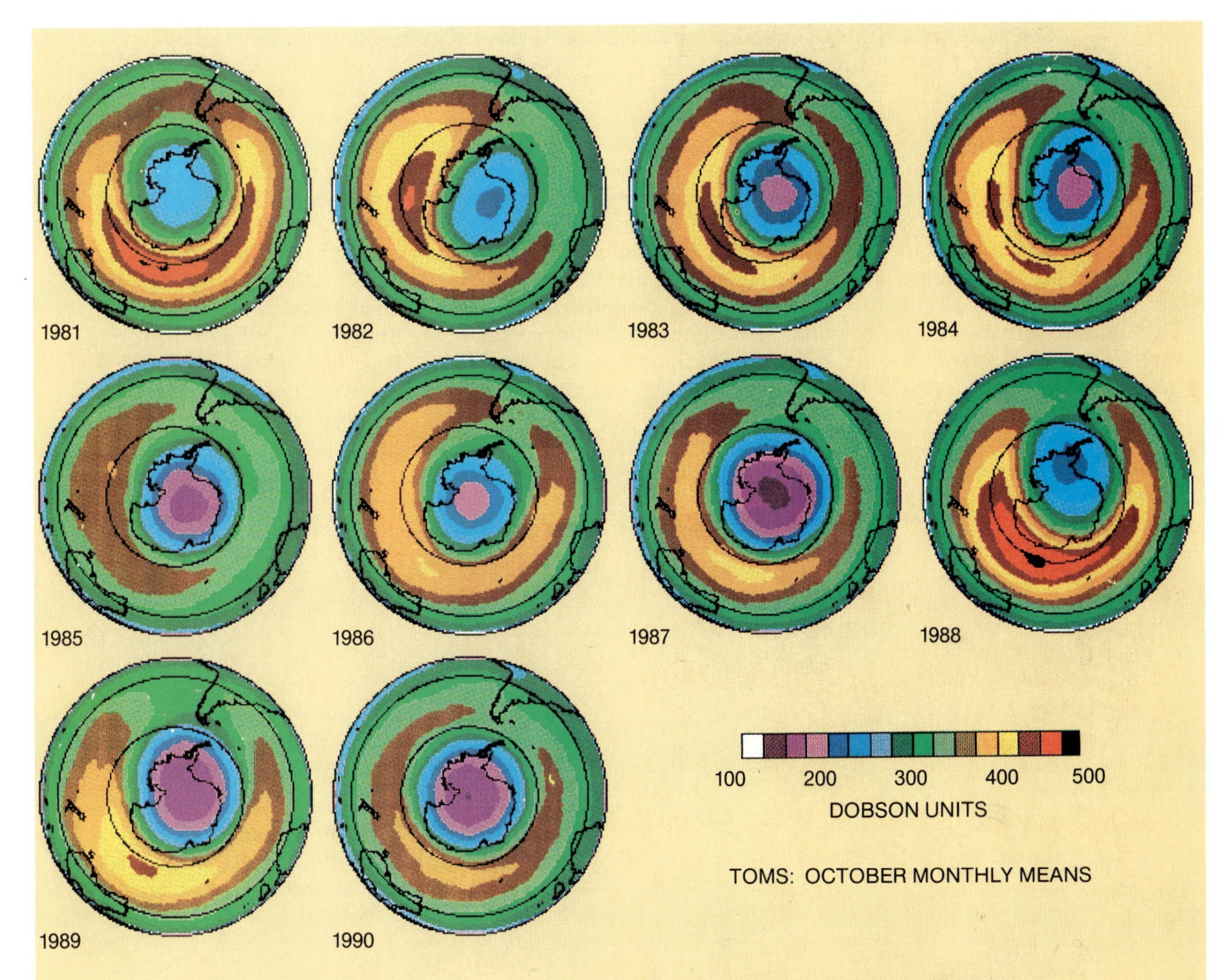

The Antarctic ozone "hole" as charted from satellite from 1981 to 1990. The hole appears each October at the start of the Antarctic spring, and its intensity has been fluctuating during the 1980s and early 1990s.

Ozone is broken apart mainly by reaction with chlorine monoxide. One molecule of chlorine monoxide can destroy many thousands of molecules of ozone. Chlorine monoxide levels rise rapidly over the poles in winter when icy cloud particles form in the main ozone layer and turn chlorine into chlorine monoxide. The chlorine monoxide becomes deactivated as the area warms in spring and the icy cloud particles melt. In October, the end of winter in the southern hemisphere, damage to the ozone is at its greatest.

Research on the breakdown of the ozone layer in the late 1980s and early 1990s has concentrated on the role of human-manufactured pollutants. The complexity of the chemical interactions going on in the atmosphere and the poor understanding of them make it difficult to draw firm conclusions. However, evidence is growing that chlorofluorocarbons used in refrigerators, aerosol cans, and the manufacture of insulating plastic foam are at least partly responsible for the decline in concentrations of ozone.

Chlorofluorocarbons were originally considered the ideal industrial chemical; they are extremely stable and nontoxic and do not break down in the troposphere. The chlorofluorocarbons, however, eventually rise into the stratosphere (because they are lightweight gases), where they encounter intense ultraviolet energy and break down, releasing chlorine. Research in Antarctica has shown that chlorine compounds, not a natural constituent of the Antarctic atmosphere, are present in sufficient quantities to produce the ozone hole. Because of their stability, chlorofluorocarbons reside in the atmosphere for long periods—up to 100 years. The problems of ozone depletion that are being recorded now are related to the much lower chlorofluorocarbon emissions of past decades. The agents of ozone destruction for most of the next century are in the atmosphere-ocean system already.

Concern over the role of chlorofluorocarbons in destroying the ozone layer has led to international attempts to limit their emission. An agreement reached

BOX 3-1

THE OZONE ISSUE—cont'd

at Montreal in 1986 as part of the United Nations Environmental Program limited emissions to 50 percent of the 1986 total by 1999. Concern that even this action may be too little and too late led to a further international agreement in London in June 1990. Industrialized countries agreed to eliminate all production of chlorofluorcarbons and related chemicals by 2000, and other countries agreed to do this by 2010. It is hoped that this will be sufficient to reduce the destruction of ozone and to enable the ozone layer to be recreated faster than it is destroyed. At the more local scale, Australia and New Zealand, which are in the vulnerable southern hemisphere close to Antarctica, have public campaigns to reduce the attractions of suntanning; children at school are encouraged to wear protective hats and visors and to sit under trees when eating lunch.

Research into "the hole in the sky" reveals the immense complexity of Earth's atmosphere and the fact that the best efforts of scientists have resulted in only a partial knowledge of how the atmosphere-ocean environment works. Moreover, the importance attached to chlorofluorocarbons in the destruction of the ozone layer highlights how human society may unwittingly alter a delicate balance that it does not fully understand.

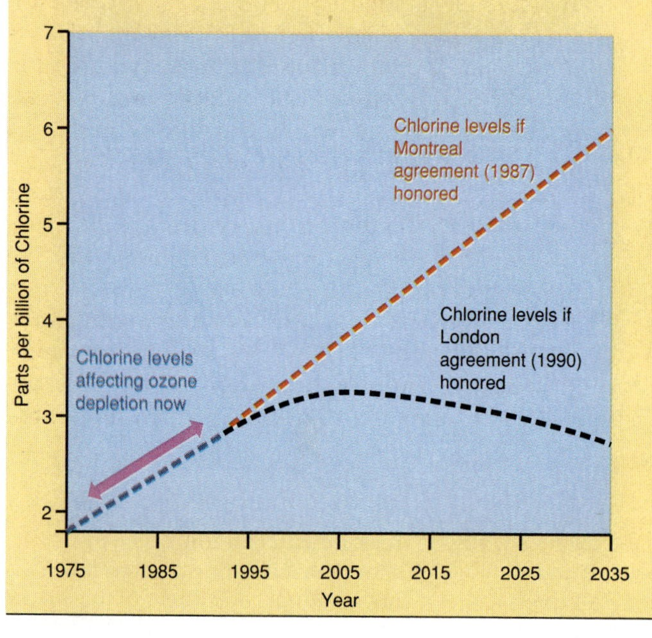

Changing levels of ozone-destroying chlorine in the stratosphere since 1975, with projections to 2035 AD. The increasing stringency of controls reflects the growing concern between the Montreal (1987) and London (1990) agreements as studies show that ozone is being depleted more rapidly.

Oxygen is the most chemically active of the major atmospheric gases. During Earth's existence it has combined with virtually all the nonliving materials at the planet's surface through the process of oxidation. Its main reactions now involve exchanges with living tissue. Oxygen is removed from the atmosphere by the respiration and decay of animals and is added to the atmosphere by plants during photosynthesis, a complex process in which green plants use sunlight, carbon dioxide, and water to produce oxygen and food molecules. The total outcome of these processes gives each oxygen molecule an average residence time of 5000 years in the atmosphere. Scientists believe that virtually all the atmospheric oxygen has accumulated during the last 3 billion years as the result of a slight excess of photosynthetic supply over removal by respiration and decay. The rate of oxygen accumulation in the atmosphere has been particularly marked in the explosive development of life during the last few hundred million years. Oxygen is also removed from the atmosphere by combustion processes, including forest and bush fires and the burning of fossil fuels. Presently, this amount is about one twentieth of that taken annually for animal respiration, and so represents a small proportion of the total being cycled.

Most oxygen exists as molecules with two atoms (O_2), but a tiny fraction exists as **ozone**, a molecule with three atoms (O_3). Ozone is produced by chemical reactions in the presence of sunlight. Although ozone occurs in the lower atmosphere, it is concentrated in a layer between 20 and 40 km (12 to 25 miles) above the surface, where atomic oxygen (O, single oxygen atoms not bonded into molecules) combines with molecular oxygen (O_2). Although it is only a small fraction of the oxygen in the atmosphere, ozone is important in absorbing incoming ultraviolet rays from the sun. Ozone protects living organisms from the harmful effects of these rays. One current concern over the future of the atmospheric environment is the growth of the "ozone hole" above the South Pole and the decrease in the amount of ozone in the protective layer of ozone around the world (see Box 3-1: The Ozone Issue).

Argon is by far the least important of the three major gases, in terms of both total volume and its contribution to environmental processes. It enters the atmosphere as a result of the radioactive breakdown of potassium within surface rocks, but builds up so slowly that changes are noticeable only over very long periods. It is seldom involved in chemical reactions with other substances and is thus said to be "inert."

The Minor Gases. The remaining gases of the lower atmosphere (excluding water vapor) make up only 0.04 percent of the total volume (Figure 3-3), of which 0.03 percent is carbon dioxide. The other gases include hydrogen, helium, neon, krypton, xenon, ozone, and methane. A number of these minor gases are more variable in quantity than the three major gases. Some of them, especially those produced by human activity, have been increasing in recent years. Carbon dioxide has received the most attention, but others such as methane, various nitrogen oxides, and chlorofluorocarbons (CFCs) are also increasing.

Carbon dioxide is the most significant of the smaller and more variable constituents of the atmosphere. It is one of the "greenhouse gases" that absorb heat energy in the lower part of the atmosphere. It also complements oxygen in interactions with living organisms, being taken up in photosynthesis and released in respiration, decay, and combustion of fossil fuels. Carbon dioxide is also released into the atmosphere by the chemical action of water on limestone rock. However, total carbon dioxide in the atmosphere is a tiny fraction of total oxygen because of the tight balance that is maintained by a series of interactions within the carbon cycle. The average carbon dioxide molecule is in the atmosphere for only 5 years before it is again involved in chemical processes at the surface. This is shorter than the time it takes to spread changes in carbon dioxide concentration throughout the atmosphere. Local and seasonal variations in carbon dioxide concentration result. Thus, the burst of photosynthesis in land plants in spring and early summer depletes the carbon dioxide in middle latitudes, but levels are restored in the late summer and fall when less photosynthesis takes place.

Other minor atmospheric gases also contribute to the greenhouse effect. Gases such as **methane** (CH_4), **carbon monoxide** (carbon bonded to a single atom of oxygen, CO), and the **chlorofluoro-**

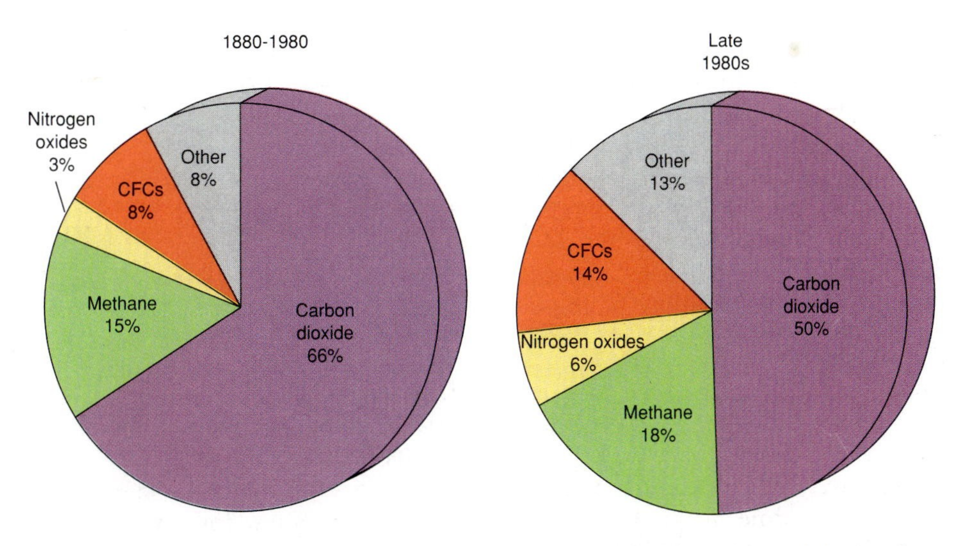

Figure 3-4 Some of the minor gases absorb long-wave rays. These increased during the 1980s at varying rates. The pie charts show *relative* proportions of the gases. Altogether, they constitute 0.4 percent of air by volume.

Table 3-1 Rising Atmospheric Concentrations of Carbon Dioxide and Methane

	10,000-300 YEARS AGO	300 YEARS AGO TO PRESENT
Carbon dioxide (ppm)	260	Rising to 350
Methane (ppb)	700	Rising to 1700

N.B. ppm = parts per million, e.g., 260 ppm is 0.026%
ppb = parts per billion, e.g., 700 ppb is 0.00007%
The use of ppm and ppb avoids long strings of zeros, making the numbers easier to read.

carbons all contain carbon and absorb heat rays. The methane and carbon monoxide come from burning vegetation, and methane is also added by the decomposition of plant matter, cattle digestion, and rice growing. The chlorofluorocarbons are produced only by human activities.

The gases that absorb energy and heat the lower atmosphere are increasing as human populations and food needs rise. The gases that are largely the products of human activity, such as methane and chlorofluorocarbons, are increasing faster than carbon dioxide (Figure 3-4). This increase suggests that there is a parallel heating of the lower part of the atmosphere. The concentrations of carbon dioxide and methane in the atmosphere were relatively constant from 10,000 to 300 years ago (Table 3-1). In the last 300 years, concentrations of carbon dioxide have risen by one third, while those of methane have risen one and a half times.

Atmospheric **sulfur dioxide** (SO_2) and **nitrogen oxides** are other atmospheric gases that are increasing as a result of human activities, particularly industrial processes. Sulfur compounds emitted in small amounts by living things react with oxygen to form the noxious gas, sulfur dioxide. The combustion of fossil fuels, however, releases much more sulfur than natural processes, and the sulfur dioxide in the atmosphere above the United States increased by 50 percent from 1940 to 1970 before falling a little because of antipollution measures. The rising levels damaged animal and plant tissues and caused the surfaces of building stones to disintegrate. The burning of forests, exhaust from automobile engines, and the use of nitrogen fertilizers produce nitrogen oxides in the atmosphere. Further chemical reactions in the atmosphere convert the sulfur dioxide to sulfuric acid and the nitrogen oxides to nitric acid. The combination of the sulfuric and nitric acids in the air causes the ground to become acidified when the acids are contained in rain and snow (see Box 6-1: Acid Rain).

Water Vapor and Aerosols. These are the most variable constituents of the atmosphere and are virtually all concentrated in the lowest 10 km (6 miles). **Water vapor** (water in its gaseous state, as opposed to the tiny liquid droplets that form clouds) is another atmospheric gas that has a major impact on weather and climate. Its concentration is more variable than that of the other minor gases and may range from almost zero to a maximum of 4 percent of the atmospheric volume. Water vapor is another gas that absorbs heat radiation, and at higher local concentrations may become more significant in the warming of the lower atmosphere than the gases containing carbon. It is also important to living organisms, since it is transported by winds over the continents and brings liquid water—a basic support for living organisms—to land areas.

Suspended particles and liquid droplets in the atmosphere are known as **aerosols.** Such particles remain in the atmosphere for months or years. They become visible when concentrated in smoke or fog but are normally too small to be seen. The solid particles include sea salt, rock dust (from volcanoes, deserts, and exposed soils), fine organic matter, and the products of combustion. Thus, they originate from a mixture of natural and human activities, but aerosol levels are particularly high over cities. Aerosols play an important part in initiating the formation of fog and clouds and may be a cooling influence by reflecting solar radiation.

In addition to occurring as a gas, water may also be present in the atmosphere as cloud or fog droplets, or as ice particles in the tops of tall clouds. The droplets of fog and clouds are tiny enough to be buoyed up by light, upward currents of air. Such water droplets also absorb heat and warm the surrounding air.

LINKAGES

Atmospheric gases interact with other Earth environments. The gases are added by the radioactive decay of minerals, by volcanic activity ***(solid-earth environment),*** *by chemical reactions with surface minerals* ***(surface-relief environment),*** *and by the living processes of plants and animals* ***(living-organism environment).*** *They are removed from the atmosphere by chemical reactions with surface minerals and by organic activity. Water vapor is cycled through ocean and atmosphere, and then over and through rocks. Aerosols are produced by wave action, volcanic activity, desert dust storms, and human activities.*

Structure of the Atmosphere

The structure of the atmosphere is a series of layers that result from heating by solar radiation and the decrease in atmospheric density away from Earth's surface. Scientists are learning more about this structure from instruments sent aloft attached to balloons and rockets, and from satellite measurements.

The processes giving rise to weather are restricted to the **troposphere,** the layer of air next to Earth's surface. This layer is heated from the ground up. As a result, temperatures generally decline with height in the troposphere (Figure 3-5). The troposphere contains 75 percent of the total weight of the atmosphere, including virtually all the aerosols and water vapor. The top boundary of the troposphere occurs where temperature ceases to decrease with height. This boundary, known as the **tropopause,** varies in height from 8 to 9 km (5 to 6 miles) above Earth's surface near the poles to 16 to 17 km (10 to 11 miles) over the equator, where the ground absorbs more solar energy and can heat a deeper layer of air. The tropopause provides an effective "lid" that prevents most tropospheric turbulence from extending upward.

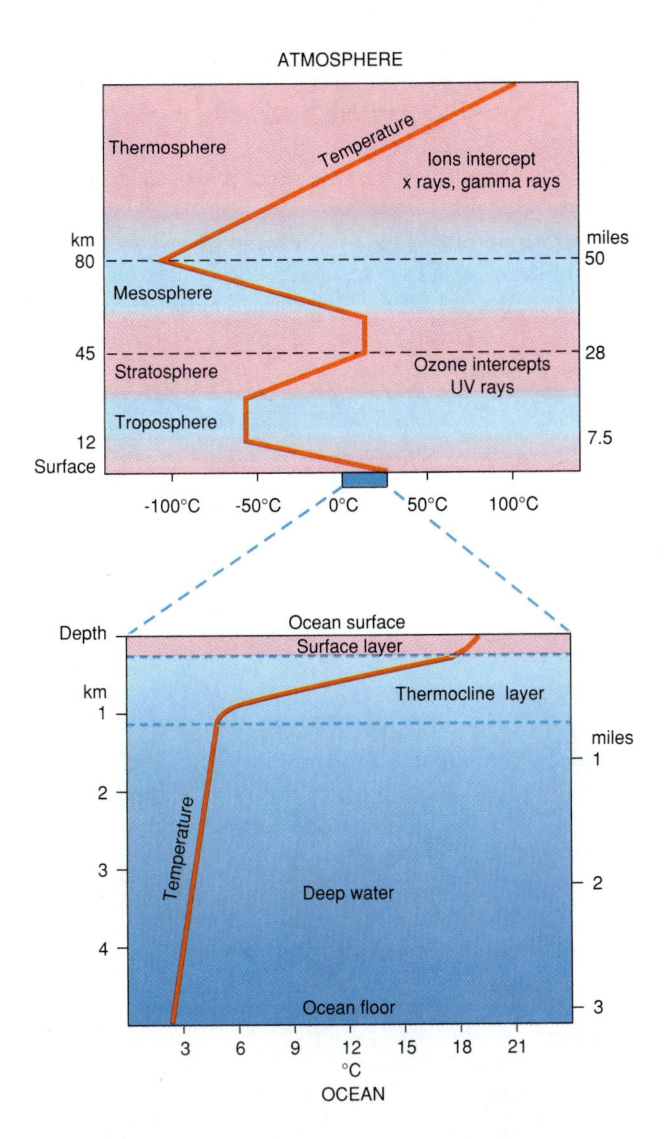

Figure 3-5 The layers of atmosphere and ocean based on temperature changes. The ocean has a warm surface layer on top of cold, deep water. In the lowest layer of the atmosphere, temperature decreases with height, then rises through the stratosphere, falls again through the mesosphere, and rises through the thermosphere.

The **stratosphere** is the layer above the tropopause. Temperatures change little with height in the lower part of the stratosphere. They rise near the top, however, as the concentration of ozone in the upper part of the stratosphere absorbs ultraviolet solar rays and heats the layer. The **stratopause** is the upper limit of the stratosphere.

The **mesosphere** lies above the stratopause and has even thinner air than the stratosphere. Temperatures fall to −90°C (−80°F) at its upper surface, known as the **mesopause.**

Above 80 km (50 miles) is the uppermost layer of the atmosphere, the **thermosphere,** which is of extremely low density. Although less than one millionth of the atmosphere's mass lies above 100 km, the extreme outer fringe is believed to extend as far as 1000 km above the surface, gradually fading to the near vacuum of space. Temperatures rise to as much as 1200°C (2200°F) as the tiny ions of the gases in this tenuous layer absorb most of the shortest-wave solar radiation (x-rays, gamma rays, and the shortest ultraviolet rays). Such rays would be deadly to living organisms if they reached Earth's surface.

These various layers within the atmosphere show that it is a complex environment. A simplified view of the atmosphere treats it in terms of a lower zone (the troposphere), in which turbulence gives rise to changing weather, and an upper zone (all the higher layers) that acts as a filter for the harmful aspects of solar radiation.

Density and Pressure in the Atmosphere

As Figure 3-2 shows, the greatest concentration of atmospheric gases is in the troposphere, and air "thins out" toward space. This is because gases are made up of molecules that are not bonded together, and can thus be compressed or may float freely where they are not confined. In the atmosphere, the downward pull of gravity compresses and concentrates atmospheric gases near the ground.

"Concentration" is often measured as **density**—the amount of a substance in a space, or, more precisely, the mass per volume (e.g., the number of molecules of atmospheric gases per cubic meter). Water has a density of 1 gram per cubic centimeter. The density of air near the ground is 0.001293 g/cm^3 (1/800 the density of water) and decreases upward. At 7 km (4.5 miles) up it falls to 0.00066 g/cm^3; at 100 km (65 miles) it is 5 ten-billionths of a gram per cubic centimeter.

The lower densities at higher levels within the atmosphere are related to another measure, pressure. Where more air molecules fill a given space, they collide more often and exert more pressure on every surface (Figure 3-6, *A*). Pressure is the ratio of force to area, and **atmospheric pressure** is the force per area exerted by atmospheric gases. Atmospheric pressure acts in all directions; the downward pressure on the roof of a house is balanced by the upward pressure of air inside the house. Pressure is lower in the upper atmosphere than just above the ground, because the atmosphere at the higher point is less compressed by gravity.

The atmospheric pressure exerted at ground level averages 1.034 kilograms per square meter (14.7 lbs per square inch). It can be measured by a mercury barometer, in which normal atmospheric pressure on the surface of the mercury supports a column of about 760 mm (29.92 in) at ground level (Figure 3-6, *B*). Meteorologists often express atmospheric pressure in units called **millibars** (mb). The average value at the ground or ocean surface is 1013.25 mb. A measurement of 760 mm on a barometer is the same as one of 1013.25 mb. Atmospheric pressure decreases with increasing height, reflecting the decreasing density of air above: at 5.5 km (4 miles) above the surface, the average atmospheric pressure is half that at the surface, and at 8.5 km (5.25 miles), it is a third. People visiting places at altitudes above 3000 meters often find that their breathing and blood circulation take time to adjust to the lower atmospheric pressure.

The vertical structure of the lower atmosphere is marked by a decrease of both temperature and pressure with height. The average temperatures and pressures at specific heights are summarized in the "Standard Atmosphere" (Table 3-2). This is used by meteorologists to compare conditions at a particular place and time with standard conditions, and by aviators for calibrating altimeters. The altimeter

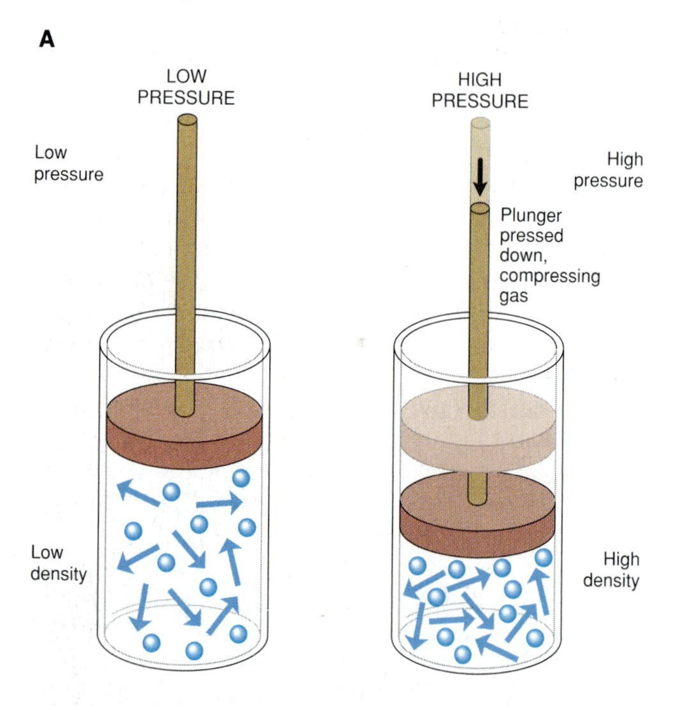

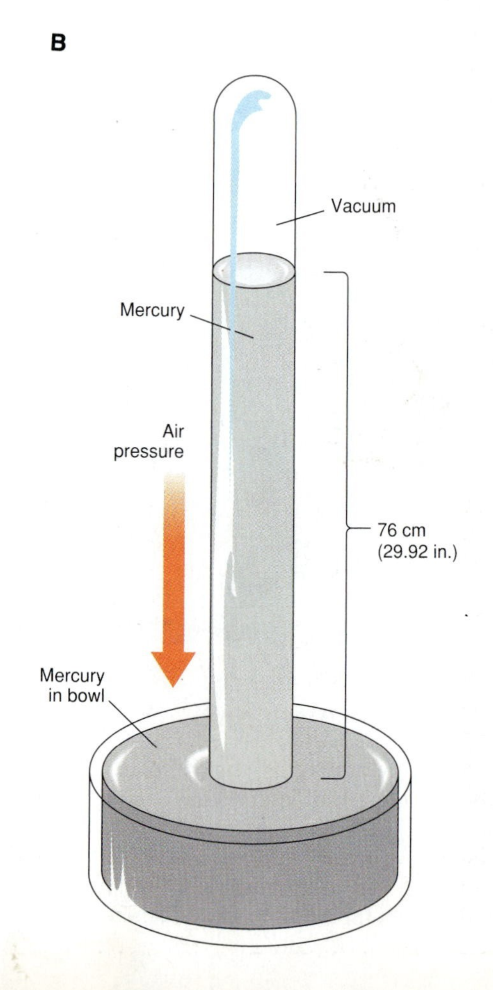

Figure 3-6 Air pressure. **A,** Increasing pressure gives rise to compression of gas molecules, greater molecular activity and higher temperature. When air is pumped into a bike tire, the pump gets hot as the air is compressed (this heat is also due to friction). **B,** A barometer. Atmospheric pressure is measured by the column of mercury supported; greater pressure pushes the mercury higher.

gives a reading of height above the surface based on the pressure at each level.

Atmospheric pressure varies from place to place at the surface because air rises in some locations and descends elsewhere. When air is heated, it becomes less dense and rises. As air rises, the force it exerts on the ground is reduced. Where air descends, a greater force is exerted at the surface, compressing air into greater density and increasing pressure. Variations in atmospheric pressure at Earth's surface are usually only a few percent above or below the average (generally from 970 to 1040 mb), but descend to 900 mb in some hurricanes and rise to 1060 mb in some high-pressure systems in the Arctic. However, these relatively small variations are important in bringing about movements of air (winds) across the surface (Chapter 5).

Table 3-2 The Standard Atmosphere

ALTITUDE (KM)	TEMPERATURE (°C)	PRESSURE (MB)	DENSITY (KG/M³)
30.00	−46.60	11.97	0.02
25.00	−51.60	25.49	0.04
20.00	−56.50	55.29	0.09
19.00	−56.50	64.67	0.10
18.00	−56.50	75.65	0.12
17.00	−56.50	88.49	0.14
16.00	−56.50	103.52	0.17
15.00	−56.50	121.11	0.20
14.00	−56.50	141.70	0.23
13.00	−56.50	165.79	0.27
12.00	−56.50	193.99	0.31
11.00	−56.40	226.99	0.37
10.00	−49.90	264.99	0.41
9.50	−46.70	285.84	0.44
9.00	−43.40	308.00	0.47
8.50	−40.20	331.54	0.50
8.00	−36.90	356.51	0.53
7.50	−33.70	382.99	0.56
7.00	−30.50	411.05	0.59
6.50	−27.20	440.75	0.62
6.00	−23.90	472.17	0.66
5.50	−20.70	505.39	0.70
5.00	−17.50	540.48	0.74
4.50	−14.20	577.52	0.78
4.00	−11.00	616.60	0.82
3.50	−7.70	657.80	0.86
3.00	−4.50	701.21	0.91
2.50	−1.20	746.91	0.96
2.00	2.00	795.01	1.01
1.50	5.30	845.59	1.06
1.00	8.50	898.76	1.11
0.50	11.80	954.61	1.17
0.00	15.00	1013.25	1.23

COMPOSITION AND STRUCTURE OF THE OCEANS

The oceans cover 71 percent of Earth's surface with water. There is an imbalance in the distribution of oceans between the two hemispheres, since they cover some 85 percent of the southern hemisphere and 57 percent of the northern hemisphere. The water fills basins produced by great movements of the planet's crust (Chapter 12); all these basins are connected to each other and to the shallower seas and gulfs at their margins. Sea level is determined by changes in the sizes of ocean basins and the amount of water they contain.

The darkness and high pressures of the deep oceans make them one of the most difficult environments to study. Since 1950, however, improving technology has added greatly to the knowledge of oceans. Previously, scientists relied on spot measurements at particular locations taken in soundings from individual ships. Today, they can use sonar techniques that make it possible to survey a wide band of deep ocean from a surface ship and, increasingly, satellites. A series of international scientific projects is overcoming the ignorance of oceans, particularly in terms of their links with the atmosphere and weather.

Just as the atmosphere is composed largely of gases but also contains some water droplets and dust particles, so the oceans are made up mostly of liquid water but also contain a variety of salts and some gases, such as oxygen and carbon dioxide, in solution. The concentration of dissolved salts in seawater is known as **salinity** and is measured in parts per thousand by weight. These minor ingredients make up a few parts per thousand of the ocean water but have important implications for living organisms.

The saltiness of ocean water is caused by the accumulation of salts from the wearing down of continental rocks. Rocks are subject to chemical change when exposed to the atmosphere, and the soluble products of such changes are carried into the oceans by rivers, glaciers, and winds. Once in the oceans, further chemical reactions may cause the dissolved minerals to be precipitated (separating from the water as solid particles), and organisms remove some of the elements for use in bodybuilding. Salts that are not removed by either of these processes remain in the ocean water and contribute to the salinity.

The average salinity of seawater is 35 parts per thousand, and the range in the oceans is 33 to 38 parts per thousand (Figure 3-7). Evaporation and ice formation increase salinity in the ocean by removing water. Precipitation and water entering the

Figure 3-7 It is easy to float in the Great Salt Lake because the greater density of the highly saline water provides buoyancy.

ocean at river mouths add fresh water and decrease salinity by dilution. For instance, the salinity of water in the equatorial Atlantic is slightly lower than that to the north or to the south because of the combination of high rainfall and fresh water flowing in at the mouths of the Amazon and Zaire rivers. Ocean waters with higher salinities have higher densities because the dissolved salts make water heavier. Lower and higher salinities than those in the open oceans are found in almost closed seas. Dilution by fresh water occurs in the Baltic Sea, where the salinity is 8 parts per thousand. Intense evaporation occurs in the Red Sea, where the salinity is 40 parts per thousand. The *relative* proportions of the major dissolved elements (sodium, chlorine, magnesium, calcium, sulfur, and potassium) do not change despite the varying salinities.

In contrast to its effect on the atmosphere, the pull of gravity has little impact on the density of seawater since water is not as compressible as gas. Seawater density is controlled by two factors, salinity and temperature. Heating of water, as of air, causes it to expand slightly, and this lowers its density. Cold water is denser than warm water. Surface ocean water is heated by the sun to a few tens of meters deep. Cold, denser water lies beneath the surface water, and the two layers are separated by a transitional layer known as the **thermocline** (Figure 3-5). Waters below 1 km deep are predominantly cold, with temperatures below 5°C (40°F).

Although the density of ocean water varies only slightly because of its salt content and temperatures, pressure increases with depth as a result of the weight of overlying water. At a depth of 1 km, the overlying water exerts a pressure of nearly 100 times the surface atmospheric pressure; the pressure at the bottom of the deepest point in the oceans, the Marianas trench, is a thousand times the surface atmospheric pressure. The rapid increase in pressure with depth in water has a major effect on the organisms living there. They have special means of coping with the changes of pressure as they change levels. Humans do not have such capabilities, and deep-sea divers have to descend and return to the surface slowly so that their bodies can adjust to changes in pressure.

The effects of human activities that alter the composition of the atmosphere receive much media attention, but less has been publicized concerning human impacts on the oceans. Oil spills from leaking tankers and from the Gulf War of 1991 get most of the adverse publicity. In addition, coastal towns pump sewage out to sea, and radioactive and other wastes are dumped at sea. People treat the oceans as a resilient and massive environment that can cope with such treatment without degradation. It is now becoming clear, however, that unrestricted development around the coasts of industrial countries can affect the quality of the ocean environment and the organisms that live in it (see Box 3-2: Pollution of the Atmosphere and Oceans).

LINKAGES

The composition of ocean water is linked to processes in other Earth environments. Streams deliver dissolved salts from the erosion of land ***(surface-relief environment)****, and undersea volcanoes erupt minerals* ***(solid-earth environment)****. Fishes and other organisms* ***(living-organism environment)*** *help to circulate materials by extracting them from seawater and then consigning them to ocean-floor deposits when they die. Sea level is determined by interactions of ocean waters with the* ***solid-earth environment*** *and the storage of water in ice sheets.*

BOX 3-2

POLLUTION OF THE ATMOSPHERE AND OCEANS

Pollution occurs when human activities lower air and water quality by adding materials to the atmosphere or surface waters. Pollution may endanger personal and environmental health or hinder the operation of natural processes. It is increasingly necessary to assess pollution's effects on people's lives and environments and to determine how fast such effects are accumulating.

Pollution of the Atmosphere

Air pollution, which is caused by most human activities ranging from farming to industrial processes, has been a significant problem since the early nineteenth century because its effects are concentrated in urban-industrial areas. By the mid-twentieth century, pollution from burning coal increased death tolls in industrial areas like the Pittsburgh region and in the cities of northern Europe. More recently, the large-scale use of cars, trucks, and airplanes has added other fossil-fuel pollutants to the atmosphere, so that summer haze resulting from air pollution has reduced visibility in the eastern United States to 25 km (15 miles) from the natural 150 km (90 miles) that is still common in much of the American West. The yellow-brown pollution domes over such western cities as Denver stand out against the surrounding clear skies (see photo). In portions of the Los Angeles basin, visibility on a good day is reduced to 14 km (8 miles). In the late 1980s the most polluted part of the world was along the Czech and former East German borders; this pollution resulted from the burning of brown coal, and it has been claimed that it was a major cause of the 1989 Czech uprising. In developing countries, the rising scale of forest burning and industrialization is increasing pollution levels. This increase is visible in the accompanying satellite image.

Atmospheric pollution resulting from forest burning. A photo of the Amazon basin with fires along Route 364. The view is looking toward the Andes, 1700 km away in the distance.

Urban-industrial pollution of the atmosphere. A pall of smog moving out from Los Angeles and curling around the Channel Islands, as viewed from the Space Shuttle.

The main air pollutants are gases such as carbon monoxide, methane, oxides of nitrogen, and compounds of sulfur. Tiny dust particles and liquid droplets (aerosols) released into the atmosphere add to natural sources from volcanic and wind activity. The atmosphere can clean itself to a large extent by chemical reactions, by dispersing pollutants in winds, and by bringing gases and particles down to Earth's surface in rain or snow. However, serious pollution events occur when atmospheric conditions lead to such a concentration of pollutants that the natural cleaning processes cannot cope. The Los Angeles basin and the deep valleys around Pittsburgh, for example, often contain large volumes of stationary air that trap pollutants rather than permitting them to be dispersed.

Air pollution has a variety of effects including acid rain, global warming, chemical reactions with building materials, and health problems. Extreme pollution causes health hazards for those

BOX 3-2

POLLUTION OF THE ATMOSPHERE AND OCEANS—cont'd

with lung ailments and irritates the eyes and skin. In the 1950s deaths from air pollution led the United Kingdom government to legislate against the burning of coal on open fires and in steam engines. This reduced the original form of smog (smoke fog), but the term is still popularly used for pollution effects that result in reduced visibility such as the buildup of car exhaust fumes.

Attempts to reduce air pollution use two main approaches. The first establishes standards of air quality by determining maximum allowable proportions of aerosols, sulfur oxides, carbon monoxide, nitrogen dioxide, ozone, and lead. Authorities may impose regulations to reduce these pollutants to levels that will not affect human health or the environment. In the United States the Environmental Protection Agency sets two levels of air quality standards for six pollutants. In each of these the primary level is the maximum exposure tolerated by human beings without ill effects; the secondary level is the maximum allowable in terms of environmental impacts. The second approach is to change technology by such means as forbidding the use of spray cans or insisting that devices such as catalytic converters be added to cars to reduce pollutant emissions.

Such controls have been applied to the United States, Canada, Japan, and the countries of western Europe. They result in reduced coal-based pollutants in urban-industrial areas, although the taller chimneys built to disperse these pollutants from power stations merely transfer the problem downwind.

Ocean Pollution

Water pollution involves streams, lakes, and oceans. It is greatest along the streams and around the coasts of urban-industrial countries, but is increasing in the developing world, where some of the worst examples occur. The open oceans are still relatively clean, but are fed by often-polluted water flowing from the continents.

The oceans form a huge reservoir where billions of tons of silt, sewage, industrial waste, and oily runoff can be "lost"—either by dilution or by sinking out of sight. However, it is important that the nature and rate of such pollution are monitored to evaluate acceptable levels of contamination. Already, unsightly plastic debris and abandoned fishing gear, together with increasing levels of toxic substances, threaten the marine environment. In many cases, water pollution takes the form of materials that poison marine plants and animals, but there is also the danger of overenriching the environment. The latter leads to unusual surface plant blooms, followed by oxygen depletion and fish kills.

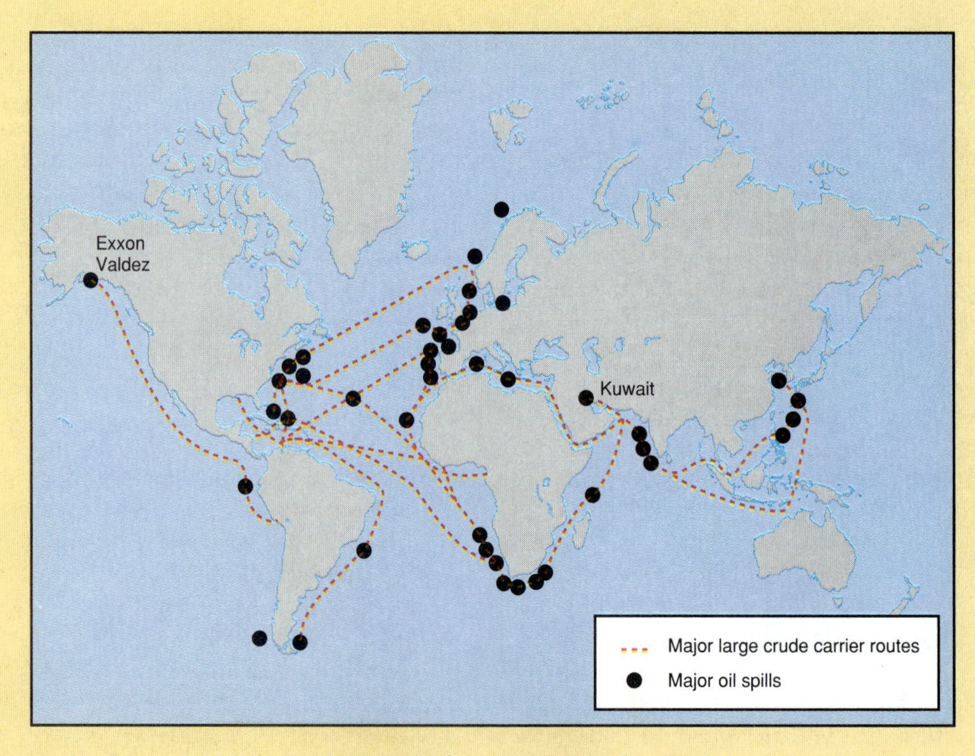

Major ocean tanker routes and oil spills.

BOX 3-2

POLLUTION OF THE ATMOSPHERE AND OCEANS—cont'd

A 1990 report from the Group of Experts on the Scientific Assessment of Marine Pollution (GESAMP), established by the United Nations, identified several trends in ocean pollution:

- Most coastal areas are polluted.
- The most widespread and serious sources of pollution are sewage and sediment from soil erosion, both the results of increasing population pressure.
- The alteration of coastal environments includes the destruction of wetlands, mangroves, coral reefs, and sand dunes as well as sewage outfall impacts.
- Oil spills and oil platform leaks are having increasing local impacts.
- Plastic litter, discarded fishing gear, and tar balls from oil spills threaten more organisms.
- Toxic metals and radioactive substances are presently less significant than other forms of pollution.

The report concluded:

We fear, especially in view of the continuing growth of human populations, that the marine environment could deteriorate significantly in the next decade unless strong coordinated national and international action is taken now.

At present, international control focuses on oil spills and the greater amounts added by the routine discharges of oil from ships cleaning their engines and tanks. The map shows how the location of major oil spills is linked to the main tanker routes. Other areas of marine pollution control have not been developed. The 1982 U.N. Convention on the Law of the Sea provided an agenda for action, and the 1985 "Montreal Guidelines" proposed methods of dealing with sewage, oily urban runoff, and other pollution sources, but little has been done to implement these proposals.

In mid-1991 an informal study was carried out on the uninhabited Ducie Atoll in the South Pacific, 4800 km (3000 miles) from the nearest continent. Almost 1000 items of trash that had been washed up on the shore were picked up from a 3-meter-wide (10 feet) beach during a 2.5 km (1.5 miles) stroll. They included 179 buoys and parts of buoys, 14 plastic bread or bottle crates, 71 glass bottles, rope pieces, plastic fragments, metal cans, toy soldiers, and light bulbs. The fact that so many items could be collected in such a remote place demonstrates the continuing human perception of the oceans as a dump.

See also Box 6-2: Acid Rain.

EXCHANGES OF MATTER BETWEEN ATMOSPHERE AND OCEANS

The atmosphere and oceans continuously exchange solids, liquids, and gases with each other and with other Earth environments. The apparently stable composition of both the atmosphere and the oceans results from a balance between these inputs and outputs, not from being static.

Among the gases exchanged between the atmosphere and oceans, oxygen and carbon dioxide are most significant; their involvement in photosynthesis provides the primary exchange of these gases between atmosphere and ocean (Figure 3-8). Single-celled plants living at the ocean surface take in carbon dioxide from the atmosphere and give out oxygen. The breaking of waves also enables water to exchange oxygen with the atmosphere.

Water is the most important liquid circulated between atmosphere and ocean. It is continuously transferred by evaporation from the ocean surface. It is then returned by rain and snow to complete part of the *hydrologic cycle* (Figure 3-9).

The solids exchanged between the atmosphere and oceans include dust and salt particles. When waves break, particles of salt in solution escape into the atmosphere. These particles are so small that they are carried by light winds and play an important role in forming cloud droplets and raindrops. The atmosphere carries other fine particles, such as dried clay. This dust gradually sinks to Earth's surface and through the oceans to form deposits on the floor. Sometimes so much dust is blown out from the Sahara that huge clouds obscure the Space Shuttle astronauts' view of the Atlantic Ocean (see Figure 16-3).

The gases of the atmosphere, the waters of the oceans, and the dust from the land provide huge reservoirs of materials whose movements and exchanges are basic to an understanding of other aspects of physical geography and to the continuation of life on Earth. Studying the exchanges of these materials between the various major Earth environments provides a dynamic basis for understanding how those environments work and are related to each other. These themes will be revisited on many occasions throughout this book.

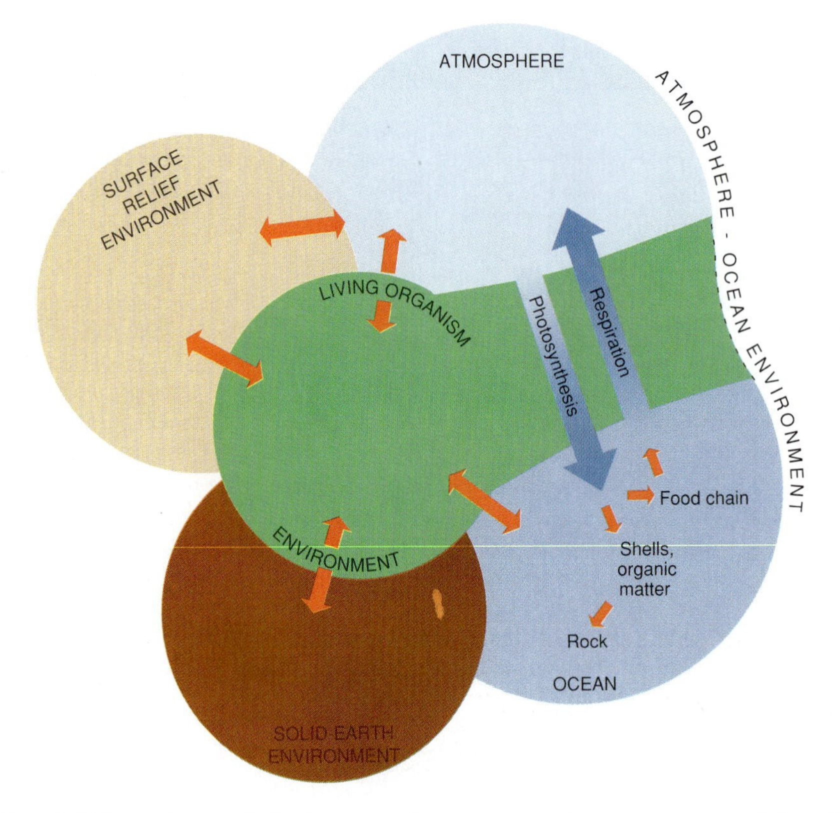

Figure 3-8 The carbon cycle in the atmosphere-ocean environment as part of the wider global cycling of carbon. Carbon is added to the atmosphere in carbon dioxide gas, but is removed in the process of photosynthesis, in which plants use sunlight to produce carbon-based chemical foods such as carbohydrates.

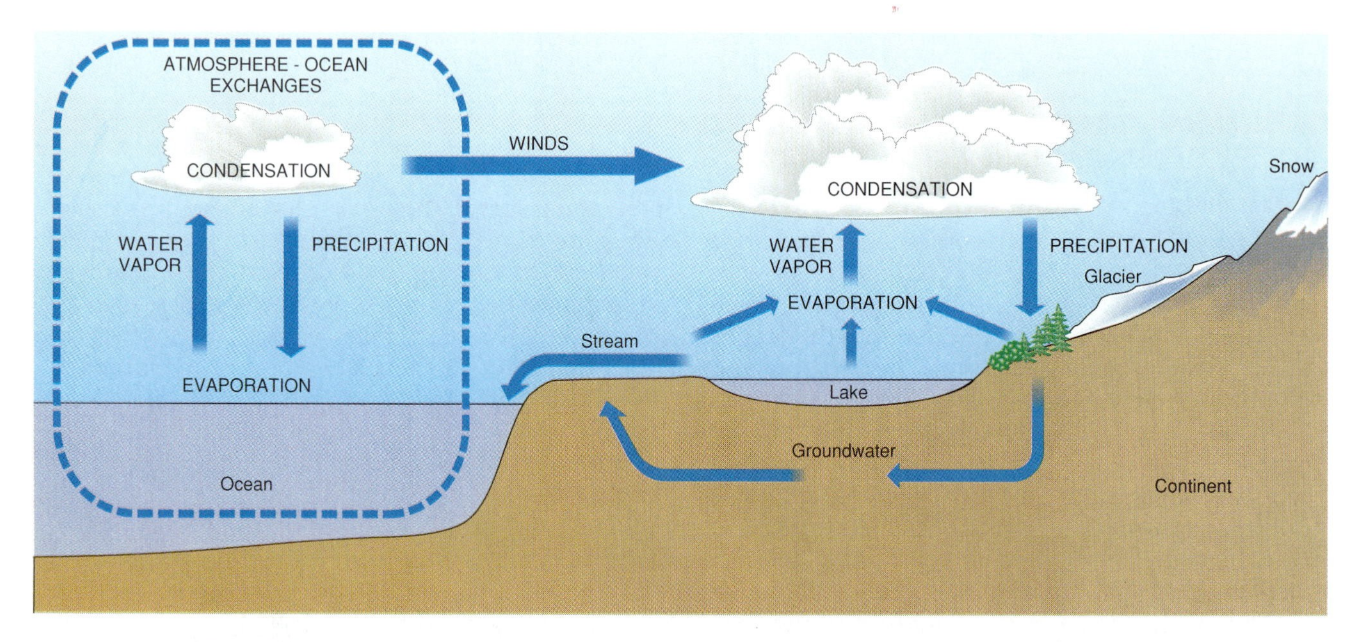

Figure 3-9 The hydrologic cycle, emphasizing the exchanges between atmosphere and ocean.

ATMOSPHERES ON OTHER PLANETS

Earth is unique among solar system planets in having both surface water and a dense atmosphere, which makes it capable of supporting living organisms. No other planet has surface water, and only Venus has such a dense atmosphere. This situation results from a variety of factors including the gravitational attraction of Earth and its distance from the sun. Earth's size gives it a gravitational force that has attracted and maintained the gaseous envelope and covering of water that is the atmosphere-ocean environment. Earth's distance from the sun, together with the presence of heat-trapping gases, produces the narrow band of temperature in which water exists as a liquid and is cycled in its gaseous and solid states.

Other planets are either too small, too near the sun, or too far from it to have a similar atmosphere-ocean environment. Earth's moon and Mercury are too small to retain an atmosphere. Venus is too near the sun, and its atmosphere traps too much heat. Mars is large enough to maintain only a thin atmosphere. Saturn, Jupiter, Uranus, Neptune, and Pluto are too far from the sun and have very different atmosphere-ocean environments from the inner "rocky" planets (Chapter 1).

Recent discoveries about the atmosphere of Mars provide a comparison that highlights some of the special features of Earth's atmosphere-ocean environment. The Martian atmosphere is 95.3 percent carbon dioxide, is much thinner than Earth's (surface pressure is less than 1 percent of Earth's), and has a mean surface temperature of −53°C (−63°F). Mars is half Earth's diameter, and has 10 percent of Earth's mass and 38 percent of Earth's gravity. Mars and Earth probably had similar atmospheric compositions to start with, but the formation of the oceans and the presence of plants modified Earth's atmosphere by increasing the nitrogen and oxygen content at the expense of carbon dioxide.

FRONTIERS IN KNOWLEDGE
Atmosphere-Ocean Composition

Scientists now understand that the atmosphere and ocean are media through which a wide range of Earth materials move. Because of the implications for possible global warming, some major research projects are currently examining the exchanges of gases that are active in absorbing heat waves such as carbon dioxide and methane. The depletion of ozone in the stratosphere is another area of intensive research.

NASA's Upper Atmosphere Research Satellite is designed to improve knowledge of the stratosphere and upper layers of the atmosphere. It carries instruments for measuring temperature profiles and concentrations of gases and will make it possible to map upper air winds and solar energy inputs. The knowledge gained will be used to assess changing conditions in these inaccessible layers and will attempt to link an improved understanding of these layers with surface weather and climate.

SUMMARY

1. The proportion of gases in the lowest 80 km of the atmosphere remains relatively constant. Dry air at this atmospheric level is composed almost entirely (99.96 percent) of nitrogen, oxygen, and argon, but includes important smaller proportions of such gases as carbon dioxide, carbon monoxide, hydrogen, helium, neon, krypton, xenon, ozone, and methane. Water vapor is the most variable of the atmospheric gases. The concentrations of some gases, notably carbon dioxide, carbon monoxide, methane, sulfur dioxide, and nitrogen oxides, are increased by human activities.
2. The atmosphere also contains aerosols, water droplets, and ice particles. These play significant roles in causing the weather.
3. The atmosphere has a layered structure made up of the troposphere, stratosphere, mesosphere, and thermosphere. The troposphere is the lowest layer and the one in which weather-forming processes occur; within this layer, temperature decreases with height. The upper layers act as filters that prevent harmful radiation from getting to the surface.
4. Density is the concentration of matter in a space (mass per volume); pressure is the ratio of force to area. In the atmosphere, pressure exerted by air is greatest near Earth's surface where air is most dense. Atmospheric pressure is measured in millibars.
5. The oceans are composed mainly of water that contains dissolved salts and gases. The proportion of dissolved salts is known as salinity and averages 35 parts per thousand in the open oceans.
6. The ocean waters are layered, with a warm surface and cold, deeper waters separated by a zone where temperature decreases sharply with depth (the thermocline).
7. Materials are exchanged between the oceans and atmosphere and between the atmosphere-ocean environment and other Earth environments. The circulation of carbon and water are two important examples of such material exchanges.
8. Earth is unique in having surface water and a relatively dense atmosphere; these conditions provide a basis for the development and maintenance of living organisms. Other planets are too small or at an inconvenient distance from the sun.

KEY TERMS

atmosphere, p. 56
molecule, p. 56
atom, p. 56
compound, p. 56
ion, p. 56
hydrogen, p. 56
helium, p. 56
nitrogen, p. 57
oxygen, p. 60
ozone, p. 60
argon, p. 61
carbon dioxide, p. 61
methane, p. 61
carbon monoxide, p. 61
chlorofluorocarbon, p. 61
sulfur dioxide, p. 62
nitrogen oxide, p. 62
water vapor, p. 62
aerosol, p. 62
troposphere, p. 63
tropopause, p. 63
stratosphere, p. 63
stratopause, p. 63
mesosphere, p. 63
mesopause, p. 63
thermosphere, p. 63
density, p. 64
atmospheric pressure, p. 64
millibar, p. 64
salinity, p. 65
thermocline, p. 66

QUESTIONS FOR REVIEW AND EXPLORATION

1. Which aspects of the composition and structure of the atmosphere and oceans are of central interest to the physical geographer?
2. List the main atmospheric gases and their properties.
3. Describe the structure of the atmosphere and oceans by reference to differences in temperature.
4. What is the difference between *density* and *pressure*? Illustrate your answer by reference to the differences between gases and liquids. Explain the changing density and pressure sequence from the bottom of the ocean to the top of the atmosphere.
5. Draw graphs of decreasing temperature and pressure above Earth's surface using the characteristics of the Standard Atmosphere (Table 3-2). What conclusions emerge?
6. What factors cause differences in ocean salinity from place to place?
7. Discuss the importance of carbon dioxide in the atmosphere, and the balancing factors that affect the level of its presence there.
8. What effects do the other major Earth environments have on the composition of the atmosphere and oceans?
9. In what ways are human activities causing changes in the composition of the atmosphere and oceans? Assess the range of demands being made on governments for action to reduce pollution and discuss the apparent order of priorities that is emerging (e.g., action on air pollution compared with action on marine pollution).
10. What features of the atmosphere and oceans are subject to change?
11. Find out about levels of atmospheric or oceanic pollution in your town or region. Have the levels changed in the last 10 years? What regulations are in force and how well do they work?

FURTHER READING

Berner RA, Lasaga AC: Modeling the geochemical carbon cycle, *Scientific American,* 54-61, March 1989. The authors emphasize the geological contribution to the carbon cycle.

Broecker WS: The ocean, *Scientific American,* 146-160, September 1983. This article describes facets of atmosphere-ocean chemistry and explains the links between the two realms.

Brune WH: Ozone crisis: the case against chlorofluorocarbons, *Weatherwise,* 3:130-143, 1990. A review of the sources of ozone depletion in the stratosphere.

Darnesh B, Paternek S: Mars, *Life,* 24-38, May 1991. This article investigates the possibility of humans colonizing Mars over the next 200 years. It provides a graphic account of the Martian environment and a comparison with Earth.

Graedel TE, Crutzen PJ: The changing atmosphere, *Scientific American,* 28-36, September 1989. This article focuses on the impact of human activities on recent changes in the composition of Earth's atmosphere.

Kahn R: *Comparative planetology and the atmosphere of Earth,* California, 1989, Jet Propulsion Laboratory. An illustrated report that compares the compositions of the solar system planets as understood in the late 1980s. Covers the subject in greater depth than the present text and refers to other sources.

Stolarski RS: The Antarctic ozone hole, *Scientific American,* 20-26, January 1988. This article explains some of the findings about the hole and discusses further issues that await research.

World Resources Institute: *World resources,* Washington, D.C., 1987, Basic Books (earlier years), Oxford University. An annual (from 1987) assessment of the global environment, its resources, and attempts to monitor and manage changes. It provides data on population, economic conditions, and the physical environment, and discusses national and international programs.

THE ATMOSPHERE-OCEAN ENVIRONMENT

California: Joshua Tree—Robert Holmes National Monument

Heat and Temperature

4

CHAPTER

When Adam had packed his burros, twilight in the clefts of the hills had deepened to purple . . . The intense heat, the vast stillness, the strange radiation from the sand, the peculiar grey light of the valley, told Adam that the midnight furnace winds would blow long before he reached his destination . . .

The heat and oppressiveness and dense silence increased toward midnight; and then began a soft and steady movement of air down the valley. Adam felt a prickling of his skin and a drying of the sweat upon him. An immense and mournful moan breathed over the wasteland, like that of a mighty soul in travail. Adam got out of the hummocky zone upon the dry, crisp, white level of salt, soda, borax, alkali, where thin, pale sheets of powder moved with the silken rustle of seeping and shifting sands. Most fortunate was the fact that the rising wind was at his back . . .

And when the midnight storm reached its height the light of the stars failed, the outline of the mountains faded in a white, whirling chaos, dim and moaning and terrible. Adam felt as if blood and flesh were burning up, drying out, shrivelling and cracking. He lost his direction and clung to his burros, knowing their instinct to be surer guided than his.

From "Wanderer of the Wasteland," by Zane Grey

Heat and temperature are central features of any study of weather. The heating of the atmosphere and oceans leads not only to a variety of temperatures in different parts of the world, but also to movements of air and ocean currents, and to the circulation of water from the oceans, through the atmosphere, and on the continents. Heat and temperature in the atmosphere-ocean environment are also central to other aspects of physical geography, including the rates of rock decomposition at Earth's surface, the various soil-forming processes, and the animals and plants that live in a region. Furthermore (as Adam found in the chapter-opening extract) extreme temperatures affect human comfort and the ability to carry out strenuous activities (see Box 4-1: The Sun, Temperature, and Human Health).

This chapter examines the nature of heat energy, the ways in which the atmosphere-ocean environment is heated, and the causes of the range of temperatures that are experienced at Earth's surface.

PHYSICAL PROPERTIES OF HEAT AND TEMPERATURE

The Nature of Heat and Temperature

Heat is the total energy of molecular movement within a body: the greater the molecular movement, the greater the heat energy. **Temperature** is a measure of the amount of heat energy present in a substance, which is a function of the molecular motion within the substance. As heat energy is absorbed, molecules move more rapidly, raising the temperature of the material.

Heat is measured in **calories**: one calorie is the amount of heat necessary to raise the temperature of 1 gram of water 1°C at sea-level pressure. (Note: A dietary *calorie* is 1000 calories.) More heat is required to increase the motion of a greater number of molecules. Thus, raising the temperature of 1 gallon of water 1° takes more heat energy (or calories) than increasing the temperature of a cup of water by 1°.

The temperature of a substance is measured by a *thermometer*. Physical geographers normally use either the Celsius or Fahrenheit scales of temperature, in which pure ice melts at 0°C and 32°F, respectively. These scales are used partly because they are familiar, and partly because the temperatures occurring at Earth's surface involve convenient numbers on those scales. However, they also involve negative temperatures, which give a false impression of the nature of temperature, so many scientists use the Kelvin scale, in which 0° is the lowest temperature that is physically possible (absolute zero), the point at which all molecular activity ceases. On the Kelvin scale ice melts at 273°. Since the Kelvin scale uses the same intervals as the Celsius scale, absolute zero is equal to −273°C (−459°F). The relationship between these scales is shown in the Reference Section.

When two bodies at different temperatures come in contact with each other, heat flows from the higher-temperature body to the cooler body. Heat energy is transferred in other ways as well. Heat, then, is a dynamic, transferable form of energy, while temperature is a measure of the presence of such energy. Temperature determines the direction of movement of heat energy from one body to another.

Different materials react differently to heating. The **specific heat** of a substance is the amount of heat required to change the temperature of 1 gram of the substance by 1°C. A gram of a substance with a high specific heat requires more heat to increase its temperature by 1° than a substance with a lower specific heat. Water has the highest specific heat of any naturally occurring substance (1 calorie per gram per degree Celsius), and metals have some of the lowest specific heats. Because the specific heat of water is five times that of dry soil, 1 calorie of heat will raise the temperature of 1 gram of soil by 5°C or 1 gram of water by 1°C; the same amount of heat increases the temperature of 1 gram of copper by 11°C, or 1 gram of gold by 30°C.

Compared with the temperature ranges elsewhere in the solar system, the range in Earth's lower atmosphere, where living organisms exist, is quite small. Temperatures in the center of the sun are thought to be several million degrees, while average temperatures in the outer planets fall below −100°C (−148°F). In Earth's lower atmosphere, temperatures normally range between about 35°C (95°F) and −20°C (−4°F), while extremes occur within the range of just over 40°C (104°F) to −70°C (−100°F). Temperatures vary from one part of the world to another. Places near the equator seldom experience temperatures below 0°C at sea level; on the other hand, places near the poles have temperatures below 0°C for most of the year. In the midlatitudes, winter temperatures commonly fall below 0°C, and summer temperatures may rival those near the equator. The range of temperatures in the atmosphere-ocean environment allows water to exist as liquid, gas, or solid. Temperature contrasts between places and changes among the three phases of water are basic factors in determining Earth's weather.

LINKAGES

There are strong links among the temperature of the lower atmosphere, movements of air, and evaporation ***(atmosphere-ocean environment).*** *Such links also determine the rates at which surface rocks are broken into fragments, and whether streams or glaciers carve the landforms* ***(surface-relief environment).*** *In the* ***living-organism environment,*** *temperature is a major factor in determining the geographic distribution of plants and animals.*

Heat Transfer

Heat energy is transferred between or through substances by three processes—conduction, convection, and radiation (Figure 4-1).

Conduction. In **conduction,** heat is transferred by molecular contact within and between bodies without movement of the bodies themselves. The more active molecules in the hotter material transfer their vibrations and heat energy to the colder material, where they enliven the sluggish molecules. This process is most important in solids. For example, a water-filled metal saucepan on an electric burner is heated by conduction as the intense molecular vibrations in the hot electric burner are transferred by contact to the cold base of the saucepan and then into the lowest layer of cold water (Figure 4-1, *A*). Conduction is less important in the atmosphere-ocean environment than in the solid-earth environment, since air and water are both poor conductors of heat. Only small quantities of heat are conducted slowly from a ground or water surface to the lowest few centimeters of air in contact with it. The poor heat conductivity of air makes it a good insulator. Storm windows or a cover of fresh snow provides good insulation because the air trapped between glass or snowflakes conducts heat poorly.

Convection. In **convection,** the liquid or gas containing the heat actually moves. Heating causes a fluid to expand and rise as the result of its lower density, while cooling causes it to contract and descend due to increasing density. A circulation is established in the liquid or gas. In the saucepan shown in Figure 4-1, *A,* the water is initially heated by conduction from the hot metal to the water in contact with the metal. Since conduction in water is poor, convection plays the major part in spreading heat energy through the water. The heated water

BOX 4-1

THE SUN, TEMPERATURE, AND HUMAN HEALTH

Human metabolism and health are affected by the temperature of the surrounding air. Exposure to low or high temperatures can be dangerous. Direct exposure to sunlight can also be hazardous, although many people think of sunbathing and acquiring a tan as healthy activities.

The human body regulates its "core" temperature to within 2°C (3 to 6°F) of 37°C (98.6°F). The "core" includes the brain, heart, lungs, and digestive tract, and these vital organs must be maintained at a constant temperature to function properly. External temperatures vary greatly, however, so the body has evolved a variety of mechanisms to sustain the current temperature.

People are most comfortable when at rest, clothed, and indoors at temperatures of 20 to 25°C (68 to 77°F). Within this range, the body maintains a core temperature of 37°C without special effort. If temperature falls below this range, shivering increases muscular activity and heat, and returns the core temperature to 37°C; blood vessels near the skin surface contract to conserve the heat of circulating blood (making the skin pale). To stay warm, it may become necessary to exercise, put on more clothing, or turn up the heat. If the air temperature rises above 25°C, perspiration begins and evaporation of the water from the skin takes heat from the skin, cooling it. Blood flow to the surface is increased to ease the transfer of heat out of the body and gives one a flushed appearance. It may be necessary to rest, shed some clothing, or turn on a fan.

Hypothermia occurs when the body core temperature falls below 35°C (95°F). Shivering becomes uncontrollable, speech suffers, and lethargy sets in. Below 32°C (90°F) the body becomes rigid and uncoordinated. Unconsciousness sets in at 30°C (86°F), and death may occur below 25°C (75°F). *Hyperthermia* occurs when the body temperature rises above 39°C (103°F), with collapse at 41°C (105°F). Severe heat stress (often known as heatstroke or sunstroke) may lead to death within a few hours. Once the body core temperature goes outside the range of 35 to 39°C, the heating regulation system of the body breaks down, and special treatment is necessary at once.

rises and is replaced by cooler water, which sinks and is heated at the base. Figure 4-1, *B*, shows how convection circulates heat upward in the atmosphere from heated ground or water surfaces. In the oceans, water cooled at the surface in polar regions sinks and flows underneath the warmer water below, forcing it slowly back to the surface.

The horizontal component of convection is often known as **advection.** An example of advection is when warm air or water is transferred from the tropics to cooler latitudes. Convection and advection are important in the circulation of heat in both the atmosphere and oceans. Conduction and convection often work together: the base of the atmosphere is partly heated by conduction, and then convection transfers the heat upward.

Two types of heat are distinguished in the study of convection in the atmosphere. **Sensible heat** affects the temperature of a heated body. When sensible heat is transferred, the change in the heat content of the body produces a temperature change.

Latent (stored or hidden) **heat** is transferred when water changes from one physical state to another. When ice is changed to water, for example, energy is needed to loosen the bonds holding the molecules together in the rigid solid: 80 calories of energy per gram are absorbed and stored in the liquid water as latent heat. The heat energy remains

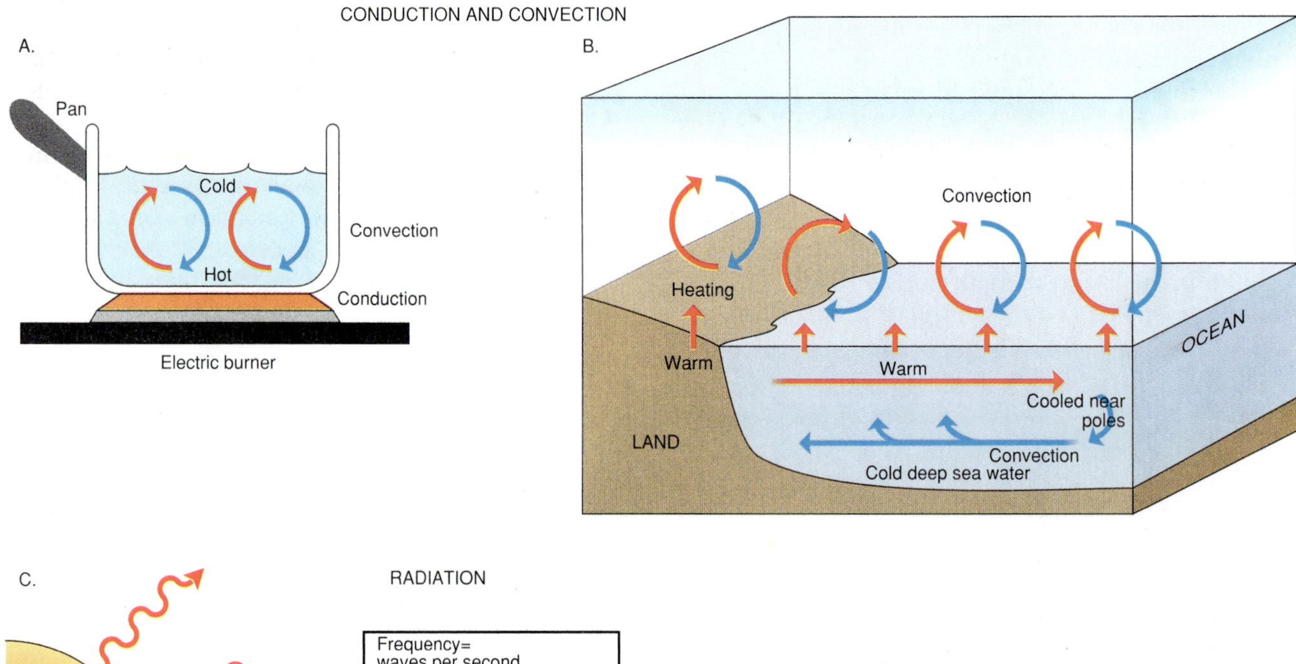

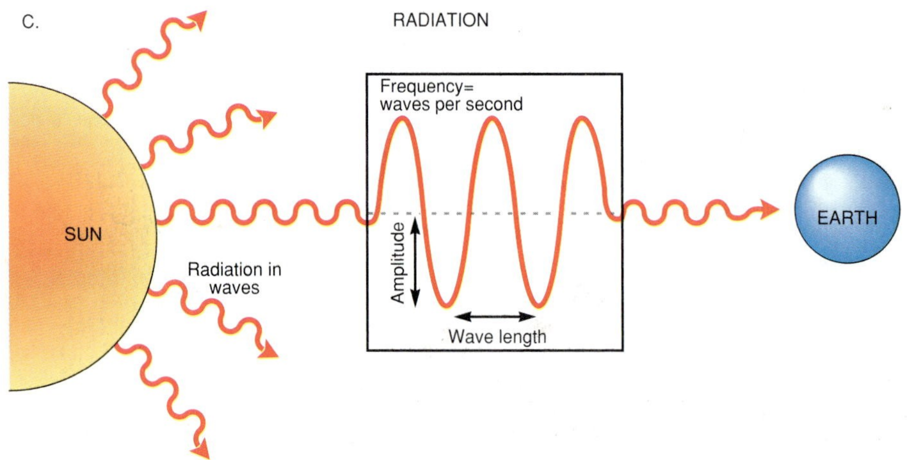

Figure 4-1 Energy transfer processes: **A,** Conduction transfers heat from a hotter to a cooler substance at the base of the pan without movement of the materials. Convection transfers heated water upward. **B,** Convection transfers heat by movement within media such as water and air. Air is heated at the base of the atmosphere and rises, being replaced by cooler air that is itself subject to heating. If water at the ocean surface is cooled, it sinks and initiates a convectional circulation. **C,** Radiation transfers heat and other forms of energy in waves that pass through some materials and are absorbed by others.

"hidden" in the sense that it does not raise the temperature of the water as the ice melts. When liquid water is converted to water vapor, even more energy is required to split apart the molecules: the amount ranges from 600 calories per gram at 0°C to 540 calories per gram at 100°C. Once again, the heat energy is stored in latent form in the water vapor. This stored heat is released only when water vapor changes back to water or when liquid water changes to ice.

The nature of latent heat can be illustrated by a common experience. Someone who gets out of a swimming pool on a hot, dry day often feels cold. This is because the water on the person's skin is evaporating, and energy from the person's body (and from the sun) is being used in the process. As the water evaporates, heat energy from the body becomes locked up in the water vapor. This loss of energy makes the person feel cold.

When liquid water changes to water vapor and there is convection in the atmosphere, latent heat energy may be transferred from the ground to higher levels in the troposphere. The air containing the water vapor rises, and as it cools it turns to water droplets, forming clouds. The formation of water droplets from water vapor changes latent heat into sensible heat, raising the temperature. This transfer of the heat locked in water vapor is what makes the circulation of water in the atmosphere such an important factor in producing the weather.

Radiation. Radiation is energy emitted by all bodies with a temperature greater than absolute zero (0°K, −273°C, −459°F). When an object absorbs such radiant energy, its heat energy increases. Thus, heat is transferred by radiation. Radiant energy moves in **electromagnetic waves** (Figure 4-1, *C*) that vary in wavelength and frequency (the number of waves per second) according to the temperature of the body emitting them. Hotter bodies emit more energy in shorter wavelengths at higher frequencies than colder bodies. In other words, the wavelength is inversely proportional to the temperature of the radiating surface. Thus, the sun emits a range of radiation that is concentrated in the shorter wavelengths, while Earth emits radiation in longer wavelengths. Radiation from the sun

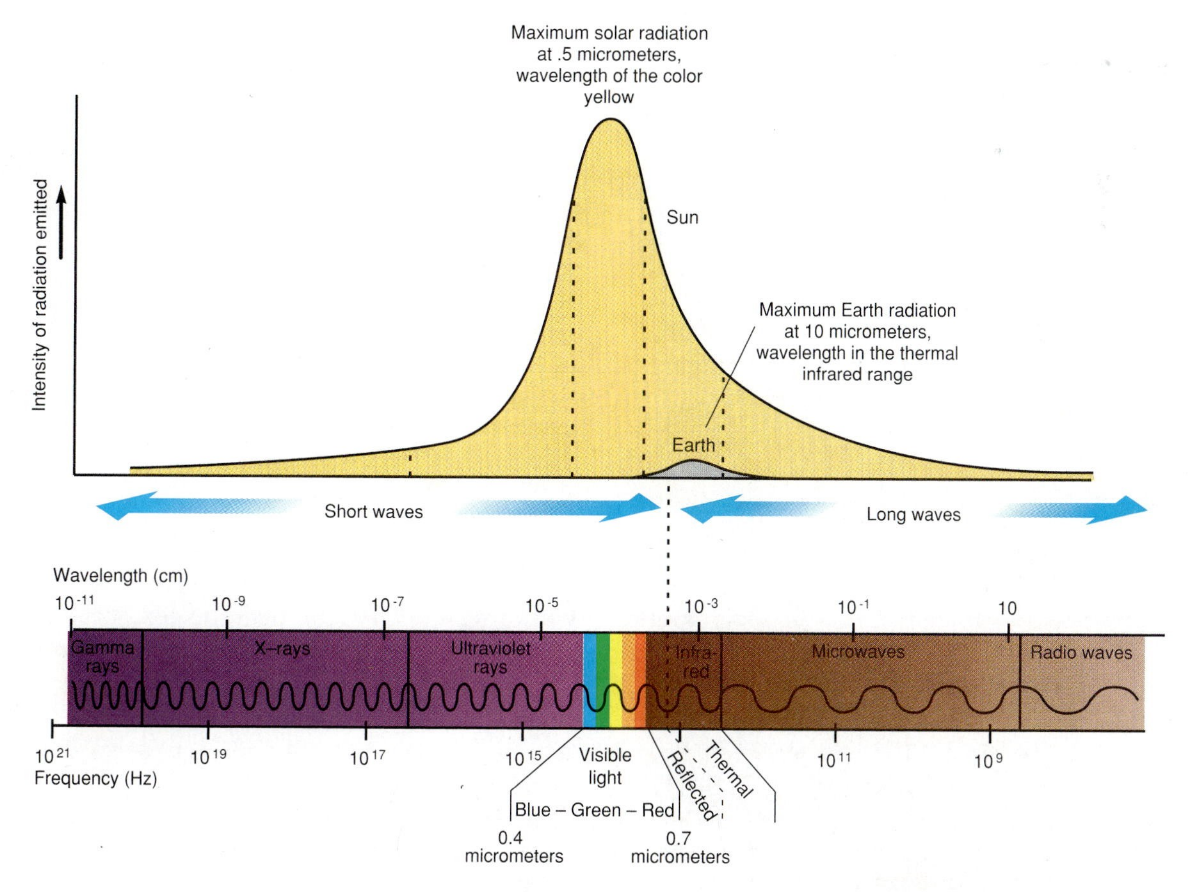

Figure 4-2 The electromagnetic spectrum of radiation wavelengths in relation to radiation from the sun and Earth. The solar radiation spectrum peaks in the visible light range and the terrestrial radiation in the thermal infrared range.

provides virtually the only source of energy for the movements of Earth's atmosphere and oceans. It is also the source of the energy converted by plants for the functioning of living organisms (Chapter 19).

The range of different wavelengths involved in the transfer of radiant energy is known as the **electromagnetic spectrum** (Figure 4-2). This spectrum is often divided into two parts for simplification. **Short-wave radiation** includes wavelengths of less than 4 micrometers (a micrometer is a thousandth of a millimeter), and is high in frequency and energy. Gamma rays, x-rays, ultraviolet rays, visible light rays, and reflected (shorter wavelength or near) infrared rays are all short-wave radiation. **Long-wave radiation** (greater than 4 micrometers wavelength) includes thermal (longer wavelength or far) infrared rays, microwaves, and radio waves. All the waves travel at the speed of light (300,000 km, or 186,000 miles, per second).

ENERGY FLOWS IN THE ATMOSPHERE-OCEAN ENVIRONMENT

The mechanisms for heating Earth's atmosphere and oceans are complex but can be understood by breaking the processes down into aspects of the flow of solar energy through the atmosphere-ocean environment. There are three aspects to be considered: the *inputs* of solar radiation, the *transfers* of this energy as it flows through the atmosphere-ocean environment and interacts with Earth's surface, and the *outputs* from Earth to space. Although the inputs are balanced by the outputs in terms of total energy, the wavelengths and paths of the two are different.

Inputs from the Sun

Insolation. The energy flow from the sun, known as **insolation** (INcoming SOLar radiATION), consists mainly of short-wave radiation, which is transferred across an average distance of 150 million km (94 million miles) of space from sun to Earth. Because the distance is so great, Earth intercepts a tiny proportion of the sun's rays (0.0005 percent or only 5 millionths of the total solar output). This tiny proportion, however, provides 99.9 percent of the energy used in the atmosphere-ocean environment. The remaining 0.1 percent comes from the planet's hot interior. The distance between sun and Earth varies over longer periods and is a major factor in producing changes of climate (Chapter 9).

The wavelength of solar radiation is determined by the surface temperature of the sun (6000°C). The wavelengths emitted most intensely are in the visible light portion of the electromagnetic spectrum. Figure 4-3 shows that the peak of solar radiation is in the narrow band of visible light: approximately 45 percent of the solar output is concentrated in this band. A further 46 percent of solar output is spread across the range of infrared wavelengths, and some 9 percent is in wavelengths shorter than those of visible light.

LINKAGES

Although the proportion of energy from Earth's interior that contributes to atmospheric and oceanic activity is small compared to solar energy, it is important to understand that Earth environments are related to both sources of energy in varying proportions. The workings of the ***atmosphere-ocean environment*** *and the* ***living-organism environment*** *are dominated by the sun; the* ***solid-earth environment*** *is dominated by Earth's interior source of energy; and the* ***surface-relief environment*** *is affected by both.*

Solar Radiation Output. Variations in the amount of solar radiation that arrives at the outside of Earth's atmosphere are determined by two major factors—fluctuations in solar output and differences in the distance between sun and Earth. These variations are small enough, however, that the insolation arriving at the top of Earth's atmosphere is known as the **solar constant.** The most recent satellite measurements of this constant give a value of 1.96 calories per square centimeter per minute.

The total radiative output from the sun has varied by only 0.1 percent during the 10 years it has been measured by satellites. It has been estimated that a 2 percent difference in the solar constant could change Earth's surface temperatures by up to 1.2°C, and a 10 percent difference by up to 6°C, but there is little evidence that the rate of solar emission will vary as much as this over the next few decades. The solar energy output can be safely regarded as unchanging over periods of a few decades.

Earth's Shape, Rotation, and Orbit. Earth's spherical shape, rotation around its axis, and orbit around the sun affect the receipt of solar energy from place to place over short periods. Earth's *spherical shape* causes insolation to be received at varying angles (and sometimes not at all) by different places on the globe (Figure 4-3). Places where the sun is more directly overhead have a greater intensity of

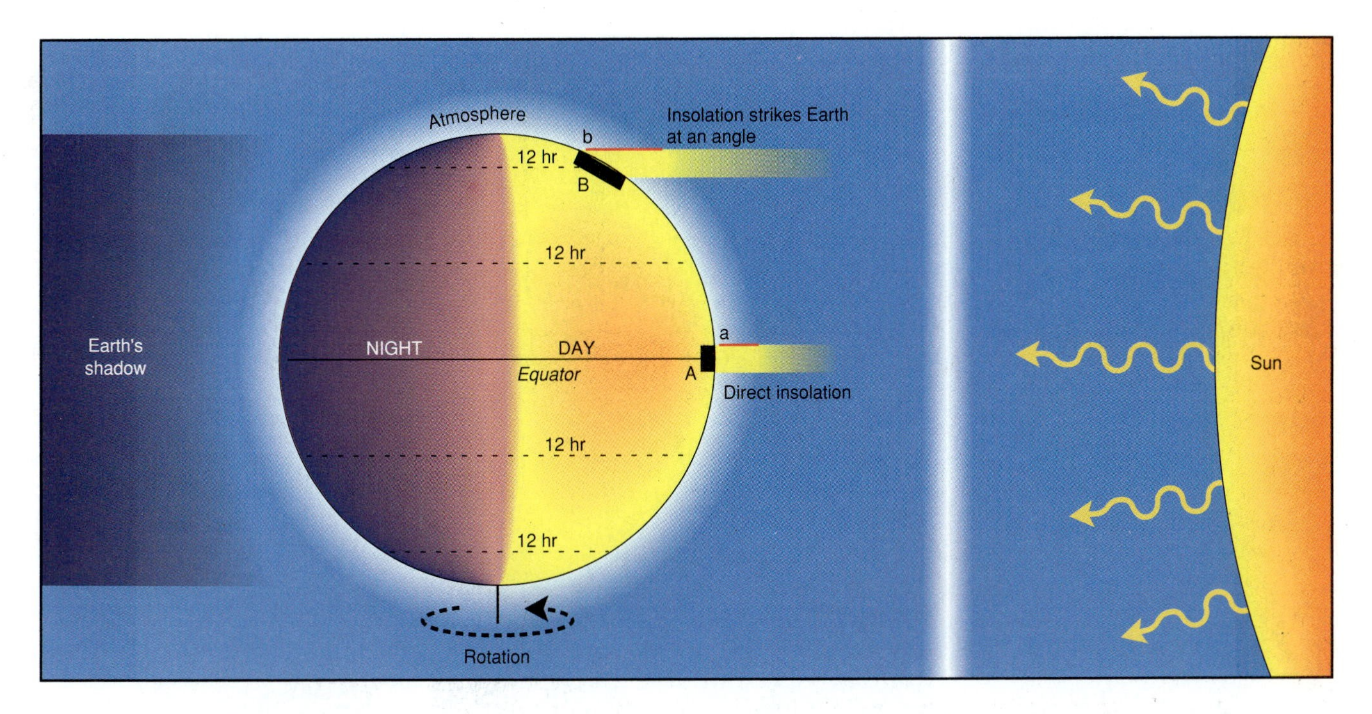

Figure 4-3 Insolation and Earth at a solstice. At the equator *(A)* the insolation is focused on a small area and the rays of the vertical sun pass through the atmosphere in the shortest distance. At the Arctic Circle *(B)* insolation is spread over a greater area and has to pass through a greater thickness of atmosphere. Earth rotates once every 24 hours, and all parts of the planet experience sunlight except for the polar regions in winter.

insolation, while those where the sun is at an angle have a lower intensity since the same amount of radiation is spread over a larger area of ground. The lower-angle rays also take a longer path through the atmosphere, which further reduces their intensity when they reach the ground at higher latitudes.

Earth's *rotation* causes daily variations of insolation around the world. The half of the globe facing the sun receives daytime insolation; the half facing away experiences night and receives no insolation. At a given location, the daily cycle of insolation begins soon after sunrise, increases until the peak at solar noon, and decreases throughout the afternoon. Earth's rotation on its axis once every 24 hours causes the insolation to be received frequently over most of the globe, although there are seasonal variations that exclude insolation from the polar regions in winter.

The *tilt of Earth's axis* of rotation gives rise to seasonal variations of insolation and of the length of day and night at different latitudes (Figure 4-4). As Earth revolves around the sun in its annual orbit, the angle of its axis to the orbit remains 66.5. As a result, the sun is vertically overhead at latitude 23.5° N on June 21st, and at latitude 23.5°S on December 21st. When the sun is vertically overhead at 23.5° N (the Tropic of Cancer), the northern hemisphere has a summer season with extra-long days, and places within the Arctic Circle have 24 hours of daylight. When the sun is vertically overhead at 23.5° S (the Tropic of Capricorn), the northern hemisphere has a winter season with shorter days, and places within the Arctic Circle receive no daylight. The two occasions when the sun is vertically overhead at the Tropic of Cancer or Capricorn are known as the **solstices.** On March 21st and September 21st the sun is vertically overhead at the equator and all parts of the world have days and nights of 12 hours. These occasions are known as **equinoxes** ("equal nights").

A minor variation in the annual heating cycle is caused by the *Earth's orbit* around the sun being elliptical (oval) rather than circular. On January 3rd, Earth is closest to the sun (perihelion), and receives 7 percent more insolation than on July 4th, when it is farthest away (aphelion). This difference alone would increase world January temperatures by about 4°C, but other effects such as atmospheric heat circulation and the predominance of oceans in the southern hemisphere minimize the impact.

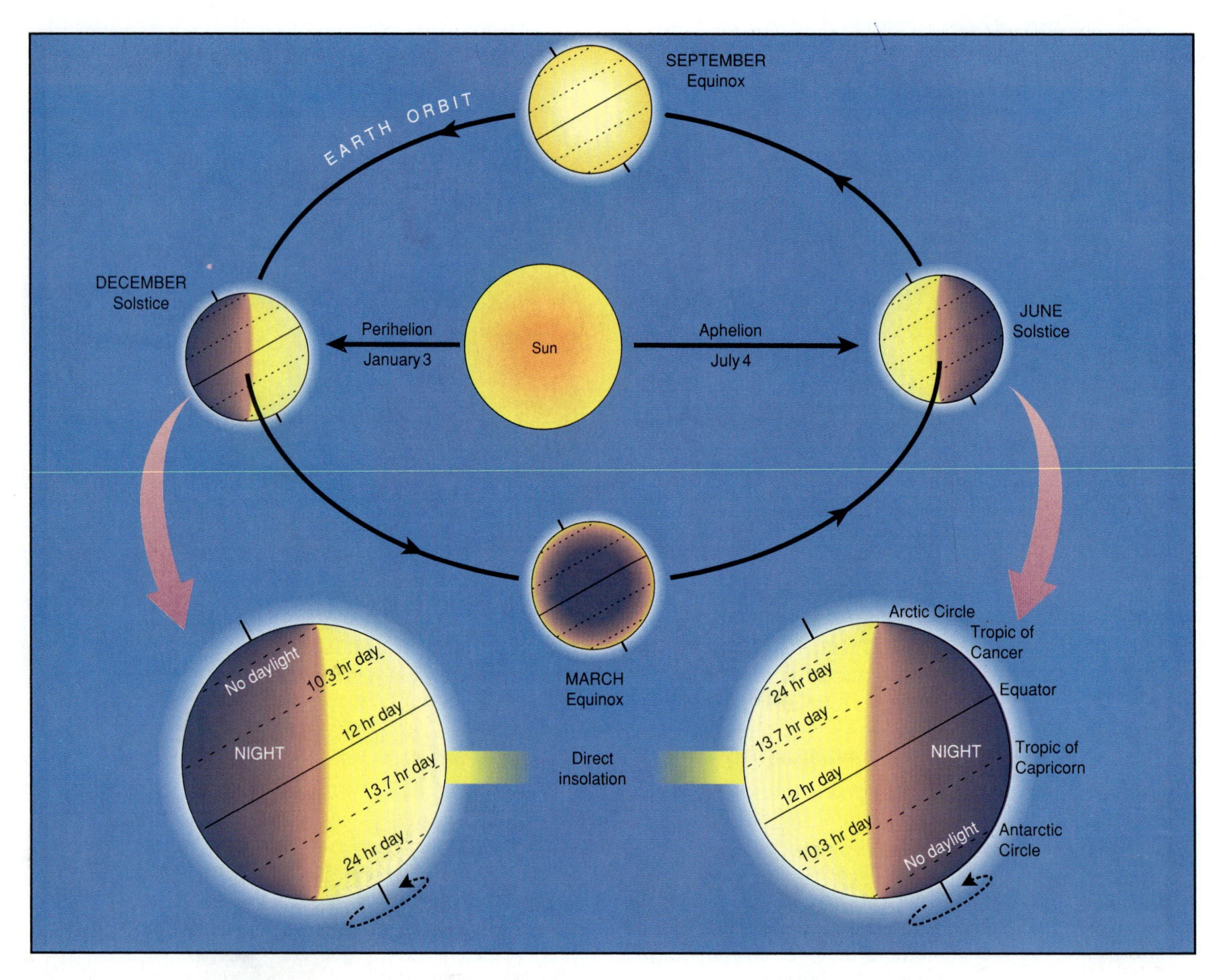

Figure 4-4 Seasonal changes in the relationship between sun and Earth. As Earth revolves around the sun, the sun's overhead position shifts from 23.5° North to 23.5° South. This produces northern hemisphere summers in June and southern hemisphere summers in December. During the summer season in a particular hemisphere, the polar area receives continuous daylight, but does not receive any during the winter.

Transfers of Insolation Through the Atmosphere

Once insolation enters the atmosphere, it is subject to a series of changes on its way to Earth's surface. The gases, dust, and clouds in the atmosphere absorb, scatter, or reflect the incoming short-wave solar radiation (Figure 4-5). Just over half is transmitted through to Earth's surface.

Absorption. Absorption is the retention of radiant energy by atmospheric gases, dust, and the water droplets that form clouds. This process results in the conversion of the sun's rays to sensible heat, raising the temperature of the absorbing material. For instance, much of the sun's shortest-wave x-ray and gamma-ray radiation is absorbed by the gas ions in the thermosphere, and heats that layer. Ozone absorbs ultraviolet rays in the upper stratosphere, and this produces another layer of higher temperature. The absorption of so much ultraviolet and shorter-wave radiation in the upper levels of the atmosphere provides a protective shield for living organisms at Earth's surface. In the troposphere the carbon dioxide, water vapor, and other gases absorb some of the incoming infrared rays. These gases are thereby heated and reradiate heat energy. The atmosphere, however, absorbs few visible light rays, acting, rather, as a transparent "window" to them.

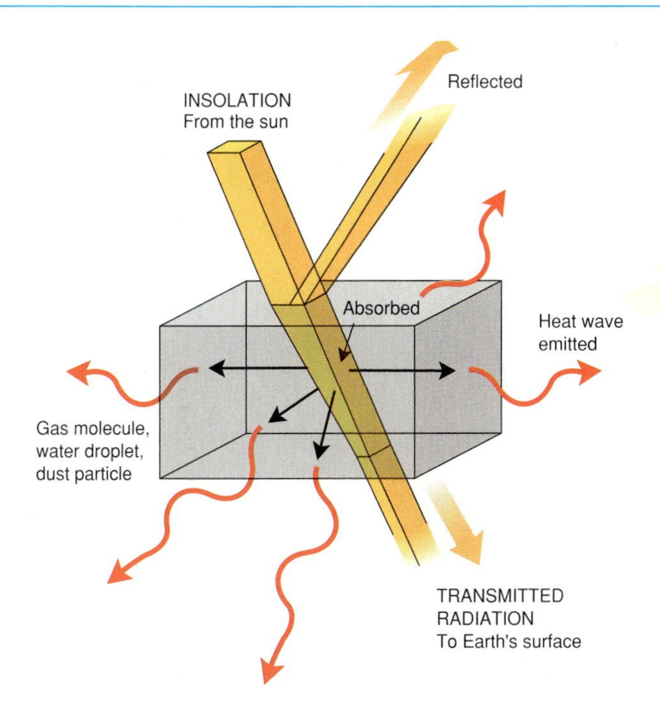

Figure 4-5 Absorption, scattering, reflection, and transmission of solar radiation. When insolation is intercepted by gas molecules, water droplets, or dust particles, part of it is reflected or scattered, part is absorbed by the substance, and part passes through. The absorbed energy heats the substance, which radiates heat.

Scattering and Diffused Light. **Scattering** occurs when radiation waves hit gas molecules or particles of dust and are not absorbed by them. The radiation bounces in all directions—upward, sideways, and downward. The result of scattering by the gases and particles in the atmosphere is that insolation reaches the ground from all parts of the sky, not just in rays directly from the sun. Scattering by gas molecules affects the shorter wavelengths of visible light to a greater extent than the longer waves. Blue light has a shorter wavelength than other colors in the visible light spectrum and is scattered more effectively, producing the blue color of the daytime sky as seen from the ground. The orange and red colors of sunrises and sunsets (Figure 4-6), however, are caused by absorption of all the blue and much of the green wavelengths as the low angle rays of the sun pass through a greater thickness of atmosphere. The colors are particularly vivid where the atmosphere contains dust and smoke particles. Scattering by dust and water droplets tends to affect all visible light wavelengths and produces a lighter blue-gray sky in polluted areas above cities. In addition to absorbing some insolation, clouds scatter visible light as it passes through, and they allow less to penetrate as they become thicker. Lighting conditions below a cloud cover are known

Figure 4-6 Red sky at dusk.

Figure 4-7 Clouds and diffused light. In this photograph the sun can be seen shining through a thin layer of low cloud, and its light is diffused as it passes through the cloud.

as **diffused light** (in contrast to direct light when the sun is visible and shadows are cast). During daylight hours, any point on Earth's surface receives energy that comes directly from the sun, is diffused through clouds, or is scattered in the atmosphere (Figure 4-7).

Reflection. Reflection occurs when the radiation is bounced off substances. Insolation is reflected directly back to space by the tops of clouds (almost one fifth of total insolation), by dust particles in the troposphere, and by Earth's surface. The **albedo** of a surface is the proportion of total incoming solar radiation that is reflected back to space. For the planet as a whole the albedo is about 0.34, which indicates that Earth reflects 34 percent of insolation.

Summary. The overall result of all these interactions between the atmosphere and the incoming solar radiation is that approximately half the radiation arriving at the outside of the atmosphere gets through to the surface. Nearly all the harmful gamma rays and x-rays, together with much of the ultraviolet and infrared radiation, are filtered out by absorption. Most of the energy in the visible light part of the spectrum gets through the atmospheric "window" directly, or in scattered form.

Solar Energy and Earth's Surface

Reflection Versus Absorption. Earth's surface reflects or absorbs the filtered short-wave insolation that reaches it. The proportions of reflection and absorption (i.e., the albedo) depend on the nature of the various surface materials covering the planet (Figure 4-8). If the sun is vertically overhead, the ocean absorbs almost all the insolation: the albedo is 3 percent in such conditions. The albedo of ocean water increases as the sun angle gets lower. With a sun angle of 30°, 6 percent of insolation is reflected; at 10° it is 35 percent; at below 5° it is 50 to 80 percent. On land the important distinction is between light and dark surfaces. The albedo of fresh snow may be up to 90 percent, but that from a blacktop road may be as little as 5 percent. When the sun is shining at a low angle on the ski slope or the beach, it is often difficult to see without dark glasses because of the strong reflection of sunlight off the snow and water. A blacktop road or sandy beach, on the other hand, absorbs solar radiation and can grow hot enough on a summer day to burn bare feet.

Insolation that is not reflected is absorbed by Earth's surface and transformed into heat energy.

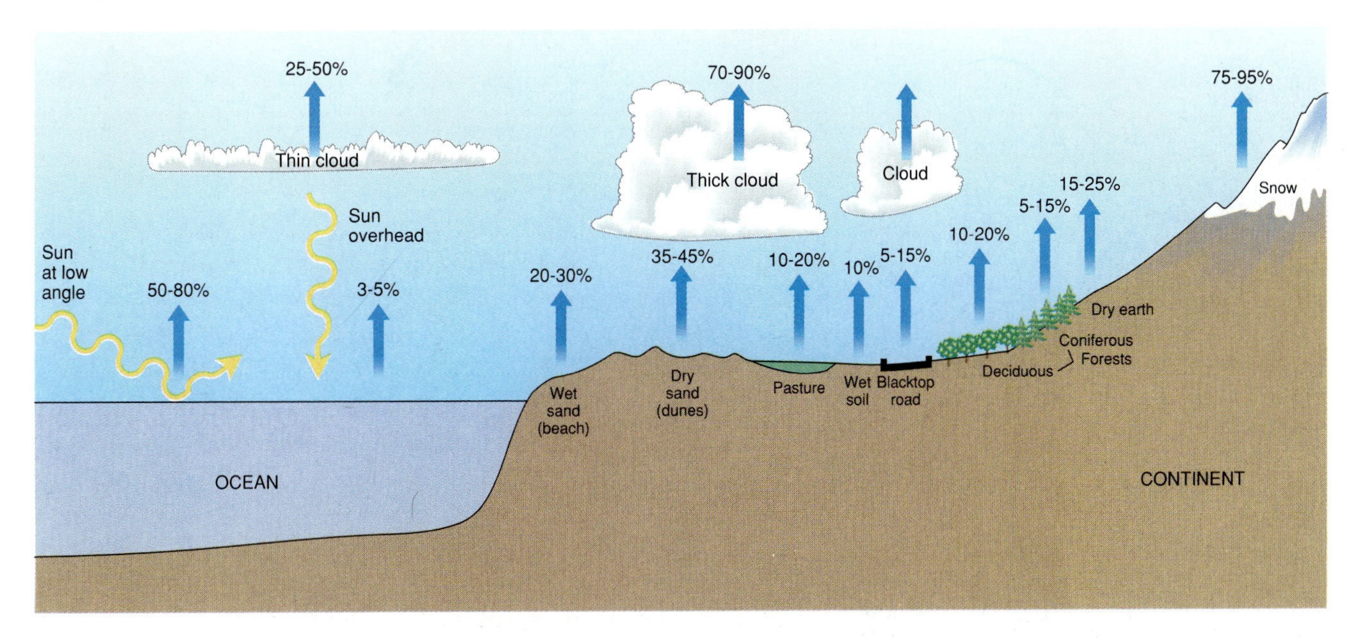

Figure 4-8 Albedos of different types of landscape. The figures are percentages of insolation that is reflected.

The oceans (under high sun angle) and dark surfaces on land absorb high proportions of the insolation as sensible heat, which raises the temperature of the water or soil. A significant proportion of the energy that might have been absorbed by the water or soil is often taken up by the evaporation of water and stored as latent heat in the water vapor produced. In desert areas, where there is little or no surface water to be evaporated, virtually all the insolation absorbed is converted to sensible heat.

Oceans Versus Continents. About 80 percent of insolation reaching Earth's surface hits the oceans, which cover 71 percent of the globe but receive a larger proportion of the total insolation since they dominate the intense radiation zone of the tropics. Some of this solar energy is transferred to latent heat in evaporated water vapor, but a great deal goes to heating enormous quantities of seawater. There is more vertical penetration by solar rays in the ocean waters than on land. Heating of the land affects depths of only a few centimeters, while in clear seawater the sun's rays may penetrate up to 40 meters (130 feet) deep. Ocean waters also mix well, and heat absorbed in the surface waters is distributed to depths of several hundred meters by this mixing. Moreover, the oceans have a higher capacity for holding heat energy because water's specific heat is higher than land's. As a result, the oceans retain 25 to 33 percent more heat than the same amount of land would retain at the same latitude. The 10-meter-deep surface layer of the oceans carries four times the thermal energy contained in the entire atmosphere.

The differences between the effects of insolation on the land and in the oceans are summarized in Figure 4-9. An important consequence of these differences is that land areas gain heat more rapidly in summer than ocean waters. In fall and winter, land areas rapidly lose all the heat that accumulated in spring and summer, whereas the oceans continue to store the heat. This makes ocean temperatures more persistent and less extreme than those of land surfaces.

Heat Transfers from Earth's Surface

Terrestrial Radiation. Absorption of solar energy heats Earth's surface (including continents and oceans) on average to 12°C (54°F). Like any body that absorbs heat, Earth radiates energy itself. However, because the temperature of Earth's surface is low compared to that of the sun, its energy transmission is less intense. Earth emits energy in the thermal infrared sector above 4 micrometers wavelength (i.e., as long waves): this is known as **terrestrial radiation**. The troposphere is heated by terrestrial radiation from the base upward to a much greater extent than by the incoming solar radiation.

The lower atmosphere is heated more over surfaces that absorb insolation rather than reflecting it. Surfaces that reflect more insolation than they ab-

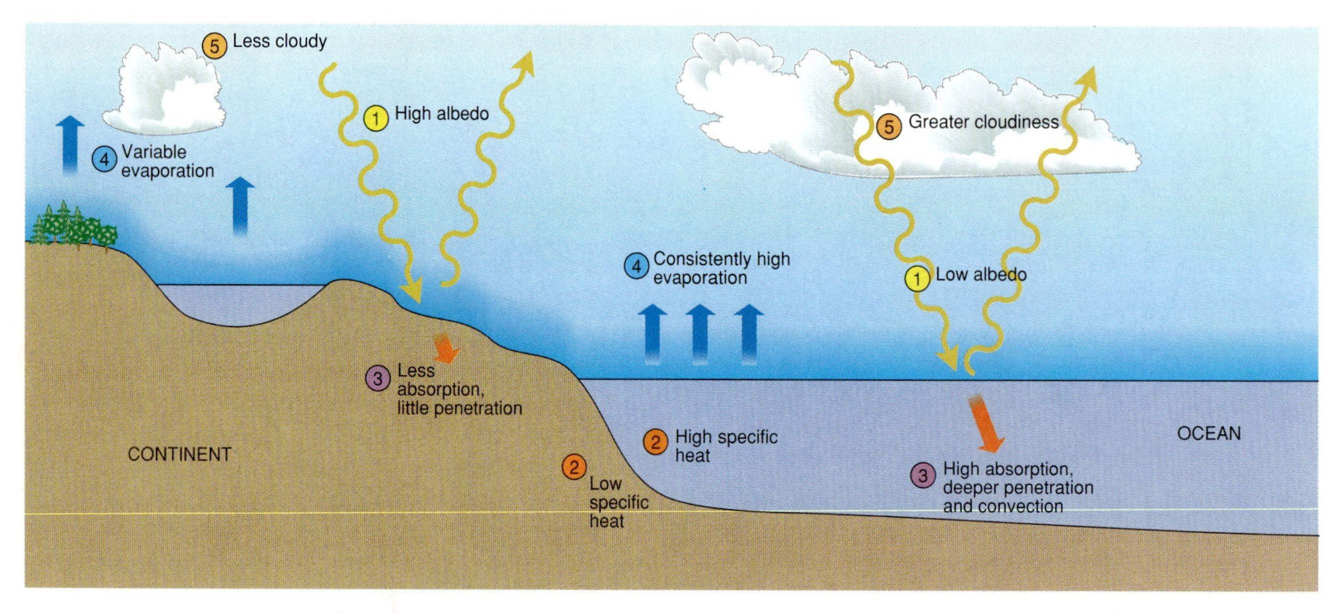

Figure 4-9 Continents and oceans: the factors that produce different heating characteristics.

sorb have less to radiate and heat the air above. Air is warmed more over blacktop roads than over grass and very little over snow. Most of the short-wave insolation is absorbed by the oceans, so they provide most of the long-wave radiation as well as the latent heat in the lower atmosphere. The oceans also maintain a more even flow from their long-term heat storage into the atmosphere than do the land surfaces, where there are greater seasonal fluctuations.

Although incoming visible-light radiation arrives at Earth's surface with little depletion after passing through the atmosphere, only a small proportion of the terrestrial long-wave radiation can escape directly to space through narrow "windows" in the atmosphere. The water vapor, carbon dioxide, and dust in the troposphere absorb most of the long-wave energy from the surface and heat up. This heating effect is less in arid regions because of the low quantities of water vapor in the air.

The gases and dust that absorb the long-wave radiation from the surface (and some from insolation) are heated, and themselves emit energy in the long-wave range. It is emitted in all directions, including upward and eventually out to space and downward back to Earth. The portion emitted downward to Earth's surface is known as **counterradiation** and helps to reinforce the heating effect near the ground. The troposphere is thus heated progressively by terrestrial radiation from the ground upward.

Conduction and Convection. Although terrestrial radiation is the main means of transferring heat from the ground into the atmosphere, conduction, convection, and latent heat transfers also play a part. Conduction transfers heat into the lowest layer of air in contact with the ground. Evaporation produces latent heat in water vapor. Convection carries the heated air and its content of water vapor upward and causes heat energy to be spread through the lower atmosphere. Again, the heating begins at ground level and affects the lower layers first and with the most intensity.

Lapse Rates and Temperature Inversion. The combination of heating from the ground up and lower atmospheric density with height (Chapter 3) leads to a decrease in temperature with ascent through the troposphere (Figure 4-10). The lower density of air at higher altitudes means that there are fewer gas molecules to absorb heat energy, so there is less heating of the atmosphere at this level. The rate at which the temperature of the atmosphere decreases with height in the troposphere is known as the **lapse rate,** and it varies depending on the rate of heating of air at the surface over land or water. In still air the rate of temperature decrease at a particular time and place is known as the **environmental lapse rate** to distinguish it from the decreasing temperature in a body of air rising through the atmosphere. On average the environmental lapse rate is 6.5°C per km (3.6°F per 1000 feet), but wide variations occur from day to day.

The decrease of temperature with height in still air varies because of differences in the heating and cooling of air in contact with the ground. When heating of air near the ground occurs, the difference

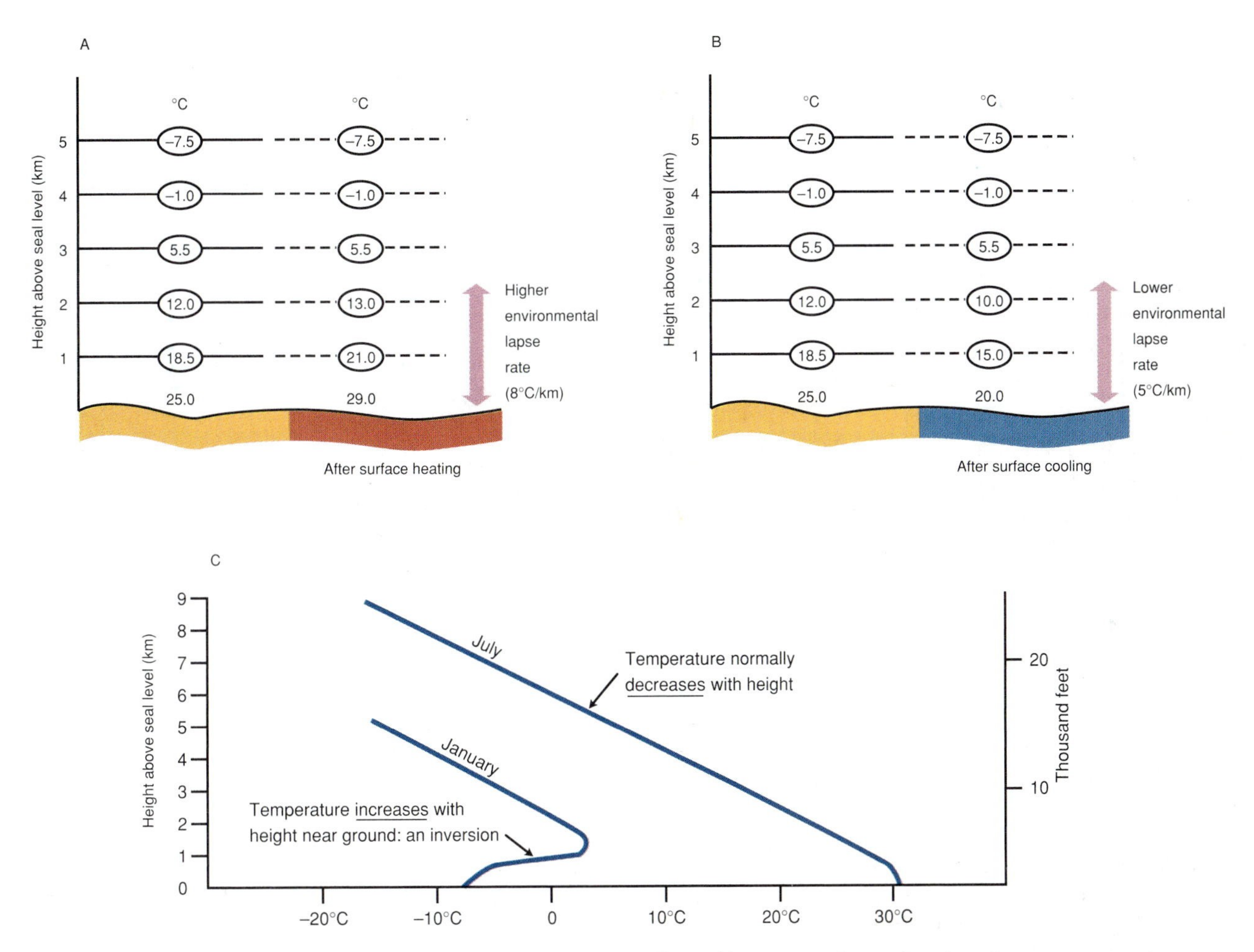

Figure 4-10 Environmental lapse rates. **A,** The effect of heating at the surface is to increase the environmental lapse rate. **B,** The effect of cooling at the surface is to decrease the environmental lapse rate. **C,** Seasonal differences in environmental lapse rates for a place in the United States Midwest. The summer graph shows how the temperature decreases with height. The winter graph shows that intense cooling at the surface has produced an inversion in the lapse rate.

in temperature between the base of the atmosphere and a point several kilometers above increases; the environmental lapse rate also increases (Figure 4-10, *A*). When cooling of air takes place near the ground, the difference in temperature between the base of the atmosphere and a point several kilometers above decreases, and so does the environmental lapse rate (Figure 4-10, *B*). It is common to find the environmental lapse rate increasing by day and decreasing by night in cloudless conditions.

In some cases cooling near the ground may be so great that the lapse rate is reversed for a few tens or hundreds of meters with cold air lying beneath warmer air. This situation is known as a **temperature inversion** because it reverses the normal pattern. The effect of a temperature inversion on the lapse rate for January is shown in Figure 4-10, *C*. Such temperature inversions are caused by rapid cooling of the lowest atmospheric layers at night in windless conditions, by dense cold air flowing beneath warmer air, or by an overlying layer of descending warm air. Temperature inversions are common in the Los Angeles basin in summer. During this season, the region's weather is dominated by descending warm air and clear skies, through which the sun's rays heat the surface. Cool, moist air is drawn in from the Pacific Ocean and gives rise to lower temperatures at the surface than at 2 km higher.

The Role of Clouds. In addition to reflecting, scattering, and absorbing insolation from above, clouds also absorb terrestrial radiation from below. Long waves are reradiated back to Earth from cloud undersides, and they are also emitted from cloud tops into space. The overall effect is to slow the return of energy to space and keep more heat in the lower atmosphere. Thus, it may be warm on cloudy days although the sun is shut out, while cloudy nights are warmer than cloudless nights since less terrestrial heat is radiated to space.

The role of clouds in heating and cooling the atmosphere was a theme of atmospheric research in the 1980s and is a major priority for study in the 1990s. It has been found that clouds reinforce the heating of the middle troposphere when latent heat is released as water vapor changes to liquid droplets. Clouds heat the troposphere most effectively over the tropical oceans and tropical rainforests, which have the greatest frequencies of the tallest clouds. In midlatitudes, the most important role of clouds appears to be their reflection of insolation, which produces a cooling effect, but they are also heated as they absorb outgoing infrared radiation. The growing realization that clouds are important to an understanding of atmosphere-ocean heating suggests that they may play a key role in climatic change (Chapter 9).

Figure 4-11 The "Blue Planet": Earth from space, showing the oceans, clouds, and continents of the southern hemisphere. Much of the atmosphere is transparent, but the clouds reflect solar rays out to space.

Outputs of Energy to Space

Earth's return of energy to space includes reflection of insolation from the clouds and Earth's surface, and radiation of long-wave energy from Earth's surface, the troposphere, and other heated layers of the atmosphere. These outputs balance the total *amount* of radiant energy inputs, but the *type* of radiation output to space is very different from that which arrives from the sun. Over 70 percent of the output is in the infrared wavelengths. The fact that some 30 percent of the output is reflected as visible light rays gives the planet a bright appearance when viewed from space (Figure 4-11), and the variations in albedo from oceans to the various land-surface types make it possible to differentiate landscape types.

The Lag Between Insolation Receipt and Heating Effect

A lag between the receipt of insolation and the heating of the atmosphere delays the timing of highest daily and annual temperatures. Insolation is most effective when the sun is highest in the sky at noon, but the hottest time of day with cloudless skies is an hour or so after noon. It takes that time for the ground to absorb insolation and transfer heat energy to the lower atmosphere by long-wave radiation, conduction, and latent heat transfer. A lag also occurs in seasonal heating and is particularly marked in midlatitudes. At these latitudes inland areas have their highest mean monthly temperatures in July in the northern hemisphere—not June when the sun is highest in the sky. Their coldest month is January, not December. Midlatitude coastal locations have an even greater lag with highest monthly temperatures in August and lowest in February. This is because the ocean takes longer than the land to heat and then to release the heat energy into the atmosphere.

THE BALANCE OF ENERGY IN THE ATMOSPHERE-OCEAN ENVIRONMENT

Inputs and Outputs of Energy

The net result of heat transfers by radiation, conduction, and turbulent heat transfer through the oceans and atmosphere is the maintenance of a complex balance of inputs from the sun and outputs from Earth known as the **heat balance** of the atmosphere (Figure 4-12). The heat balance is important in keeping temperatures in Earth's atmosphere at the right level for animals and plants to exist, and in preventing wild fluctuations, which would produce more frequent weather extremes.

The input of insolation is straightforward: the very short-wave radiation is absorbed in the upper atmosphere and the visible light rays that are not reflected away to space by clouds bring energy to

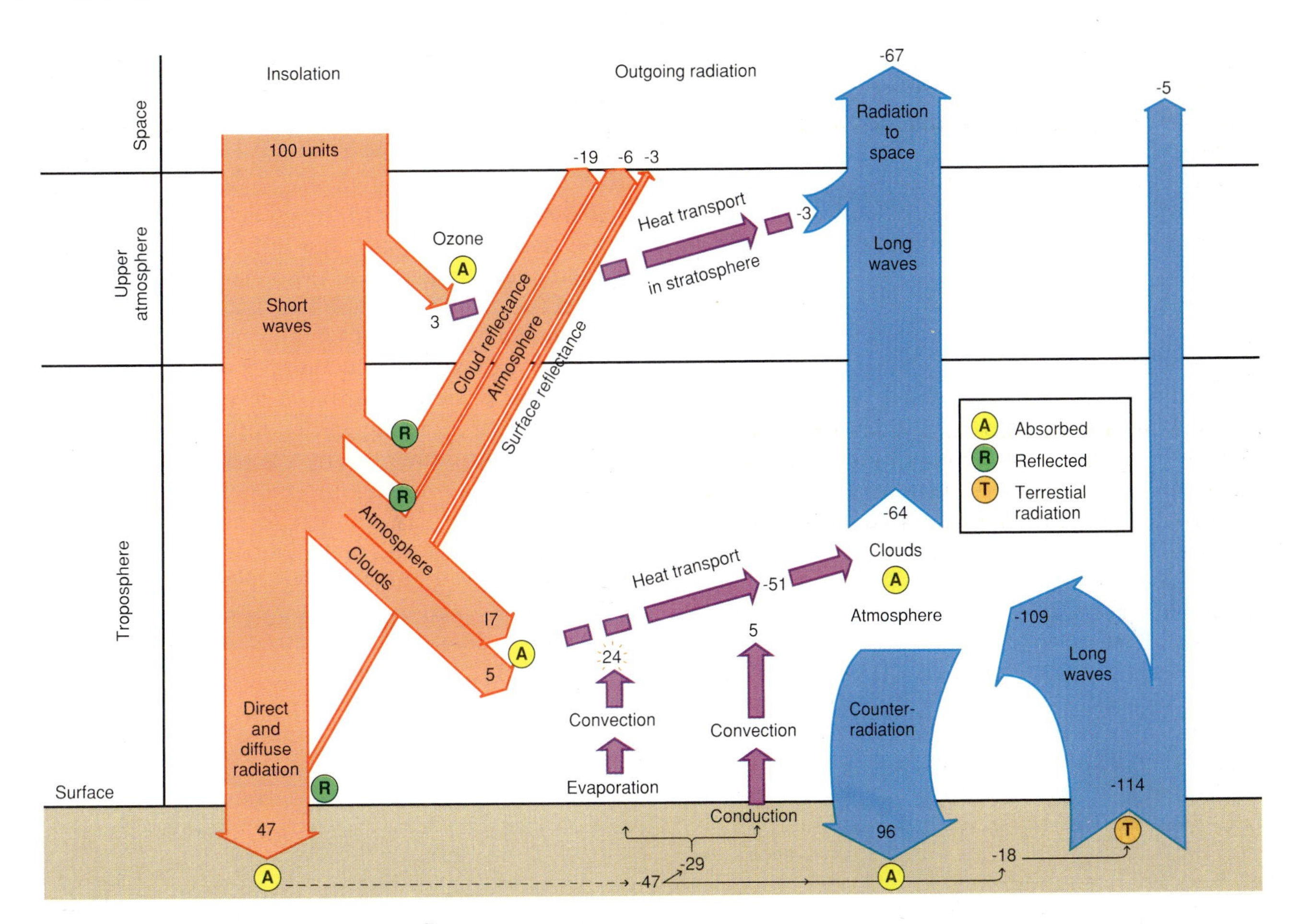

Figure 4-12 The heat balance of Earth's atmosphere. The total insolation is treated as 100 units. Of this, less than half gets through the atmosphere to the ground, although 25 of the units that do not reach the ground are absorbed on the way and help to heat layers of the atmosphere directly. Reflection causes an immediate loss of 28 units to space. The 47 units absorbed by the ground are radiated as long waves (18 units, of which 5 go directly to space), conducted into the lower layers (5 units), or consumed in the process of evaporation (24 units). Air heated by the ground and containing evaporated water vapor is convected upward until the latent heat is released in condensation. The right-hand side of the diagram shows how heat is circulated in the lower atmosphere by being radiated from Earth, absorbed by the atmosphere, and counterradiated. This cycle is the greenhouse effect.

the surface. The output component of the balance is more complex, with several strands. First, most terrestrial radiation is absorbed in the lower atmosphere, from where it is reradiated back to the ground or out to space. Energy is transferred back and forth within the lower atmosphere. Second, a small proportion of the energy (5 units on the diagram) is transferred by conduction from the ground into the air in contact with the ground. The air that is warmed by conduction becomes less dense and is lifted by convection, by the convergence of air, or by forced ascent over mountains or colder air bodies. This circulates the warmth up through the troposphere. Terrestrial radiation and conduction-convection can be grouped as "dry" energy transfers upward from the ground. Third, the remaining transfer (24 units) is carried out by insolation being absorbed in the process of evaporation and forming water vapor containing latent heat. The air containing water vapor rises, cools, and condenses in clouds, releasing latent heat. Such "wet" transfer releases heat energy in the middle and upper troposphere.

Present worries over global warming are based on the potential impact of changes in the balance between insolation and radiation to space. Increases in the materials that trap terrestrial radiation in the troposphere, such as carbon dioxide, other gases that absorb long-wave infrared radiation, dust from industrial activity, and water vapor have attracted much recent attention because they could result in such warming (Chapter 3). Attention has been drawn to this concern through the popular image of the "greenhouse effect."

The Greenhouse Effect

The atmosphere acts in some ways like a giant greenhouse. In a greenhouse (Figure 4-13) the glass lets in solar short-wave radiation but prevents most long-wave rays from escaping. Once inside the greenhouse, the short-wave radiation is either reflected out or absorbed by plants, soil, and atmospheric gases. When warmed, these materials radiate long waves that the glass absorbs and reradiates—mainly back inside.

In the atmosphere, short-wave solar radiation passes through "windows," is absorbed at the ground surface, and is reradiated in longer waves. The water vapor, carbon dioxide, and other so-called "greenhouse gases" prevent the long-wave radiation from escaping directly to space by absorbing it, and so cause the lower atmosphere to be heated. The **greenhouse effect** refers to this process and is a positive feature of Earth's atmosphere. It produces surface air temperatures that are 33°C

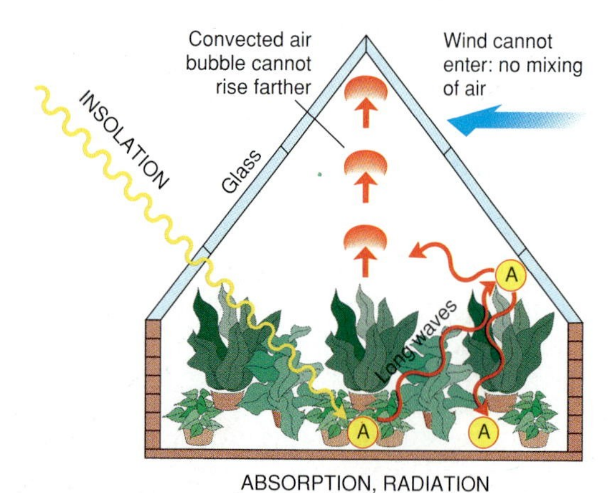

Figure 4-13 A greenhouse. The interior is heated as the short-wave rays pass through the glass and are absorbed by surfaces inside. These radiate heat waves that cannot escape through the glass. The atmosphere resembles a greenhouse by letting insolation through and then trapping the long-wave energy that is radiated from the surface. However, the greenhouse prevents exchanges of air with the rest of the atmosphere.

higher than those that would occur if there were no gases to trap the terrestrial radiation. In the absence of the greenhouse effect, the average surface atmospheric temperature would be −18°C (0°F), instead of 15°C (59°F).

So far, the analogy of the greenhouse fits. However, there are several major differences between a greenhouse and Earth's atmosphere. One difference is that the greenhouse glass prevents the transfer of sensible and latent heat by convection into the atmosphere above and the replacement of this air by cooler and drier air at ground level. The glass ensures that the air in the greenhouse functions as an isolated entity, whereas the atmosphere is a mass of continuously mixing and moving bodies of air. The balance of energy in the atmosphere-ocean environment is the result of the whole range of energy transfers; a greenhouse exploits only one part of the system that functions in the atmosphere.

Another difference is that a greenhouse owner can control the greenhouse environment, whereas human beings cannot control what happens in the atmosphere. If the owner is not careful, the air in the greenhouse will heat up until the plants are killed. If the temperature inside the greenhouse gets out of control, the runaway greenhouse effect occurs. The owner prevents this by opening a door or windows to let out some of the heat and thus achieves the correct balance. When it is cold in winter, the owner heats the greenhouse. The atmosphere, on the other hand, has its own control system, and the balance is maintained by the incoming radiation equaling the outgoing radiation. This is a very complex process, and humans cannot control it. If a runaway greenhouse effect were to occur in the atmosphere, nothing could be done to avert it.

The term *greenhouse effect* has mistakenly come into common usage to describe the *runaway* greenhouse effect in the atmosphere-ocean environment. The temperature of the global atmosphere may be rising because of the increased concentration of greenhouse gases, dust, and water vapor that trap heat in the troposphere. These increases are attributed to human activities such as burning fossil fuels. However, the complexity of feedback mechanisms in the atmosphere-ocean environment and the limits of possible human actions to control what happens in that environment (other than reducing the rise in greenhouse gases) make it difficult to assess the threshold at which irreversible changes will begin, the extent of the possible impacts, or a clear solution. Because more and more human activities appear to be contributing to the likelihood of a runaway greenhouse effect, and because the extent of such activities is increasing, governments are showing signs of serious concern about this issue and are discussing possible actions. In particular, more research on the subject is being funded. The U.S. Global Change Research Program is attempting to monitor changes in the environment and to advise governments on the amount of change to be expected.

HEATING OF OTHER PLANETS

A comparison of Earth with its nearest neighbors in space makes it possible to appreciate the significance of the atmosphere-ocean environment and of the heating level and balance within it. *Earth's moon* has no atmosphere, and its surface is heated only when the sun shines on it. It has no mechanism for trapping heat above the surface, and there are extreme temperature contrasts between the lit and dark sides. *Mars* is farther from the sun than Earth is and receives less insolation. Its thinner atmosphere and lack of clouds transmit insolation to the surface with a minimum of filtering, but little is trapped in the lower atmosphere, so temperature changes between day and night are large and rapid. *Venus* is closer to the sun than Earth, and is about the same size as Earth. Venus's greater warming and dense atmosphere with high levels of carbon dioxide have produced a runaway greenhouse effect in which the insolation is totally absorbed by moisture and gases in the atmosphere, and the cloud cover prevents a balance of radiation back to space. Surface temperatures are estimated to be around 500°C (900°F). Any oceans that once existed boiled off, and the additional water vapor increased the extent and thickness of cloud cover. This in turn intensified the trapping of heat in the Venusian atmosphere.

TEMPERATURE DIFFERENCES ON EARTH

Isotherms and Temperature Gradients

Temperatures measured at particular locations may be recorded and plotted on maps. **Isotherms** are lines joining places of equal temperature; their locations are estimated based on the locations of sites of known temperature (Figure 4-14). The world maps of temperature at Earth's surface in January and July (Figure 4-15) are drawn on the basis of temperatures averaged over some 30 years. They show the differences of temperature between tropical and polar regions.

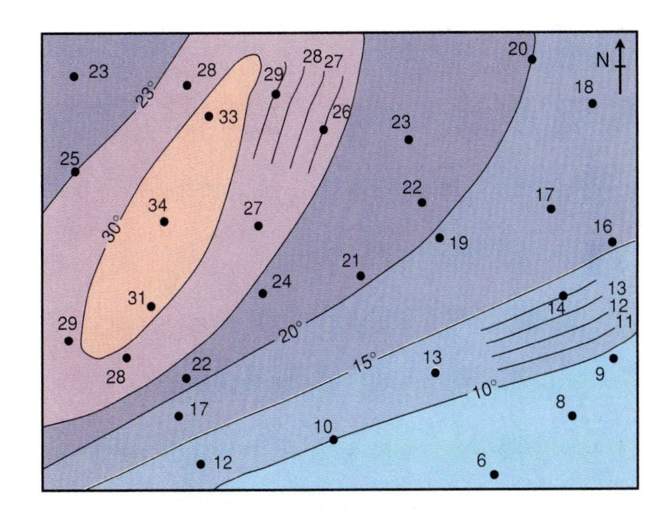

Figure 4-14 Isotherms joining places of equal temperature on a horizontal surface. Temperature is recorded at specific locations, and the isotherm lines are drawn through points at the correct ratio between the measuring locations. For instance, the 20° isotherm is drawn two-fifths of the way between 22° and 17°.

A **temperature gradient** is the rate of temperature change with distance, and may be expressed as degrees per 100 km. A global temperature gradient exists between the hotter tropics and cooler polar regions. In winter the difference between the tropics and the pole is greatest and the maps show a greater number of isotherms drawn at the same interval (Figure 4-16). This reflects a steeper temperature gradient. The lapse rate, described earlier in this chapter, is a vertical temperature gradient in the atmosphere. Figure 4-17 shows world temperatures averaged for June 1988, as observed by an NOAA satellite.

Factors Affecting Temperature Distribution

The pattern of temperature distribution shown on world maps is determined by interactions among the factors involved in heating the atmosphere-ocean environment. The factors include *latitude* and its effect on incoming solar energy supply, the varied nature of *surface materials* with their different absorption/reflection ratios, the nature of *mountainous relief*, and the *movements of air and water* that transport heat energy. Large *urban-industrial areas* have become distinctive "heat islands," showing that human activities can also influence temperature distribution.

In general, isotherms on the world maps run parallel to the lines of latitude, indicating that temperatures decrease from tropics to polar regions.

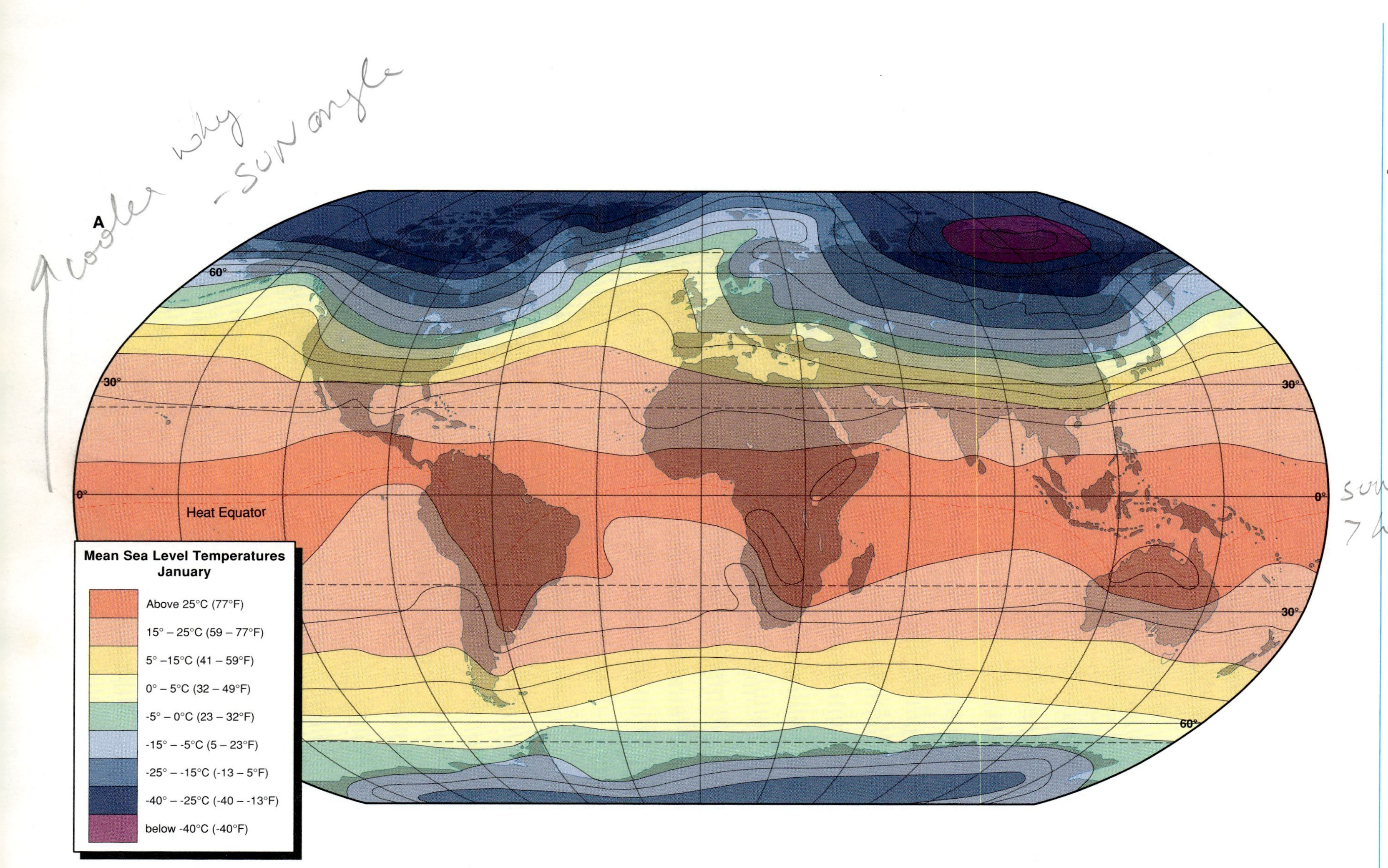

Figure 4-15 World isotherm maps for January *(A)* and July *(B)*. These isotherms are based on average temperatures over 30 years. The "heat equator" connects the points of highest temperature at each longitude.

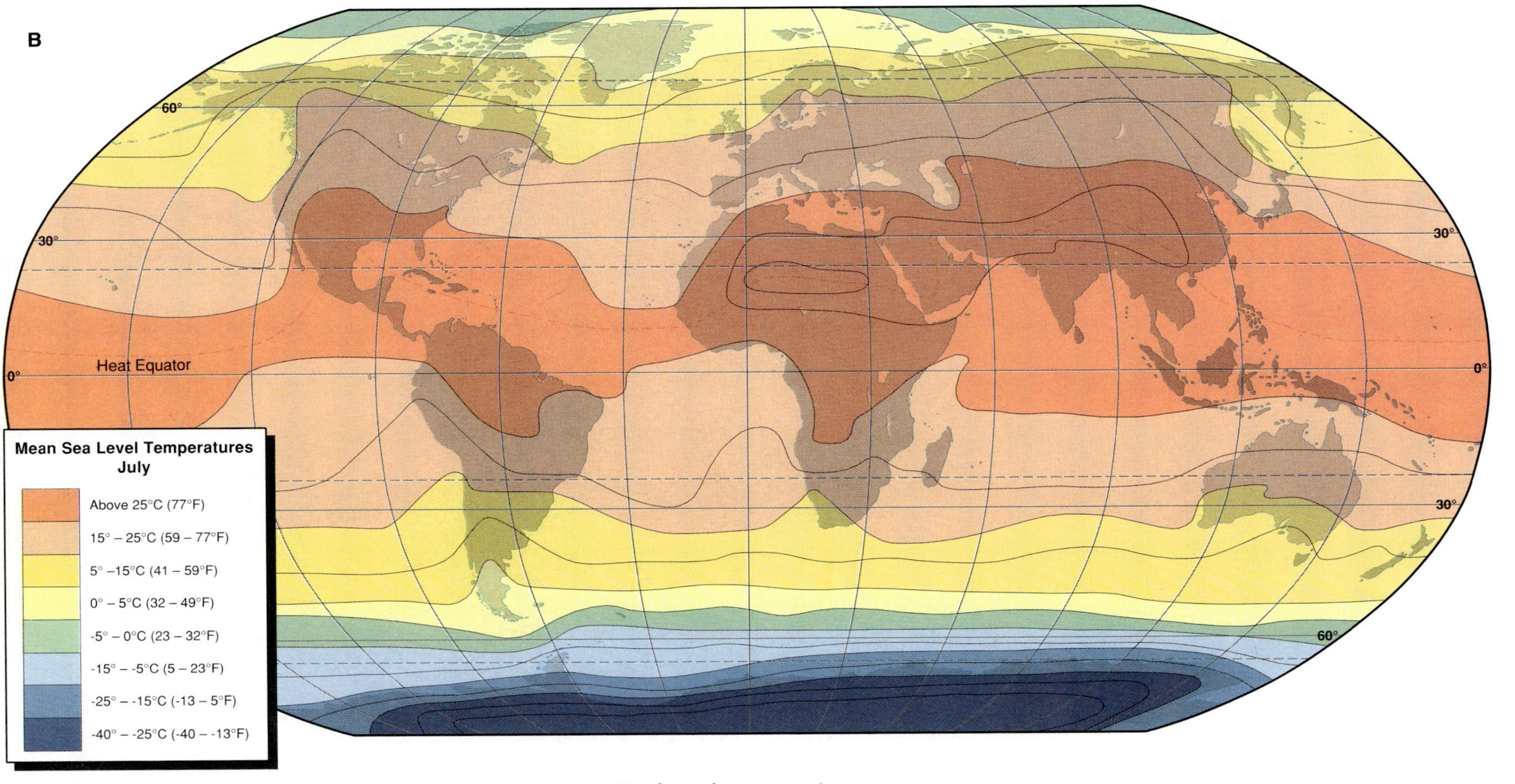

For legend, see opposite page

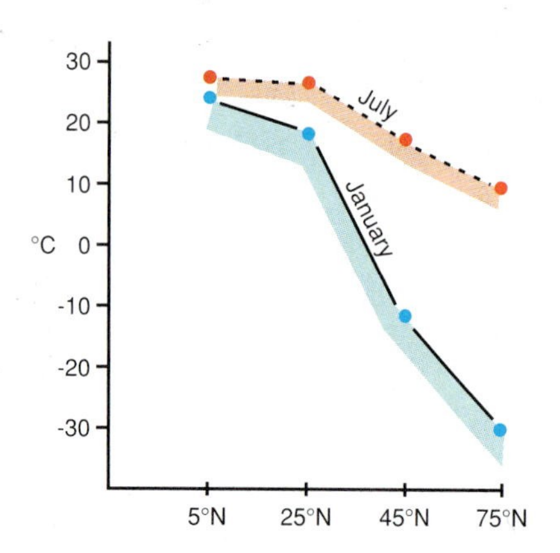

Figure 4-16 Temperature gradients from tropics to polar regions in the northern hemisphere: summer and winter compared.

This is what would be expected on a globe with a uniform surface, since the greatest intensity of heating is received where the sun's rays are perpendicular to the surface (Figure 4-3). Earth's tilted axis causes the hottest and coldest zones to shift with the seasons. The sun is vertically overhead in the area bounded by latitudes 23.5° N and 23.5° S, which defines the **tropics.** The hottest zone in July is along the northern tropic, and the hottest zone in January is along the southern tropic. The polar regions have summer days of 24 hours of sunlight and winter days of 24 hours of darkness, and this produces greater extremes than if Earth were not tilted.

World maps show that in winter the isotherms over the oceans bend toward the poles, indicating

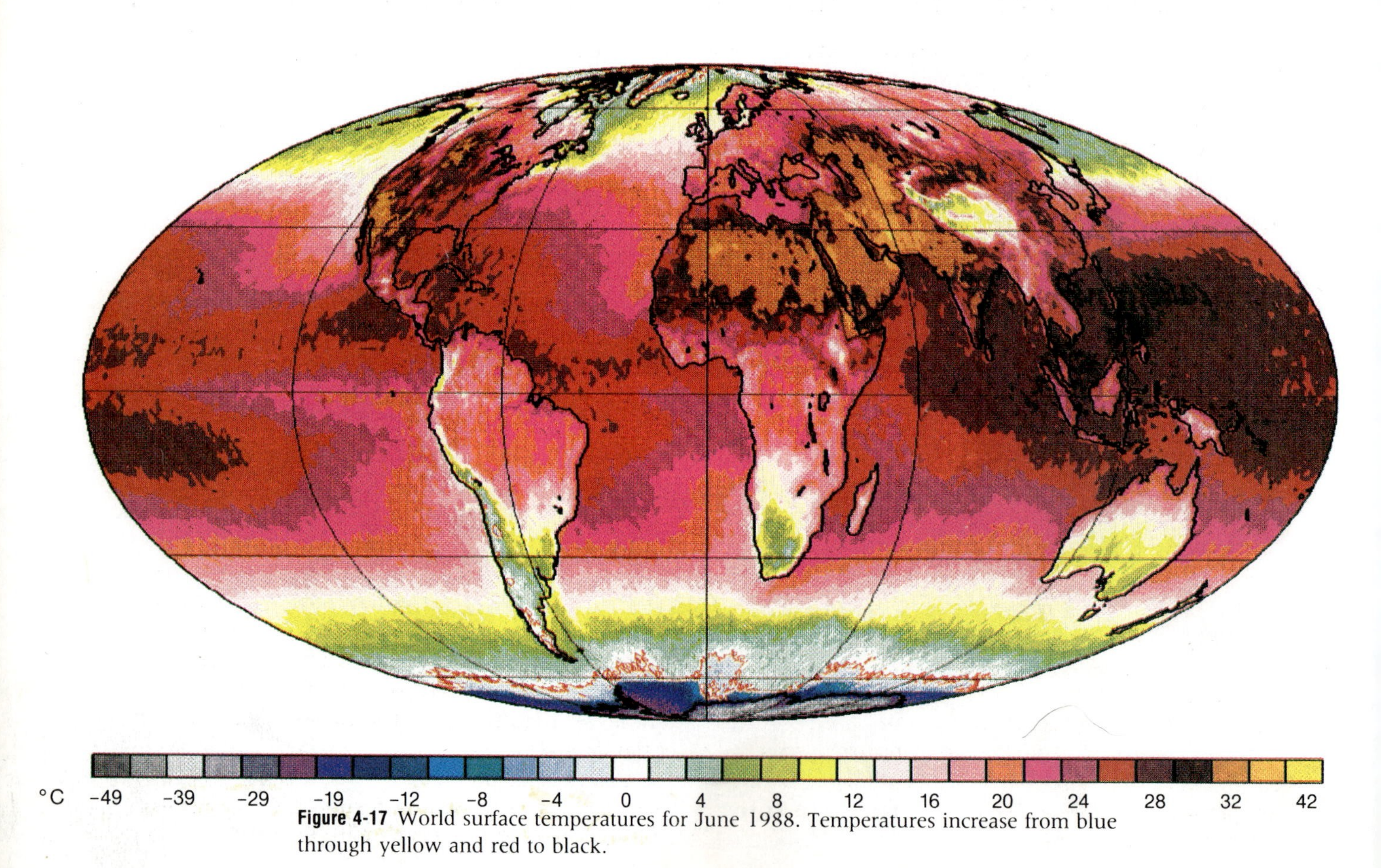

Figure 4-17 World surface temperatures for June 1988. Temperatures increase from blue through yellow and red to black.

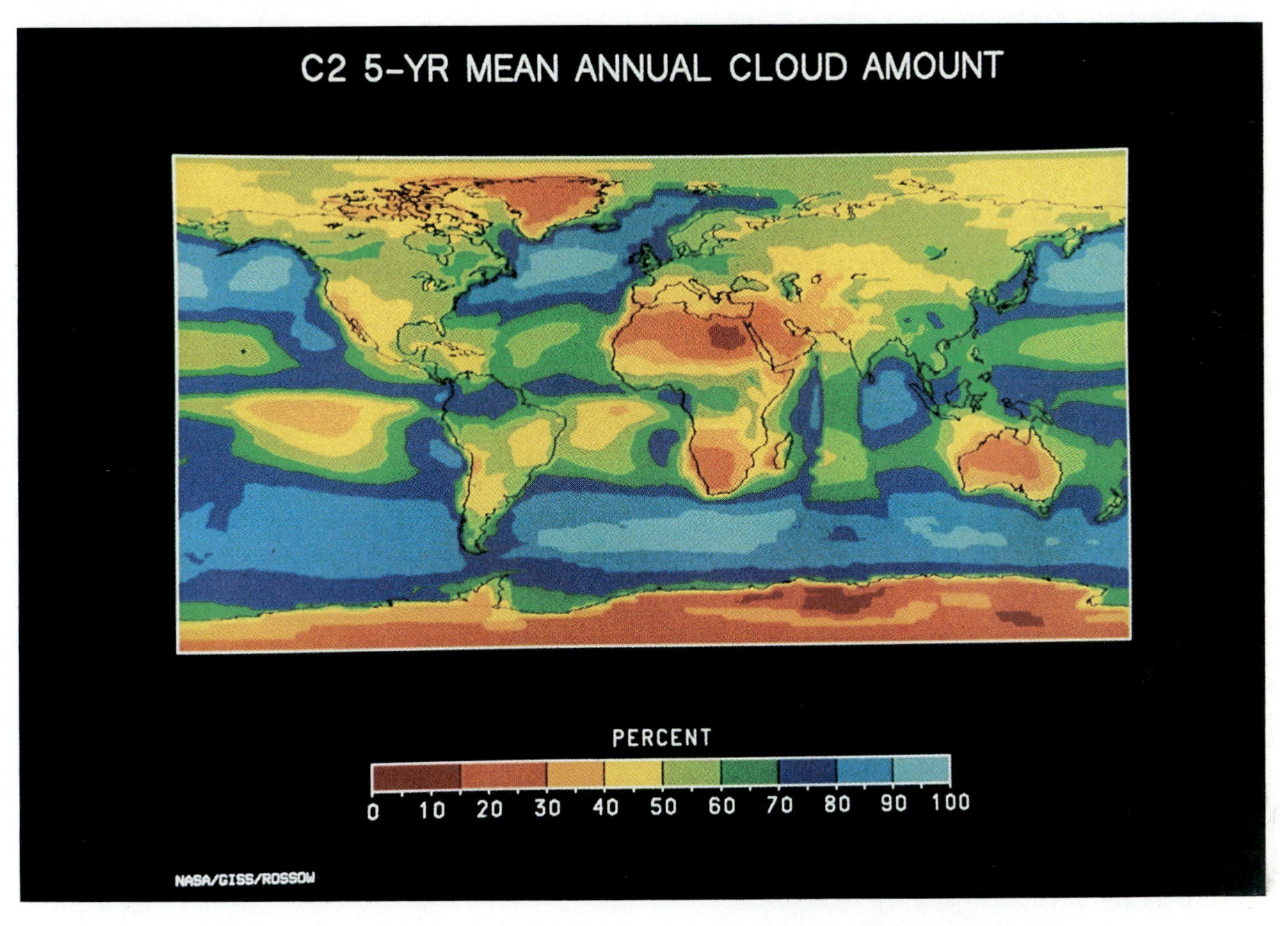

Figure 4-18 World map of cloud cover, based on the average for 1984-1988. The lowest percentages of cover are shown in red, the highest in blue.

that the air is relatively warmer over the oceans than over the continents for equivalent latitudes. In summer the isotherms over the land bend toward the poles, indicating that land areas are warmer. The overall effect is that much larger seasonal temperature contrasts are experienced over continents than over oceans. Islands and places on coasts have cooler summers and warmer winters than places in the interiors of continents because they are close to ocean influences. The *continentality* of a place is gauged by its high range of temperature from summer to winter compared with other places on the same latitude.

The differences in seasonal temperatures between the continents and oceans result from the contrasting reactions of the two to insolation (see Figure 4-9). This contrast leads to the oceans absorbing and storing more heat than the continental surfaces. It takes longer to heat or cool ocean water than it does to heat or cool soil and rock, and so continents and the air above them heat up to higher temperatures and cool down more rapidly to lower temperatures than the oceans.

Two other factors that cause differences in temperature between continents and oceans are relative cloudiness and the circulation of warm and cold ocean water. Ocean areas are more cloudy than the continents because of the moisture evaporated from the surface and condensed above them, and this affects surface temperatures (Figure 4-18). The cloud cover lowers insolation but prevents radiation beneath the cover from escaping to space. This reduces daily and seasonal temperature ranges, particularly in the cloudiest regions near the equator and in the midlatitude storm belts. The circulation of warm and cold currents in ocean surface waters also affects the temperatures of adjacent coasts; places on coasts where there are warm currents offshore have higher temperatures than those where there are cold currents. For instance, summer temperatures on the coasts of southern California are lower than those on the coasts of the southern Atlantic states.

A large-scale outcome of the differences between continents and oceans in response to insolation is that the northern hemisphere, which has a greater

Figure 4-19 Human activities and their output of heat. In this satellite image, white represents city lights, red depicts forest and agricultural fires, and yellow shows gas flares.

proportion of continental surfaces, has warmer summers and cooler winters than the southern hemisphere. Isotherms are more nearly east-west in the southern hemisphere than in the northern hemisphere because more of the southern hemisphere, especially in midlatitudes, is occupied by ocean.

Altitude and *aspect* also modify surface temperatures. The combination of these factors is very significant in mountainous regions. Temperature decreases with **altitude,** or height above sea level, because the atmosphere gets thinner. As density decreases, so does the quantity of greenhouse gases and hence the air's capacity to be warmed. **Aspect** is the orientation of a place with respect to the sun. Sun-facing slopes are warmer than those that are shaded from the sun. Sun-facing slopes in the European Alps are marked by the lush growth of vines, higher altitudes of forest growth, and higher levels at which permanent snow is reached. Measurements of temperature on adjacent north- and south-facing slopes at latitude 45° N show little difference between the two opposite-facing slopes in summer but an immense difference in winter when the sun shines at a low angle. In winter, south-facing slopes receive more insolation and reach higher temperatures than horizontal surfaces at the same altitude.

Human activities have not had a major impact on the world isotherm maps to date, since their contribution to atmospheric heating is tiny compared to solar heating. However, with the growth of urban-industrial areas, such as the megalopolis region of the northeastern United States stretching from Boston to Washington, D.C., the isotherm pattern may become somewhat modified (Figure 4-19). Such areas absorb more incoming insolation and produce more heat energy from human activities than do rural areas. In the depth of winter the most densely populated parts of these midlatitude urban-industrial regions contribute more heat to the atmosphere from human sources than they receive from the sun.

Temperatures in a particular place are not determined merely by the interactions of insolation and surface conditions at that spot. The fluid media of atmospheric gases and ocean waters are constantly in turbulent movement and transfer sensible and latent heat from tropical regions toward polar regions, and exchanges occur between continents and oceans. Ocean surface temperatures are cooler than those on the continents in summer and warmer in winter. Ocean heat is transferred to the overlying air and then by onshore winds to coastal regions; coastal towns have temperature ranges that closely follow those of the adjacent oceans. Anchorage, Alaska (62° N), has warmest-month average temperatures of 12°C (58°F) and coldest-month temperatures of −12°C (13°F), an annual range of 24°C (45°F). Conversely, places in the center of midlatitude continents and isolated from such oceanic influences have hot summers followed by freezing winters: Vekhoyansk, in central Siberia (66° N), is the coldest place on Earth in winter outside of Antarctica, with an average January temperature of −45°C (−49°F), but has an average July temperature of 15°C (59°F); its annual range is 60°C (108°F).

The world isotherm maps are based on average conditions and do not record the changes associated with daily and seasonal variations of insolation. Diurnal (day-night) changes are the most significant fluctuations in the tropics, where seasonal variations are small. The seasonal changes are particularly significant in midlatitudes, where there is the greatest contrast from summer to winter.

GLOBAL DISTRIBUTION OF SOLAR ENERGY

The energy from the sun that heats the atmosphere-ocean environment is not distributed evenly over Earth's surface. There is a major contrast between the tropical and polar regions, a contrast enhanced by seasonal changes. In the tropics, more heat arrives than is radiated to space. Polar regions, on the other hand, radiate more than they receive from the sun. Tropical areas have a net gain, or surplus, of radiation, while polar areas have a net loss, or deficit (Figure 4-20). However, the tropical regions do not

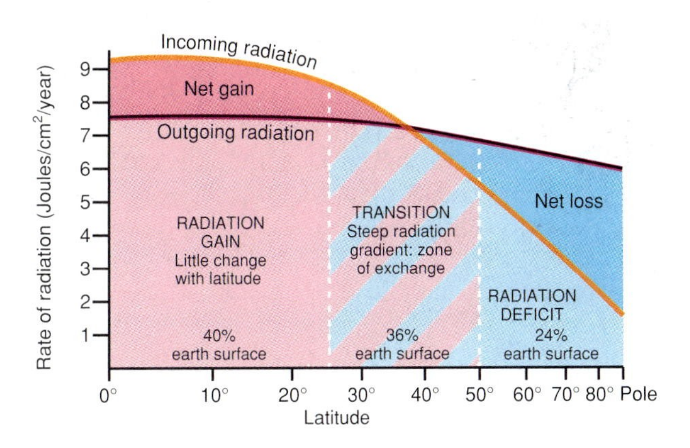

Figure 4-20 The surplus of heating in low latitudes, and deficit in high latitudes. The spacing on the horizontal axis is proportional to the area within each 10° of latitude.

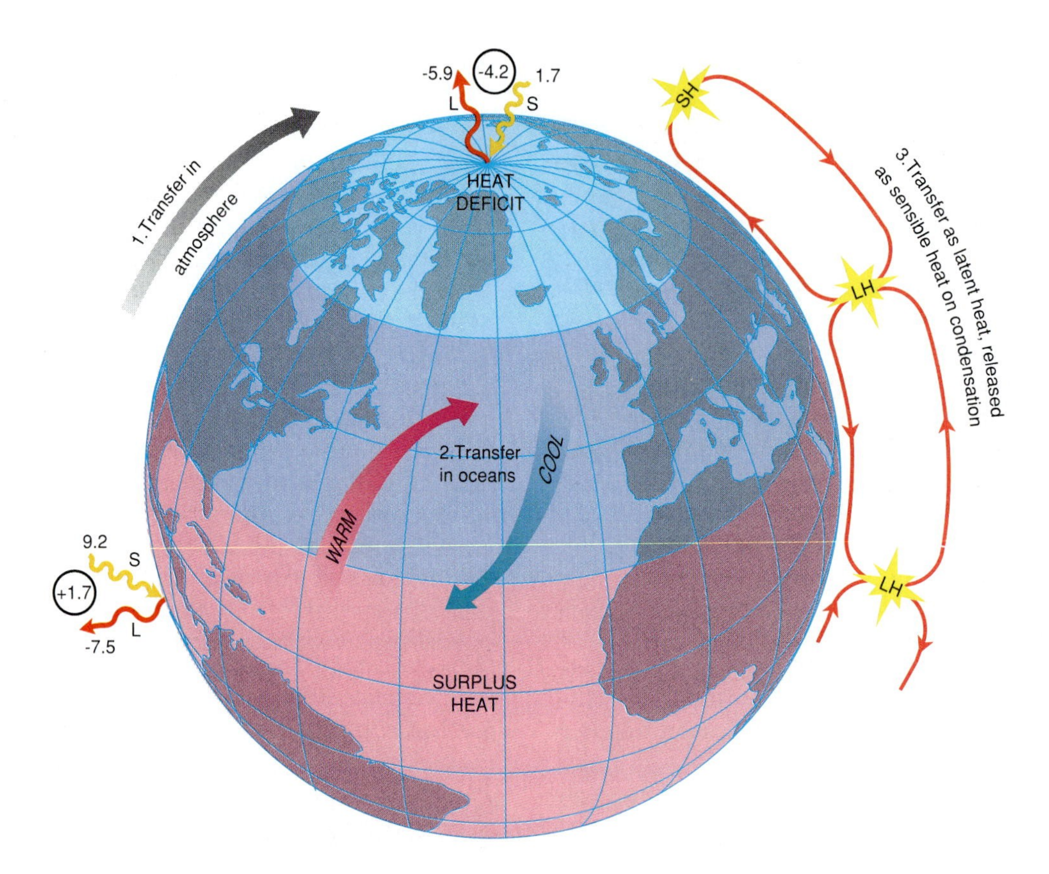

Figure 4-21 Mechanisms of sensible heat (SH) and latent heat (LH) transfer from low to high latitudes. The inputs of insolation and outputs of terrestrial radiation for equivalent areas are shown for equator and pole in units of energy. There is an excess at the equator and deficit at the poles, but these are balanced by flows from the tropics toward the poles.

get progressively hotter, and the polar regions do not grow continuously colder. A balance operates to maintain the temperatures in different parts of the world.

This balance is made possible by massive horizontal transfers of energy through the atmosphere and oceans. These transfers can be viewed as taking heated air and water from the tropics to midlatitude and polar regions. After cooling by radiation the air and water are returned to the tropics for reheating. This is the essence of the atmospheric heat engine.

Three main sets of processes are involved in these transfers (Figure 4-21). The most important is the transfer of sensible heat from the tropics toward polar regions by convectional (and advectional) movements of warmed air. On the way, the midlatitudes are heated, and some of the warmed air gets to the polar regions. In high latitudes the air from tropics and midlatitudes is cooled and sinks, especially in winter when there is virtually no insolation.

The second set of processes is the movement of

ocean waters, which carry out a similar function to the atmosphere. Tropical oceans absorb a large amount of intense radiation. The heated surface waters are transferred poleward in ocean currents such as the Gulf Stream. The water's heat is transferred to the overlying air in lower and midlatitudes. Recent satellite-based estimates of energy transport in the atmosphere-ocean environment suggest that ocean transport may be nearly 75 percent of the total at 20° N and half of the total at 30° to 35° N. During winter in middle latitudes, ocean currents may contribute more heat to the air above them than do the sun's rays.

The third set of processes occurs when water vapor evaporated from Earth's surface carries latent heat with it to the point of condensation. The zone of maximum evaporation is in the tropics, and there are two main zones of latent heat release. Warm, moist air is forced to rise near the equator, where condensation releases the latent heat, helping to drive the circulation of air in the tropics. Moist tropical air is also carried into the middle latitudes, where it is forced to rise over the denser, cold air around the poles. This uplift leads to cooling, condensation, and the release of latent heat.

The relative contribution of such transferred heat energy as compared to direct insolation increases toward the poles. Those living in the Seattle area (47° N), for instance, may have puzzled over the fact that in winter, dull and cloudy days are often warmer than sunny days. This is because most of the winter heat in Seattle is brought by air warmed over the North Pacific Ocean. The clouds and humidity associated with this ocean air help to keep the warmth near the surface. Even when it is shining, the winter sun is at a low angle and its rays are of low intensity at the ground. Prolonged sunny periods at this time of the year are often accompanied by cold northerly airflows.

FRONTIERS IN KNOWLEDGE

Heating the Atmosphere-Ocean Environment

Energy flows through the atmosphere-ocean environment have become a topic of major public interest because of concern about global warming. Studies being undertaken in this field concern the role of oceans in the transport of heat, how oceans transfer heat into the atmosphere, and how polar regions function in the exchange of heat. One major research program is the Tropical Ocean—Global Atmosphere Program (TOGA), which is making a special study of the links between the heating of oceans and atmosphere.

How human activities affect the heating of the atmosphere is also being investigated and is of special interest to physical geographers. All transformations of energy produce heat, and the more human activities involve such transformations, the more they will contribute to global warming. The burning of fossil fuels is still the main human source of greenhouse gases, but studies are being made of other sources, such as methane. Furthermore, it is necessary to learn how increases of population and different types of land use affect the production of greenhouse gases and the heating of the atmosphere.

Other scientists are investigating the variability of solar energy output and the potential impact of variations. The satellite that has been monitoring solar output for 10 years reentered the atmosphere earlier than anticipated in 1989 but is being replaced in the early 1990s.

INVESTIGATIONS in Physical Geography

The Oceans: Carbon Sink or Carbon Source?

Research into the processes of contemporary climatic change examines interactions in and among Earth environments. To explain the process of global warming, for example, scientists must trace the continuous movement or *flux* of carbon between the atmosphere and the ocean, and its wider exchange among the atmosphere-ocean, living-organism, and solid-earth environments, all at a global scale (see Figure 3-8). The researchers must also separate the natural flux from that caused through human activity, and then predict the pattern in each component over the next few decades and centuries. It is a huge scientific problem. Its solution will involve observing chemical, physical, and biological reactions at all scales in Earth environments, from individual molecules in laboratory experiments to models of vertical mixing in oceans and the transport of gases in the atmosphere. The scale of international scientific cooperation needed to address the problem is also daunting. The issue is considered worthy of such attention in part because of the possible role of atmospheric carbon dioxide in promoting a runaway greenhouse effect and global warming.

A "bloom" of coccolithophores (a type of plankton) shows as a cream band off the northwest coast of Scotland. Coccolithophores take up carbon from the atmosphere, using some of it to build a skeleton. Phytoplankton are a major sink of carbon in the oceans.

The Intergovernmental Panel on Climatic Change published a report in 1990 describing the state of human knowledge about global warming and what is still left to find out. This panel of experts is certain of only two things: that there is a natural greenhouse effect, and that human activity is substantially increasing the concentration of carbon dioxide and other greenhouse gases in the atmosphere. The panel considers that "there are many uncertainties in our predictions, particularly with regard to the timing, magnitude and regional patterns of climate change, due to our incomplete understanding of . . . sources and sinks of greenhouse gases . . . clouds . . . oceans . . . [and] polar ice sheets."

It is clear from the limited knowledge that exists that the oceans are a key control in the global movement of carbon. Ocean waters contain nearly 50 times as much carbon as exists in the atmosphere, and around 20 times as much as exists on land in plants, animals, and soil. About half the carbon released to the atmosphere by human activity is unaccounted for in the global carbon budget and is thought to be absorbed by the oceans. Without the oceans, therefore, atmospheric carbon might be increasing at double the current rate; without human inputs the concentration might actually be dropping.

International research on climatic change is focusing on the oceans, and in particular, on how carbon gets into and out of them. One large program is called the Joint Global Ocean Flux Study (JGOFS), set up by a committee of the International Council of Sci-

A satellite image showing a front between a body of warm water *(dark)* and a body of cold water *(light)* in the North Atlantic. The eddies along the front allow some mixing of the two water bodies. Data like these help to guide research ships working in the ocean.

entific Unions (ICSU). JGOFS is a part of the larger International Geosphere-Biosphere Program organized by the Union and launched in 1986. The aims of the larger program mirror the goal and challenge of Earth Systems Science described in Chapter 2. These broad aims have been translated into separate research themes, and a series of interdependent research programs has grown up to address them. One of these is JGOFS.

The basic aim of JGOFS is to understand and express numerically the role of the oceans in regulating the global composition of the atmosphere. In detail, the program includes experiments to examine the processes by which chemical elements, particularly carbon, are moved from one state to another in the atmosphere-ocean environment and to examine the periods of time and distances over which this occurs. Within JGOFS, teams of scientists are researching the transfer of carbon between the atmosphere and ocean at the sea surface; between the surface, mixed layer of the ocean, and the deep ocean; and between the deep ocean waters and the sediments on the sea floor.

There are two main controls on the transfer of CO_2 across the sea surface, and scientists are as yet unsure which is more important. The first control is water temperature. Cold water draws CO_2 from the atmosphere to the ocean, and warm ocean water releases CO_2 to the atmosphere. There is therefore a spatial pattern of carbon exchange related to latitude and a temporal pattern related to the seasons; the ocean's absorption of carbon is greater near the poles and in the winter. The second control is biological and involves tiny organisms called phytoplankton. These organisms take up carbon in the process of photosynthesis and release it in the process of respiration (Chapter 19). In some cases the carbon is used to form the skeletons of the phytoplankton. In contrast to the effects of temperature, biological uptake of carbon is greatest in the summer and has a strong daily pattern. Scientists must understand how both temperature and phytoplankton control the transfer of carbon, how the two processes interact, and, most importantly, how each process might be expected to change in a warmer atmosphere. At the moment, the ocean seems to be a "net sink" of carbon, in that it absorbs more carbon than it releases overall. In a warmer atmosphere it may become a net source, releasing more carbon to the atmosphere than it absorbs, thus accelerating global warming.

Remote sensing is an important tool in investigating the spatial and the temporal variability of both phytoplankton and ocean temperature. The phytoplankton give a distinctive visible color to the ocean surface, and sensors in infrared wavebands can measure sea surface temperature. Using the two data sets together shows the conditions of temperature in which the phytoplankton thrive. Remotely sensed data have been used in JGOFS to guide the sampling activities of research ships in the Atlantic. The satellite data were beamed out to ships within a day of their reception, showing the scientists on board where and when to concentrate their sampling. The satellite data are therefore valuable in themselves, and are a vital part of the more detailed spatial measurements needed to complete the picture of carbon exchange.

SUMMARY

1. Temperature is a measure of the hotness of a substance. It is determined by the heat energy applied to the substance and the specific heat of the substance.
2. Heat energy is transferred from one substance to another by conduction, convection, and radiation.
3. Temperatures in the atmosphere are the result of heating by the sun and the distribution through the atmosphere-ocean environment of the heat received.
4. The spectrum of radiant energy from the sun is mainly short-wave and is filtered during passage through the atmosphere. The upper atmosphere filters rays that are harmful to living organisms, leaving visible light and some infrared wavelengths to penetrate to ground level.
5. Insolation absorbed by Earth's surface is partly radiated as long waves, which are absorbed by water vapor, carbon dioxide, and other greenhouse gases and heat the troposphere from the base upward. The heated Earth surface also conducts a little heat into the atmosphere in contact with it, and more heat energy is transferred upward as latent heat.
6. Earth's atmosphere behaves in some ways like a greenhouse in trapping heat. Transfers of heat in the atmosphere, however, are not bound by glass walls, and there is greater exchange by vertical and horizontal movements.
7. The *runaway* greenhouse effect occurs when more heat is trapped in the atmosphere than is released to space, as in the case of Venus; it is not yet certain that this is occurring on Earth.
8. The excess of heat received in the tropics is distributed to cooler parts of the globe by movements in the atmosphere and oceans and by the transfer of latent heat in water vapor.
9. Temperature at a particular place is related to latitude, the distribution of oceans and continents, altitude, aspect, land use, and the effects of the movement of warmed air and waters. Seasonal variations in insolation are caused by the tilt of Earth's axis relative to its plane of orbit around the sun.

KEY TERMS

heat, p. 76
temperature, p. 76
calorie, p. 76
specific heat, p.76
conduction, p. 77
convection, p. 77
advection, p. 78
sensible heat, p. 78
latent heat, p. 78
radiation, p. 79
electromagnetic waves, p. 79
electromagnetic spectrum, p. 80
short-wave radiation, p. 80
long-wave radiation, p. 80
insolation, p. 80
solar constant, p. 80
solstice, p. 81
equinox, p. 81
absorption, p. 82
scattering, p. 83
diffused light, p. 84
reflection, p. 84
albedo, p. 84
terrestrial radiation, p. 85
counterradiation, p. 86
lapse rate, p. 86
environmental lapse rate, p. 86
temperature inversion, p. 87
heat balance, p. 88
greenhouse effect, p. 90
isotherm, p. 91
temperature gradient, p. 91
tropics, p. 94
altitude, p. 97
aspect, p. 97

QUESTIONS FOR REVIEW AND EXPLORATION

1. Explain how the sun and Earth radiate energy in different wavebands.
2. Give examples of atmospheric processes that demonstrate heat transfer by conduction, by convection, and by radiation.
3. How do the composition and structure of Earth's atmosphere affect the passage of insolation through it?
4. Describe atmospheric colors at sunrise, midday, and sunset, and visibility during different weather conditions. Relate them to the scattering of insolation and the atmosphere's content of aerosols and water vapor.
5. Explain the different absorption properties and albedos of wet and dry soils, trees, ice, fresh snow, and asphalt roads. Suggest how these different types of surface might affect the temperature of the overlying air on a still, sunny day.
6. Compare Earth's heat balance with that of Earth's moon, Mars, and Venus.
7. Explain why an increase in carbon dioxide in the troposphere could lead to a rise in temperatures at Earth's surface. What other changes could have a similar or opposite result?
8. What is the difference between the "greenhouse effect" and a "runaway" greenhouse effect? What could be done to prevent the latter affecting Earth?
9. Polar regions have an overall heat deficit; the tropics have an overall heat surplus. Why do temperatures in these regions remain approximately the same over time? How does this affect temperatures in midlatitudes?
10. Because of differences in albedo, specific heat, and other characteristics, the temperature of air over land changes much more rapidly than that over ocean. Why does this result in greater daily and yearly *extremes* of temperature over the continents?
11. Account for seasonal differences in temperature among a place on the equator, one over the North Atlantic Ocean, and one in central Siberia.
12. Rewrite the extract that opens this chapter as a scientific description of how intense atmospheric heat affected Adam.
13. What are some of the difficulties facing attempts to utilize solar power in domestic and industrial situations?

FURTHER READING

Asimov I: *Understanding physics: motion, sound, and heat,* 1969, Mentor: New American Library. A clearly written explanation of basic physical principles of heat and temperature.

ERBE Science Team: First data from the Earth Radiation Budget Experiment (ERBE), *Bulletin of the American Meteorological Society,* 67:818-824, 1986. The results of satellite measurements of solar and terrestrial radiation.

Foukal PV: The variable sun, *Scientific American,* 34-41, February 1990. Variations in solar output over time.

Hamakawa Y: Photovoltaic power, *Scientific American,* 87-92, April 1987. New technical developments and prospects for solar energy.

Ingersoll AP: The atmosphere, *Scientific American,* 162-174, September 1983. Summarizes some of the major features of heating the atmosphere.

Lindzen RS: Some coolness concerning global warming, *Bulletin of the American Meteorological Society,* 288-299, 1990. Raises questions about the links between rising carbon dioxide levels and global warming.

Quayle R, Doehring F: Heat stress: a comparison of four different indices, *Weatherwise,* 3:120-124, 1981.

Winds are the atmosphere in motion. They may come from any point of the compass. Their strength varies from the destructive force of hurricane and tornado winds to very light breezes. Gusty conditions bring moment-to-moment variations in wind strength and direction on a local scale. Air can also be quite still. The winds described in the opening extract concern patterns of air movement at regional and global scales, but winds also occur on a local scale.

Winds affect people's lives in many ways, including lowering temperatures by "windchill" (see Box 5-1: Human Comfort—The Windchill Factor). Strong winds may destroy homes, woodland, and standing crops; winds propel sailboats; and wind can be harnessed as a source of power (Figure 5-1). Winds also affect other Earth environments. They control ocean waves that affect coastal landforms (Chapter 17), produce dune landforms in sand that is not anchored by vegetation (Chapter 16), and affect the forms of vegetation (Chapter 19).

Both winds and ocean currents are important in circulating heat through the atmosphere-ocean environment. This chapter first examines the nature and causes of winds and ocean currents. It then establishes the geographic patterns of movement within the global atmosphere-ocean environment. This study includes the winds of the upper troposphere and the movements of water in the deep oceans, because these unseen flows influence surface winds and ocean currents.

BOX 5-1

HUMAN COMFORT—THE WINDCHILL FACTOR

Strong winds may decrease human comfort at low temperatures, while gentle breezes may increase comfort at high temperatures. Moving air increases heat loss from the body by conduction. In calm conditions a body at rest is surrounded by a layer of still air that acts as an insulator, since air is a poor conductor of heat. As wind speed increases, that layer gets thinner and the rate of heat loss from the body rises. In cold weather, wind can increase the danger of frostbite and hypothermia.

Winter weather reports often indicate the "windchill factor." The table of values was first compiled in the 1940s by polar scientists and has been refined on the basis of subsequent experience. At an air temperature of 6°C, a light breeze will reduce the effective temperature by a few degrees. A strong wind at 6°C, however, reduces the effective temperature to −10°C, and exposed parts of the body lose heat at a rate experienced in calm conditions at −10°C. The effect becomes greater as the air temperature falls.

Although other factors, such as humidity, body heat production, and the type of clothing worn also affect body temperatures, the windchill factor is a useful guide to taking protective measures in harsh winter weather.

	THERMOMETER READING °C						
WIND (KM/HOUR)	5.0	2.5	0	−2.5	−5.0	−7.5	−10.0
CALM	5.0	2.5	0	−2.5	−5.0	−7.5	−10.0
LIGHT (15) BREEZE	4.0	−5.5	−7.0	−9.5	−12.0	−15.0	−18.0
30	−7.0	−10.0	−14.5	−18.0	−21.5	−25.0	−28.5
45	−10.0	−14.0	−18.5	−21.5	−24.5	−27.5	−30.5
STRONG WINDS 60	−13.0	−16.5	−20.0	−23.5	−27.0	−30.0	−33.5

VERY COLD
DANGER
EXTREME DANGER
LITTLE FURTHER EFFECT ABOVE 60 kmph

The windchill equivalent temperatures, measured as the effect of wind on an unclothed body.

Figure 5-1 Wind is necessary for many sporting activities. It also blows the trees and flags in the background.

PATTERNS OF FLOW IN WINDS AND OCEAN CURRENTS

Winds are horizontal movements of air relative to Earth's surface. **Ocean currents** are horizontal flows of surface water in the oceans; they are faster than the similar movements called *drifts*. World maps of surface winds and ocean currents (Figure 5-2) show that their directions of flow are closely related. Winds in the tropics blow toward the equator from the northeast in the northern hemisphere, and from the southeast in the southern hemisphere (Figure 5-2, *A,B*). On either side of the equator the surface ocean water also flows from east to west. These are **easterly winds and currents** since they blow and flow from the east (Figure 5-2, *A,C*). In the zones where the Tropics of Cancer and Capricorn cross the oceans, both winds and currents move in circular patterns. In the regions poleward of 40° of latitude, there are prevailing **westerly winds and currents**—i.e., that blow and flow from the west. In the southern hemisphere these are known as the "Roaring Forties," and there are similar patterns of wind and ocean currents in the northern hemisphere that are also linked with ocean currents. Wind patterns over the continents shift seasonally, as flows from continent to ocean in winter are reversed in summer (compare Figure 5-2, *A*, and 5-2, *B*). What causes these movements of air and water, and how are their flows linked?

CAUSES OF WINDS

Winds have a speed (kilometers or miles per hour) and a direction (e.g., westerly, easterly). The speed is determined by the forces that set the air moving—the differences in density that induce movement from higher to lower density in convectional patterns—and those that slow it, such as friction as air blows over a hilly land surface. Wind direction is determined partly by differences in density, partly by Earth's rotation, and partly by the effect of friction. The land masses surrounding the ocean basins and forming high mountains also affect wind directions.

Solar Heating and Winds

Solar heating is the main cause of density differences in the atmosphere. Air warmed above the land or sea expands (i.e., its volume increases) and thus has a lower density than cold air. Some surfaces absorb more insolation and so are better at heating the air above them than others because they radiate more energy. This results in uneven heating and a patchwork of warmer and cooler sectors in air near the ground. Lower density air in warmer sectors rises, reducing atmospheric pressure on the surface below. Pressure on the surface increases where cool, dense air descends. The horizontal pressure contrasts that result are shown on weather maps by drawing **isobars**—lines that join places of the same atmospheric pressure.

Winds are caused by air flowing over the surface from high to low pressure. The difference in pressure per distance between the high and low centers is known as the **pressure gradient.** It creates a force that acts on air between the pressure centers. The force is known as the **pressure gradient force**. It determines the strength and initial direction of the wind. If there is a steep pressure gradient—a large difference in pressure over a small distance—the winds will be strong. The spacing of isobars on weather maps reflects the pressure gradient; isobars are closer together where the pressure gradient is steeper and the winds are stronger. The direction of a wind is determined by the positions of high and low pressure centers.

The effect of pressure on air movement is evident on a local scale where the heating contrast between land and sea often produces daytime breezes blowing from the sea (Figure 5-3). During the day, air over land heats more rapidly than that over sea. The heated air over land expands and rises, creating a zone of lower pressure at the surface (Figure 5-3, *A*). Cooler air from the area of higher pressure over the sea flows in and replaces the rising air, resulting in a sea breeze. The air drawn in from the sea is replaced by descending air a few kilometers offshore, and convection is established. The strength of the winds increases with the difference in heating between land and sea and is often greatest in mid-afternoon. The circulation reverses at night, when air over the land cools more rapidly than that over

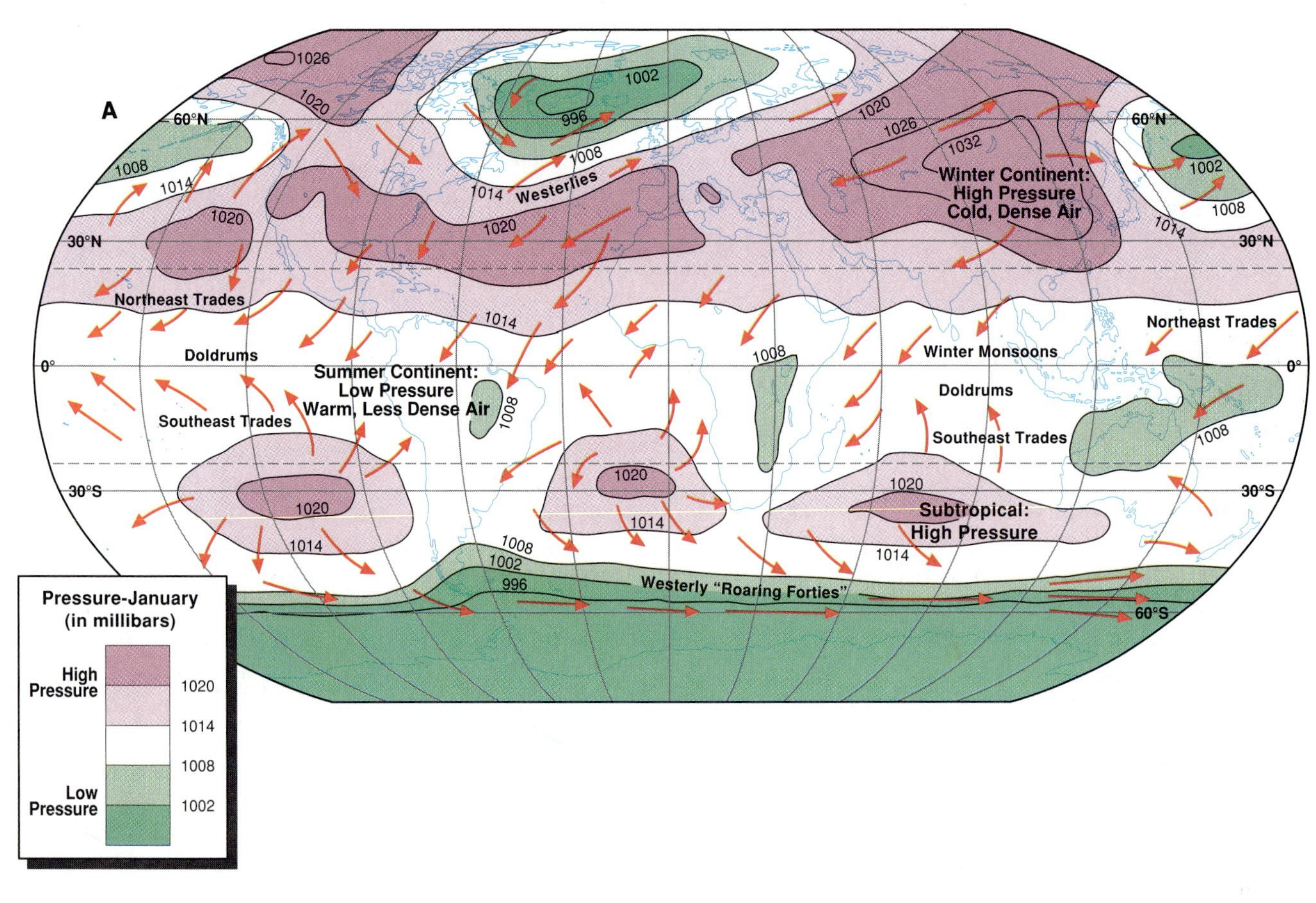

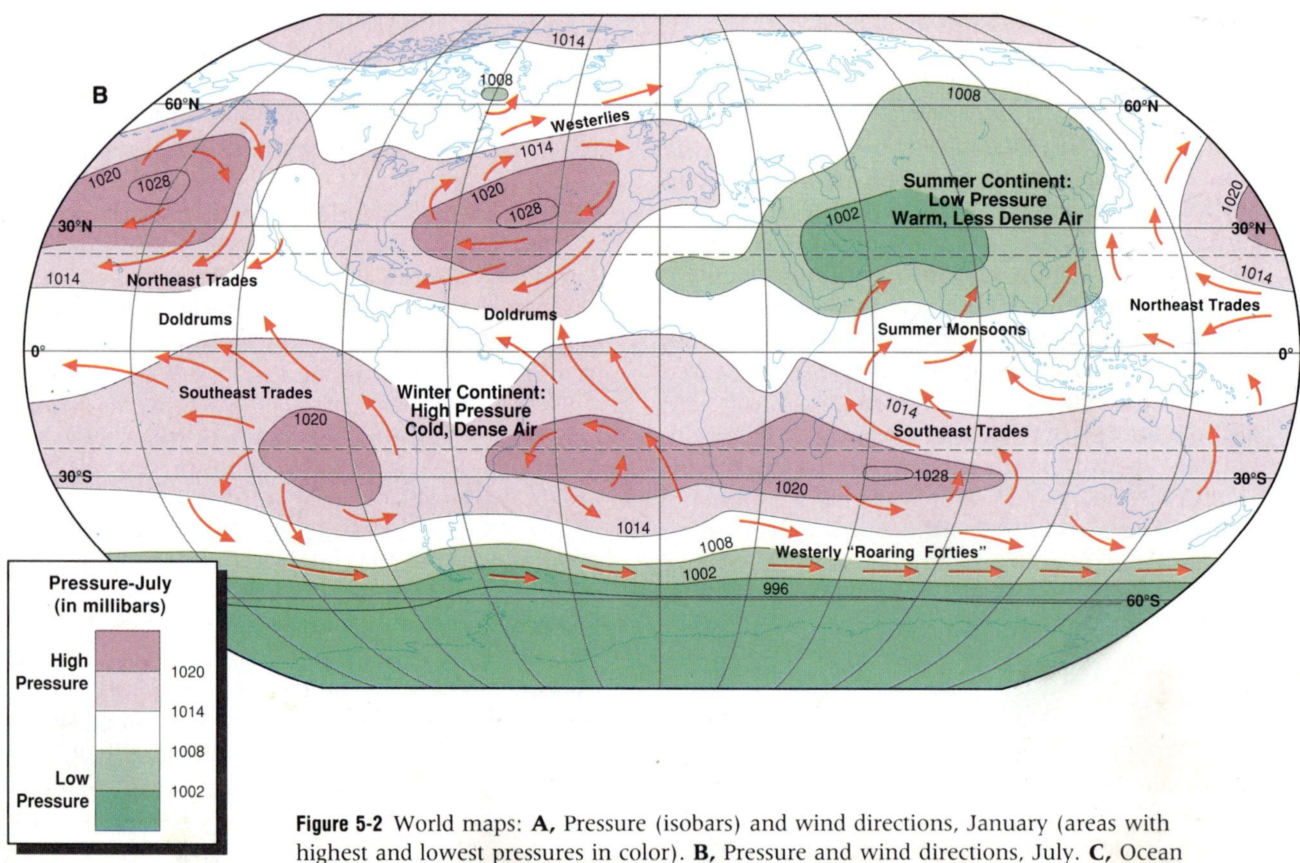

Figure 5-2 World maps: **A,** Pressure (isobars) and wind directions, January (areas with highest and lowest pressures in color). **B,** Pressure and wind directions, July. **C,** Ocean surface currents, January. Warm currents (red) and cold currents (blue).

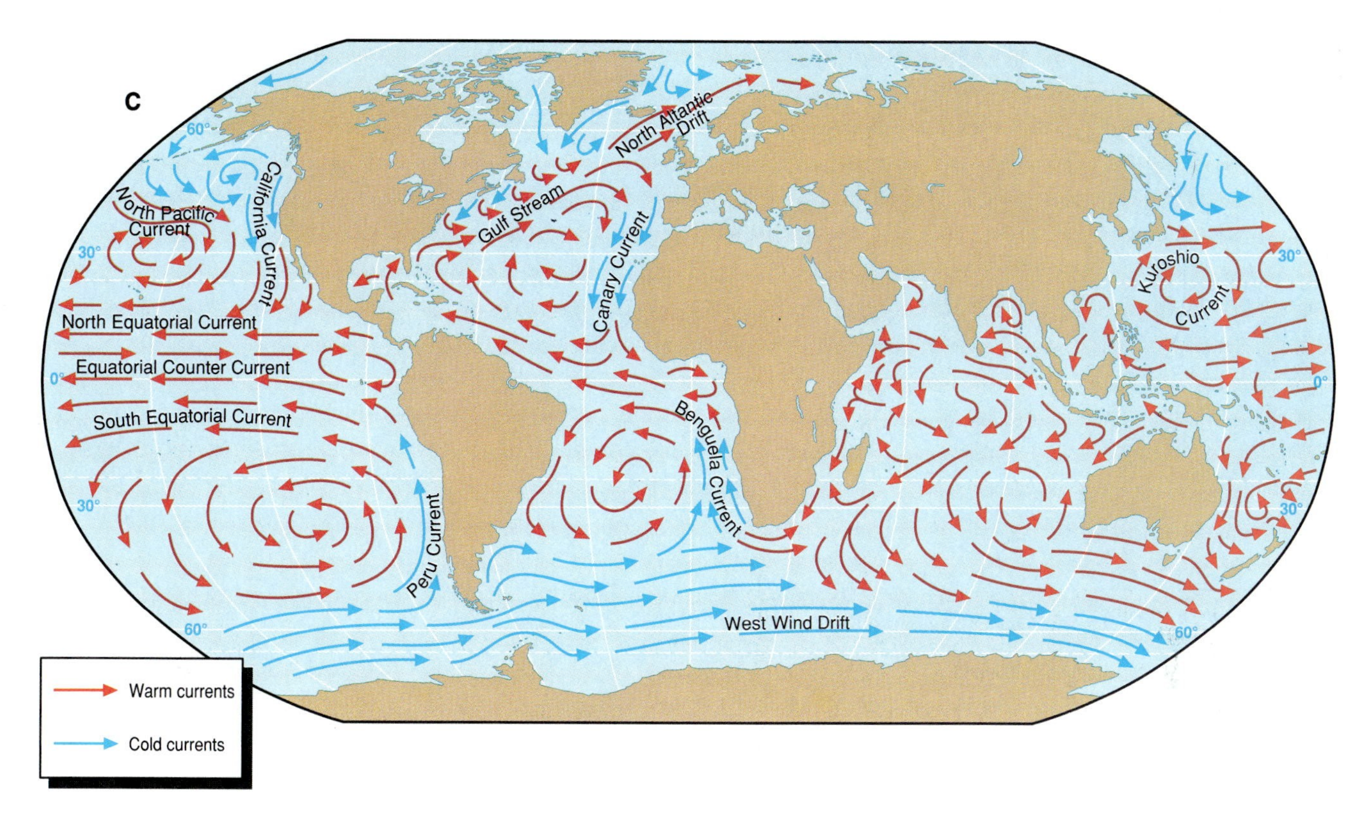

For legend, see opposite page

the sea (Figure 5-3, *B*). Surface air flows toward the sea, but the winds are usually less strong than in the daytime, since the nighttime temperature contrast between land and sea is not as great as that in the day.

If the pressure gradient force were the only factor affecting wind strength and direction at the global scale, there would be a convectional circulation between the tropics and poles. Air would rise over the heated tropics, flow aloft toward the poles, descend on cooling, and flow back over the surface toward the tropics. This would make much of Earth uninhabitable, since the surface flow would be cold air. Fortunately, this does not occur. Rather, the wind patterns form a series of distinct circulations, with some places receiving surface airflow from the tropics (Figure 5-2, *A, B*). Nevertheless, it is important to keep in mind that the basic impetus for large-scale movement in the atmosphere is produced by the contrast in heating between the tropics and polar regions.

The Effect of Earth's Rotation on Winds

The directions of winds, initially determined by differences in air density, are strongly modified by Earth's rotation. A full understanding of how Earth's spin affects the directions of movements in its envelope of gases is a matter of complex physics, but it is possible to summarize the major effects.

The atmosphere and oceans are fluids in contact with the solid Earth, but they are not fastened to it. Movements in the atmosphere and oceans have to be viewed as separate from the spinning Earth beneath. The effect of Earth's rotation on objects moving through the atmosphere above the surface is often illustrated by the example of a rocket launched in northern Canada and aimed due south at New York City. While the rocket is in the air, Earth's rotation moves New York City to the east, clear of any danger. The north-south path of the rocket is maintained, but its path relative to Earth's surface bends to the right as observed by those it passes over. In the southern hemisphere the path of a rocket would bend to the left.

The effect of Earth's spin on the direction of movements of air in the atmosphere has three aspects. First, air in motion appears to be deflected to the right in the northern hemisphere, and to the left in the southern hemisphere. Second, the deflection has no effect on the speed of the wind but the speed of the wind affects the amount of deflection: the higher the wind speed, the greater the deflection. Third, deflection is zero at the equator and greatest at the poles, with a gradation between.

Since the impact of Earth's rotation on movement in the atmosphere is so complex, some rule-

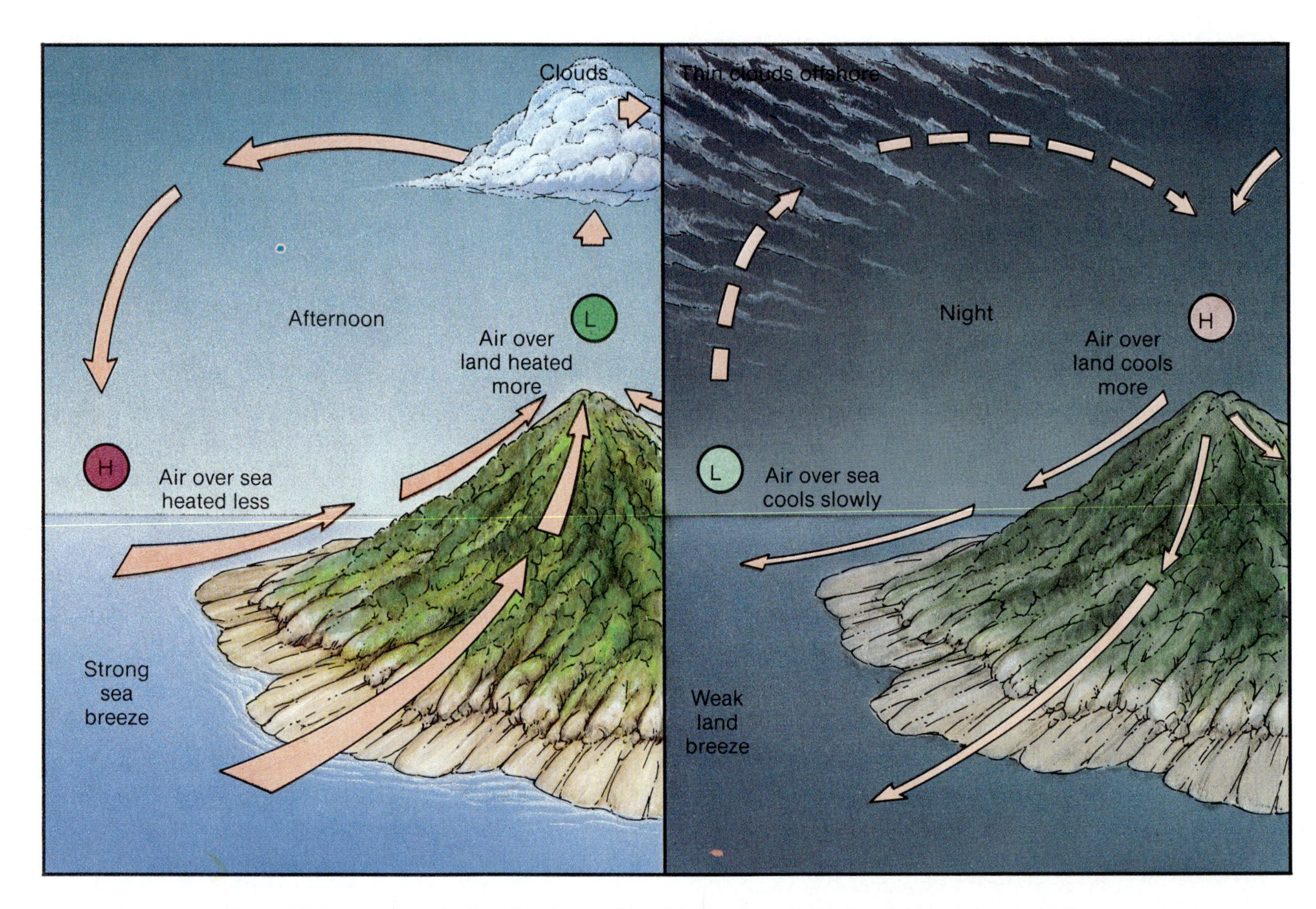

Figure 5-3 Sea breezes during the day and land breezes at night. Land heats more rapidly than sea during the day, and so transfers heat more rapidly to the air above. The heated air above the land expands and rises, lowering the surface pressure and establishing a pressure gradient from sea to land. Air flows along the pressure gradient toward the land as a sea breeze. Over the sea the air that moves toward the land is replaced by air descending from above and a convectional circulation forms. At night the flow is reversed because air over land cools more rapidly than air over sea. The cooled air is denser and flows downslope and out to sea. The night pressure gradient is not so steep as that during the day and so the land breezes are not so strong.

of-thumb "laws" have been invented for understanding and applying the effects. In 1835, Gaspard C. de Coriolis proposed a "force" to explain the observed reality. The Coriolis force is opposite to the direction of windflow determined by the pressure gradient. Coriolis force is not a true force such as gravity, but aids in understanding the way in which Earth's rotation affects apparent wind direction from the point of view of an earthbound observer. For this reason, the term **Coriolis effect** is now commonly used. Because of this effect, winds are apparently deflected to the right of their expected path in the northern hemisphere, and to the left in the southern hemisphere. The strength of the Coriolis effect depends on wind speed, which varies with the pressure gradient force. The Coriolis effect is caused by Earth's spin effect, which has a greater influence at higher latitudes (which are almost parallel to the plane of rotation) and none at the equator.

The way in which the Coriolis effect works can be visualized by imagining a wind caused by a pressure difference between two places (Figure 5-4). The air begins to flow in response to the pressure difference and at first travels at right angles across the isobars. As it accelerates toward the speed dictated by the pressure gradient, the Coriolis effect increases and its role is greatest when the wind reaches the maximum speed dictated by the pressure gradient. By this stage the wind is turned fully to the right (in the northern hemisphere) and blows parallel to the isobars. A wind that blows parallel

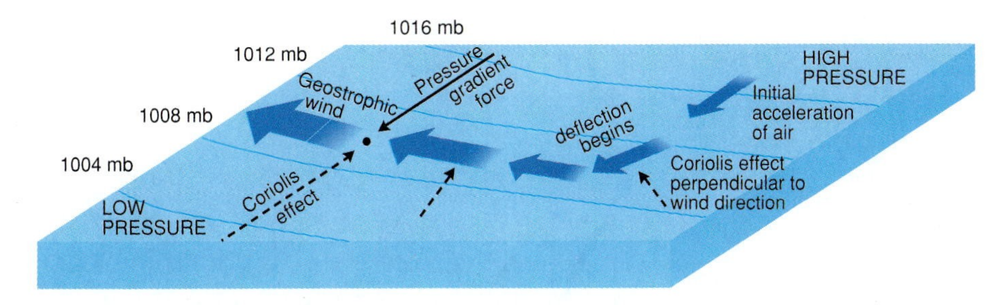

Figure 5-4 Coriolis force and a geostrophic wind. The sequence shows the Coriolis effect changing the direction of the wind (viewed from the ground) as the wind accelerates in response to the pressure gradient force. The Coriolis effect always acts at right angles to the direction of flow. It increases in strength as the wind speed increases until it balances the pressure gradient force and causes the wind to blow parallel to the isobars instead of across them.

to the isobars is known as a **geostrophic wind.** C. H. D. Buys Ballot (1857) suggested a rule that illustrates the geostrophic wind phenomenon: when a person stands with his or her back to the wind in the northern hemisphere he or she has higher pressure to the right and lower pressure to the left.

Figure 5-5 shows the surface *(A)* and upper air *(B)* maps of winds and pressure lines at one moment over North America. These maps reveal how far the effect of Earth's rotation helps to explain the relationship of winds to pressure at Earth's surface and higher up in the troposphere. The map of the higher level *(B)* shows geostrophic winds blowing parallel to the isobars at high speeds associated with close-spaced isobars. An airplane flying in the same direction as this high-level wind (i.e., with its tail to the wind source) has high pressure on the right and low pressure on the left in the northern hemisphere. The Coriolis effect, inducing a geostrophic wind, explains the direction of flow at this level. At the surface *(A)*, however, the winds are slower and are not geostrophic. They flow away from high pressure areas and toward low pressure areas, crossing the isobars, but not at right angles to them. The effect of Earth's rotation explains the surface wind directions partly, but not entirely: surface friction must also be taken into account.

Winds and Friction with Earth's Surface

Winds in contact with Earth's surface are slowed by friction, which reduces the effect of the pressure gradient force by up to half over hilly land. The Coriolis effect is reduced to a greater extent than the pressure gradient force, and this relative change in force strength affects the direction of the winds. Winds blow across the isobars instead of parallel to them because of the greater relative strength of the pressure gradient force (Figure 5-6). In the northern hemisphere, surface high-pressure areas are marked by descending air at the center and clockwise, diverging flow outward. There is a counterclockwise, converging flow of air into surface areas of low pressure, accompanied by rising air at the center. In the southern hemisphere the outflow from high-pressure areas is counterclockwise, and the flow into low-pressure areas is clockwise.

Winds over hilly land blow in directions at a higher angle to the isobars than those over a flat sea, because the greater friction of the land increases the relative influence of the pressure gradient. The surface map of Figure 5-5*A* shows how light winds blowing across the Appalachians or Rockies cross the isobars at high angles, while strong winds out in the North Pacific blow almost parallel to the isobars. The opening extract about winds in the Great Plains illustrates the low frictional effect of flat land.

The layer of the atmosphere near the ground that is affected by surface friction is known as the **friction layer.** The depth of this layer is determined by the turbulence in moving air caused by the different types of terrain it is in contact with. On average it is 1 km deep but is deeper over mountains and shallower over plains and oceans.

The Effects of Surface Relief on Winds

Some local winds are caused by differences in relief that set up centers of high or low pressure, as in the case of the land and sea breezes discussed earlier (see Figure 5-3). In upland areas it is common to

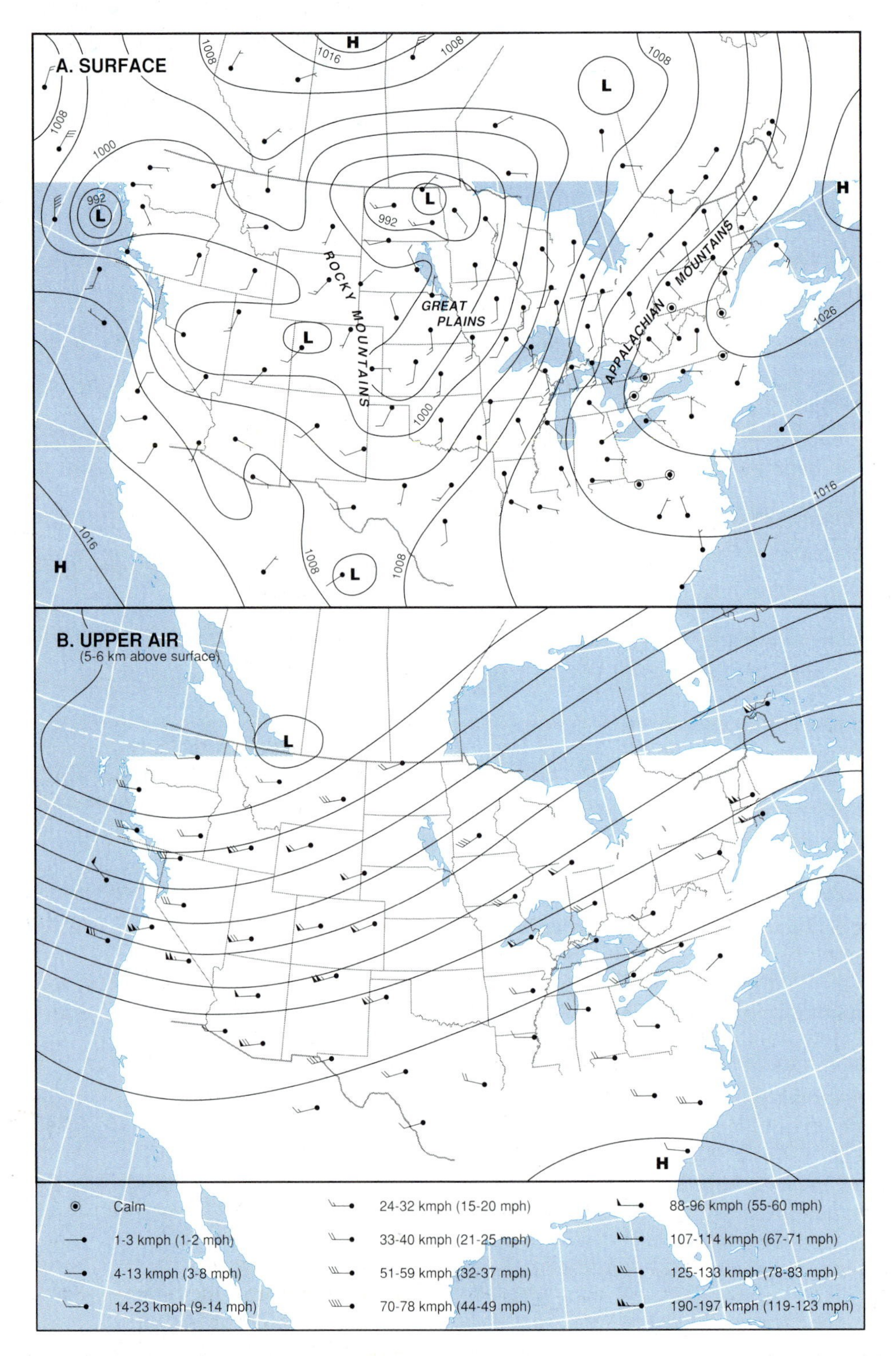

Figure 5-5 Maps of atmospheric pressure and winds for North America: the surface *(A)* and the upper troposphere where the atmospheric pressure is approximately half that at the surface *(B)*. These patterns occurred on a day in October and are taken from the Daily Weather Report forecast maps. The key shows wind direction and strength by symbols: the arrows indicate direction, and the number of "feathers" the wind speed. Wind speed is often given in knots, or nautical miles per hour: a knot is approximately 1.8 kmph (1.2 mph).

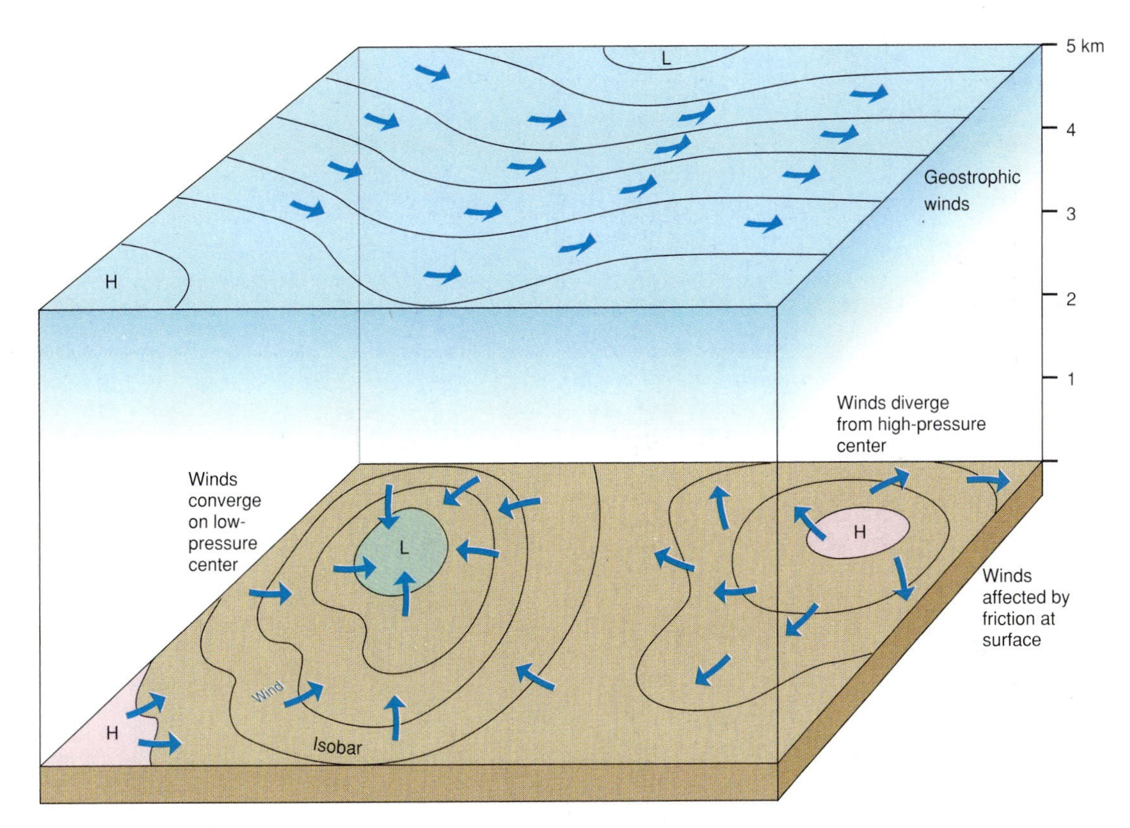

Figure 5-6 Surface friction and winds. Winds blowing across Earth's surface are slowed by friction with the ground. Winds at higher levels are not affected by friction and blow parallel to isobars. This diagram is a northern hemisphere case and can be compared with the maps of Figure 5-5.

have winds blowing upslope in the afternoon and downslope at night (Figure 5-7). Daytime heating of the upper slopes causes the air above them to expand and rise by convection, while air from the valley floor moves upward to replace the air that is rising. At night rapid radiation from the upper slopes leads to cooling of the air, making it more dense. The cold air then drains downward to the valley floor. Where the valley floor is confined, cold air may accumulate and lower the temperature more rapidly than in overlying air. This may produce a temperature inversion. Cold winds draining off the ice sheets in Greenland and Antarctica are also features of the surrounding areas. Cold winds draining into a valley floor can be disastrous for fruit crops. In the Napa Valley of California, the effect of cold night winds was combatted by heaters and "smudge pots," but environmental legislation has forced their replacement by windmills that circulate and mix the cold air on the ground with warmer air above (Figure 5-8).

On a larger scale, wind direction is affected by major relief features acting as barriers. Most mountains are insignificant in relation to the thickness of the troposphere, but the highest ranges reach fully half this thickness and play a major regional role in air movements. The Himalaya-Tibet mountains (8 to 10 km or 5 to 6 miles high) exclude cold Siberian airflows from the Indian subcontinent in winter. A massive temperature contrast is established between the extreme cold of Siberia to the north and the subtropical conditions of the Indian subcontinent to the south, which strongly affects the climates of the two regions (Chapter 8).

At a regional scale mountains have the effect of channelling winds through the valleys between them. An example of this is the cold *mistral* wind that blows down the Rhône valley in France. Cold air flowing from a high-pressure area over northern Europe toward a low-pressure area over the Mediterranean Sea is channelled along the valley, producing strong, cold winds at the southern end that can harm early flowering fruit trees in spring. The *Santa Ana* winds in southern California occur when high pressure over the interior desert causes warm, dry winds to be concentrated through the narrow valleys toward the Pacific coast. Such burning-hot winds may carry desert dust.

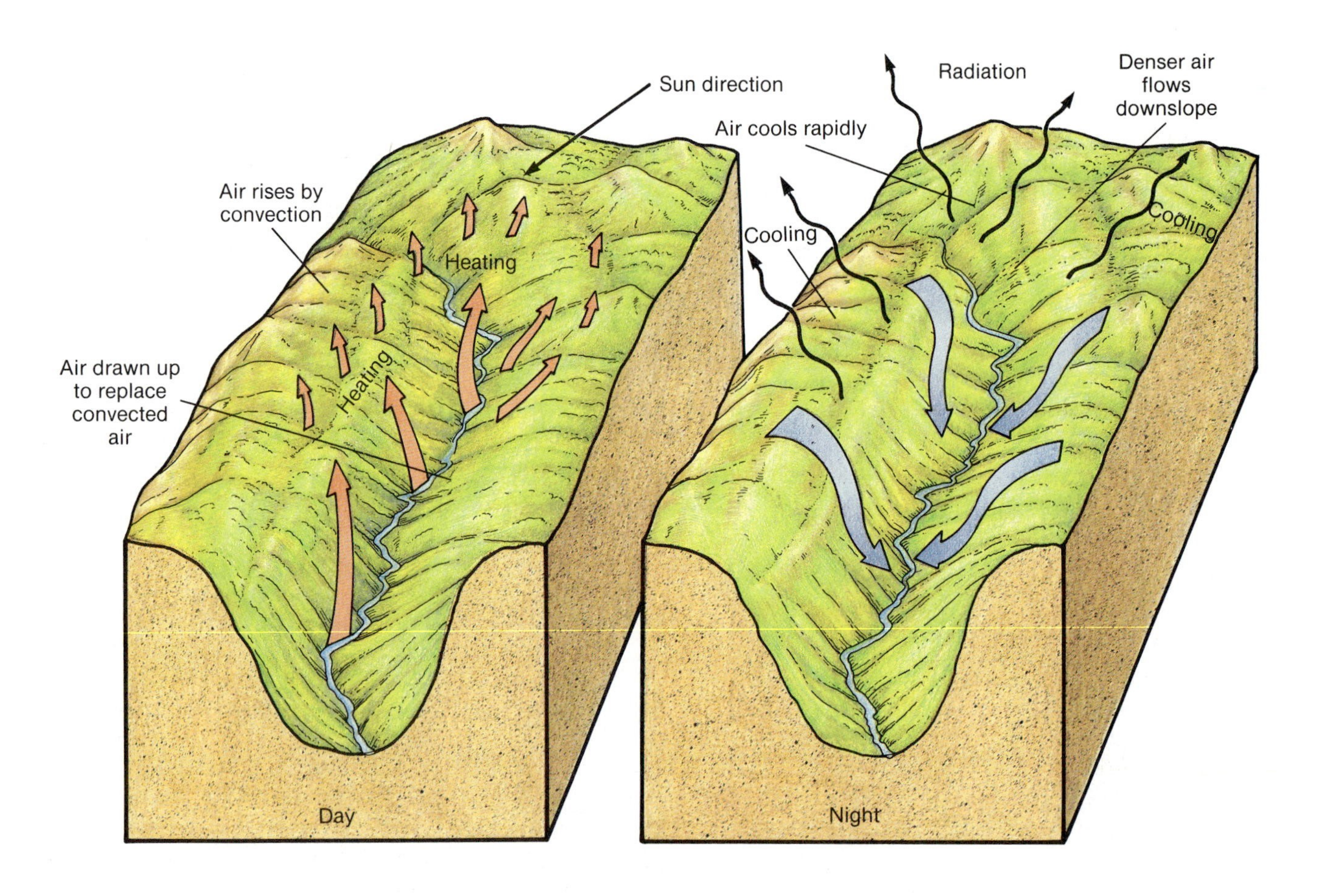

Figure 5-7 Mountain winds. During the day *(A)* air is heated on the mountainsides, expands, and rises. Air from the valley floor is drawn up to replace the rising air and upslope winds occur. At night *(B)* the cooling of air occurs first on the upper slopes, and the denser air drains down to the valley floor.

Figure 5-8 Cold air draining into the Napa valley floor in central California can kill the early flowers of the grapevines. The older heaters have been replaced by fans that circulate the air and mix the cold surface flow with warmer air above, thus preventing frost.

CAUSES OF OCEAN CURRENTS

Ocean current speeds and directions are caused by similar forces to those that cause winds, but there are some major differences. First, the speed and direction of surface ocean currents are largely determined by winds blowing over the surface, although density differences and the Coriolis effect also play a part. Second, the confinement of oceans to depressions in Earth's crust causes the water to circulate within the boundaries provided by the continents.

Solar Heating, Salt Content, and Ocean Currents

Density differences in the oceans provide a starting point for considering movement within ocean waters. These differences result from a combination of solar heating and variations in the salt content of the water (Chapter 3). Warmed water expands and has a lower density than cold water. Water with a high salinity is denser than that with a low salinity. As in the atmosphere, the density differences in the oceans provide an initial impetus for ocean current flows. When surface water is heated in the tropics, it moves poleward in currents such as the Gulf Stream off eastern North America.

The densest water is both cold and has a high salinity. It sinks to the bottom of the ocean, leaving less dense water at the surface. As newly cooled water sinks, it pushes beneath the older water gradually, raising it back toward the surface. This process establishes a vertical circulation from the top to the base of the oceans. In the shallower surface layer of the ocean, overturning of water also occurs as a result of the turbulence that is part of the water movements. If this vertical circulation related to density differences were the only factor involved, the oceans would be marked by a convectional, north-south circulation between the tropics and polar regions. In fact, as with winds, circular patterns of surface flow predominate, and deep ocean movements are not merely due to convection.

The Effect of Earth's Rotation on Ocean Currents

In the oceans Earth's rotation also affects the movements of water, and there is a surface circulation based on deflecting the waters to the right in the northern hemisphere and to the left in the southern hemisphere. This occurs, for instance, in the circular flows of the northern and southern halves of the Atlantic Ocean (Figure 5-2, *C*). These circular patterns in oceans are termed **gyres.** They are not, however, truly products of the Coriolis effect, since the shapes of the ocean basins and the influence of surface winds are more important in affecting the surface water flow than Earth's rotation.

The Effects of Winds on Ocean Currents

The directions of ocean current flow are affected by winds blowing across the ocean surface. In the uppermost layers of the oceans, the surface winds are more important than Earth's rotation or differences in seawater density in determining the direction of flow. Thus, the dominant westerly flow of ocean currents in midlatitudes is caused by the prevalence of westerly winds. Near the equator, the trade winds blowing from the east push water toward the west.

The wind-induced flow of water causes the most rapid movements to occur at the surface. An ocean current typically flows at 2 percent of the wind speed above it. The speed of water flow drops off with depth and below about 100 meters, the water is unaffected by the surface flow. As water depth increases, the effect of the wind decreases and the relative effect of Earth's rotation becomes more significant. The **Ekman spiral** (Figure 5-9) demonstrates these changes. The Coriolis effect deflects the *surface* flow to the right (northern hemisphere) or left (southern hemisphere) of the wind direction by some 45 degrees or less, but the *overall* water movement is at approximately right angles to the dominant wind direction. This has important economic implications where winds blow parallel to the shore, such as over the southward-flowing ocean current off California or over the northward-flowing current off southern Peru. Water movement in both is away from the coast, bringing up cooler waters from greater depth to replace the surface waters. Such deep waters are rich in nutrients after traveling around the world and accumulating organic materials that have fallen into them from above. The upwelling regions produce some of the world's richest fisheries.

Ocean-Basin Boundaries and Ocean Currents

The form of ocean basins is also significant in directing ocean-water movements. The continents are barriers to the ocean currents and obstruct the pattern of flow that might have been established on a smooth planet covered totally by ocean. Examination of Figure 5-2, *C*, shows that the shape of the land around the midlatitude North Pacific Ocean tends to favor a circular pattern, whereas the wes-

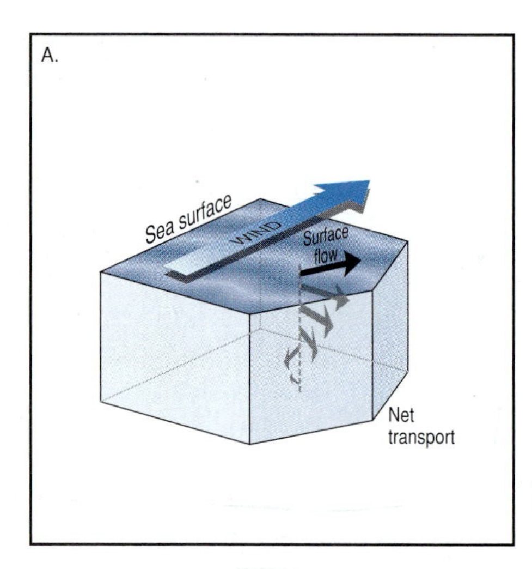

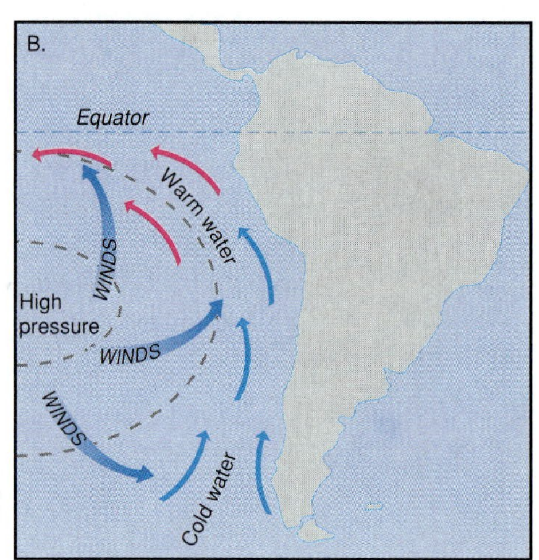

Figure 5-9 The effect of surface winds on ocean currents. **A,** Winds drag the surface waters forward at 2 percent of the wind speed. Earth's rotation deflects the flow to the right in the northern hemisphere. As depth of water increases, the movement is less and the deflection increases. This effect is shown in the Ekman spiral. Overall, water is moved at right angles to the wind direction. **B,** Winds and currents in the Pacific Ocean off South America, where deflection is to the left (southern hemisphere). The winds cause surface water to flow away from the coast and bring up cold water from depth.

terly flow in the North Atlantic Ocean continues north of Europe. In the southern hemisphere the absence of land south of 40° latitude allows a west-east flow around the globe, since there are no continents to direct the flow to north or south. The major factor governing surface water flow in this region is the prevailing winds.

The ocean basins are mainly north-south in orientation because of the arrangement of continents. This causes the dominantly east-west flows of the tropics and the west-east currents of midlatitudes to split and be directed along the coasts in northerly or southerly flows known as **boundary currents**. Narrow deep, warm currents flow rapidly poleward along the western margins of oceans, and shallower and broader cold currents flow more slowly toward the tropics along the eastern margins (Figure 5-10).

The movement of air and water involves massive exchanges of energy. Some attempts have been made to tap this energy for human use.

THE INFLUENCE OF UPPER-AIR WINDS AND DEEP-OCEAN CURRENTS

Earlier in this chapter upper-air and surface winds were treated as separate entities in order to explain the factors that cause them to blow. Research since World War II has demonstrated a number of important linkages between these two sets of winds. Further study of the oceans has also made it possible to construct the linkages between surface and deep flows of water.

Upper-Air and Surface Winds

On the global scale, transfers of energy by winds and currents produce zones of concentrated ascent and descent of air that are part of a larger pattern of convection on the global scale. Ascent occurs over surface zones where airflows come together, or **converge**, and descent occurs where surface airflows move apart, or **diverge**. Surface atmospheric pressure in a zone of converging and rising air is relatively low; it is relatively high in a zone of descending and diverging air.

In ascent, the pull of Earth's gravity is overcome by several special mechanisms that propel the large-scale uplift of air. First, converging surface air streams of similar temperatures result in air rising along the zone of meeting. This occurs close to the equator, where warm and moist winds from the northeast in the northern hemisphere and southeast in the southern hemisphere confront each other along the **intertropical convergence zone (ITCZ)**. Second, convergence of contrasting types of air causes the warmer (less dense) to ascend over the cooler. This is common in midlatitudes, where polar and tropical air meet and create a zone of steep temperature gradient known as the **polar front.** This zone occurs all around the globe between 40°

Figure 5-10 The effect of a cold current. San Francisco beach on a fine day in August. The only bathers are hardy people in wet suits.

and 60° of latitude in both hemispheres. Third, the release of latent heat raises temperatures in the air some distance above the ground. This occurs where air is forced to rise, as above the ITCZ and the polar front. In both cases the uplift that begins with convergence is enhanced by the release of latent heat during cloud formation. The locations of the sites of uplift are thus indicated by the equatorial and midlatitude bands of cloudiness (Figure 5-11).

Strong high-level winds, known as **jets,** occur near the top of the troposphere in zones where there is a very steep gradient of temperature and pressure (Figure 5-5, *B*). The most continuous of these zones occurs in the midlatitudes where tropical and polar air converge, setting up extreme variations in upper air temperature and pressure over small distances. This zone features the **polar-front jet** at 10 to 20 km (6 to 12 miles) above the junction of the polar front with the ground (Figure 5-12). Wind speeds of 500 kmph (300 mph) have been observed in the core of the jet, although average wind speeds are less than half this extreme. Jet wind velocities are highest in winter and lowest in summer in response to changing temperature and pressure gradients. Aircraft save time and fuel by flying along the margins of these high-speed winds (the core is too turbulent) when travelling in the same direction. Other jets occur when steep temperature and pressure gradients exist over the subtropical high-pressure zones (the westerly subtropical jet) and over the equator (easterly equatorial jet).

The path of the polar-front jet forms wave patterns, known as **Rossby waves,** as it circles the globe. There are generally four to six of these waves around the globe in the high and midlatitudes of each hemisphere (Figure 5-13). The wave forms go through a cyclic pattern of development that includes a phase of low-amplitude waves with the jet blowing mainly west-to-east (Figure 5-13, *A*) followed by increasingly meandering (more wavy) phases that become more north-south in flow direction (Figure 5-13, *B, C*). As the meanders develop, lobes of cold air on the polar side of the jet are known as *troughs*, and lobes of warm air on the tropical side are known as *ridges*. In the troughs, cold air aloft pushes toward the equator, as warm air flows poleward in intervening ridges (Figure 5-13, *C*). Finally, the troughs of cold air are cut off south of the jet and the warm ridges north of the jet (Figure 5-13, *D*), forming part of the processes that transfer heat from tropics to polar regions. Cycles of Rossby waves each last for several weeks, but variations in their duration cannot be predicted at present.

The polar-front jets are the strongest winds known, and are sufficiently powerful to control the wind patterns at the surface. They are responsible for the extent and intensity of converging and di-

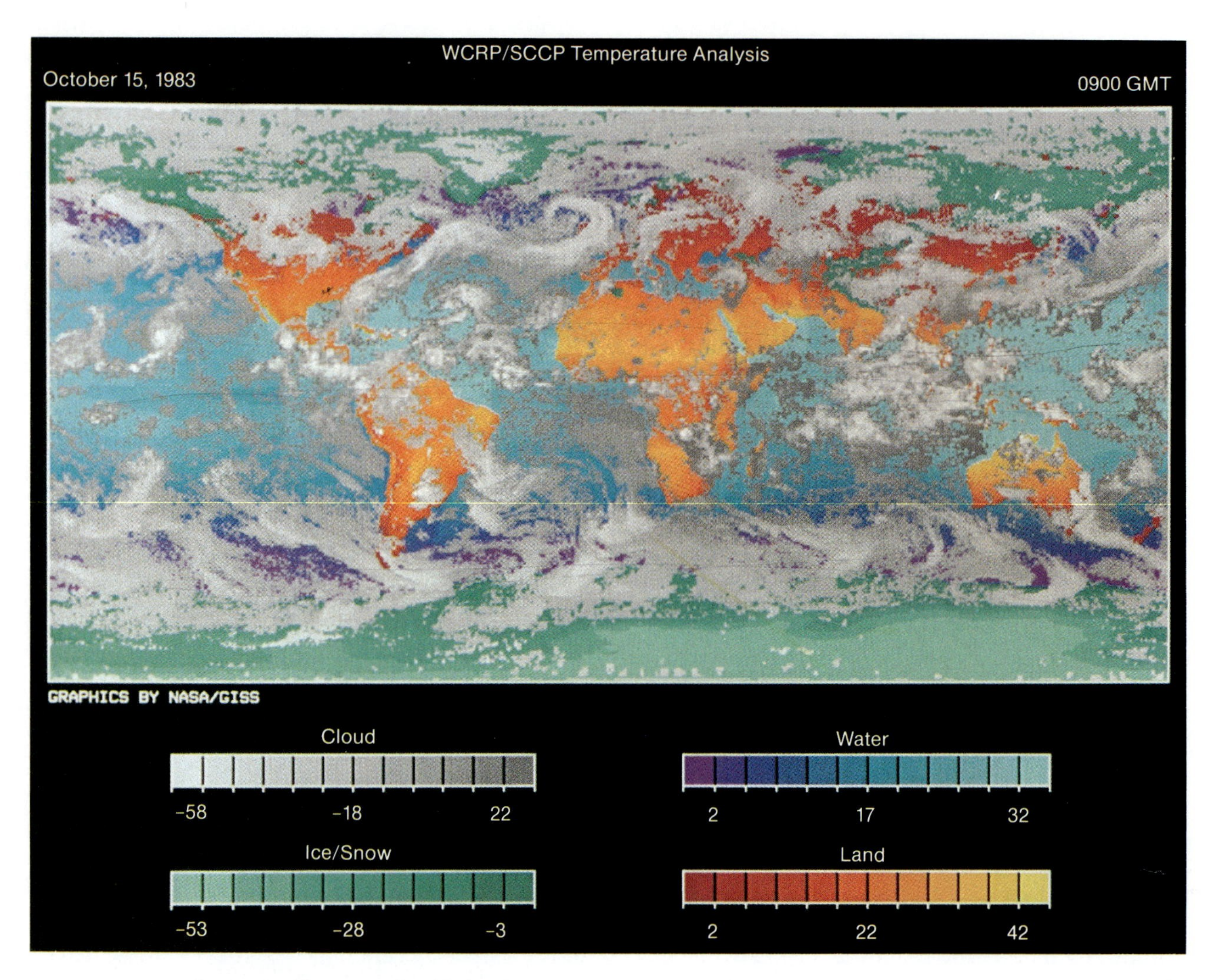

Figure 5-11 The main cloud belts close to the equator and between 50° and 60° North and South, seen from a weather satellite. The cloud belts occur along zones of uplift in the atmosphere.

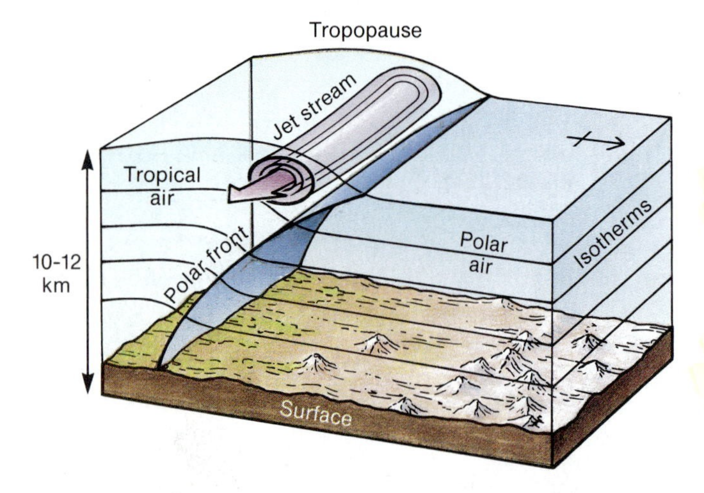

Figure 5-12 The polar-front jet is a powerful wind, often blowing at 300 kmph and more in the central core, while winds in the margins blow at 120 kmph. The jet occurs above the polar front, a major zone of temperature and pressure change between tropical and polar air in middle latitudes.

verging movements of air along the polar front at ground level. Within the wavelike jets of the upper troposphere, zones of converging and diverging air are linked to diverging and converging air at the surface. As shown in Figure 5-14, air is forced down toward Earth's surface at zones of convergence in the jet stream. High-pressure systems with diverging winds form at the surface as a result of this downward movement of air. Convergence above is linked to divergence near the ground. Where there is divergence within the jets, air from the surface is pulled upward through low-pressure surface systems that are marked by converging surface winds. The force of surface winds depends on the strength of pressure differences near the surface, which, in turn, result from the strength of flow in the upper troposphere.

Zones of descent occur between the major zones of ascending air over the equator and midlatitudes. One such zone occurs in the subtropics above the

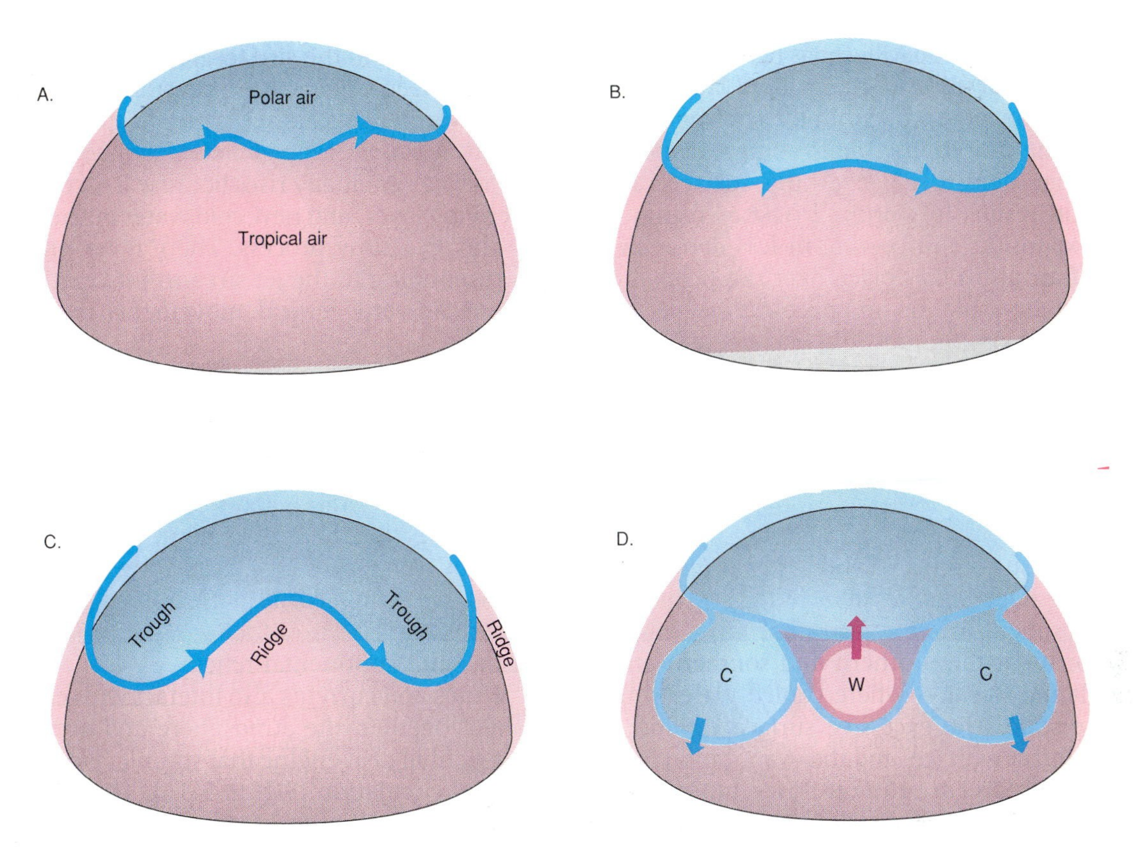

Figure 5-13 Rossby waves in the polar-front jet: a model of their cyclic development. The sequence may take 4 to 6 weeks, but its timespan varies greatly. **A,** A strong westerly geostrophic flow, with little mixing of tropical and polar air aloft. **B,** The waves grow larger. **C,** Ridges and troughs become pronounced, with north-south and south-north flows. **D,** The waves break down, transferring large "pools" of polar and tropical air across the polar front. Stage A is resumed.

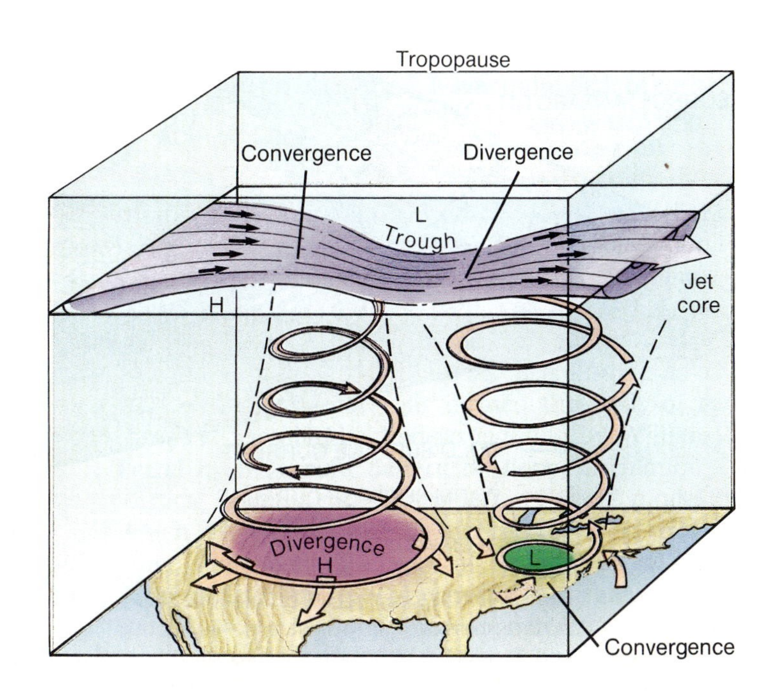

Figure 5-14 The polar-front jet and its impact on surface winds and atmospheric pressure patterns in middle latitudes. Converging airflow in the jet forces air down toward the surface, producing high-pressure centers with air that diverges at the surface in clockwise patterns (northern hemisphere). Diverging airflow in the jet draws air upward from the surface, where there is low pressure and inflowing winds in counterclockwise patterns (northern hemisphere).

a satisfactory basis for explaining circulation in the tropical atmosphere, air movements in midlatitude and polar regions are different. The tropics, midlatitudes, and polar regions are three distinctive circulation environments. The three-dimensional patterns of global atmospheric circulation are made clearer by comparison with the surface winds and upper-air flows shown in Figure 5-17. This illustration is referred to in the following discussion.

The *tropical circulation* affects some 40 percent of Earth's surface. The concept of a simple Hadley cell based on convection fueled by heating at the equator and cooling aloft has been developed further by recent investigations of the tropical atmosphere. The modern view of the tropical circulation focuses on the convergence of trade winds at the surface and the release of latent heat above the equator. It also recognizes the importance of the subtropical jet some 15 km (10 miles) above the surface at approximately 25 to 30° North and South. Air flows from above the equator and the jet forms above a major surface pressure gradient at the outer margin of the tropics. Some of the air is forced down toward the surface beneath convergence in the jet flow, while the rest is conveyed poleward. The air that descends is warmed by compression and becomes drier. On reaching the surface in the **subtropical high-pressure zones,** it spreads out. Some of this surface air flows equatorward as it moves toward lower pressure areas along the intertropical convergence zone. This flow is extremely consistent and produces the trade winds.

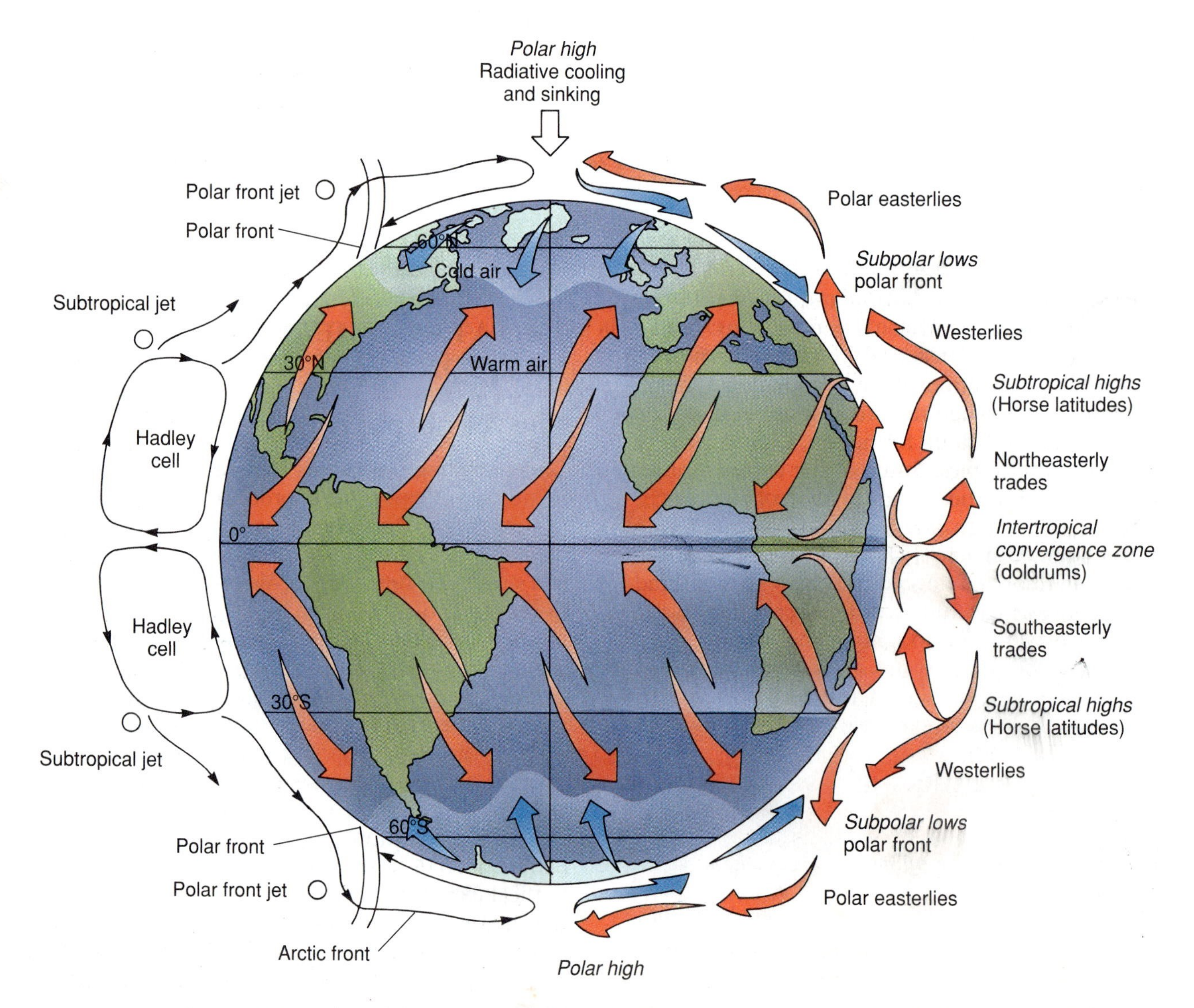

Figure 5-17 The global wind circulation. The surface pattern can be compared with the vertical movements of air in the troposphere. The three major sectors (tropics, middle latitudes, and polar regions) are distinguished.

Trade winds are surface flows of air from the subtropical high-pressure zones toward the equator. The trade winds are particularly consistent in strength and direction over the Atlantic and Pacific oceans, where the area they influence varies little from one season of the year to another. This consistency is due to the strength and continuity of the subtropical high-pressure belts over the oceans. The diminished role of the Coriolis effect close to the equator prevents the trade winds from becoming involved in circulating weather systems.

The trade winds made it possible for the early European explorers, such as Columbus in 1492 AD, to cross the Atlantic. The northeast trades blew Columbus westward from Spain and brought him to the Caribbean. On the return to Europe he sailed northward, being blown by the westerly winds around the northern flank of the high-pressure zone. This circular ocean route around the subtropical high-pressure zones later became a major commercial passageway involving a triangle of trade between Europe, West Africa, and the Americas.

Near the equator, the low-pressure zone where the trade winds converge—the intertropical convergence zone—is frequently marked by ascending air and little horizontal movement at the surface. This zone was termed the **doldrums** by sailors becalmed there in the days of sailing ships. The ITCZ shifts seasonally north and south with the passage of the vertical sun.

The greatest northward extent of the ITCZ occurs over southern Asia and western Africa. In both areas the land in the northern hemisphere faces sea in the southern hemisphere. Heating of the continent in summer draws in air containing large quantities of moisture from the equatorial and southern oceans. As the southeast trades cross the equator into the northern hemisphere, the Coriolis effect causes them to divert to the right and blow from the southwest. These seasonal winds are known as **monsoons** (see Chapter 8 for a discussion of the effect this process has on climate).

The trade winds and the intertropical convergence zone play important parts in the circulation of heat energy and in fueling the global circulation. During their passage over the tropical oceans, the trades pick up large quantities of evaporated moisture. The latent heat stored in this moisture is released in the clouds as the air ascends over the intertropical convergence zone.

The second of the three distinctive sectors of the global atmospheric circulation is in the middle latitudes between the tropics and 60° north and south: the *midlatitude circulation* affects some 40 percent of Earth's surface. This zone is one of surface conflict between converging tropical and polar air and is subject to steep horizontal gradients of temperature and pressure along the polar front. The main impetus for air movements is the meeting of tropical and polar air along this zone, and the activity of the polar-front jet aloft. The surface winds are known as the **westerlies** (Figure 5-17). However, surface winds in the midlatitude circulation are more variable in direction than the trade winds of the tropics because they are involved in weather centers of high and low pressure with their diverging and converging wind patterns, respectively. These circulating systems develop along the polar front in latitudes where the Coriolis effect exerts a strong influence and move eastward. Winds in circulating systems may blow from any direction, but the westerly component is dominant in systems that move from west to east. The westerly winds are most consistent in strength and direction in the southern hemisphere, where the "Roaring Forties" blow almost entirely over the sea. (These winds mostly blow between 50° and 60° of latitude but have retained the name given them by early navigators.)

The third major section of the global atmospheric circulation affects latitudes poleward of 60° North and South: the *polar circulation* affects 20 percent of Earth's surface. This circulation is weak, since the region lacks motivating forces such as heating of the ground, high-level jets, or the release of latent heat. High-latitude areas lose more heat by radiation to space than they gain from the sun. The deep cooling of the atmosphere extends upward into the stratosphere in winter and leads to general sinking of air around the poles, forming the polar highs. Dense, cold air builds up and flows out from the polar regions at the surface. The **polar easterlies** shown on the map of global circulation (Figure 5-17) are less constant than either the trade winds or the midlatitude westerlies.

The arrangement of continents also affects the extent and influence of polar winds in the two hemispheres. Antarctica acts as a single cold core for the southern hemisphere throughout the year, but its influence is partly ameliorated, and prevented from spreading, by the warmer surrounding ocean. The contrast between the Antarctic cold and the warmer surrounding ocean, however, produces a strong temperature gradient along the polar front and a strong westerly wind circulation. In the northern hemisphere, the Arctic Ocean core of high pressure and cold air is augmented in winter as intense cooling extends the high pressure into northern Siberia and northern Canada. Cold air flows southward from these cells in winter to affect extensive areas on the middle latitude continents of the northern hemisphere.

The atmosphere and oceans are locked together by exchanges of matter and energy as was shown in Chapter 3. These exchanges include the **hydrologic cycle** (Figure 3-9), a set of processes that move water from place to place on Earth. Water is transferred from storage in the major oceans into the atmosphere as water vapor, and is then changed back into water droplets to form clouds, from which it returns to Earth's surface as rain and snow. The work of the hydrologic cycle on the land surface and in the living-organism environment is covered in Chapters 14 and 19; the focus in this chapter is on the cycle's role in the formation of weather as water vapor, clouds, rain, and snow. The processes that are part of the circulation of water through the atmosphere are linked closely to the transformations and transfers of solar energy through the atmosphere-ocean environment.

Human activities are strongly affected by the weather elements related to the hydrologic cycle. The water vapor in air affects personal comfort, especially on hot days; temperatures of 33°C (90°F) when air has the high humidity typical of summers in Washington, D.C., are very uncomfortable, but with the low humidity of air in Phoenix, Arizona, such temperatures are more bearable. Rain and snow provide water that can be used for domestic and industrial uses, and may determine whether a farmer will have a profit or loss. Visibility is strongly influenced by clouds and fog. The quotation opening this chapter demonstrates how a writer builds a seasonal scene by describing weather elements that are based on the water factor.

This chapter is organized around the major weather-forming aspects of the hydrologic cycle. It follows the sequence of processes from the movement of water vapor out of storage in the oceans to become an atmospheric gas, until it returns to the ground and ocean as rain or snow. At each location on Earth's surface these processes result in a water budget of inputs and outputs that determine the amount of water available for local use, and a discussion of water budgets closes the chapter.

SURFACE WATER STORAGE

The oceans contain 97.2 percent of Earth's water (Figure 6-1), and this huge body of water has an important role in the heating and movements of the atmosphere, as was shown in Chapters 4 and 5. The remaining 2.8 percent of Earth's water exists as fresh water on the continents. Of this, over three fourths is stored in glaciers and ice sheets, and almost all the rest is held in rocks below the surface. Only 0.6 percent of the fresh water is contained in lakes, inland seas, and rivers; 0.05 percent in the soil; and 0.04 percent in the atmosphere.

Water moves rapidly through the atmosphere in relatively small amounts. The average amount of water in Earth's atmosphere would be sufficient for only 10 days of rain or snow if no water were added from ocean storage. Most of Earth's water remains stored on the surface, particularly in oceans, but also in seas, lakes, and ice masses, or underground in rocks. It may remain there for several thousand years without being involved in movement through the hydrologic cycle.

FROM SURFACE WATER TO ATMOSPHERIC GAS

Water begins its journey through the atmosphere when it is converted from liquid to gas by evaporation or transpiration. **Evaporation,** the change

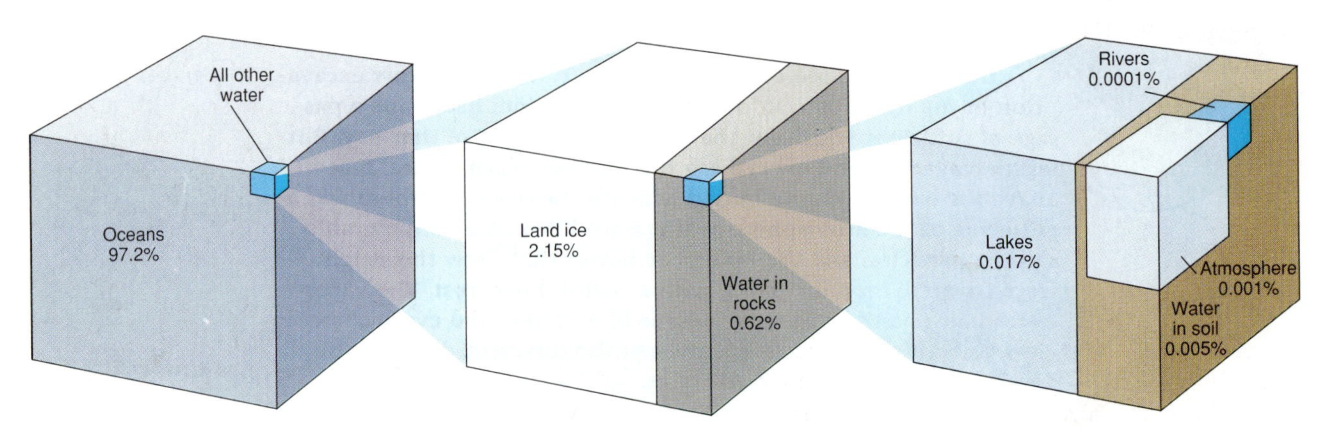

Figure 6-1 The storage of water at Earth's surface. Percentages shown are by weight. A tiny proportion of the total is involved in movement through the atmosphere at one moment. Lakes include inland seas.

from liquid water to the gaseous form (water vapor), takes place from water surfaces exposed to the air. The bonds linking the molecules in liquid water are broken apart so that each molecule is released into the atmosphere as a free gas molecule. The process of breaking the bonds takes a lot of energy. The evaporation of 1 gram of water (1 cubic centimeter, 5 of which fill a teaspoon) requires the same amount of energy needed to heat 6 grams of liquid water from freezing to the boiling point. The energy that powers evaporation is not "used up," but exists in the gaseous water as latent heat (Chapter 4). The immediate and local effect of this use of energy is to cool the air in contact with the water surface. Most of the time, sufficient heat energy is present in the atmosphere for evaporation to occur from water surfaces.

Transpiration occurs when plants pass water to the leaf surfaces, and it is evaporated into the atmosphere. Plants draw up water from soil through their roots and use it in the manufacture and transport of their own foods. The water passes up the stem and into the leaves, where it is transpired as vapor into the atmosphere. Up to 98 percent of the water taken up by plant roots is transpired. For every gram of dry plant matter produced, 200 to 1000 grams of water is transpired. The proportion transpired affects the water demands of particular crops. Sorghum, for instance, transpires 275 parts of water for every part of dry plant matter produced. For wheat the figure is 507, and for alfalfa it is 1068. That means sorghum can be grown in drier conditions, while alfalfa requires much more water. The total volume of water released varies as well. A corn plant transpires 2.4 liters (.74 gallon) of water vapor per day, and an oak tree up to 675 liters (185 gallons). Transpiration takes place from multiple layers of leaves in a forest and may equal evaporation from an open water body. Evaporation and transpiration over land are commonly referred to collectively as **evapotranspiration.**

Rates of evaporation and transpiration (or their combined total) vary in relation to the presence of surface water, the availability of heat energy, and the capacity of the air to accept more water vapor. Evaporation is uninterrupted over oceans or lakes. Over land, water is not always available, since the soil surface may dry out, reducing the rate of evaporation.

As water vapor enters the air, it adds to the molecules of other gases already there, but acts separately from them. The amount of water vapor that can coexist with the other atmospheric gases depends on the temperature of the atmosphere, since that controls the relationship between evaporation and the change from water vapor back to liquid water. At higher temperatures molecules become more excited and more water molecules escape from water surfaces to become water vapor in the atmosphere. The capacity for evaporating water increases with the temperature of the air. The maximum possible occurs at sea level in the tropics where the pressure exerted by water vapor may constitute around 40 mb of the average 1013 mb of atmospheric pressure—almost 4 percent. When as much water vapor is changing back to water droplets as is being evaporated, the air is said to be **saturated.** Air that is not saturated is often known as "dry" air. The rate of evaporation into a body of air that is not saturated will exceed the rate of formation of water droplets in that body of air.

Evapotranspiration is most intense on hot days and in warmer parts of the world. Extremes of atmospheric water vapor have been recorded in the hot tropics (just over 4 percent by volume of total atmospheric gases) and in the winter conditions of northern Siberia (virtually zero). Winds are important in maintaining evaporation rates by moving saturated air away from a water or plant surface and bringing dry air into contact with that surface. The map of evaporation rates around the globe (Figure 6-2) illustrates these points. The highest evaporation rates take place over the tropical oceans because of intense insolation, high atmospheric temperatures, and constant water availability. The trade winds blow dry air from deserts over the oceans, and evaporation is especially rapid into that air. On average, water to a depth of 1.2 meters is evaporated annually from the oceans (and is replenished by precipitation and by streamflow). This depth varies from almost 2 meters in the tropics and 2.5 meters over the Gulf Stream to less than 0.2 meter in high latitudes. On the continents, only the tropical forest areas rival ocean areas in evapotranspiration rates; other areas tend to have lower rates because of the lower availability of water.

LINKAGES

*The exchanges of the hydrologic cycle link the atmosphere and ocean together and the **atmosphere-ocean environment** with the **surface-relief environment** and the **living-organism environment.** Evaporation from water surfaces combines with transpiration from plants to provide water vapor in the atmosphere. On cooling, water vapor condenses on particles that enter the atmosphere following volcanic activity and the wind acting on broken rock materials on the continents. Precipitation as rain or snow brings water to the continents, where it fashions landforms and sustains living organisms.*

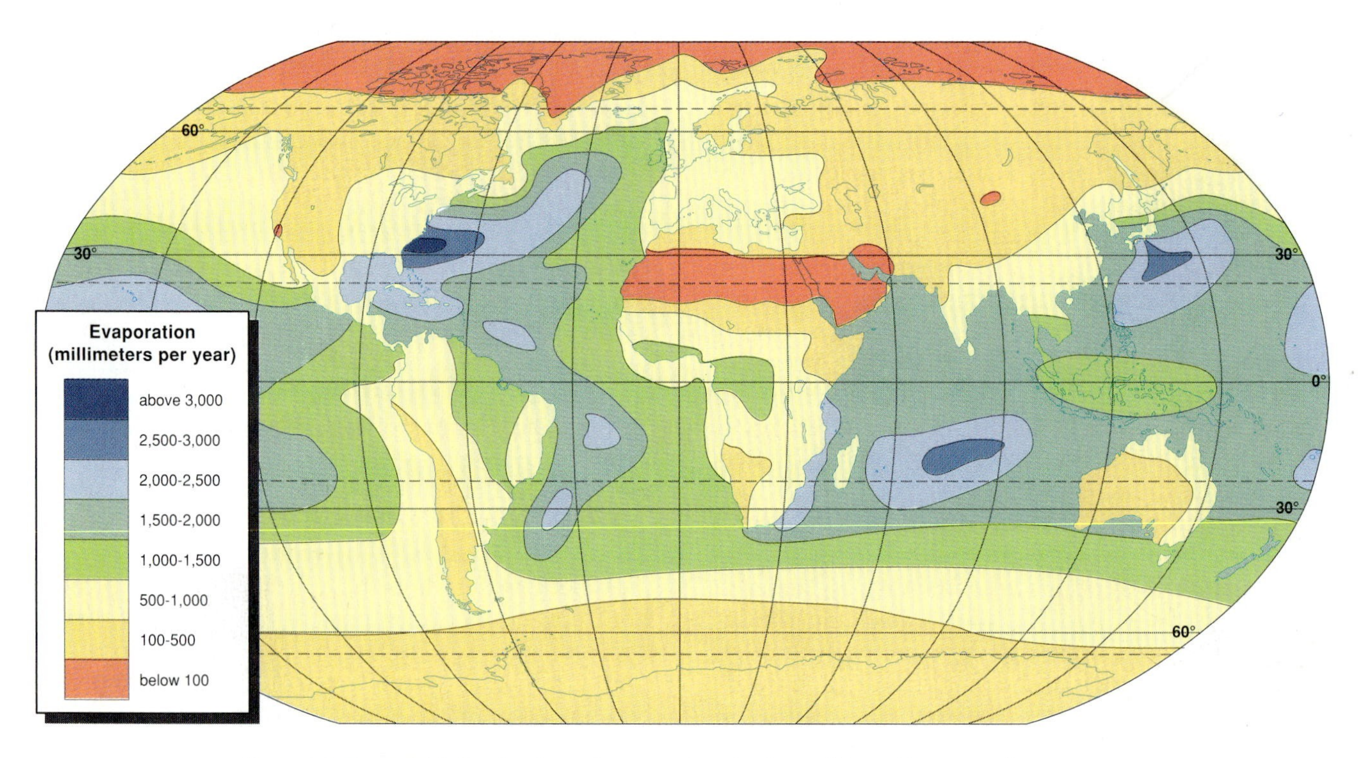

Figure 6-2 The world distribution of evaporation (estimated in millimeters of water per year). The highest rates are over warm ocean currents (Figure 5-2, *C*), and the lowest over continents.

WATER VAPOR IN THE ATMOSPHERE

Water vapor, like many gases, is invisible, but it affects personal comfort and plays an important part in the weather. The amount of water vapor in the atmosphere is known as the **humidity.** A combination of high temperature and high humidity slows evaporation from the skin and may lead to heat exhaustion (see Box 6-1: Humidity and Human Comfort). Water vapor is an important greenhouse gas, so humid air traps heat near the ground.

Humidity is measured in several ways, which relate the weight (or mass) of water vapor to the weight or volume of the air that contains it. **Absolute humidity** is the mass of water vapor per volume of air (grams per cubic meter). Figure 6-3 shows quantitatively that the amount of water vapor that can be held in a body of air increases with temperature. At 10°C (50°F), air can hold 10 grams (0.4 ounce) of water vapor per cubic meter, at 20°C (68°F) it can hold 17 grams (0.7 ounce), and at 30°C (85°F) it can hold 31 grams (1.25 ounces). It is noteworthy that the amount increases more rapidly at higher temperatures.

Relative humidity is the amount of water vapor in the atmosphere, expressed as a percentage of the total water vapor the air can hold at a particular temperature and pressure. The maximum amount of water vapor in the atmosphere depends on pressure as well as on temperature, but temperature variations are generally more significant to weather processes. At 20°C and sea-level pressures, 100 percent relative humidity is 18 grams per cubic meter; 50 percent humidity is 9 grams per cubic meter (Figure 6-3). The amounts of water vapor at a location decrease with altitude because both atmospheric temperature and pressure decrease with increasing height above the ground. Dehydration is a problem for mountain climbers.

Relative humidity is sometimes measured by comparing the temperatures of two thermometers, one of which has its bulb covered by a moistened wick (wet bulb), while the other is a conventional thermometer with no wick (dry bulb). The thermometer reading is lower on the wet bulb because water evaporating from the wick cools the thermometer. A person taking readings of these thermometers has to make sure air is passing over the wet bulb, since the water being evaporated from it could increase the local humidity around the bulb; some systems involve whirling the thermometers before taking the reading. The difference between the temperatures on the two thermometers is greatest when the relative humidity is lowest. When relative humidity is high and evaporation is balanced by the formation of liquid droplets, the dry bulb reading may be equal to that on the wet bulb.

BOX 6-1

HUMIDITY AND HUMAN COMFORT

Wind has its greatest effect on human comfort at low temperatures (Chapter 5); humidity has its greatest impact when high relative humidity is combined with high temperatures. Humidity is most significant in these conditions because air at high temperatures may contain more water vapor and because the body is attempting to lose heat by evaporative cooling in perspiration—in which the body's sensible heat is absorbed by evaporating water. If the air is hot but has a low relative humidity, sweating and evaporative cooling are effective, and the body feels more comfortable. When the relative humidity in summer rises over 80 percent, the air feels sultry because perspiration evaporates from the skin with difficulty; if the humidity rises a few percentage points higher, the air becomes saturated and it cannot take up any more moisture. In the southwestern United States, by contrast, the relative humidity is usually less than 50 percent despite the high temperature, and the air does not feel sultry, since evaporation from the skin is easier.

If high humidity prevents heat loss by evaporative cooling, the body core temperature may rise, leading to irritability, reduced ability to work, and in extreme cases, heat exhaustion, heatstroke, and death. It is estimated by the National Oceanic and Atmospheric Administration (NOAA) that some 150 people in the United States die each year as a result of illnesses induced by high atmospheric heat and humidity. Very low humidity can also bring discomfort if water is not available to replace fluids lost by sweating.

The National Weather Service reports a "heat index" to advise people of potentially dangerous weather. When high humidity is combined with high temperature, the "apparent temperature" rises, as shown on the table. At low humidities the apparent temperature may be less than the air temperature because of the evaporative cooling effect. Increasing the humidity of a room in winter increases the comfort level because less heat is lost from the body by evaporation.

However, once the relative humidity of air has risen above 50 percent at 21°C (70°F), or 30 percent at 35°C (85°F) or above, the apparent temperature exceeds the air temperature and discomfort increases. When the apparent temperature goes over 40°C (104°F), sunstroke or heat exhaustion is likely and prolonged exposure may end in heatstroke (hyperthermia).

When the air becomes hot and humid, relief can be obtained by going outdoors where a breeze may increase evaporative cooling. In hot countries people commonly have a siesta during the hottest part of the day.

RELATIVE HUMIDITY (%)	AIR TEMPERATURE (°F) 70	75	80	85	90	95	100	105	110	115	120
0	64	69	73	78	83	87	91	95	99	103	107
10	65	70	75	80	85	90	95	100	105	111	116
20	66	72	77	82	87	93	99	105	112	120	130
30	67	73	78	84	90	96	104	113	123	135	148
40	68	74	79	86	93	101	110	123	137	151	
50	69	75	81	88	96	107	120	135	150		
60	70	76	82	90	100	114	132	149			
70	70	77	85	93	106	124	144				
80	71	78	86	97	113	136					
90	71	79	88	102	122						
100	72	80	91	108							

PROLONGED ACTIVITY LEADS TO FATIGUE

PROLONGED ACTIVITY LEADS TO HEAT EXHAUSTION

HEAT EXHAUSTION LIKELY

HEAT STROKE IMMINENT

Humidity and temperature: the Apparent Temperature Index. This provides an indication of personal comfort based on temperature and relative humidity.

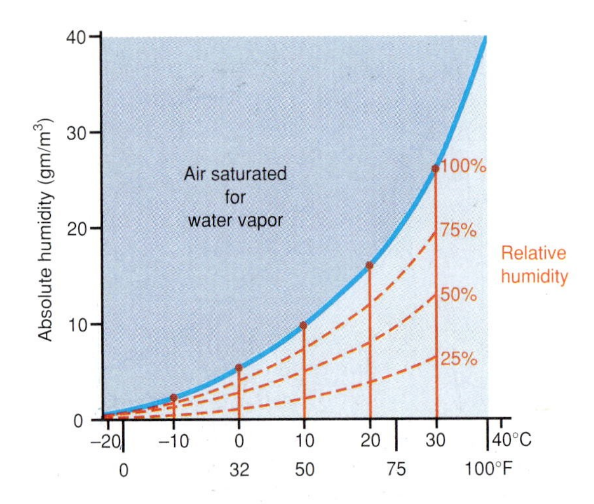

Figure 6-3 Water vapor in the atmosphere under varied temperatures. The curved graph line shows the maximum amount of water vapor that air can hold at each temperature, the *dew point.* Above the line the air is saturated and will not be able to add further water vapor without loss by condensation. Below the line the proportion of the maximum possible at each temperature gives the relative humidity.

CONDENSATION

Condensation Processes

The Basic Mechanism. Water vapor changes to liquid water by **condensation.** This process takes place on a large scale in the atmosphere and converts the invisible gas to masses of water droplets that form visible features such as clouds and fog. Condensation must occur before precipitation can take place, but additional processes are necessary to allow condensed water droplets to grow large enough to be precipitated.

The relative importance of evaporation and condensation depends on the amount of water vapor already in the air, together with the temperature and air pressure. Condensation becomes more important than evaporation once air becomes saturated with water vapor. If more water vapor enters the air after the saturation point is reached, some of the vapor condenses. Alternatively, if the temperature or pressure of the body of air is lowered, condensation exceeds evaporation. The maximum temperature at which a relative humidity of 100 percent is reached in a body of air with a given absolute humidity (the red lines in Figure 6-3) is known as the **dew point.** In a cooling air parcel, condensation begins to exceed evaporation when the temperature falls to the dew point.

Condensation is more rapid when the atmosphere contains large numbers of dust, smoke, or salt particles that provide solid surfaces on which condensation can take place. These particles form the centers of water droplets and are known as **condensation nuclei.** The most effective condensation nuclei are larger particles and those that have a special attraction for water; the latter are known as **hygroscopic nuclei.** Nuclei are 10 times more prevalent over land than over the oceans. Experiments in laboratories have shown that air without such nuclei may be cooled below dew point without condensation occurring—a state known as *supersaturation.* At the other extreme, condensation may take place with a relative humidity as low as 78 percent when large numbers of hygroscopic sea salt particles are present.

Large-Scale Condensation in the Atmosphere. Three main types of large-scale cooling and condensation occur in the atmosphere and produce fog (Figure 6-4) or clouds. First, the air is cooled when *advection* transports a body of warm air over a cold surface. The cooled air becomes saturated and fog forms as condensation takes place in the air above the cold surface. Second, *radiation* from the ground at night results in the loss of surface heat and rapid cooling of air close to the surface.

Third, a body of air at ground level may be forced to rise by *lifting* into air above it that is not only cooler, but of lower pressure. The rising body of air expands under conditions of lower pressure in the surrounding atmosphere, but the amount of energy in the parcel does not change. As the air body expands, this heat energy is spread through a greater volume of air, so the air temperature drops (Figure 6-5). The mechanism is known as **adiabatic expansion** (*adiabatos* is Greek for "impassable," indicating that heat energy is not lost from the rising body of air). The decreasing temperature of the rising body may lead to condensation and the formation of clouds.

Once water droplets have formed from water vapor, they continue to grow by further condensation. Eventually, they reach a maximum size for such droplets, but are still tiny compared to raindrops (Figure 6-6). The small size of the water droplets means that they are light enough to be held well above the surface by the upward air currents that have produced the adiabatic expansion, cooling, and condensation.

Figure 6-4 Fog at Golden Gate Bridge, San Francisco. Such fogs are a common feature of the Pacific coast at San Francisco and are caused by warm, moist air flowing over the cold ocean current offshore.

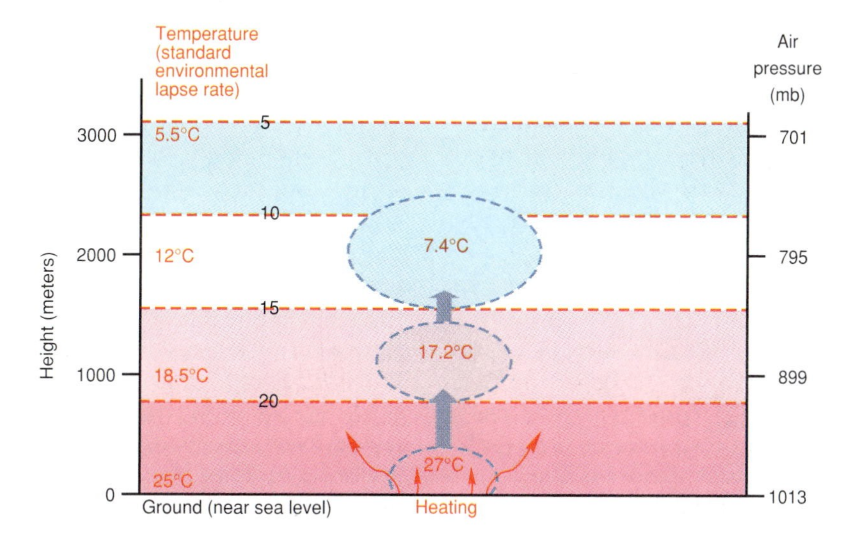

Figure 6-5 Adiabatic expansion and cooling of air. A body of air near the ground is heated, expands, and rises. When it reaches lower-density air 1 km above the surface, it has expanded and its temperature has dropped by 9.8°C—faster than the decrease in temperature in the surrounding air. By the time it gets to 2 km above the surface the rising body is markedly cooler than the surrounding air.

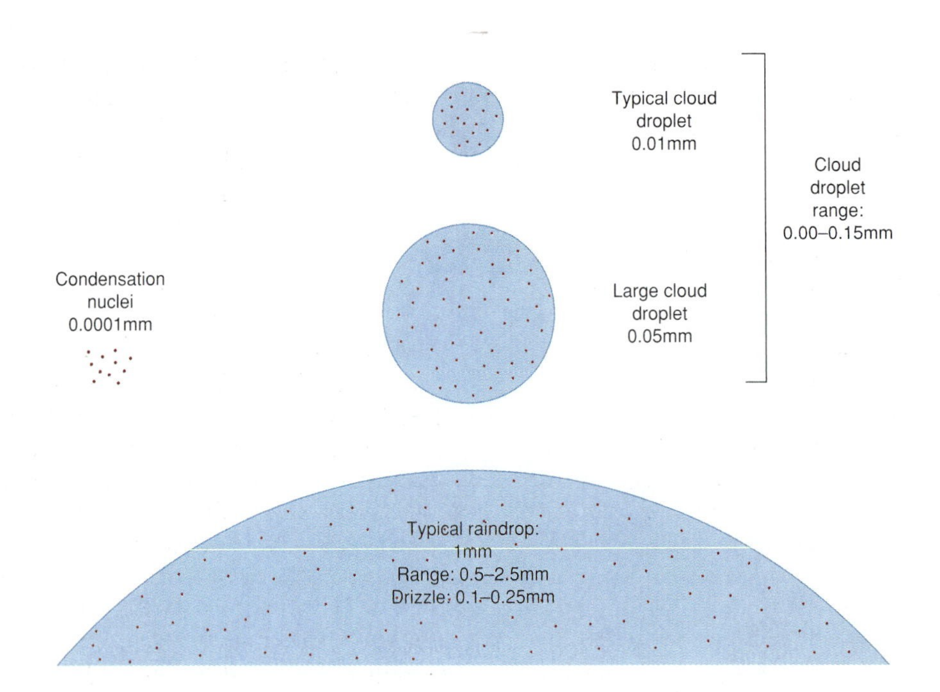

Figure 6-6 Rain drops, water droplets (cloud, fog), and condensation nuclei: a comparison of sizes (radii).

Dew, Frost, and Fog

Condensation in still air in contact with cooling ground at night produces **dew** (liquid droplets) or **frost** (frozen crystals) on the ground surface. Dew occurs on surfaces cooled below saturation temperature, or dew point. Condensation on walls and windows is a form of dew. Frost occurs on surfaces cooled below freezing point as feathery crystal patterns on glass, or fernlike *hoar frost* on grass and trees. Frost is not the same as the solid freeze feared by Florida citrus growers, since the latter affects water inside the plant tissues and often causes crops to die.

When there is a slight movement in the air above the ground, cooling may be carried upward for several tens of meters. Condensation in this layer in contact with the ground produces fog. In the official meteorologic definition, **fog** reduces surface visibility to under 1 km; dense fogs may be a hazard to motorists, seagoing vessels, and landing airplanes. Most fogs are caused by advection or radiation.

Advection fog is produced when warm, humid air passes over a cold surface that has a temperature lower than the dew point of the moving air. The coastal fogs of the San Francisco area of California and the fogs over the Grand Banks of Newfoundland both occur where warm, moist air flows over cold ocean currents. Advection fogs are common over land in winter if warm, moist air moves over a snow-covered landscape. Another type of advection fog occurs in upland areas when air cools on the upper slopes, becomes denser than its surroundings, and flows into a valley bottom (Chapter 5). There the cold air mixes with the moist air above a stream, lowers its temperature, and causes condensation. The wind velocities associated with advection fog are at most 20 to 30 km per hour (12 to 20 mph). Greater wind speeds cause increased turbulence that lifts the mass of condensed moisture off the ground as low clouds.

An *upslope fog* is somewhat similar to advection fog. It occurs when moist air is blown upslope, as from the lowlands of Manitoba, Canada, to the higher Alberta Plains. If adiabatic cooling is sufficient, saturation occurs, and a moving fog forms.

Radiation fogs are most likely to form during calm atmospheric conditions, such as occur in the fall and winter seasons of eastern North America or in the winters of the Central Valley in California. Warm days give rise to evaporation into the lower layers of air, and the clear night skies provide a long period of cooling. Radiation from the surface at night produces cooling of the ground, and the ground in turn chills the lowest few centimeters of the atmosphere by conduction. With a light breeze, condensation spreads through the lowest 15 to 100 meters (50 to 300 feet). At wind speeds greater than 10 kmph, the cooled air is mixed with warmer air above and condensation is less likely to occur. The radiation mechanism works slowly and may require several hours to produce the initial condensation. Once begun, the condensation spreads rapidly and visibility decreases. The following morning the sun quickly warms the atmosphere and the fog evaporates.

Fog-forming conditions in urbanized areas are often associated with pollution, and the combination is known as *smog* (see Box 3-2: Pollution of the Atmosphere and Oceans). Such conditions may be caused and then maintained by the presence of a temperature inversion. This is most common when cool, moist air moves over a warm surface beneath warm, overlying air and causes advection fog: the inversion lasts longer in these conditions than when it is due to radiation at night.

Clouds

Clouds are masses of water droplets, ice particles, or both suspended in the atmosphere above Earth's surface. Nearly all clouds form as the result of adiabatic expansion in a rising body of air, although advection may bring about low clouds if the winds are strong enough to create lifting by surface air turbulence. When viewed from the ground, clouds appear to be solid and stable masses, but a time-lapse film of clouds moving across a landscape reveals that they are in a constant state of change. Water vapor in air rising underneath a cloud condenses as it enters the cloud. Such condensation may be balanced by evaporation into the drier surrounding air at the tops and sides of the cloud. Thus, clouds are dynamic features and exist only as long as condensation exceeds evaporation. Study of the factors that cause the air body to rise, of the level of humidity in the rising air, and of the conditions in the surrounding air leads to an understanding of the different cloud shapes observed.

Atmospheric Stability and Instability. A variety of conditions lead to the uplift of bodies of air. Air is forced to rise when it is heated at the surface, converges with other air or meets a mountain range, or when cold air pushes in underneath. Heating of air at the surface causes a bubble to rise by convection. A slow-moving or stationary mass of dense, cold air acts as a sort of wedge, forcing faster-moving warm air to climb over it (see Figure 7-5). All these processes force a body of air to rise through the overlying atmosphere and undergo cooling by adiabatic expansion. They frequently act in concert, reinforcing the uplift; a mountain range or convectional uplift may coincide with a cold wedge of air.

The rising body of air interacts with the surrounding air in either of two ways. First, it may rise so far and then stop, even falling back toward earth. When rising air becomes cooler and denser than the surrounding air and falls back, this is termed the condition of **stability.** Any cloud that develops under these conditions will be limited in height, possibly forming a thin layer or small, isolated heap cloud. The second possibility is that the rising air body remains warmer and less dense than the surrounding air, in which case it continues rising. This condition is known as **instability** and increases the likelihood of condensation and the formation of tall clouds. The conditions of stability and instability control the nature and form of clouds.

In the atmosphere as a whole, temperature generally decreases with height from the surface to the tropopause at the environmental lapse rate (see Figure 4-10). It averages a decrease of 6.5°C per kilometer (3.5°F per 1000 feet), but this varies with time of day and surface heating and cooling. On a cloudless day, daytime heating increases the temperature of the lower air layers more than those higher up, and so increases the environmental lapse rate. Conversely, with clear skies at night, cooling by radiation near the surface lowers the temperature of the lower layers and decreases the environmental lapse rate. If the surface cooling is very intense or warm air flows above the lower layer, the environmental lapse rate may be composed of a lower section in which the temperature decreases with height, overlain by a section in which it increases, and then another where it decreases again; this causes an inversion in the lapse rate.

A rising body of air must negotiate its way through the atmospheric environment, and at each level its progress depends on the relationship between its temperature and that of the surrounding atmosphere. The rate at which temperature decreases *within* a rising parcel of air experiencing adiabatic expansion is different from the rate in the *surrounding* atmospheric environment. The **adiabatic lapse rate** is the rate at which temperature decreases in a rising body of air as it expands and its heat is redistributed. It is constant at 9.8°C per kilometer (5.4°F per 1000 feet) for unsaturated air. As noted, the environmental lapse rate is more variable, but averages 6.5°C per kilometer. The adiabatic lapse rate of 9.8°C per kilometer (approximately 1°C per 100 meters) means generally that the temperature of a rising body of air decreases more rapidly than that of the surrounding air and soon reaches a level where it is colder, and therefore denser, than the air around it. At that point, the air body stops rising and falls back toward earth (Figure 6-7, *A*). This stable sequence precludes the growth of large clouds and precipitation. However, when

Figure 6-11 Cumulonimbus clouds over the Mekong delta, Vietnam. The combination of strong heating and surface water gives rise to impressive vertical growths in these clouds, as seen from the Space Shuttle.

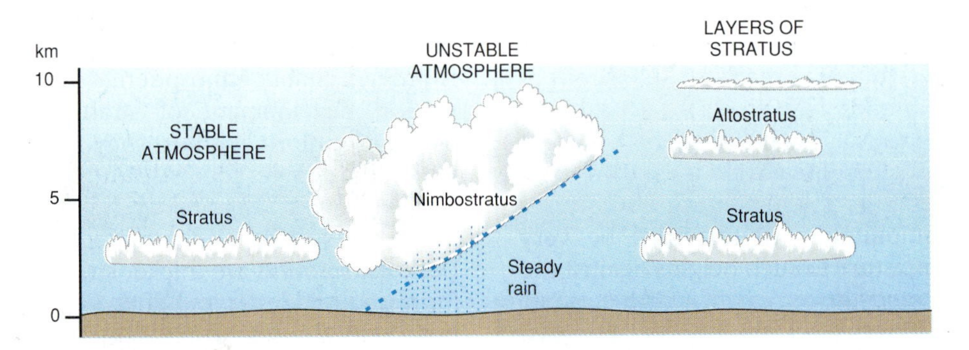

Figure 6-12 Stratiform clouds. Layers of stratus are formed by widespread ascent of warm, moist air over cold air or by the decay of cumulonimbus clouds.

tion. The crystals fall until they sublimate (change directly from ice to water vapor).

Prediction of cloud formation and growth—an important facet of weather forecasting—relies on a knowledge of conditions in the atmosphere several kilometers above the surface. Meteorologists launch balloons to measure conditions in the upper troposphere and radio back readings of temperature and humidity. Such balloons reach heights up to 18 km (11 miles), where they burst because pressure within them exceeds that of the surrounding air by more than they can withstand. Weather satellites are also used for acquiring such data.

The upper troposphere readings are used to calculate environmental lapse-rate patterns, so that meteorologists can predict how bodies of rising air will react when they reach condensation levels. For example, if an inversion layer is present, it may provide a "lid" that prevents further upward movement of air until powerful updrafts break through and give rise to violent weather.

PRECIPITATION

Types of Precipitation

Precipitation is the deposition of atmospheric water on Earth's surface. Over most of the globe this occurs as rain (Figure 6-13), but the water falls as snow in colder regions. Water drops fall as **drizzle** or **rain,** defined by drop size (Figure 6-6). In cold conditions some of the rain freezes at the surface to give **freezing rain. Snow** falls as ice crystals grouped into flakes, the sizes of flakes depending on the water vapor available during crystal growth. Small flakes are more common in very cold conditions, while in warmer conditions wet snowflakes stick together and may have diameters up to 5 cm (2 inches). Other forms of precipitation include hail and sleet. **Hail** is rounded or jagged lumps of ice that may have a structure of concentric layers like an onion. **Sleet** consists of raindrops that have frozen as they fall through very cold air. Sleet and hail bounce, freezing rain does not.

The size of precipitation, from drizzle to raindrops and hailstones, determines the rate of fall and thus the intensity of impact at Earth's surface. Very heavy rain with large drops falling at rates of several hundred millimeters (a few feet) per second washes away surface soil when it hits the ground. Light drizzle has small drops that may fall for several hours and have no such effect.

Precipitation types can be linked to the clouds from which they fall. Drizzle often falls from stratus clouds. Rain and snow fall from nimbostratus and cumulonimbus clouds. Hail falls exclusively from cumulonimbus clouds that have the strong convective updrafts needed to carry the hailstones up and down and so add layers of ice. Rain does not fall from cirrus clouds because they are too thin and the ice sublimates to water vapor when it falls.

Mechanisms of Precipitation

Cloud droplets stay above the ground because upward air currents prevent them from falling. Growing cloud droplets to a size that will be precipitated requires special mechanisms. There are two complementary theories to explain the physical processes that produce precipitation (Figure 6-14). The processes often work together, but involve different ways of growing the droplets.

Figure 6-13 Rain falling from a cumulonimbus cloud over Nag's Head, North Carolina.

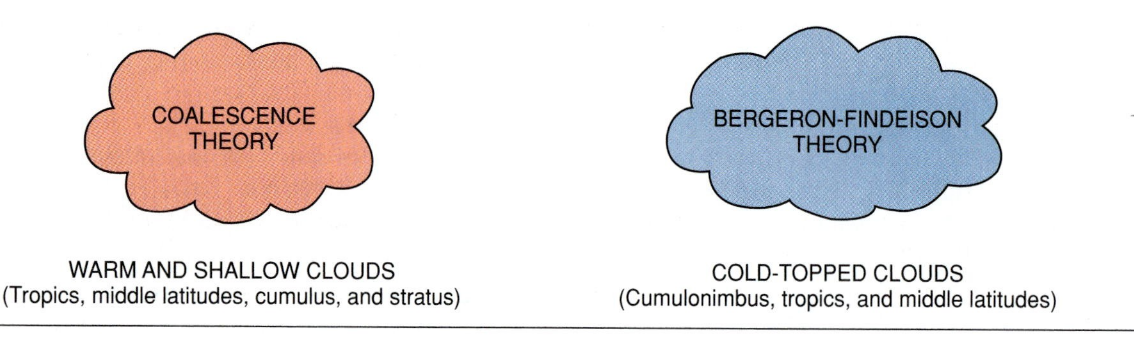

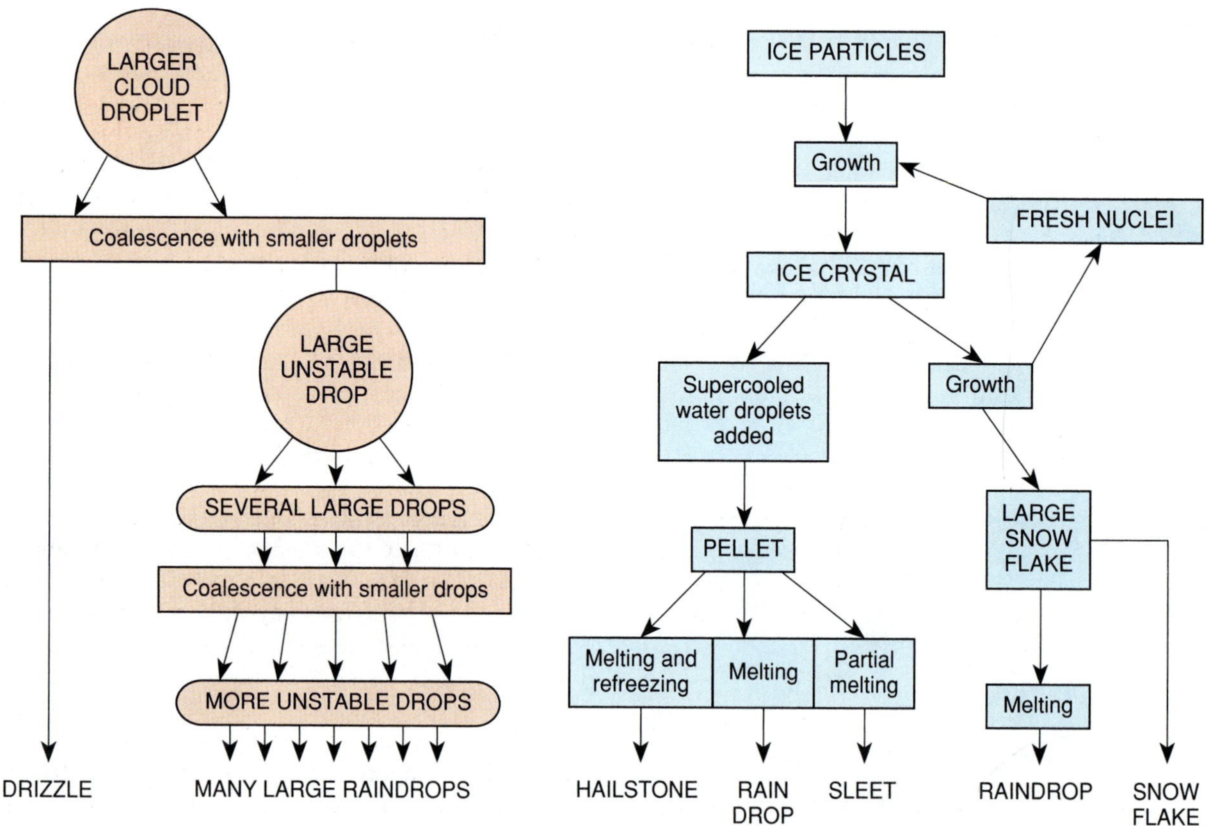

Figure 6-14 Theories of precipitation compared. The two main theories in terms of the processes involved in the production of precipitation.

The **Bergeron-Findeison theory** is named after the Swedish meteorologists who developed it in the 1930s. This theory depends on the growth of ice crystals in the tops of clouds to form particles that are large enough to fall to the ground. When cloud temperatures are between 0°C and −20°C, water occurs in a mixture of gaseous, liquid, and solid forms. Liquid water existing at subfreezing temperatures is said to be **supercooled.** It is common in clouds when there is a shortage of condensation nuclei or freezing nuclei. Freezing nuclei include tiny particles of clay, but are less common than condensation nuclei. More importantly, these special freezing nuclei function only at temperatures below −9°C (16°F). Once ice particles begin to form in the top of a cloud at around −10°C, they grow rapidly by attracting supercooled water droplets and uncondensed water vapor. By the time the temperature has fallen to −20°C (7°F), all the cloud droplets have frozen. The ice crystals may splinter to form additional particles, which also grow until they are heavy enough to fall. On falling, ice crystals often combine in loose aggregates (snowflakes), which melt to raindrops at the 0°C level in the atmosphere. The Bergeron-Findeison theory can be applied to clouds in which the upper layers are below the freezing point: that includes most cumulonimbus clouds outside the tropics, and the tallest cumulonimbus in the tropics.

The second precipitation process is described by the **collision-coalescence theory.** Coalescence (fusing together of water droplets) occurs in clouds where the lowest temperatures do not permit the formation of ice, as well as in those where it does. It is particularly common in clouds with prolonged updrafts of warm and moist air, such as clouds of moderate height above the tropical oceans. It also occurs in stratus clouds in the tropics and midlatitudes. Water droplets in these clouds range to about 0.05 mm in radius, the size of the drop being related to the size of the condensation nucleus. The larger droplets collide and coalesce with slower-moving smaller droplets that are carried by updrafts or downdrafts within the cloud. Smaller droplets are also drawn into the wake of the larger droplets.

Most of the raindrops produced by collision-coalescence are small (drizzle), but large drops may form in tall clouds, especially when the ice formation in the upper cloud is combined with collision-coalescence lower down. As the snowflakes and hailstones drop, they melt and grow by collision-coalescence to form the largest raindrops. The largest drops become unstable and break up as they fall. Smaller clouds without ice in their upper parts usually produce smaller drops up to a maximum of 2 mm diameter because there are no ice particles in the cloud, and collision-coalescence acts on its own. However, in exceptional conditions, such as occur in strong updrafts over Hilo, Hawaii, warm clouds may produce raindrops 4 to 5 mm in diameter.

Measuring Precipitation

Precipitation is measured by depth of rain or snow, or by depth falling over a period of time. Rain is gathered in a rain gauge, which is a cylindrical container with a funnel-shaped opening at the top. The volume of water collected is measured as millimeters (or inches) collected in a standard measure. Rain gauges used by meteorologists have mechanisms for automatically recording and storing information about the volume of rainwater falling within a specific time period.

Snowfall depths are expressed in either of two ways, which yield different values for the same amount of snow. To compare snowfall to rainfall (as in the precipitation map in this chapter), the depth of snow covering an area is assessed and converted to an equivalent of liquid water. Typically, the value measured for frozen snow is 10 times greater than the liquid equivalent for the same amount of water. The snowfall measurements usually reported and forecasted to the public, on the other hand, are unconverted measurements of the actual depth of fallen snow, or the depth of snow drifts where wind has piled the snow higher.

Where a recording site has poor access—as in the Cascade Mountains east of Seattle—automatic rain and snow-depth gauges are linked to transmitters that send the information to satellites, which then radio it back to ground receiving stations.

The World Distribution of Precipitation

The world map (Figure 6-15) on pp. 148-149 shows that precipitation is not evenly distributed around the globe. Sufficient precipitation is essential to the survival of living organisms on the continents, and so the geographic distribution of precipitation is of vital significance. Zones of high and low precipitation can be distinguished and are a major factor in classifying climatic types (Chapter 8). Precipitation also varies in quality as well as in quantity (see Box 6-2: Acid Rain).

Areas of High Precipitation. High precipitation totals occur in the zones of surface air convergence near the equator (i.e., along the intertropical convergence zone) and along the polar front. In these zones moist air is forced to rise, producing bands of clouds and precipitation (Figure 5-11). At least two thirds of the world's precipitation falls on the tropical zone. The combination of high temperature, high rates of evaporation into the trade winds as they blow over the oceans, and high rates of evapotranspiration over tropical rainforests produces great quantities of water vapor in the low-latitude atmosphere, and this is turned into precipitation. In summer, this zone is drawn northward into southeastern Asia in an area known as the monsoon lands. Heavy rainfall is caused by moist air being drawn over the land from the oceans (Chapter 8). The second zone of clouds and rain occurs along the midlatitude polar front, where warm, moist air is forced to rise over denser cold air.

Variations occur within these two zones of high rainfall. One distinction is between the oceans and continents. Cumulus clouds over the tropical oceans produce precipitation by collision-coalescence when they grow to only 1 to 3 km deep from cloud base to top, and so rain falls over the oceans in frequent but small amounts. Cumulonimbus clouds also occur as parts of weather systems over tropical oceans and produce heavier rains. Precipitation over the land is less regular because there may be less water vapor in the atmosphere, and convectional uplift relies more on local conditions of heating, moisture supply, and wind. Over the continents a greater proportion of precipitation

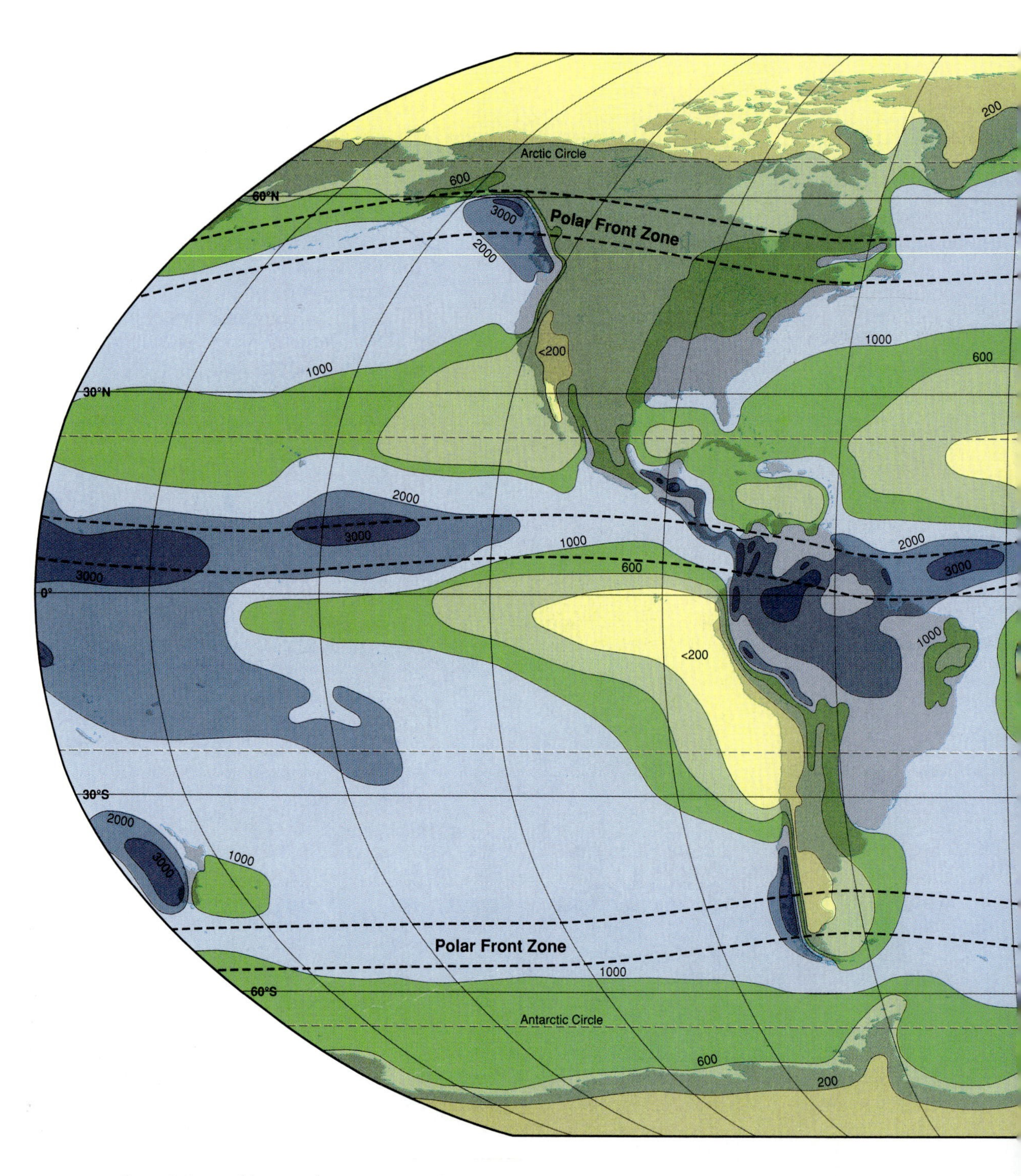

Figure 6-15 World map of average annual precipitation. The intertropical convergence zone and the polar fronts are indicated to emphasize their links to the main zones of precipitation.

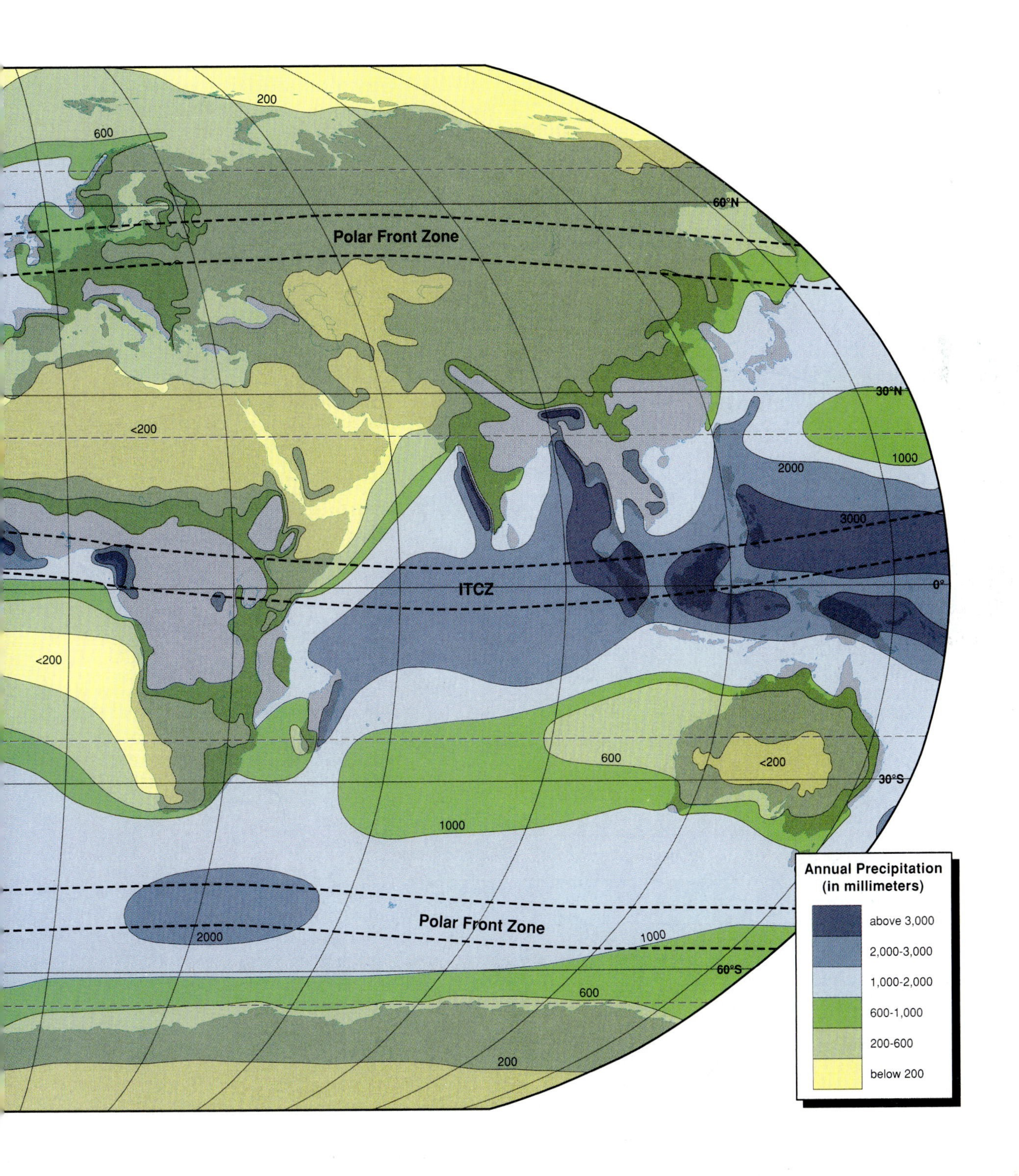

200
600
60°N
Polar Front Zone
30°N
<200
1000
2000
3000
ITCZ
0°
<200
600
<200
30°S
1000
Annual Precipitation (in millimeters)
above 3,000
2,000-3,000
1,000-2,000
600-1,000
200-600
below 200
Polar Front Zone
2000
1000
60°S
600
200

BOX 6-2

ACID RAIN

Acid rain is precipitation that has an enhanced acidity. It is often known as "wet acid deposition," since its greatest impact is on the ground in soil and water.

Acidity is defined by the number of hydrogen ions present in a solution and is measured by the pH scale, where a value of 3 denotes one part of hydrogen in a thousand ($1/10^3$) and 7 denotes one part in 10 million ($1/10^7$). The scale is shown in the accompanying table. Neutral pH is 7.0, higher values are alkaline, and lower values are acidic. Natural rainfall has an average acidity of 5.5 because of the carbon dioxide dissolved within it, which makes it a weak acid. Locally, rain may vary from a pH of 4.0 after a volcanic eruption to 7.5 downwind of a cement factory, where calcium carbonate dust in the atmosphere makes rain alkaline. Acid rain is defined as rain with a pH of less than 5.5 where human activities have contributed to the acidity. Values for the northeastern United States are shown on the map.

Acid rain is caused by emissions of sulfur and nitrogen compounds into the atmosphere by the combustion of fossil fuels, particularly in power stations or vehicle engines. The sulfur and nitrogen released in these processes unite with atmospheric oxygen in the presence of sunlight to form compounds called oxides. The oxides may be deposited as solid particles near the source in a process known as dry deposition. Farther from the source, the oxides remaining in the atmosphere are converted by further chemical reactions into sulfuric and nitric acids. These are deposited by precipitation up to several hundred kilometers downwind. The acidity downwind of manufacturing and electricity-generating areas has increased in recent years. Tall power station chimneys, built as an initial response to air pollution in the early 1970s, reduced local dry deposition of oxide compounds but increased distant wet deposition of acid rain.

Acid rain has been blamed for a variety of effects, including fish kills in upland lakes, loss of tree leaves, destruction of building materials, and hazards to human health. Of these, the effects on lakes in upland areas such as the Adirondacks in upper New York State and parts of eastern Canada and Scandinavia have been most obvious. The thin acidic soils of these areas transmit the additional acid from rain and snow to the streams and lakes. It is more difficult to directly link the effects on trees to acid rain, since other chemical processes may be involved, and the role of acid rain in combination with these processes is not fully understood. Other human activities, such as farming practices that do not replace soil nutrients, and the planting of evergreens for lumber, may also enhance acidification. It is thought that the most significant effect of acid rain may be on crop yields, but few data are available.

Acid rain has become a public issue, especially for people subject to deteriorating environmental conditions downwind of sulfur and nitrogen emissions. It has even affected international relations between the United States and Canada because Canadians living in eastern upland areas blame the power utilities of the Ohio Valley and the northeastern United States for increasing local acidification in their area. In Europe some smaller countries have established international confer-

pH Value		Common substances	
14	STRONG ALKALI	Caustic soda	
13			
12			
11		Household ammonia	
10		Soap solution	
9		Sea water	
8			
7	NEUTRAL	Distilled water	
6		Milk	
5		Pure rain	ACID RAIN
4		Wine	ACID RAIN
3		Orange juice	ACID RAIN
2			
1	STRONG ACID	Battery acid	

The pH scale.

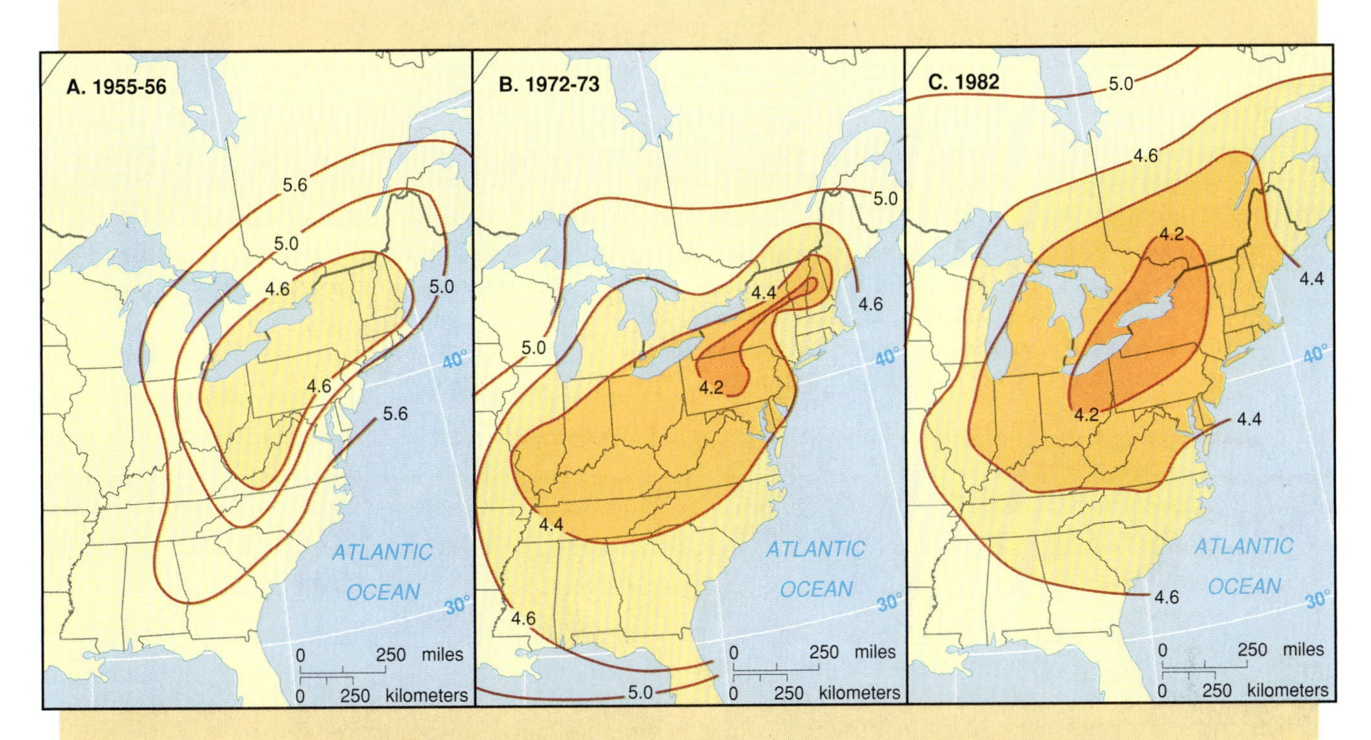

The distribution of acid precipitation (pH values) in eastern North America, 1955-1982, showing the increase in intensity and geographic coverage (A,B, Park after Likens; C, U.S. National Atmospheric Deposition Project.)

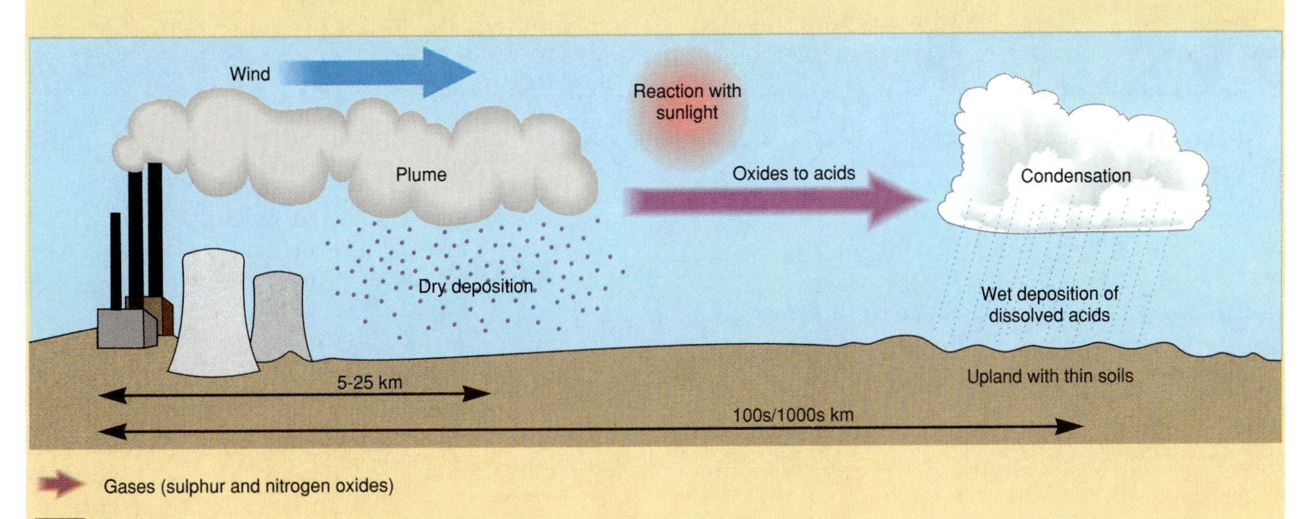

The processes that give rise to acid rain.

ences to highlight the issue and to demand government action by the main source countries like Britain, Germany, and the eastern European states.

Acid rain's status as a public environmental issue has led to the passage of regulations aimed at reducing it in several nations. The main thrust of these regulations is to reduce emissions of contributing substances, either by using low-sulfur fuels or by adding filters to remove sulfur from the emitted fumes. Such expenses can be borne in developed countries, to which acid rain has been largely confined up to the early 1990s. However, the increase of industrialization and vehicle use in developing countries may spread acid rain to other parts of the globe and be more difficult to control because of the costs involved.

comes from tall cumulonimbus clouds because of the intense convectional uplift resulting from surface heating.

Areas of exceptionally high-intensity precipitation occur where moist trade or monsoon winds blowing from the oceans are forced to ascend over mountain ranges. For this reason the coastal ranges of the northern Andes and Central America (facing trade winds), and the hills of southwest India and southeast Asia (facing monsoon winds from the southwest in summer) stand out on precipitation maps.

In the midlatitudes rain is produced where warm, moist air is forced to rise over cold polar air. The highest totals occur where air, already rising for this reason, has to cross mountains. The rising air converts water vapor to deep clouds and heavy precipitation as it rises over mountainous coasts in the northwestern United States, western Canada, southern Alaska, and northwestern Europe. East coasts in midlatitudes have their highest totals when warm, moist air is drawn into the continents from the oceans in summer.

Areas of Low Precipitation. Low precipitation is one of the main factors causing **aridity,** which is a lack of surface water. Aridity occurs where there is an excess of evapotranspiration over precipitation and where other sources of water cannot compensate for the deficit that results. Low precipitation is characteristic of continental subtropical areas beneath the high-pressure zones in which air is descending and diverging at the surface. The Sahara, the western half of Australia, and southwest Africa are prime examples of low precipitation beneath subtropical highs. The southwestern United States and the coasts of Peru and northern Chile are narrower zones where such arid conditions are restricted in area by mountain ranges. The aridity of these coastal areas and of the ocean offshore is often enhanced by cold surface ocean water; the cold air above the cold water decreases the environmental lapse rate and increases stability, making rain less likely.

In midlatitudes the driest areas are in the interiors of continents. Heavy precipitation on the coastal areas leaves little water vapor in the air passing through the interiors of large continents distant from the oceans. Such air has a low absolute and relative humidity. The low relative humidity may be enhanced by descent on the eastern side of mountain ranges standing in the way of the westerly flow of air in these latitudes—as in Nevada (see Figure 6-8) or the Great Plains area east of the Rockies. Low precipitation on the lee sides of mountain ranges causes the places to be known as **rain-shadow** areas. The Patagonian desert in southern Argentina is another example.

Low precipitation totals are generally a characteristic of high-latitude polar regions as the result of a combination of little surface water (limiting evaporation), low air temperatures (reducing the amount of water vapor in the air), and outflowing cold air at the surface (preventing the entry of moist air from lower latitudes).

Variability in Precipitation Over Time

Precipitation varies not only from place to place, but also in its timing. It is not always produced in the same amounts at the same place from year to year or from season to season. This variability is particularly significant for human activities. Fluctuations in precipitation have attracted most publicity in areas such as the Sahel desert-margin zone along the southern edge of the Sahara in Africa, where slight changes in rainfall have a dramatic impact on human life.

Precipitation may vary over periods of several years, so that groups of wet and dry years can be identified. The graphs for coastal areas of northern and southern California in Figure 6-16 show periods of wetter years from 1903-1918, 1937-1944, 1967-1974, and 1979-1985. In between were periods of lower-than-average precipitation, giving southern California annual precipitation totals below 250 mm (10 inches) for several years. In contrast, the northern Californian totals seldom fell below 750 mm (30 inches), and the dry periods that occurred did not differ from the mean as much as dry periods to the south did.

Precipitation may also vary from season to season. This is particularly noticeable in the tropics where seasons are defined most clearly by variations in precipitation, whereas midlatitude seasons are defined by variations in temperature. The seasonal shift of the intertropical convergence zone with the position of the vertical sun takes the belt of precipitation north of the equator in June and south of the equator in December. However, the amount of rain in the rainy seasons of these regions varies considerably from year to year, especially over land. Seasonal precipitation also occurs in midlatitudes. On the west coasts of continents there is usually a winter maximum of precipitation because that is the season when the storms bringing rain from the ocean are strongest. The interiors and east coasts, however, tend to have summer rainfall maxima because of heating of the land that draws moist air in from the oceans.

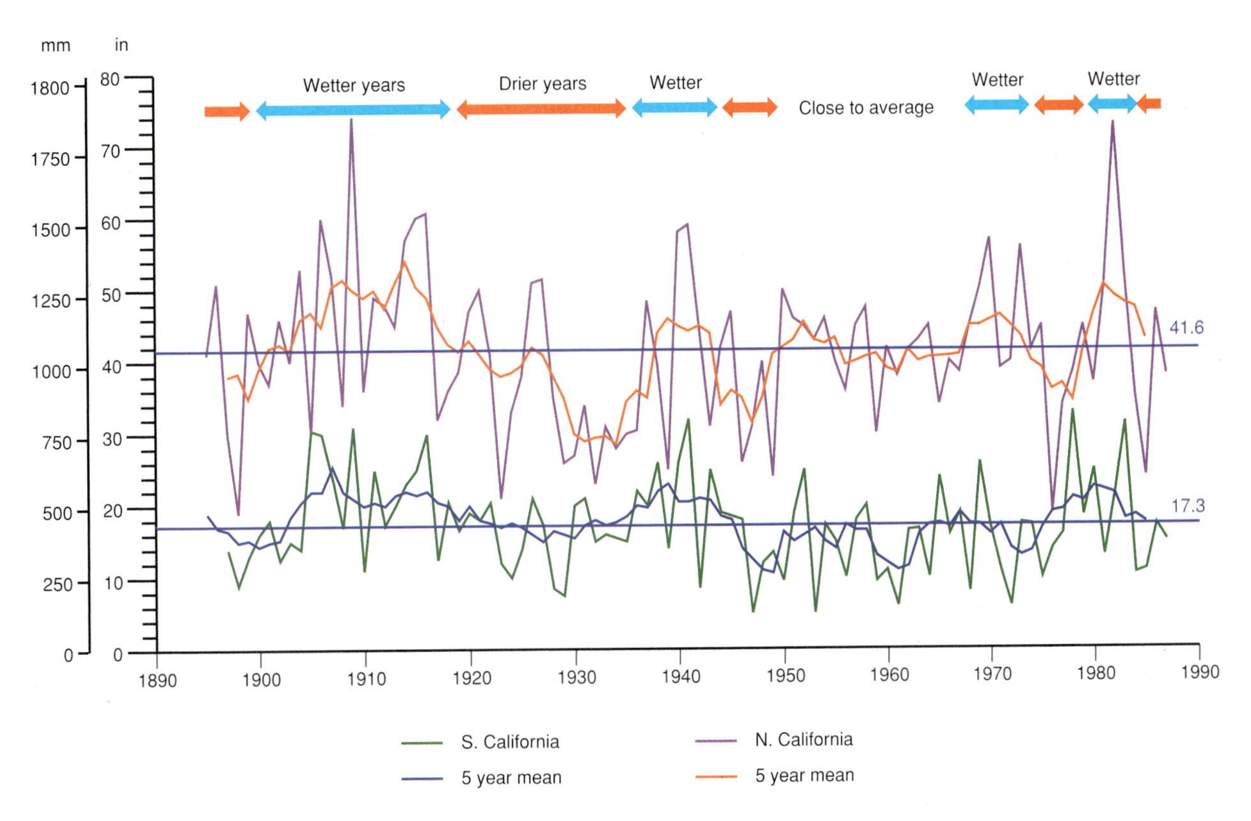

Figure 6-16 Annual precipitation, 1895-1989: northern and southern California compared in graphs. The 5-year mean graph has been drawn to emphasize the regularity of fluctuations of groups of wetter and drier years.

> LINKAGES
>
> *The distribution and timing of precipitation over the globe have significant effects on related processes in most Earth environments. Within the* ***atmosphere-ocean environment,*** *precipitation is linked to the release of latent heat in condensation and to the general circulation of the atmosphere; it is also linked to the sea-surface temperatures of the oceans and their impact on environmental lapse rates. Precipitation is a prime factor in supplying agents of erosion (streams, glaciers) in the* ***surface-relief environment.*** *It is also a major determinant of the type, density, and productivity of plants and animals in the* ***living-organism environment.*** *The tropical rainforests in particular play a major role in providing evapotranspiration to fuel further condensation and precipitation in the air above.*

Prospects for Rainmaking

Rain and snow are so important for human activities that attempts to cause or increase precipitation have figured in traditional cultures through the ages. Modern attempts to make rain are based on a scientific understanding of the processes at work. Those that make use of a fuller comprehension of cloud physics seem to have achieved some success. However, no attempt so far has definitely caused rain to fall in a particular place and at a particular time. One of the problems faced by all rainmaking projects is that the scale of atmospheric movement is very large, and the conditions that give rise to precipitation are extremely sensitive to small variations in temperature and humidity. This makes it difficult, if not impossible, for humans to manipulate the atmospheric environment.

The two main approaches used in recent years are related to the theories of precipitation processes. The Bergeron-Findeison theory of ice formation in clouds has persuaded scientists to seed clouds containing supercooled water droplets (at $-5°C$ to $-15°C$) with freezing nuclei such as silver iodide or dry ice (solid carbon dioxide). These substances are used because they have similar crystalline structures to ice, and it is expected that ice will form around them. The dry ice is released from airplanes flying through suitable clouds, and the silver iodide from planes or ground-based burners sited beneath strong upward currents.

The second method, known as "warm seeding," is based on the collision-coalescence theory. A cloud is sprayed with liquid containing hygroscopic nuclei such as ammonium nitrate. Within a minute increased condensation causes droplets to grow from 0.02 to 0.05 mm in diameter. The collision-coalescence process leads to raindrops of 5 mm in diameter being formed within 20 minutes.

During the California drought of 1976-1977, cloud seeding was attempted in the northern part of the state. In July 1977, a contract for $127,000 was awarded by the state government to try to increase rainfall. The ultimate goals of the rainmaking were to raise groundwater levels, to provide winter runoff, to reduce fire hazards, and to increase water supply in northern California. Southern California would also benefit from this effort in the north, since it depended on water transported from the north.

From July 20th through September 28th, 47 aircraft-borne cloud-seeding operations were carried out, and it was judged that 20 of these resulted in some increased precipitation. Another contract for $289,000 was awarded in December 1977 to initiate cloud seeding in the far north of California as part of an emergency drought relief program. Air heated by 44 flame chambers carried silver iodide aloft from hilltops. Three aircraft also participated. On this occasion the operations had only just begun when heavy rains set in that were caused by a natural weather system. The program had to be suspended to avoid potential legal claims following floods and avalanches.

Environmentalists have expressed concern about the pollution of rainwater if silver iodide is used extensively as a seeding agent, since it is brought down to earth in raindrops. A bill was passed in Pennsylvania to prohibit its use, but analyses have shown no significant changes in silver iodide concentrations in grassland where the chemical has been used in cloud seeding. If cloud seeding becomes operational (as opposed to the current experimental phase), the public will have to be involved in decisions about the extent to which such programs will be pursued.

There are other potential problems. Scientists have tried several approaches to alleviate major water shortages in the lower Colorado River basin. Some of the projects have focused on increasing the precipitation on the mountain ranges that supply the river's headwaters, rather than on trying to create rain in the dry areas where the proper cloud conditions rarely occur. Unfortunately, increased snowfall on these mountains creates a greater avalanche hazard to winter sports centers in the region.

WATER BUDGETS

The processes that make up the hydrologic cycle also influence the natural availability of water at a particular place. The inputs of precipitation and the outputs of evaporation, transpiration, flow through soils, and flow over the land determine whether a given locale will have a surplus or deficit of water. Calculations of these inputs and outputs are important in estimating domestic and industrial water supply, flow in rivers, flooding, and the need to irrigate crops.

Over Earth as a whole, precipitation and evaporation balance at around 880 mm (35 inches) of each per year. However, there is an excess of evaporation over the oceans, and of precipitation over the continents. A simple equation:

$$\text{Evaporation (E)} = \text{Precipitation (P)} + \text{Surface runoff (R)},$$

works out as follows for the two:

$$\text{For oceans, } E = P + R = 1200 + 100 \text{ mm} = 1300 \text{ mm}$$

$$\text{For continents, } E = P - R = 710 - 240 \text{ mm} = 470 \text{ mm}$$

Note: 240 mm of runoff from the continents equals 100 mm of river water entering the oceans because of the different areas of the continents and oceans.

The inputs and outputs of water at a particular location on Earth's surface may be viewed as a budget (Figure 6-17). Precipitation provides the inputs, which have to be compared against immediate losses by evapotranspiration. If precipitation exceeds evapotranspiration, the extra water goes into soil and rock storage underground; the process of replacing water in soil and rock is known as *recharge*. If the ground storage is full, the *surplus* water will run off over the surface into streams. If evapotranspiration exceeds precipitation, water is extracted from the ground store by plant roots faster than it is replenished or is evaporated from the surface. This continued *usage* of water can only occur as long as there is water available in surface water

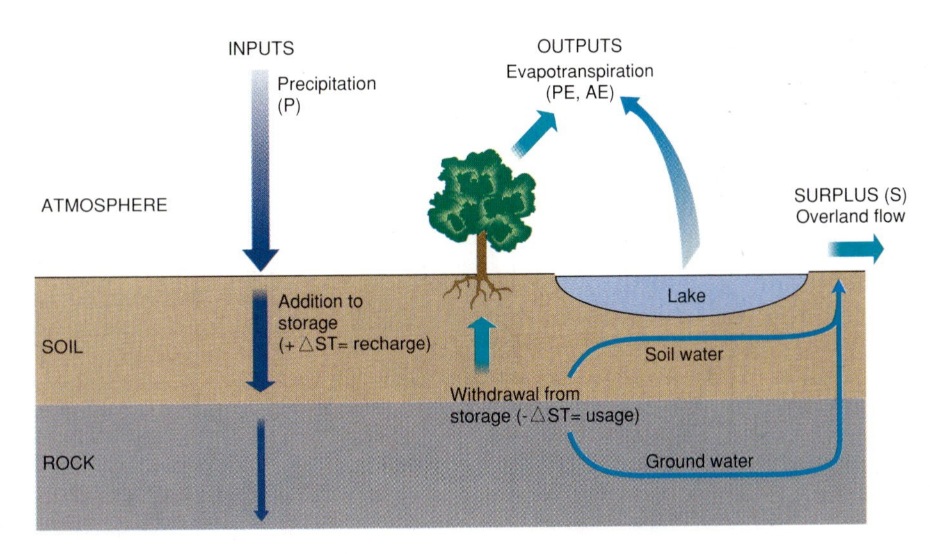

Figure 6-17 The water budget at Earth's surface. The terms are explained in the text.

bodies or in the ground; once these water stores are emptied, there is a *deficit,* and evaporation ceases. The amount of change in water stored in soil and rock is shown by the use of the Greek letter Δ (delta). Recharge is $+\Delta ST$ (storage increases by ΔST); usage is $-\Delta ST$ (storage decreases by ΔST). **Potential evapotranspiration** is the amount of evapotranspiration that would occur if water were always unlimited at the surface and at plant-root level. **Actual evapotranspiration** is that which occurs in practice: it will be equal to the potential when water is available for evapotranspiration, but will be less in a dry season.

Figure 6-18 shows how a simple water budget is calculated and a summary graph drawn. First, the monthly totals for precipitation *(P)* are compared with those for potential evapotranspiration *(PE)* and the difference is entered *(P − PE).* In months when *P* is greater than *PE* (wet seasons), more water falls than is removed by evapotranspiration, and the area gains moisture. When *P* is less than the *PE* rate (dry seasons), precipitation is insufficient to maintain the maximum evapotranspiration. The area may have water stored in the soil or rocks near the surface, and this may make up the difference as long as it lasts. In the simplified case illustrated it is assumed that the maximum storage is 100 millimeters; this would normally be worked out by a more complex formula for each situation.

It can be seen from the table that the dry months from May to September at Houston use up the stored water until there is a deficit *(D):* actual evapotranspiration *(AE)* is less than *PE* from July to September. When the wet months begin in October, the excess builds up stored water so that by January the ground store is full and no more can be accepted. The surplus *(S)* then runs off into streams. Over the year, *AE* is about 90 percent of *PE* at Houston.

The graph plots the figures for *P, PE,* and *AE* as lines. Areas between the lines can be shaded to highlight periods of surplus, deficit, usage, and recharge. Water budget graphs for three places across the United States are shown in Figure 6-19 and illustrate three different relationships between *P* and *PE.* San Francisco (*A*) has most of its precipitation in the winter 6 months from October through March, when *PE* is low and recharge occurs, leading to a surplus that contributes to surface runoff from February to March. In the summer, low precipitation and high *PE* combine to create a major deficit period, and farmers in central California need to irrigate their crops. In Nebraska (*B*), the main period of precipitation coincides with the highest *PE.* There is a small late-summer deficit and a short period of water recharge, but little surplus water for runoff in winter (the period of low precipitation). In Boston (*C*), precipitation occurs throughout the year, and there is a very short period of deficit in late summer. Water passes into and out of ground stores, and there is a long period of water surplus and surface runoff in winter.

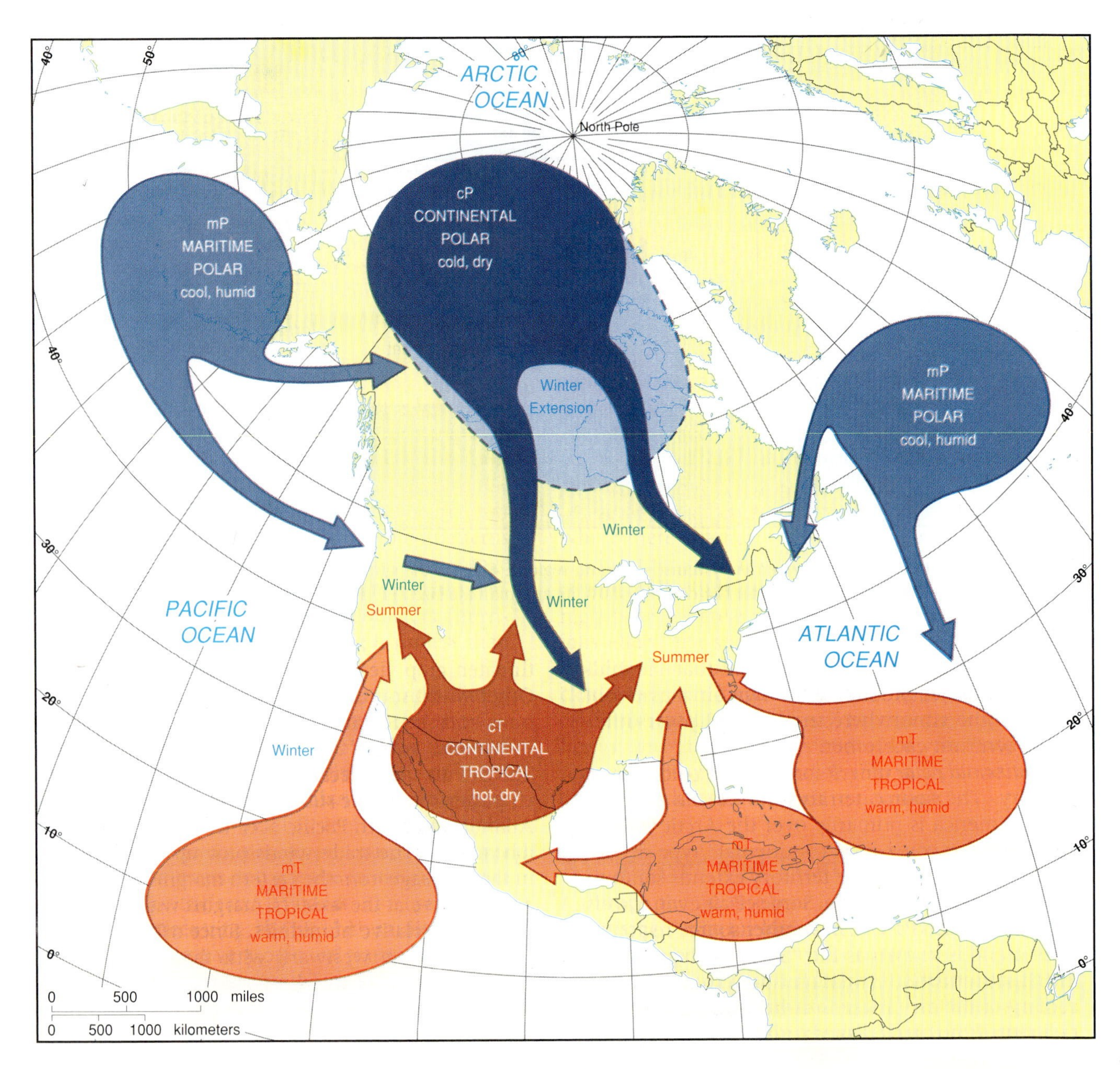

Figure 7-2 Air masses of North America. Source regions and paths.

moisture and warmth from the ocean. Such modification is slow and requires a passage of several days over the water to have much effect. The Great Lakes of North America are large and provide a source of heat and moisture that modifies a cP air mass passing over them in winter, but the effects are not great enough to convert it to mP. These modest changes do affect the weather, however. The surface layer of air over the lakes is heated by the warmer water, producing a steeper environmental lapse rate and encouraging instability in the air above (Figure 7-3). Heavy winter snowfalls often occur on the eastern and southern sides of the lakes as a result.

Maritime polar air may also result from the cooling of mT air as it travels over the poleward sectors of the oceans. Such cooling brings stability to the air mass by reducing the environmental lapse rate, but the high relative humidity of the air mass makes it liable to condensation if it is forced to rise over mountains or a colder air mass. Maritime polar air masses are frequently involved in confrontation with cP or mT air, or with other mP air masses of slightly different temperature. Maritime polar air is a feature of the midlatitude and high-latitude regions of the oceans and adjacent continents.

Table 7-1 **Air Mass Types and Characteristics**

		PERCENT OF WORLD		CHARACTERISTICS	
HEMISPHERE	AIR MASS	JANUARY	JULY	WINTER	SUMMER
NORTHERN	cP	10.7	2.5	Very cold, dry	Cool, dry
	mP	10.6	14.6	Cold, humid	Mild, humid
	cT	6.1	7.1	Hot, dry	Hot, dry
	mT	31.5	25.8	Warm, humid	Warm, humid
SOUTHERN	mT	21.5	21.8	Warm, humid	Warm, humid
	cT	2.4	4.9	Hot, dry	Hot, dry
	mP	13.8	16.5	Cold, humid	Mild, humid
	cP	3.4	6.8	Very cold, dry	Cool, dry

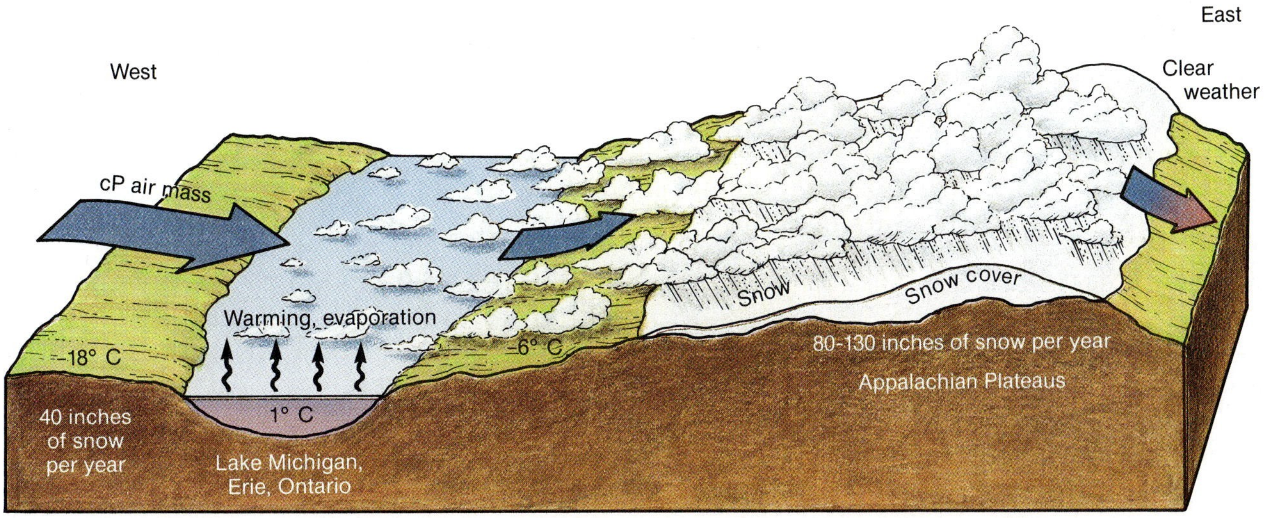

Figure 7-3 Modification of an air mass. The passage of cP air over the Great Lakes in winter is marked by heating of the lower air and evaporation. This produces instability in the air and leads to clouds and precipitation over land to the east.

These four types of air mass are regularly experienced in North America. Some meteorologists recognize only three basic types of air mass—cP, mT, and cT—and view mP as a modification of other types, subject to many variations. Other meteorologists add to the four by distinguishing arctic, antarctic (included here as varieties of cP), and equatorial (included here as a variety of mT) air masses. Regardless of how they are classified, all air masses move away from their source areas and are modified slowly as they move to areas with different surface conditions.

The movement of air masses from their source regions has the effect of transferring large bodies of heated air from the tropics toward the poles and of returning cooled polar air toward the tropics. Air masses are thus important in the global processes of heat transfer. On a more local scale, they affect the weather in ways that cannot be explained solely by the seasonal shifts in the global distribution of solar radiation. Variations in winter weather in North America are often caused by cold snaps of cP air or unseasonably warm spells of mT air.

Air masses tend to move along certain paths, such as those shown in Figure 7-2, and come into contact with each other along **fronts**, which are narrow zones of changing conditions where one air mass meets another air mass of different temperature and humidity. Since the mixing of differing air masses is relatively slow, the contrasting conditions on either side of the front may continue for a few days. The term *front* was first used by meteorologists in 1917, when people were very conscious of the almost static battlefronts in eastern France. Atmospheric fronts can be seen as zones where converging air masses come into conflict with each other.

The interactions between contrasting air masses produce much of the weather commonly experienced in midlatitudes. When two air masses of differing temperatures, humidities, or both come together, the relatively warm air rises over colder, denser air. Fronts are classified on the basis of the dynamic relationships between the air masses. If warm air is moving faster than a sluggish mass of cold air, it will rise over the cold air along a sloping surface at its leading edge; this is known as a **warm**

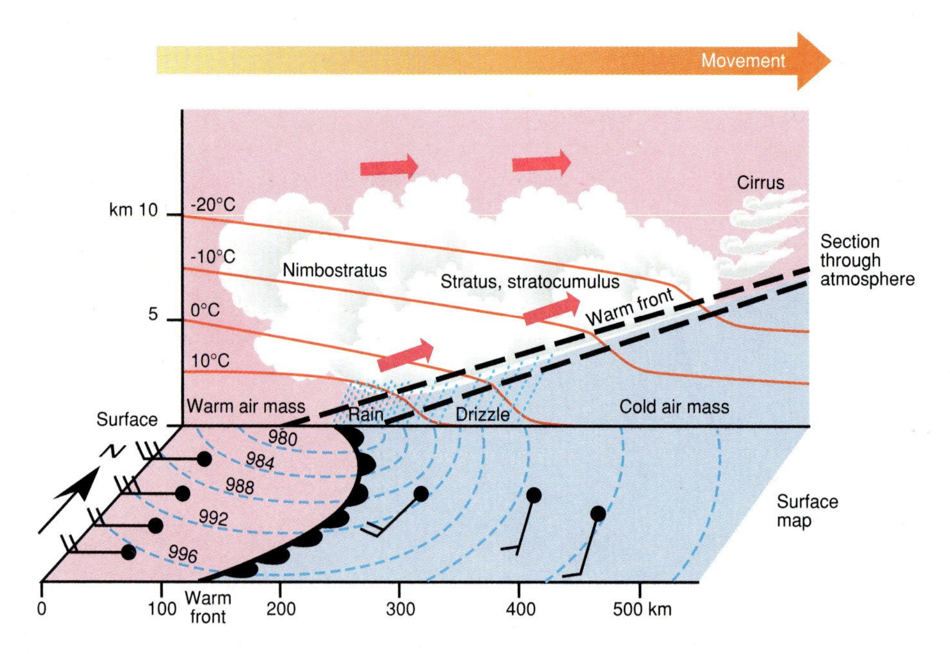

Figure 7-4 A warm front. A warm air mass is forced to rise over the slow-moving cold air in front of it. If the warm air mass is unstable, cloud development will be thicker and rainfall heavier.

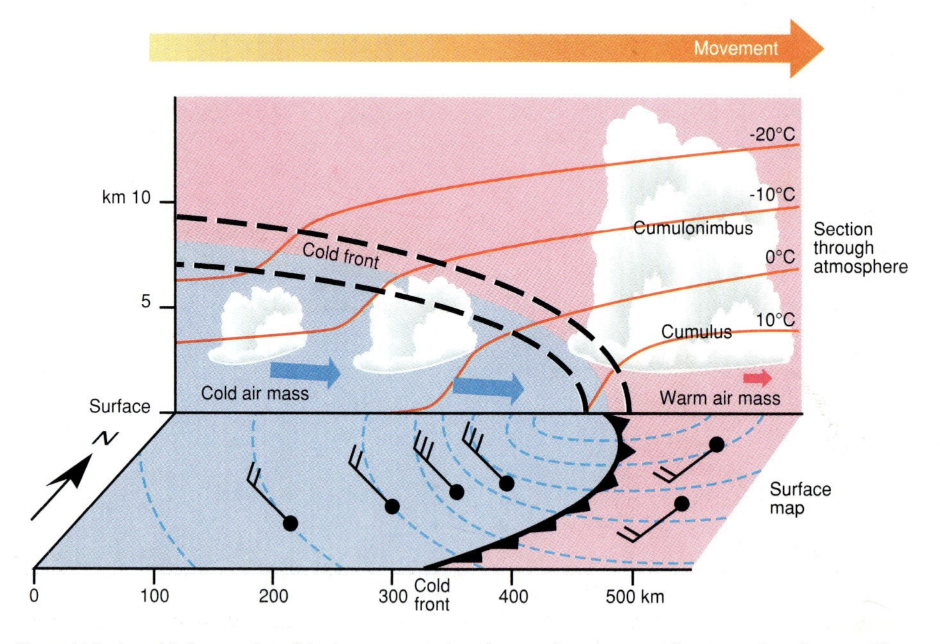

Figure 7-5 A cold front. A cold air mass pushes beneath a warm air mass that is travelling more slowly and produces a steep boundary that forces warm air to rise almost vertically.

front (Figure 7-4); if the cold air is actively pushing beneath warmer air, the front will be steeper and is known as a **cold front.** Warmer air is uplifted in both circumstances, and both fronts produce cloud cover and precipitation according to the conditions in the air masses involved. In stable air the clouds forming along fronts may be shallow in height because the moist air ceases to rise and condense. If the warm air is unstable there will be a deeper development of cumulonimbus clouds and more violent storms (Figure 7-5). Instability is often the case along a cold front due to its greater steepness.

Fronts are not features of the surface weather in the tropics because there are fewer contrasts between air masses in converging airflows. Similar mT air masses meet on either side of the intertropical convergence zone. When cT and mT air converge, the less dense dry cT air may rise over the denser humid mT air but produces little noticeable weather.

LINKAGES

*Air masses take their characteristics from contact with specific types of surface condition. The surface types include the oceans **(atmosphere-ocean environment),** cold or dry land surfaces **(surface-relief environment),** and vegetation cover **(living-organism environment).** The tropical rainforest areas, for instance, transpire so much moisture that the air above takes on mT characteristics.*

Thunderstorms

Thunderstorms occur most frequently of all weather systems. They are formed of single or multiple cumulonimbus clouds and produce heavy rain, hail, thunder, lightning, and strong downdrafts of air. Some 16 million thunderstorms occur around the world each year, and as many as 2000 are in

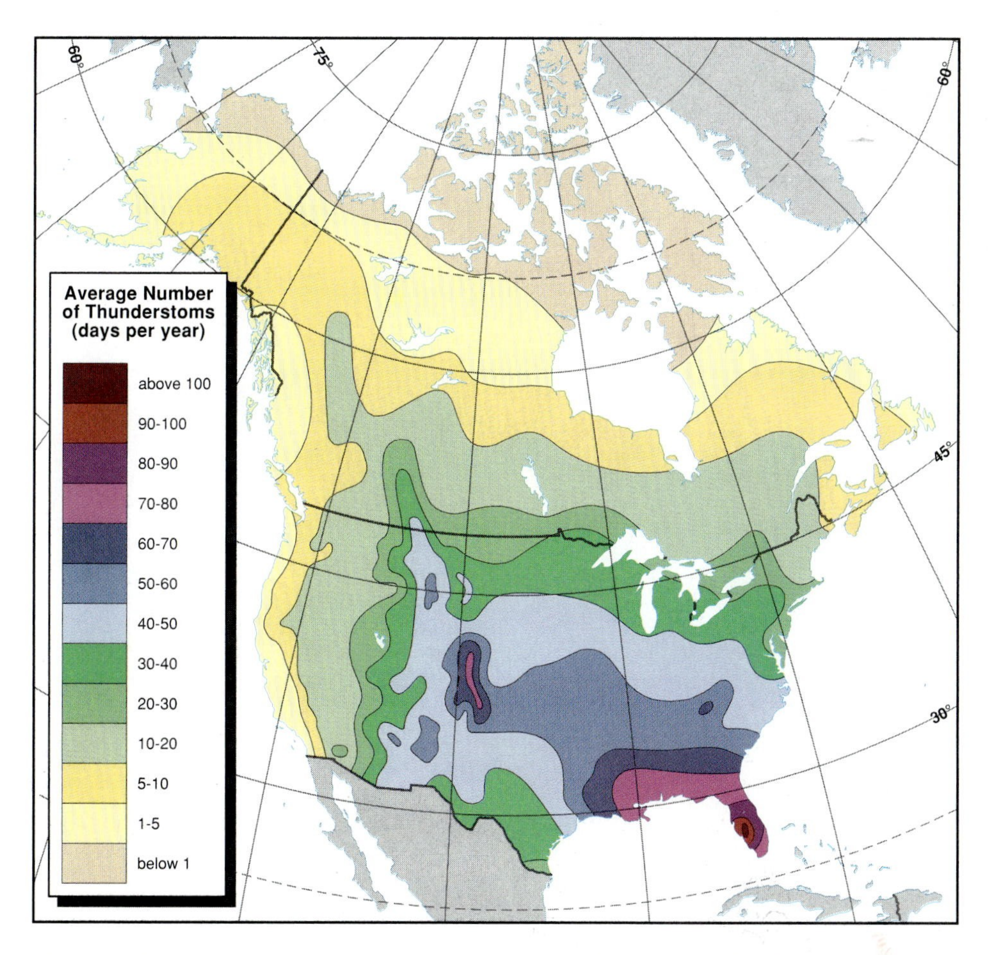

Figure 7-6 The frequency of thunderstorms in the United States and Canada: average number of days with thunderstorms per year. Thunderstorms are most frequent in the warm humid southeast and in the central Great Plains.

progress at any one moment. They are particularly frequent in equatorial regions. Kampala, Uganda, averages 242 thunderstorm days per year. Outside the tropics they are less common. Florida has the most (80 to 100) in the United States, followed by the southern Rocky Mountains area, as is shown in Figure 7-6. The frequent occurrence of thunderstorms in most parts of the world makes them a significant part of the mechanisms for transferring heat energy through the atmosphere (by the release of latent heat during condensation).

Thunderstorms develop in atmospheric environments where there is a violent upward movement of moist air and where large volumes of water vapor are converted to cloud droplets. This occurs when heating near the ground leads to the uplift of bodies of humid air in a very unstable atmosphere. Such instability is commonly caused by a large contrast between warm surface air and cold upper layers of air. Strong uplift along a steep front between air masses or over mountainous terrain may act in combination with surface heating to trigger a thunderstorm. Once a body of air begins to rise in these unstable conditions, it continues upward, and condensation produces towering cumulonimbus clouds. A complex group of convective cells may be involved; new ones form next to older ones that are dying out as the storm moves across the land (Figure 7-7).

Three stages can be identified in thunderstorm cell development. In the *initial stage,* cumulus clouds are formed in updrafts of air and continue to grow upward to form cumulonimbus clouds. Conversion of water vapor to cloud droplets releases latent heat, and this reinforces instability and the convectional ascent of air (Chapter 6). Air is drawn in at the surface, and strong diverging winds aloft may aid growth by sucking moist air upward. In

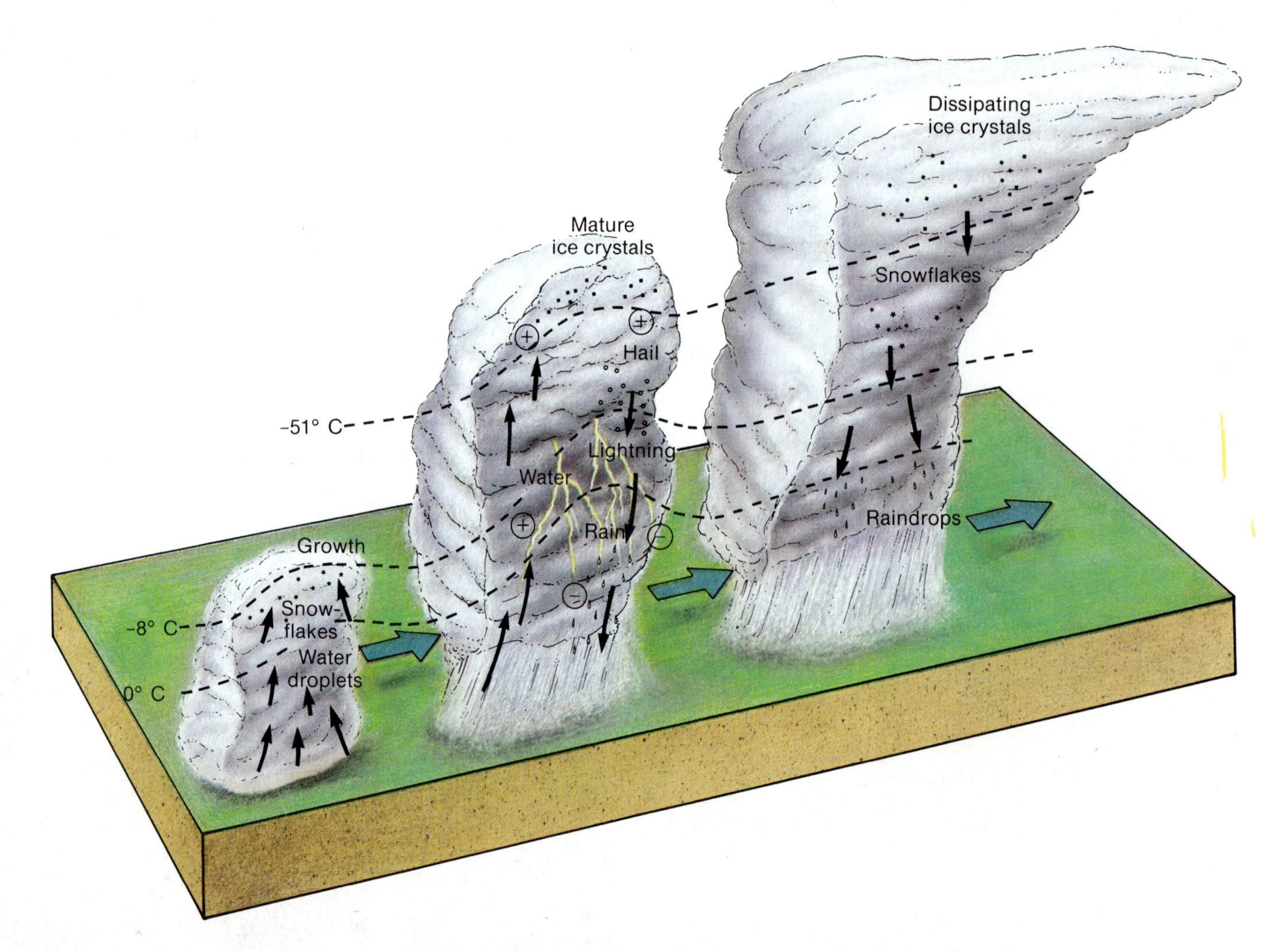

Figure 7-7 The development of a thunderstorm cell. The cutaway clouds show how the updrafts produce tall clouds and icing conditions; the downdrafts that follow produce lightning and heavy rain. A thunderstorm often includes several cells that grow and dissipate as the storm travels along.

the *mature stage,* the cloud top rises to a level where cloud droplets and water vapor are converted to ice particles. The top is spread out by high winds to form a diffuse, anvil-like feature just below the tropopause, where the warmer air above prevents the air and moisture from rising further. Precipitation starts as the cloud grows and reaches a maximum after the ice phase begins to form. A strong downdraft of air is initiated as falling ice crystals and raindrops drag air downward. Lightning and thunder are caused by the combination of drafts with the presence of water in liquid, vapor, and ice forms; they follow soon after the main precipitation begins. Eventually the downdraft becomes so strong that it shuts out the updraft of air and destroys the thunderstorm cell. During this *dissipating stage* the cell loses its supply of moisture and energy and disintegrates. The thunderstorm will continue to exist if new cells are added at its margins.

The downdraft of air beneath the clouds during the period of heaviest thunderstorm rain may form a powerful wind known as a **downburst.** Large downbursts can last for up to 30 minutes and have surface winds blowing at up to 210 kmph (130 mph). They may cause damage in a swath over 4 km (2.5 miles) long. Smaller downbursts have been blamed for a number of commercial aircraft accidents.

Thunderstorms often produce hail, thunder, and lightning. *Hailstones* are precipitated from thunderstorms in which the ice crystals experience several phases of falling to levels where they partially melt before being lifted again to the freezing sector of the cloud. The ice particles grow layer by layer and can be as large as 5 cm (2 inches) in diameter when they fall to the ground (Figure 7-8).

Continental interiors where cold and warm air masses meet favor the formation of hail. The upper Mississippi valley may have up to 10 devastating hailstorms per year, but fewer occur in the southeastern United States despite its greater number of thunderstorms. Hail can be particularly damaging to soft fruit, such as grapes, cherries, and peaches, and also to other standing crops. It causes annual crop damage estimated at over $1 billion in the United States. In the summer of 1991 car dealers in Mississippi gave large reductions on new car prices because of hail damage.

Various attempts have been made to prevent the formation of hail in thunderclouds. At one time, making a loud noise, such as firing cannons or ringing church bells, was thought to be effective. A more scientific approach has been adopted in recent years with some success. Rockets filled with silver iodide were shot by Soviet scientists into clouds likely to produce hail. The theory was that more, smaller hailstones would form and melt before reaching the surface. However, tests of this theory in the United States in the 1970s were inconclusive, and federal funding for the hail suppression program ended in 1979.

Lightning and thunder both result from the existence of violent updrafts and downdrafts in cumulonimbus clouds. **Lightning** is a visible flash (Figure 7-9) that results from electrification within a thunderstorm. A thundercloud usually has concentrations of positive charges in its upper parts and of negative charges at its base that develop as the updrafts and downdrafts increase. It is thought that the charges build up by friction between the drafts of air—much like a static charge results from rubbing feet on a rug. At Earth's surface a positive

Figure 7-8 Large hailstones. Hailstones can damage crops and other surface features.

Figure 7-9 A lightning flash from a cumulonimbus cloud.

charge develops beneath the cloud and surges up tall buildings, trees, and other high points. The atmosphere is a poor conductor of electricity, and the difference between the positive (surface) and negative (cloud) has to build up to as much as 100 million volts before lightning occurs, momentarily connecting the two by an electric arc. Lightning may occur between oppositely charged parts of the cloud, or between the cloud base and Earth's surface. Once a lightning stroke has discharged the electrical field, charges may build again to create a new field. **Thunder** is the sound generated by this discharge of electricity. Pressure waves caused by the heating and rapid expansion of the air along the lightning path cause sonic booms as they move outward. Thunder can be heard up to 10 km away from its source, but its timing is increasingly separated from the lightning flash as distance between the storm and hearer increases, since sound waves travel more slowly than light waves.

Lightning is often underestimated as a killer, and figures given by public agencies are usually lower than the real death toll since small numbers of people are involved in each event and not all the records are gathered. Most lightning deaths occur outdoors in rural areas, and a fourth involve those who seek shelter beneath trees. Males are killed four times as often as females; those who play golf, farm, swim, or boat are particularly at risk because these activities are performed in the open.

Forest fires, such as those that devastated Yellowstone National Park in 1988, are often caused by lightning, and buildings can be damaged as well. The Empire State Building in New York City is struck by lightning up to 70 times each year, but like all tall buildings, it is protected by having lightning rods attached to external metal conductors or internal scaffolding so that the current is carried to Earth. Little else can be done to avert damage by lightning, although "chaff seeding" of aluminum-coated nylon threads has succeeded in discharging clouds likely to produce lightning during a rocket launch from Cape Canaveral, Florida. This option is too expensive to be used for a large number of thunderstorms. Lightning occurrence should be an important factor in building design in areas prone to thunderstorms. In 1991 the Florida Department of Transportation had to replace a weigh station on Interstate 75 that had been built in the mid-1980s but had been abandoned after repeated lightning strikes. Although its location was in Marion County, which records the highest number of lightning strikes in the United States, lightning was not considered in the design of the weigh station, and this caused a waste of several million dollars.

Tornadoes

Tornadoes are small but extremely violent rotating storms, in which a distinctive funnel-shaped column, or vortex, descends from the base of a thunderstorm to the ground. This vortex is commonly 100 meters or so in diameter at its base, whirls usually in a counterclockwise direction in the northern hemisphere, and has winds that may exceed 400 kmph (250 mph). The pressure at the center of a tornado funnel is often 100 mb less than the surrounding air. Each funnel is typically short-lived and moves across the ground in an erratic path at 50 to 60 kmph (35 mph) for up to 15 km (10 miles), although some have traveled as far as 200 km (125 miles), carving a swath of damage hundreds of meters wide. In a similar way, waterspouts form over the sea or a large lake but are usually less energetic and last for a shorter time.

The formation of a tornado requires special conditions, which occur in severe thunderstorm cells when convective updrafts of air are very strong. Such updrafts are generated by an extreme temperature and pressure contrast between the surface and upper troposphere air, producing a very steep environmental lapse rate and a sharp increase of wind speed with altitude. The difference in wind speed begins a rolling motion in the lower atmosphere, and the updraft of air caused by the strong environmental lapse rate shifts this spiralling air into a vertical position high above the ground. A vertical cylinder of air 10 to 20 km in diameter builds upward and downward and strengthens the updraft (Figure 7-10). At a particular stage—not fully understood since it occurs only about half of the time after a cell reaches this state—the rotating cylinder narrows and spirals downward to the ground as a funnel (Figure 7-11). The narrowing increases the wind speeds to violent levels. Tornado winds churn up the surface and even destroy buildings. The conditions that cause a tornado are not maintained for more than a few minutes. When the funnel dissipates, the debris it carries is dropped. However, other tornadoes may form from the base of the same cloud, producing a stop-and-go progress across the land.

Tornadoes are particularly common in the central United States (Figure 7-12) and occur much less frequently elsewhere in the world. This is because the central lowlands of the United States are relatively flat and provide a unique zone for the confrontation of continental polar and maritime tropical air; this area is known as "tornado alley." The Gulf states are the main center of tornado activity in February, but the zone shifts northward to west of the Great Lakes by June. This shift

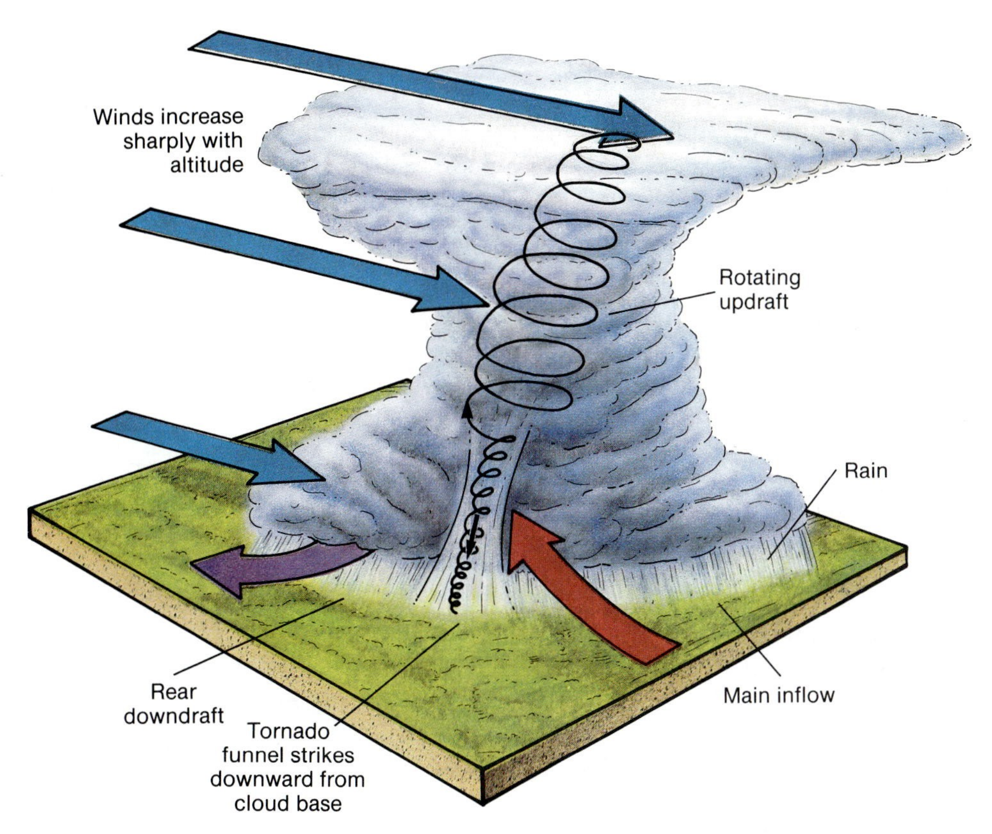

Figure 7-10 Features of a tornado. A rotating motion is generated in strong updrafts of air where there is a sharply increasing wind velocity with height. Once developed, it will extend down to the ground.

Figure 7-11 A tornado funnel moving cross country.

may contain 50 to 80 towering cumulonimbus clouds rising up to 17 km above the surface, piercing the tropopause in places. When the ITCZ is farthest from the equator, some of these convective centers may deepen into easterly waves, spinning tropical storms, or hurricanes, all of which will be described later.

Cloud clusters are responsible for most of the rain in tropical areas, and much of it occurs in very heavy falls; one half of total rainfall at a location commonly falls on 10 to 15 percent of rainy days. The difference between an adequate and a poor rainy season may be accounted for by just two or three cloud clusters. When the intertropical convergence zone coincides with an island as happens in the East Indies for much of the year, the uplift resulting from the convergence of moist air is intensified by daytime heating over the island and produces strong sea breezes and a regular daily pattern of cloud cluster development. Late-afternoon downpours of rain are followed by dry nights and clear mornings during which the clouds begin to build up.

Subtropical High-Pressure Zones

The subtropical high-pressure zones form semipermanent belts over the oceans centered at latitudes 30° North and South (see Figure 5-17). Sea level pressures within these zones average around 1020 mb over the North Atlantic, North Pacific, North Africa, South Atlantic, South Pacific, and southern Indian Ocean. In between the high-pressure zones in these latitudes are narrower zones of low pressure that provide north-south passages for air movement; such north-south movement is toward the equator on the eastern margins of the "highs" and toward midlatitudes on the western sides. The belts of high pressure give rise to the dominantly easterly (trade wind) circulation on their equatorial margins and the westerly flows on their poleward edge.

The subtropical high-pressure zones are often omitted from a list of tropical weather systems. Since they seldom move, they cannot be described as atmospheric "disturbances" (as can cloud clusters, easterly waves, and hurricanes), and the weather they bring about is relatively calm. However, these high-pressure zones are huge—covering 3500 to 5000 km (2000 to 3000 miles) east to west and over 1500 km (1000 miles) north to south. They play a major role in the weather of the large areas they dominate and in the global circulation.

The high-pressure zones are produced by air descending below the westerly subtropical jet on the poleward side of the Hadley cell (see Figure 5-19). The descent of air is strongest on the eastern sides of the high-pressure zones and in winter. The compression of descending air in the high-pressure zones results in warming but does not always extend to the surface. The warm air often spreads out over a surface layer of air, which is cooled over the cold currents on the eastern ocean margins. The warm air overlying the cooler air produces a marked temperature inversion. These conditions make it difficult for tall clouds and rainfall to develop where the inversion is most pronounced over the cold currents, since the cooler air cannot rise through the warmer air. Convection affects a deeper layer of the atmosphere above the western parts of oceans. The effect of subtropical high-pressure zones is different over land, where insolation from the vertical sun through cloudless skies gives rise to high surface daytime temperatures that can reach up to 45°C (115°F) in the Sahara.

Easterly Waves

Easterly waves are disturbances of the pressure patterns in the region of the trade winds that blow around the subtropical high-pressure zones between 5° and 20° of latitude. An easterly wave forms when converging air becomes subject to weak turning by the Coriolis effect, which is not very strong in these latitudes. A low-pressure trough forms, in which the isobars bend away from the equator, giving a wave form. Rising air in this trough produces clouds and rain, mostly east of the trough.

Easterly waves are 2000 to 4000 km (1200 to 2500 miles) across from east to west, last for 1 to 2 weeks, and travel 6 to 7 degrees of longitude per day. They are common from June to September in the tropical western Atlantic and Caribbean, and some 50 such waves cross this area each year. Some of them continue on and cross central America into the eastern Pacific. Similar weather systems occur in varied forms over the Pacific Ocean, where easterly waves often make up a transitional state to a tropical storm.

The movement of an easterly wave within the trade-wind flow helps to intensify rainfall on the east-facing slopes of islands and coasts in their path. For instance, the east-facing slopes of the Pacific Ocean islands have some of the world's highest rainfall totals, while the west-facing lee coasts remain relatively dry.

Tropical Cyclones and Hurricanes

Hurricanes are one of a group of tropical ocean weather systems known collectively as **tropical cyclones**; the crucial defining factor is that wind speeds in a tropical cyclone average over 115 kmph (75 mph) for at least 1 minute. They are known as **hurricanes** in the western North Atlantic, as **typhoons** in the western North Pacific, and simply as cyclones in the Bay of Bengal and northern Australia (Figure 7-15). Most of these storm systems form between 10° and 20° of latitude, and 70 percent occur in the northern hemisphere. They are among the most devastating of natural phenomena; although they are about one third the size of large midlatitude cyclones, they are much more intense. Less intense but related weather systems in the same latitudes are known as **tropical storms** if the average wind speeds are between 60 and 115 kmph (40 and 75 mph), and as **tropical depressions** if they do not exceed 60 kmph (40 mph). Pressures in the centers of tropical cyclones fall below 950 mb and often close to 900 mb. The pressure difference between the edge and core of a hurricane is similar to that in a tornado, but the smaller size of the tornado produces a much greater pressure gradient and therefore higher winds.

Tropical cyclones occur several times a year in summer or early fall, and they vary in number from year to year. The Florida-Caribbean area has between four and fourteen per year, but several times this number occur in the western Pacific Ocean. Those that reach tropical storm intensity are rare

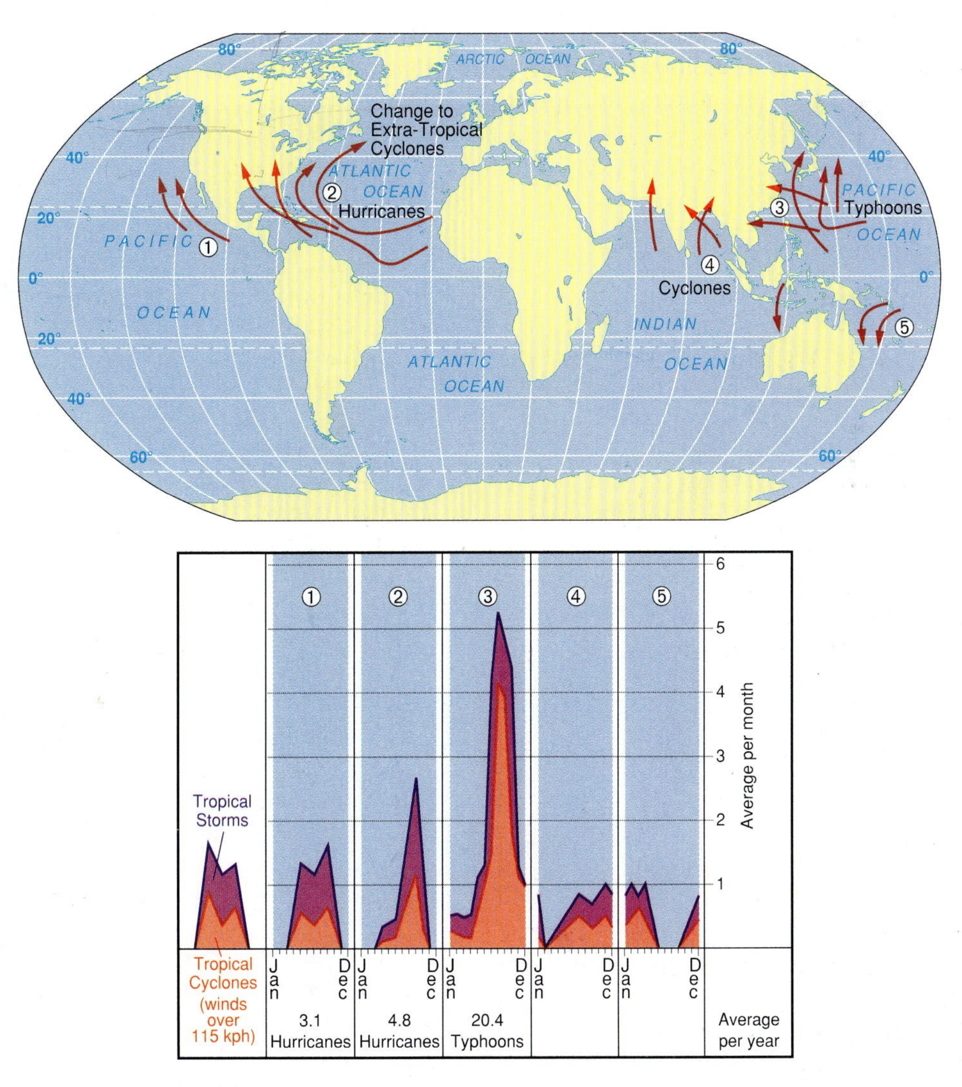

Figure 7-15 Tropical storms and tropical cyclones: worldwide distribution and seasonal occurrence of regular storms and their more severe manifestations. Note that the greatest occurrence is in the central and western Pacific.

Tropical storms that develop into hurricanes follow paths westward across the Atlantic to the south of the subtropical anticyclone. At the western margin of the ocean, hurricane tracks diverge (Figure 7-18). Some carry straight on through the Caribbean, some turn north and remain offshore, and some turn northwest, hitting the Gulf coast, Florida, or the southern Atlantic coast of the United States. Forecasting this path is one of the most important tasks of the monitoring teams, who use information gathered by satellites and aircraft.

Hurricanes continue as long as the conditions of surface inflow of hot moist air and divergence aloft continue. They dissipate when they move over land and are separated from the warm water, or when the sea temperature cools. When a hurricane gets to 30° of latitude, colder air is drawn in and fronts develop; it changes to a midlatitude cyclone. A common feature of hurricanes passing over land is that tornadoes are generated on the margins.

Individual tropical cyclones can bring death and destruction to coastal areas. In May 1991, a severe tropical cyclone hit Bangladesh. The combination of high winds, heavy rainfall, and especially the surge of water several meters above the normal level from the Bay of Bengal that drowned the Ganges delta, devastated the whole country.

Hurricane Hugo (September 10-22, 1989) provides an example of the progress of a hurricane and how people and agencies react. It was the strongest storm to strike the United States since Hurricane Camille in 1969. At one point east of Guadeloupe winds in the core were measured at over 250 kmph (150 mph), and wall pressures fell to 918 mb; these characteristics rated Hugo in the highest category of hurricanes. It was not quite so intense when it struck the South Carolina coast, but it caused record storm surges of over 6 meters (20 feet). A *storm surge* is a rapid rise in sea level caused by storm winds pushing water toward a coast.

Hurricane Hugo began as a cloud cluster moving off the coast of Africa (Figure 7-18). On September 10 it became a tropical depression 200 km (125 miles) south of the Cape Verde Islands and travelled due west along latitude 12° North. By the time it reached a point 1700 km (1100 miles) east of the Leeward Islands late on September 13 it was classified as a hurricane, and was moving west at 32 kmph (20 mph). It continued to move west, with internal winds increasing as the forward movement slowed. A hurricane watch was posted for Puerto Rico and the Virgin Islands on September 14, and this was raised to a hurricane warning two days later. The hurricane hit St. Croix in the Virgin Islands early on September 18 and destroyed or damaged over 90 percent of the island's buildings; much of the island was stripped bare of its vegetation. Hugo then turned to the north, skirting the northeastern tip of Puerto Rico, and began to weaken. However, as it moved out over the Atlantic off Flor-

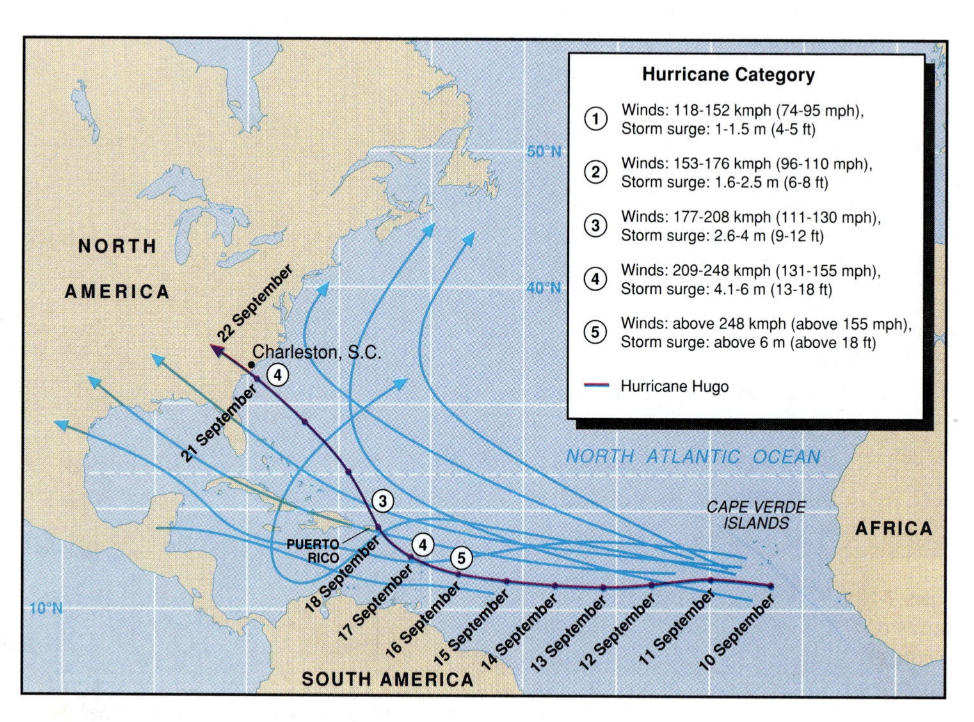

Figure 7-18 Hurricane paths in the North Atlantic Ocean, and details of the path taken by Hurricane Hugo in 1989.

ida it picked up strength and by September 20 had a well-defined eye again. When it passed over the warm Gulf Stream waters, the winds increased and central pressure fell further, while its forward speed increased to 40 kmph (25 mph). People living on the barrier islands and beaches from Georgia to southern North Carolina were evacuated. The eye crossed the coast near Charleston early on September 22 (Figure 7-19) with sustained winds estimated at over 200 kmph (135 mph), several inches of heavy rain, and a 6-meter storm surge inundating large areas. By sunrise on that day the main force of the hurricane was spent, although some tornadoes were reported and heavy rains continued into West Virginia.

Hurricane Hugo became the nation's most expensive hurricane, costing an estimated $9 billion in direct damage and economic losses on the U.S. mainland, Puerto Rico, and the Virgin Islands. The homes of over 200,000 American families were damaged. Some 49 people were killed by the hurricane. The relatively small loss of life reflected the efforts invested in combining improved monitoring and path prediction techniques with public education, timely warning, and evacuation procedures. The latter must be coordinated so that the area evacuated is as small as safety will allow, and as few people as possible are diverted from their normal occupations.

It is significant that most of the deaths from tropical cyclones (96 percent) occur in the developing countries of Asia, while most of the property damage occurs in the United States. The United States possesses the most sophisticated hurricane monitoring and warning system, but the amount of expensive real estate constructed in hurricane-prone coastal areas of the "Sunbelt" has grown rapidly since 1950. Fewer people in the United States are killed by hurricanes now than earlier in this century, but the cost of damage to property continues to rise. The two major cyclones experienced by Bangladesh in 1970 and 1991 each caused over 250,000 deaths in an area where the density of population has increased rapidly in the last 30 years; although property damage was relatively low in dollar values, it often represented all the survivors had.

LINKAGES

The impact of weather systems on other Earth environments may be illustrated by hurricanes. In addition to their impacts on human activities, they may flatten forests ***(living-organism environment)****, change the features of shorelines, and enhance river erosion* ***(surface-relief environment)****.*

Figure 7-19 Hurricane Hugo as it approached the South Carolina coast on the evening of September 21st. The image was constructed from information gathered by a defense satellite. In this presentation, clouds and precipitation appear yellow and atmospheric moisture is seen as a darkening over the ocean.

MIDLATITUDE WEATHER SYSTEMS

Midlatitude weather systems reflect the atmospheric environment of which they are a part. In the midlatitudes, contrasting air masses meet along fronts, and the polar-front jet plays a major part in controlling the surface weather. The Coriolis effect significantly influences the formation of rotating weather systems in these latitudes.

Midlatitude Cyclones and Anticyclones

A sequence of alternating low- and high-pressure areas move around the globe from west to east between latitudes 35° and 60°. **Midlatitude cyclones** and **midlatitude anticyclones** show up on surface weather maps (e.g., Figure 5-5, *A*) in patterns of roughly circular isobars. Each of these weather systems gives rise to its own pattern of weather. It was the study of these midlatitude weather systems that laid the basis for modern meteorology and weather forecasting.

Midlatitude cyclones and anticyclones contrast with each other. Cyclones are characterized by surface low pressure (often down to 970 mb over the oceans, but rarely below 950 mb); anticyclones are characterized by surface high pressure (commonly up to 1040 mb over continents, but seldom above 1060 mb). Cyclones have moderate to strong winds; anticyclones generally have weak winds. Cyclones involve surface convergence of air and uplift at the center; anticyclones involve surface divergence of air following subsidence in the center. In cyclones the convergence is linked to the conflict of contrasting air masses along fronts, leading to uplift and the formation of clouds and rain; in anticyclones it is more common to have clear skies and no rain.

Midlatitude Cyclones. Most midlatitude cyclones form along fronts, and they are often called *frontal lows.* Frontal lows are particularly important features of the surface weather conditions in midlatitudes. There are also nonfrontal lows, but they form a minority of low-pressure weather systems in the region.

The development of a frontal low can be viewed in terms of a generalized ideal, or model, in which there are several stages. This model helps to explain the workings of a frontal low, but no particular system is perfectly standard. The total sequence of stages in the model may take from 1 to 7 days. The northern hemisphere case is described here.

The first stage begins with contrasting air masses—typically cP and mT air—coming into contact as they move in parallel and opposite directions along a straight front between the air masses (Figure 7-20, *A*). The second stage begins when interaction between the opposing winds leads to a curving front that creates a *frontal wave* up to a few hundred kilometers across. These waves look like an ocean wave in profile on weather maps, with circular isobars surrounding a low-pressure cell at the wave's crest (Figure 7-20, *B*). A warm front and a cold front now bound a central *warm sector.* As winds blow in a counterclockwise spiral toward the low-pressure center, the warm front moves poleward along the leading edge of the warm sector, and the cold front pushes in behind and beneath the warm sector.

The wave is at its largest in the mature cyclone (Figure 7-20, *C*). As the frontal low moves east, the faster-moving cold front tends to catch up with the warm front and lift the entire warm sector off the ground, forming the occluded stage (Figure 7-20, *D*). A single front remains on the ground where the cold front has undercut the warm sector and lifted the warm front off the ground. This is known as an **occluded front** (or, in Canada, as a TROWAL, for **TRO**ugh of **W**arm air **AL**oft) and is shown on surface weather maps by combining the two frontal lines and symbols. The occluded stage ends the frontal contrasts and the low-pressure cell "fills in" as surface pressure rises. Wind strength declines and the fronts and cyclone disappear. A remnant of the low-pressure system remains in the air aloft. The average life of a midlatitude cyclone is approximately 3 days.

The typical weather sequence produced by a frontal low crossing a particular location from west to east is the passing of (1) cold-sector air, (2) a warm front, (3) warm-sector air, (4) a cold front, and (5) cold-sector air at the back. The *cold sector* ahead of the low often has winds blowing from the south, but forms a sluggish mass of air that the system pushes against. *Warm fronts* (Figure 7-5) have gentle slopes (0.5 to 1 degree) and are characterized by the widespread and gradual ascent of warm air. High-level cirrus clouds may signal the arrival of a warm front 12 hours in advance. These give way to cirrostratus and altostratus clouds as the warm front approaches and the barrier between the warm and cold air comes closer to the ground. If the conditions in the warm air behind the front are unstable, vertical movement of air forced along the warm front gives rise to thick, dark nimbostratus clouds and steady precipitation. At other times the gradual ascent over the warm front does not overcome a predominant stability in the air, and relatively thin layers of cloud form light-gray sheets of stratiform clouds from which only fine drizzle, if any precipitation, falls. As the warm front passes over an area, there is a change of wind direction from southeasterly to southwesterly. The *warm sector* often contains mT air, and its arrival brings a rise in temperature and a clearing of the skies. Local showers may occur from tall cumulus clouds if this air is unstable. The *cold front* (Figure 7-5) typically has a steeper profile than the warm front, and this produces stronger upward movements of air as cold-sector air pushes from the back beneath the warm-sector air. Where the warm air is unstable, cumulonimbus clouds and thunderstorms form. Tornadoes may occur in the special conditions described earlier in this chapter. The *cold sector* at the rear of a cyclone is often marked by individual cumulus or cumulonimbus clouds in a northwesterly airstream.

Small *secondary lows* may form along the cold front behind a frontal low. This occurs most often over the ocean in winter when the moist air is warmed, generating local instability and ascent of air. Such lows may grow rapidly, becoming part of

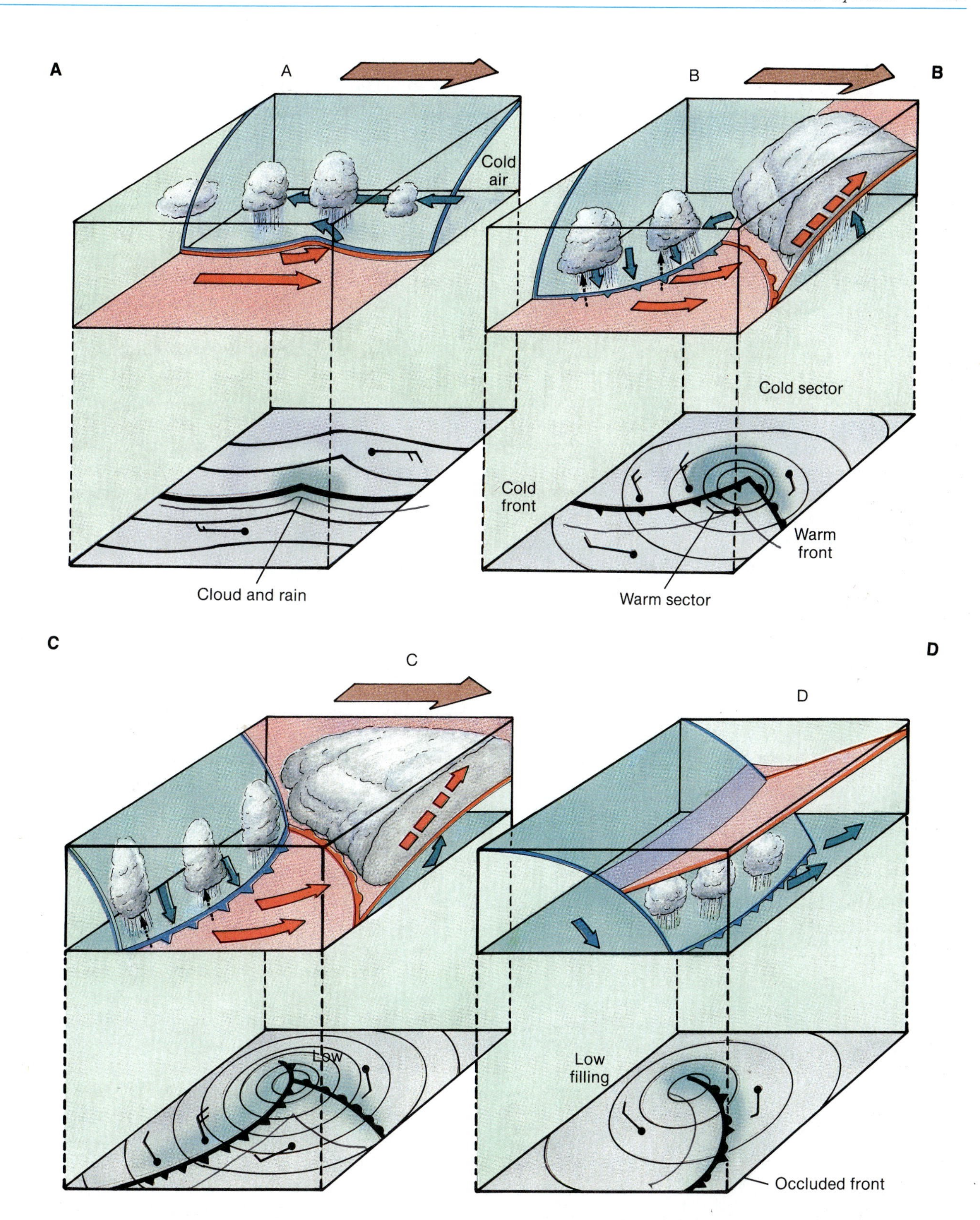

Figure 7-20 The evolution of a midlatitude cyclone in the northern hemisphere. **A,** Initial stage: cold and warm air masses flow parallel to the front between them and in opposite directions; a small wave begins to form. **B,** The wave enlarges with recognizable warm and cold fronts and low pressure at their juncture. **C,** Mature stage. **D,** Occluded stage. The cold front pushes beneath the warm-sector air and lifts it off the ground. Only colder air remains at ground level.

the main circulation with fronts. The strong winds caused by a steep pressure gradient produce stormy conditions.

The general model of conditions associated with midlatitude cyclones can be applied to each individual case shown on daily weather maps or satellite images. As with many natural systems, each cyclone is unique in some way and only partially replicates the model (Figure 7-21). However, the model has stood the test of time as increasing sources of information have confirmed the general pattern it describes.

Nonfrontal lows are, by definition, low-pressure weather systems that form without the presence of a front. They are produced by a variety of atmospheric conditions. One such set of conditions occurs when winds blow over a mountain range, such as the Rockies. As air is squeezed into a smaller space between the higher ground and the tropopause, it is compressed vertically. When the air descends on the leeward side of the mountain range, the depth of air increases and so does the speed of flow and the Coriolis effect. A cyclonic spin is imparted to the expanding air, giving rise to a low-pressure trough. This type of nonfrontal low is known as a **lee cyclone.** Lee cyclones form east of the Rocky Mountains and draw air down from the mountains, which warms and dries out as it descends in *chinook* winds.

Figure 7-21 A satellite view of an occluding midlatitude cyclone centered east of Britain. The circular pattern of clouds is the result of the reduced condensation as the cyclone "fills up." In the mature stage, the cloud bands are thicker.

Thermal lows form when intense surface heating produces expansion of the air and lowers surface pressure. Such lows do not move, do not have fronts, and are shallow if the rising air is stopped at an inversion layer. They are common in areas such as Arizona and northern Mexico.

Midlatitude Anticyclones. Midlatitude anticyclones occur in two major forms. The first type are smaller anticyclones occurring between the cyclones that dominate the westerly flow beneath the polar-front jet. These anticyclones provide a respite from cloudiness, rain, and high winds, but pass through rapidly, often within a day. The descending and outflowing air gives rise to clear skies, gentler air movements, and absence of frontal conflicts between air masses.

The second type of midlatitude anticyclone is much larger and may dominate the weather by staying over an area for several weeks. Such weather systems appear to deflect the passage of midlatitude cyclones and smaller anticyclones around them to the north and south, and are often termed "blocking anticyclones." These conditions in summer bring prolonged periods of sunshine, intensifying the warmth of cT air, so that high temperatures and drought may result. In winter the loss of heat through clear skies during long nights lowers the temperatures in cP air to extreme levels. If the air is almost still, fog may result and polluting dust and gases may accumulate. A combination of low clouds with the fog and pollution may cause the bright skies to give way to anticyclonic gloom.

Midlatitude Weather Systems and the Polar-Front Jet. Midlatitude cyclones and anticyclones were identified during the early part of the twentieth century as the main weather systems in midlatitudes. In the middle of the century it became clear that their occurrence is connected to the polar-front jet in the upper troposphere. The westerly movement of cyclones and smaller anticyclones is controlled by this jet and its Rossby-wave cycle (Chapter 5). This forms a "conveyor-belt" type of mechanism, in which the strong upper wind drags along the surface systems. The waves in the jet stream produce alternating sections of converging and diverging air, which force air down toward the surface or draw it upwards. This produces the flow of surface midlatitude cyclones and smaller anticyclones when the jet is in its west-east pattern. The jet drags the surface systems along beneath it at about 30 kmph (20 mph) in summer and about 50 kmph (30 mph) in winter. The difference in speed reflects the differing temperature gradients between tropics and poles in the two seasons and the consequent pressure gradients. Greater extremes of temperature between the tropics and polar regions in winter induce larger differences in air pressure, causing the jets to be stronger and the Rossby waves to move more rapidly around the globe.

The movement and development of a midlatitude cyclone can be linked to the troughs and ridges in the jet. In the developing stage of the cycle (Figure 7-20, *A-C*) the low-pressure center is to the south of the jet and air is drawn up to the diverging flow within the jet. This link continues as the cyclone develops and until the cold front catches up with the warm, resulting in occlusion (Figure 7-20, *D*). At this stage the low's center moves north of the jet, is cut off from the upward flow of air, and ceases to exist.

Frontal lows affecting North America are commonly generated off the east coast of Asia, and are generally occluded by the time they reach the west coast. They may be regenerated following descent of the westerly air flow east of the Rockies, or off the east coast where contrasting air masses meet at the junction of Labrador and Gulf Stream currents. This pattern is typical of the first stage of the Rossby-wave cycle (Figure 5-13, *A*). When the Rossby waves become more extreme (Figure 5-13, *C*), they carry the cyclone "conveyor belt" in nearly north-south courses beneath the jet. The cores of the large waves become dominated by large blocking anticyclones.

THE POLAR ATMOSPHERE

The polar atmosphere is characterized by the cooling and sinking of air, and in winter this results in polar high-pressure systems and outflowing easterly winds. The lack of weather observations in polar regions has made it difficult to identify weather systems that can be compared to those of the midlatitudes or tropics. However, as was pointed out in Chapter 5, these systems differ in kind from the interlinked pressure and wind systems of the middle and lower latitudes. They are subject to static, rather than dynamic, atmospheric conditions. In summer, the polar regions are invaded more frequently by midlatitude weather systems, especially when the waves in the polar-front jet are at their most extreme.

FORECASTING THE WEATHER

A **weather forecast** is an attempt to predict the weather at a place for the next few hours or days. Midlatitude weather systems such as fronts, cyclones, and anticyclones are the main features of weather maps, since they are the basis of models that forecasters use to predict weather. In the United States there is particular interest in forecasting severe weather conditions such as hurricanes, tornadoes, and downbursts. Before the era of weather forecasting in this century people attempted to read the skies for information about impending weather (see Box 7-1: Weather Lore).

The U.S. government annually funds billions of dollars to the National Weather Service. The four main functions of the National Weather Service are to provide severe weather warnings, weather observations and forecasting, education, and aviation briefings. In addition to the general forecasts broadcast on TV and radio and published in newpapers, specialized reports are provided to farmers, pilots, and those who work in other specific sectors of the economy.

The business of forecasting weather is an excellent example of the scientific method in action. Weather data (temperature, pressure, wind speed and direction, cloud forms, rain) are collected and plotted on maps, and general atmospheric conditions are analyzed using weather-system models based on the principles covered in this chapter. The visual models described here are converted to numerical computer models. Forecasts are made on the basis of the computed speed, direction, and internal characteristics of these systems. Since the continuous flow of data indicates that these are changing (sometimes in unexpected ways), the forecasters must continuously update their forecasts. The systems themselves in midlatitudes last only for a few days (except for some large anticyclones, which may persist for a week or so), and forecasts are limited to that period. Forecasting weather for longer periods will depend on a better understanding of how the polar-front jet functions and helps to produce surface weather systems. Like other applications of scientific principles, the forecasts are not perfect, but better technical facilities and understanding have resulted in great improvements in the past 30 years.

Ground-Based Data Collection. The familiar weather map of TV and newspapers is produced each day from the weather maps compiled by the U.S. National Weather Service. The task of compiling the maps begins with gathering the relevant weather information summarized for each weather station across North America and adjacent regions.

BOX 7-1

WEATHER LORE

The weather has been a subject of observation and speculation throughout human history. Before the days of scientific measurement and modeling, people worked out a number of "common-sense" rules that can be seen to have some link with a fuller understanding of the workings of the atmosphere. This popular weather lore is often expressed in short sayings. A few of the sayings common in English-speaking midlatitude countries are discussed here; similar sayings from other cultures could be collected and analyzed for their validity.

"Red sky in the morning, shepherd's (or sailor's) warning; red sky at night, shepherd's (or sailor's) delight."

Low-angle solar rays at the start and end of the day cause the sky to be reddened as the blue light is scattered away by the dust and haze in dry, slow-moving air. Such conditions typify anticyclones and their fair weather. Red dawn skies in the east show that an anticyclone has passed over an area and is likely to be followed by a cyclone with its fronts and wet weather. A red sunset to the west signals the approach of an anticyclone.

"Hen's scratchings and mares' tails make tall ships carry low sails."

The "hen's scratchings" and "mares' tails" are two names for high cirrus clouds that signal the advance of a warm front and low-pressure stormy weather systems. These two signs may give 15 to 20 hours warning of arriving wind and rain, so sailing ships had ample time to reduce the sails as the winds rose.

This saying is complemented by, *"The lower they get, the nearer the wet"* (reflecting the approach of the warm front and its lowering cloud base), and *"Long foretold, long last; short notice, soon past"* (linking the size of the weather system and its potential effect to the time it takes to build up). A related saying is: *"Ring around the moon, rain by noon; ring around the sun, rain before night is done."* Here, the shining of Earth's moon or the sun through thin, high-level cirrostratus cloud also heralds the arrival of a warm front some 12 to 18 hours away.

This saying reflects anticyclonic conditions with little air movement:

"When the dew is on the grass, rain will never come to pass."

Dew forms on grass when air near the ground cools at night in conditions of little air movement. Such conditions are typical of an anticyclone, in which clear skies allow radiation losses at night, but rain does not form.

This information is recorded in a shorthand format using the symbols shown in Figure 7-22. Data from land-based weather stations are supplemented by records from several thousand aircraft and merchant ships. Specially sited weather ships are being replaced by automatic weather buoys.

Worldwide, a set of primary observations is made four times each day—every 6 hours based on Greenwich Mean Time (midnight, 6:00 AM, noon, and 6:00 PM). Noon Greenwich Time is 7:00 AM Eastern Standard Time (and 6:00 AM when daylight saving operates), so daily weather maps published in the United States show conditions at that time. At each reading air temperature, cloud type, cloud cover, cloud-base height, humidity, wind speed, wind direction, atmospheric pressure, visibility, and precipitation are recorded. Upper air measurements are taken by radiosondes, which are instruments for recording humidity, pressure, temperature, and winds in the upper atmosphere that are carried upward by large balloons.

Weather observations from around the world are collected at three major international centers, Washington, D.C.; Melbourne, Australia; and Moscow. Forecasters need to have access to the world view, since events in one sector of the atmosphere affect those in other parts of the globe. The coded reports are converted into local station summaries and are plotted on maps. This process is carried out by computer. Isobars are drawn at 4 mb intervals, and fronts are added by comparing the patterns of pressure, temperature, and precipitation. Maps of highest and lowest temperature and of precipitation totals for 24 hours are also included in the Daily Weather Map record (Figure 7-23). In addition, the forecaster requires a map of conditions in the upper troposphere. The 500 mb chart, showing windspeed and direction at approximately 5.5 km above the surface (e.g., Figure 5-5, *B*), is also provided on the Daily Weather Map and can be related to the surface movements of weather formations.

Weather Satellite Data Collection. Since 1960 forecasters have been aided in their work by an increasing worldwide network of weather satellites, which have augmented observations from surface weather stations. The present generation of weather satellites includes the NOAA (National Oceanic and Atmospheric Administration) series, which orbit Earth on a polar route at lower altitudes, and the GOES (Geostationary Operational Environmental Satellite), which orbits (at a height of 36,000 km, or 22,000 miles) at the same speed as Earth turns. As the older satellites come to the end of their useful life, they are replaced by new versions with en-

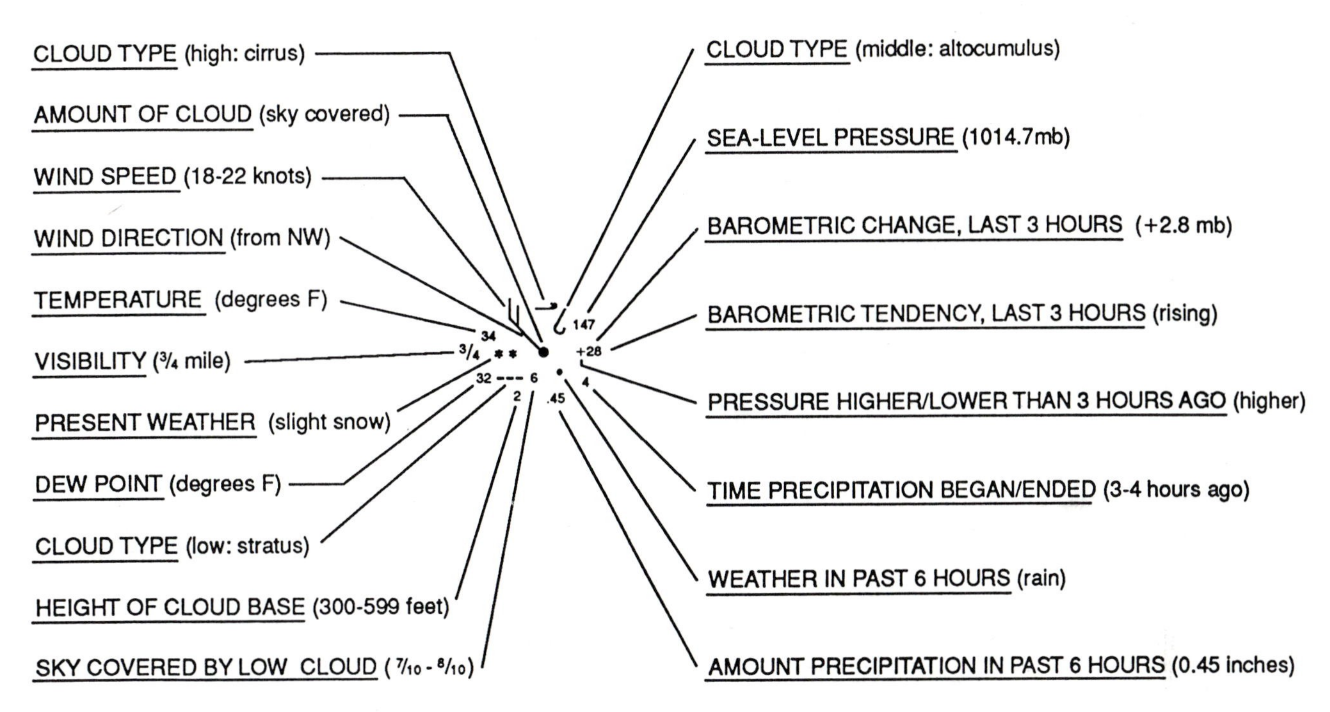

Figure 7-22 A weather station record: this forms the basis of daily weather maps.

hanced instrumentation capable of obtaining better data. The technological basis for weather forecasting is further enhanced by some of the world's most powerful computers, which collect and analyze observations from increasing numbers of stations and levels in the atmosphere.

Processing of satellite images taxes the most powerful computers because of the sheer volume of data provided by satellite images that are updated every 30 minutes. Furthermore, the atmosphere is so vast and subject to so many influences that it is difficult to describe everything affecting its movements in the mathematical equations that are the basis of computer analysis. Computers have not yet replaced human weather forecasters. Images obtained from satellites are routinely used as a visual source of information alongside the maps generated by the computers from ground-collected data. The impact of further improvements in the time scale of forecasts could be immense (Table 7-2), and this is one reason for developing the science of meteorology and gathering further information about the atmosphere.

Making the Forecast. The forecaster now has a range of maps and satellite images at hand and also has information on the rates at which the weather systems have been moving. If a front is moving at a particular speed, it may be expected to continue at this rate for several hours more. This approach to forecasting is known as the **persistence method.** However, movements of weather systems are subject to changes in velocity, so the persistence method is regarded as reasonably accurate only for periods of 6 to 12 hours. Another way of extending trends is known as the **continuity method**, in which a frontal low's progress may be forecast by reference to the development stages discussed under "Midlatitude Cyclones and Anticyclones" earlier in this chapter. If a frontal low has been observed for 2 or 3 days, its pattern for the next 24 hours may be suggested, depending on whether it is approaching maturity or beginning to dissipate (Figure 7-20). These approaches may sound fairly simple, but the forecaster also takes account of a wider range of factors, including special local conditions of relief or the presence of large water bodies. Despite the use of sophisticated modern technology, forecasters still must make the forecast on the basis of their own knowledge and experience of the patterns displayed by the plotted weather data and the evidence from satellite images.

Improving Weather Forecasts. Some developments that will increase the accuracy of weather forecasts include the use of Doppler radar (Nexrad), the implementation of the Automated Surface Observation System, and LIDAR (Light Detection and Ranging) systems that combine radar and lasers. The Doppler radars being installed at 175 sites across the United States in the early 1990s provide a major upgrade from the early radar systems that have been used to monitor rainfall. The new radar scans the atmosphere from the ground upward and produces three-dimensional images that make it possible to detect the sort of wind shifts that occur

Climate is the long-term behavior of the atmosphere-ocean environment. Weather varies over hours or days; climate concerns periods of years. The climate at a location is not merely the "average weather"; it is also defined by the diversity of temperature, moisture, and atmospheric pressure conditions experienced through the seasons and over a number of years. For instance, Chicago and Valentia Island (on the southwest coast of Ireland) have almost the same annual *average* temperatures, but Chicago has a much greater *range* of monthly temperatures (Figure 8-1). Chicago is subject to humid summer heat with thunderstorms and to freezing, snowy winters, while Valentia Island experiences cool summers and few winter frosts (although it is 10 degrees of latitude farther north than Chicago).

The everyday usefulness of information about climate in human decision making is illustrated by the Pilgrim Fathers (opening excerpt), who based some early decisions about settlement and crops on a rudimentary knowledge they had gleaned of the climate and natural environment of eastern North America.

Physical geographers are particularly interested in recognizing climatic differences around the world at different scales from world regions to local areas. Climate is basic to an understanding of other aspects of physical geography, such as the study of landforms, the formation of soils, and distributions of living organisms. Differences in climate strongly influence glaciers, rivers, and other processes that break down rocks (Chapters 13 to 17) and help to determine the type of soil (Chapter 20). The climatic factor is very significant in the consideration of global differences in landforms (Chapter 18) and natural vegetation (Chapter 21). Climate also has significant impacts on human activities. One major set of problems currently facing physical geographers is to determine the relative roles of natural climatic change and human intervention in issues affecting the lives of millions of people. The phenomenon of desertification—the degrading of land quality on the margins of dry regions—is one such issue.

The study of climate is also compelling for physical geographers because one of their tasks is to apply a broad understanding of the atmospheric environment to the planning and management of a wide variety of human endeavors. Physical geographers are employed for this purpose by a variety of corporations and agencies. During World War II the planning of large-scale amphibious invasions and jungle combat required an improved knowledge of climate. Advances begun then have continued in such aspects as designing and constructing buildings for use in varied climates from the tropics to tundra and in determining the best sites and orientation of airport runways. The space program required greater knowledge of climatic conditions for rocket launchings and Space Shuttle landings. The potential of warfare in polar conditions and the Gulf War of 1991 require military equipment modifications for these extreme environments (Figure 8-2).

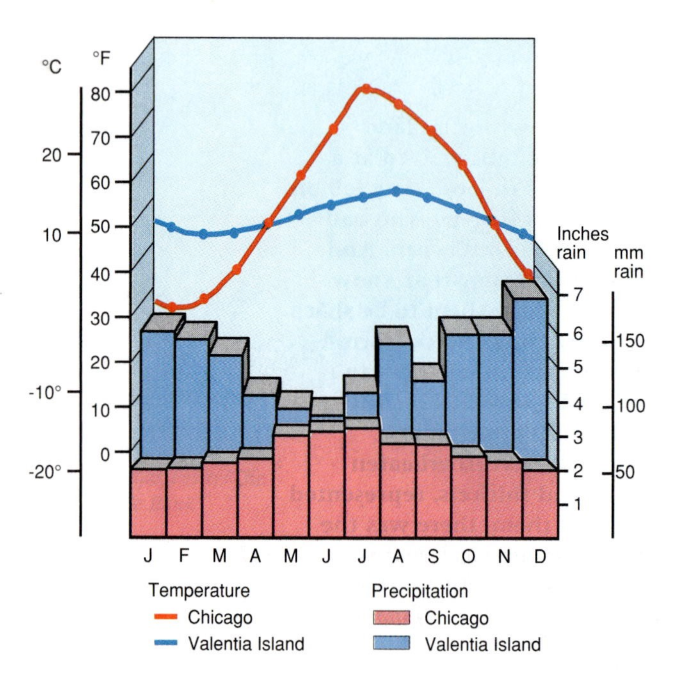

Figure 8-1 Climatic data for Chicago (41°56′ N) and Valentia Island (51°50′ N). Both have an average temperature just over 10°C (50°F), but Chicago's annual range is 26°C (49°F) and Valentia Island's is 8°C (15°C). The precipitation figures show contrasting seasonal regimes.

Figure 8-2 A tank that had to be repainted and have engine modifications for use in the 1991 Gulf War.

United States federal agencies concerned with the collection and use of climatic data are many and varied. The facilities of the National Oceanic and Atmospheric Administration (NOAA) in the Department of Commerce include the National Climatic Data Center at Asheville, North Carolina, which holds information ranging from private eighteenth century weather diaries to modern observations made worldwide at the rate of 100 million per year. The Department of Defense requires climatic information for its operations and works closely with the National Climatic Data Center. Other federal agencies with an interest in the application of climatic studies include the Department of Agriculture, the Department of Education, the Department of Housing and Urban Development, the Department of the Interior, the Department of Transportation, the Environmental Protection Agency, and the National Aeronautics and Space Administration. The length of this list emphasizes the importance of climate studies for public policy and planning.

This chapter examines how distinctive types of climate are recognized at different geographical scales from the global to the local, and describes the main types of climate. Chapter 9 considers how global climates change over time. The two chapters together are central to the consideration of all Earth environments.

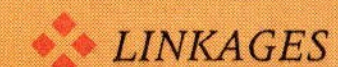
LINKAGES

*Climate has major linkages with the other surface Earth environments. Different climatic environments give rise to distinctive processes of rock breakdown and transfer that give rise to landforms in the **surface-relief environment,** and also affect the formation of different types of soils and the distribution of plants and animals in the **living-organism environment.***

CLIMATIC ENVIRONMENTS

Each place on Earth has a distinctive climate that is different from all others. To conduct climate studies of larger geographic areas, geographers identify similarities between places and group them in a **climatic classification.** A climate classification is often based on selecting a limited set of weather data (e.g., annual or monthly values of temperature and precipitation) and grouping places with similar records into *climate regions*. Boundaries are drawn between different regions. Climate classifications based on the use of observed data are said to be **empiric** classifications. Climatologists can also define climate regions by studying the factors that cause weather and climate. Such a classification is said to be **genetic** because it seeks to understand the origins of observed conditions.

A **climatic environment** (Figure 8-3) is defined

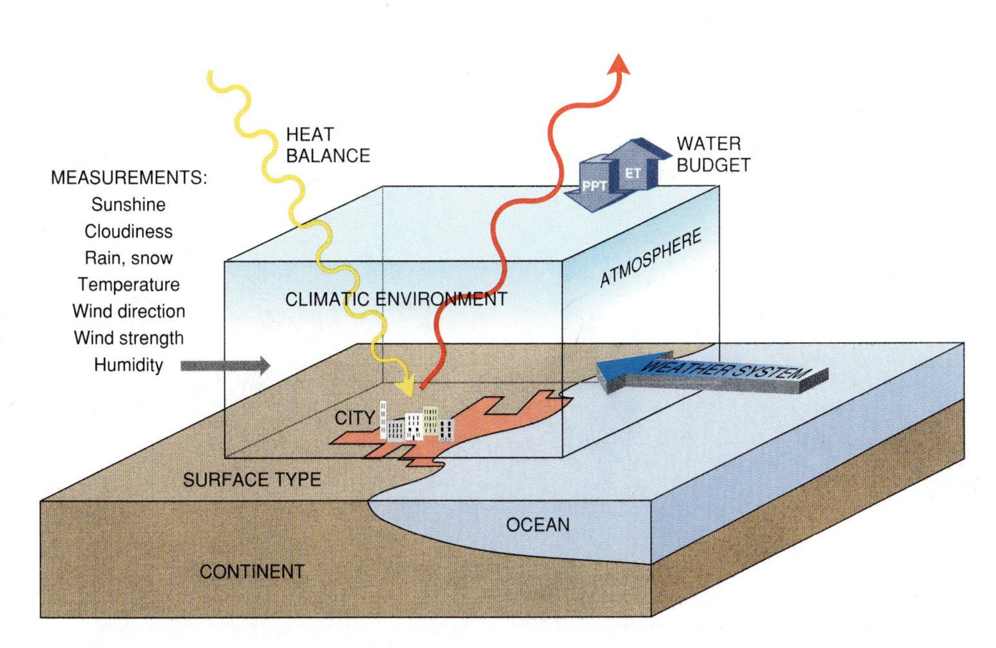

Figure 8-3 A climatic environment and its characteristics. Weather data are recorded and averaged over a period of years; the four sets of interacting conditions—heat balance, water balance, surface type, and weather systems—explain these weather conditions.

by a distinctive group of genetic factors—the heat balance, water budget, surface type, and weather systems—that provide the environmental context of the climate. Empiric data show how the different combinations of genetic factors work out in practice and can be used to identify subdivisions within each major environment.

Climatic environments occur at different geographical scales (Figure 8-4). At the "macro" scale, the world is divided into major climatic environments based on the workings of the global atmosphere-ocean environment; at this scale the whole of western Europe or eastern North America may be regarded as one type. At the "meso" scale, the variations of climate within the major environments form subdivisions, often on the basis of relief or vegetation cover. Examples are mountain, plateau, forest, and grassland climatic environments. At the "micro" scale the emphasis is on the local surface or land-use type—a field, a forest, or part of a town.

Climate Classification Using Observed Data

In using empiric data, the basis for a climatic classification is determined, data are selected, and the boundaries are drawn between regions. Some em-

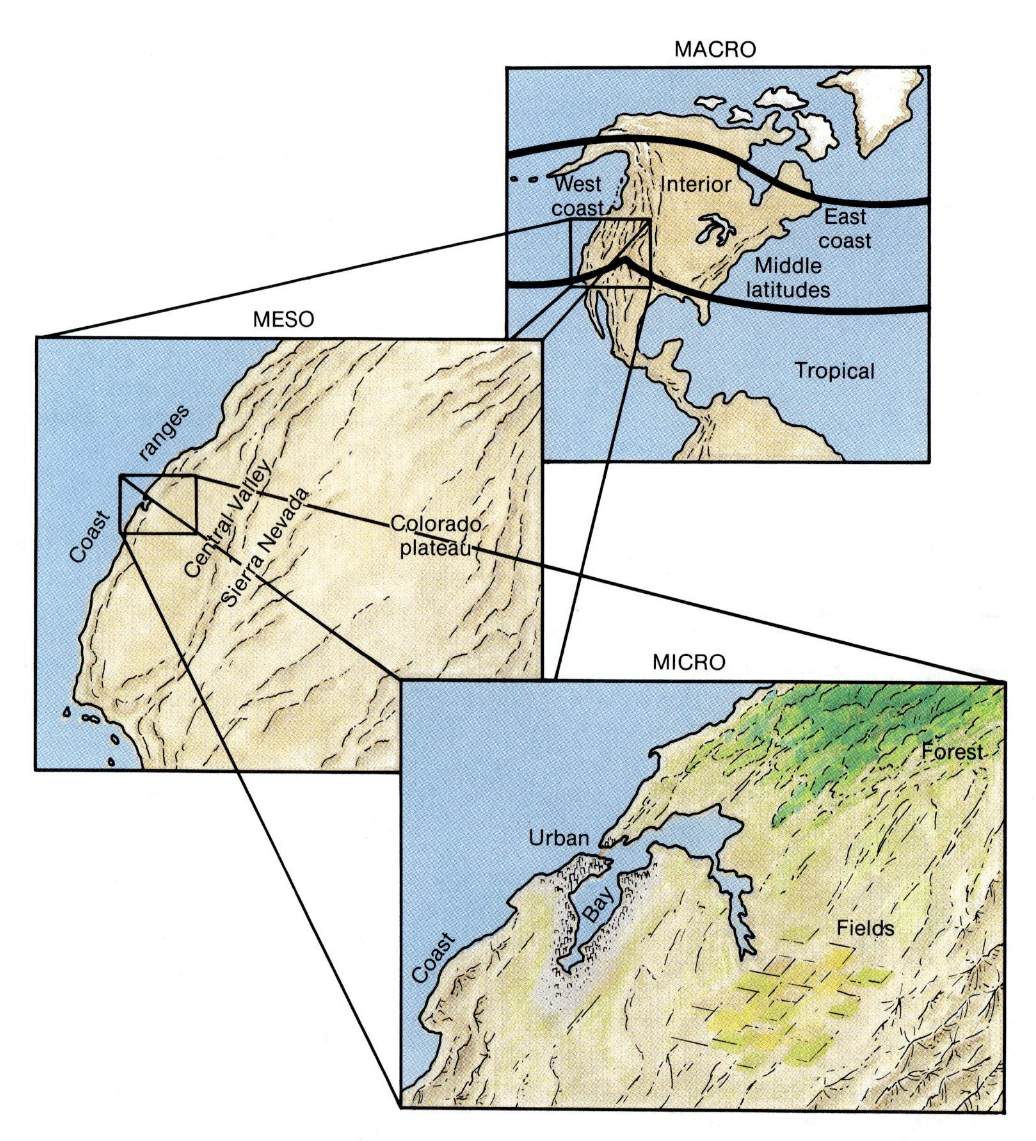

Figure 8-4 Climate and geographical scale. The continent is the unit at the macro scale; major relief features form the mesoclimatic units within a continent; local land-use differences provide the basis for many studies of microclimate—those shown can be split further into individual fields or even streets.

piric classifications are based on climate-caused differences observed in phenomena such as vegetation, soil, and crop types. In 1918 Wladimir Köppen proposed a system that has continued to be widely used. From studies of plant geography, Köppen took the boundaries of vegetation zones and picked average monthly and annual amounts of temperature and rainfall that coincided with them.

The main categories of the **Köppen classification** are summarized in Table 8-1. Four of the five categories (A, C, D, and E) are based on temperature zones divided by three significant temperatures (coolest monthly temperatures of 18°C [64°F] and −3°C [26.6°F to 50°F], and warmest monthly temperature of 10°C [50°F]). The fifth category (B) is defined by annual precipitation that is less than evaporation; subdivisions in this category are based on temperature. Of these, the three equatorward categories (A, C, and D) are subdivided on the basis of the seasonality of precipitation (all-year, summer, or winter) and the most poleward categories on the basis of summer and winter extremes. Köppen also identified a sixth "highland" (H) category for high mountain regions where the internal conditions were too variable for designation in one of the vegetation-based categories.

The world map of Köppen's climate regions (Figure 8-5) is used widely for relating climate to other

Table 8-1 The Köppen Classification of Climate

CRITERIA	MAJOR CATEGORY	SUBDIVISIONS
Coolest month 18° C (64.4° F) and above	**A** **TROPICAL RAINY**	f - rain all year (6 cm or over in driest month) m - monsoon (major contrast between very high rains in summer season and marked winter dry season) w - dry winter, summer rain (less than 6 cm in driest month)
Annual evaporation greater than precipitation	**B** **DRY**	Boundaries between humid and dry climates, and between BS and BW climates determined by formula (see below) W - arid; subdivided further by temperature (BWh = average annual temperature of 18° C or over; BWk = average annual temperature of under 18° C) S - steppe (semi-arid)
Coolest month less than 18° C but −3° C (26.6° F) or more; warmest month at least 10° C (50° F)	**C** **MESOTHERMAL CLIMATES** (Midlatitude rainy with mild winter)	f - wet all year: Cfa - hot summers; Cfb - warm summers; Cfc - cool summers w - dry winter (driest month with less than 10% of wettest summer month precipitation): Cwa - hot summers; Cwb - warm summers; Cwc - cool summers s - dry summer (driest month with less than 33% of the wettest month and less than 4 cm precipitation)
Coldest month less than −3° C; warmest month at least 10° C	**D** **MICROTHERMAL CLIMATES** (Midlatitude rainy with cold winters)	f - wet all year: Dfa - hot summers; Dfb - warm summers; Dfc - cool summers; Dfd - extreme winter cold w - dry winter (driest month with less than 10% of wettest summer month precipitation): Dwa - hot summers; Dwb - warm summers; Dwc - cool summers; Dwd - extreme winter cold s - dry summer (driest month with less than 33% of the wettest month and less than 4 cm precipitation)
Warmest month less than 10° C	**E** **POLAR CLIMATES**	T - warmest month average 0-10° C F - warmest month average less than 0° C
	HIGHLAND CLIMATES	Subdivisions too small for world map

Formula for B climates (P = mean annual precipitation in centimeters, T = mean annual temperature in °C):
BW, arid: $P < T + 7$ e.g., for T = 20°, $P < 27$ cm; for T = 10°, $P < 17$ cm
BS, steppe: $P < 2(T + 7), > T + 7$ e.g., for T = 20°, P = 27-54 cm; for T = 10°, P = 17-34 cm
[Humid A, C, D climates: $P > 2(T + 7)$ e.g., for T = 20°, $P > 54$ cm; for T = 10°, $P > 34$ cm]

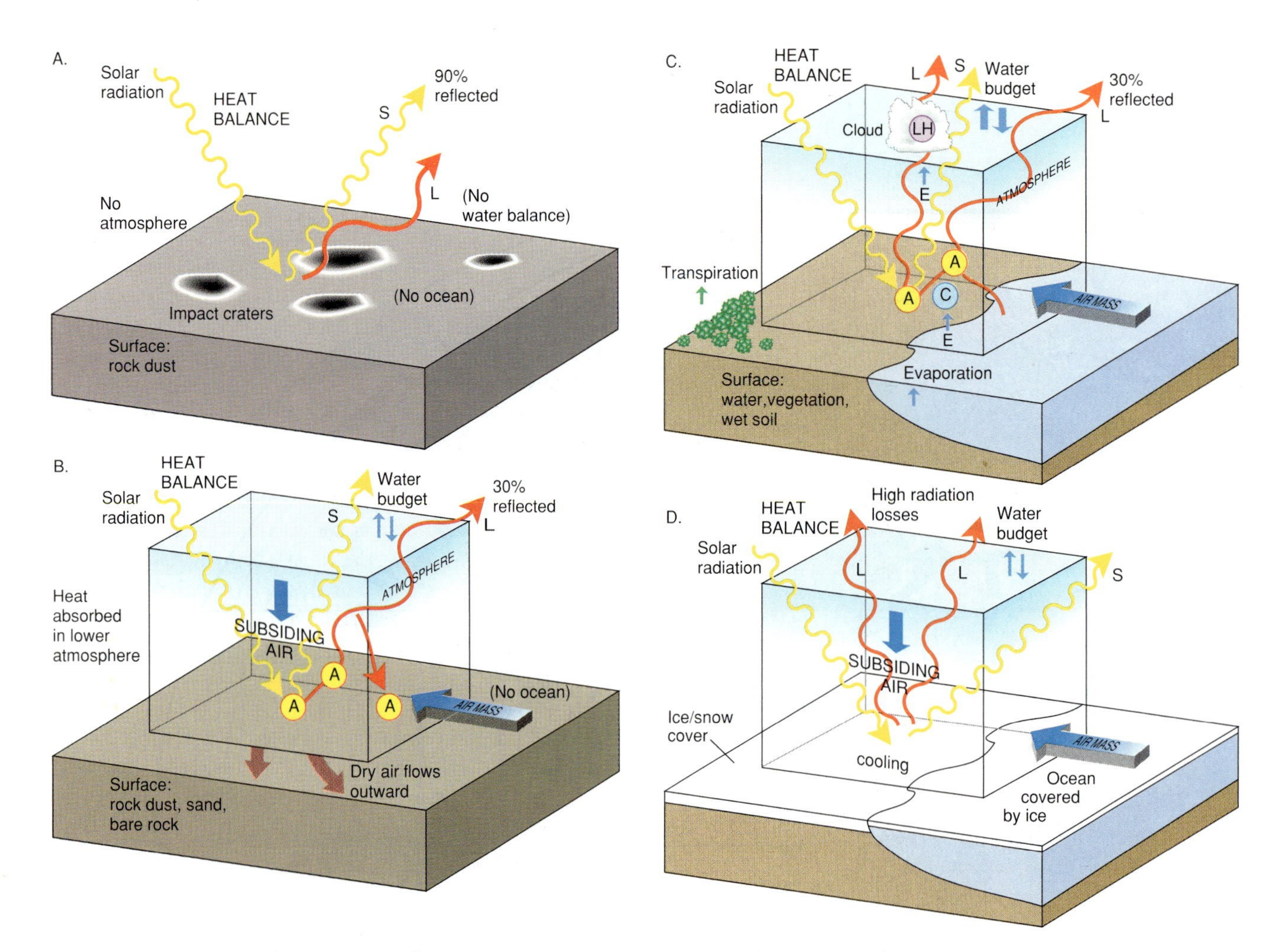

Figure 8-6 Contrasting climatic environments. **A,** No atmosphere, dry surface. **B,** Atmosphere, dry surface, moderate to high insolation. **C,** Atmosphere, moist surface, moderate to high insolation. **D,** Atmosphere, ice or snow cover, low insolation.

natural phenomena. In using Köppen's classification it must be remembered that he utilized just two measures of climate—monthly averages of temperature and precipitation—and ignored others such as wind, cloud cover, humidity, and diurnal (daily) extremes. This provides an incomplete view of the whole climatic environment of each region, so Köppen's scheme is best regarded as a starting point for climate studies. The fact that he began with vegetation boundaries and assumed climate was intricately linked to these is a debatable basis for classifying climate because other factors apart from climate affect vegetation distribution, and much of the natural vegetation has been replaced by farming. Moreover, the divisions Köppen adopted are limited to land areas, and therefore cover only 29 percent of the globe.

Climate Classification Based on Causal Factors

When climate is classified based on the causes of climatic conditions, the different types are determined by the range of dynamic factors shown in Figure 8-3. The advantages of a genetic basis for climate classification are that it arises from a broad consideration of processes in the atmosphere-ocean environment and can be applied more easily to questions of climatic change than systems based on specific empiric criteria or related to one phenomenon such as vegetation boundaries.

Four simplified sets of climatic environmental conditions experienced on Earth and the inner planets, as shown in Figure 8-6, foster an appreciation of the significance of the factors in a genetic

Figure 8-7 Astronaut John W. Young, Apollo 16 (1972) commander, collects samples of moon rocks. His spacesuit provided a personal atmosphere and protected him from solar radiation.

classification. Other planets are included in this comparison because they illustrate conditions that are not found on Earth and emphasize Earth's climatic distinctiveness. The four cases show a progression from simpler to more complex conditions, first without an atmosphere and then adding an atmosphere with dry, moist, or icy surfaces. The level of complexity builds through these cases.

Earth's moon and Mercury have a *dry surface and no atmosphere* (Figure 8-6, *A*). Surface temperatures depend solely on the insolation received and its absorption by rocks. The moon's albedo is 93 percent, so virtually all the sun's rays are reflected back to space. The small fraction of energy that is absorbed and radiated by the moon escapes directly to space because there is no atmosphere to trap it. Figure 8-7 shows the moon surface and the blackness of the "sky" without an atmosphere. There is no moisture on Earth's moon or Mercury.

Earth's continental desert regions are characterized by a *dry surface with moderate-to-high insolation* (Figure 8-6, *B*). This case covers the dry (B) climates of the Köppen system on Earth and applies to most of Mars. The atmosphere causes scattering, reflection, and absorption of the insolation. Mainly cloudless skies let in a high proportion of solar radiation and allow terrestrial radiation to escape directly back to space. Heat gained by absorption at the surface and in the lower atmosphere is reduced by the high albedo of sand and bare rock and the low quantity of water vapor in the atmosphere (which diminishes the greenhouse effect). Sensible heat is more important than latent heat in energy transfers. In these conditions, the daily temperature regime is one of extreme variation from hot days to cold nights. Although desert regions are dry relative to the rest of Earth, some water vapor exists in the atmosphere there because air masses transport it.

The most common type of climatic environment on Earth is a *moist surface with moderate to high insolation* (Figure 8-6, *C*). In addition to the oceans, three of Köppen's categories—tropical rainy, mesothermal, and microthermal (A, C, and D)—exhibit such conditions. Much of the insolation reaching the surface in moist areas is taken up by evaporation and transpiration in addition to absorption. The energy supplied by the sun is transferred up through the atmosphere as both sensible and latent heat, the proportion depending on the relationship between precipitation and evapotranspiration. The use of energy in evaporation and its absorption by water vapor and other greenhouse gases in the lower atmosphere may result in more moderate ground temperatures than are found in desert lands receiving the same amount of sunlight. However, the moderating effect of moisture from vegetation is seasonal, since transpiration is reduced during dry or cold periods. Overall, the climatic environment over a moist surface is characterized by a greater complexity of atmospheric processes and exchanges and less emphasis on a simple balance of incoming and outgoing radiation.

Ice and snow surfaces with low insolation (Figure 8-6, *D*) equate generally with Köppen's polar (E) climates. Once a snow cover is established, it tends to persist, reflecting insolation and cooling the air above by conduction even when it is sunny. Such surfaces have high albedos and reflect over 70 percent of insolation back to space, leaving little to be absorbed. The small amount of water vapor in the low-temperature atmosphere restricts the greenhouse effect. Although such surfaces cover relatively small parts of the planet at present, they have covered more in the past, and the extent of snow cover on Earth is acknowledged as an essential factor in predicting the direction of climatic change.

north as San Francisco when the aridity extends northward in summer.

Temperature Fluctuations. The combination of intense insolation through cloudless skies and adiabatic warming of the air leads to high daytime temperatures. Shade temperatures commonly exceed 30°C (86°F). The world's highest shade temperature of 58°C (136°F) was recorded at Azizia in the Libyan Sahara, and 57°C (134°F) has been measured in Death Valley, California. Temperatures of 82°C (180°F) occur in desert air above exposed bare rock or sand surfaces out of the shade. However, these temperatures would be even higher were it not for the high albedos of sand and bare rock, which reflect away a high proportion of insolation, and the low content of water vapor in the air, which diminishes the amount of heat that the atmosphere can absorb. Terrestrial radiation continues through the clear skies at night, and the air in contact with the surface cools rapidly, giving diurnal ranges of as much as 40°C (72°F). The low nighttime temperatures can result in dew (despite low daytime relative humidity) and winter frosts.

Trade-Wind Climatic Environments

When it grew light next morning, a thick mist lay over the coast of Peru, while we had a brilliant blue sky ahead of us to westward. The sea was running in a long quiet swell covered with little white crests, and clothes and logs and everything we took hold of were soaking wet with dew. It was chilly, and the green water around us was astonishingly cold for 12 degrees south. There was a rather light breeze, which had veered from south to south-east.

And the wind came. It blew up from the south-east quietly and steadily. Soon the sail frilled and bent forward like a swelling breast, with Kon-Tiki's head bursting with pugnacity. By the late afternoon the trade wind was blowing at full strength. It quickly stirred up the ocean into roaring seas which swept against us from the stern. We knew that from now onwards we should never get another onshore wind or a chance of turning back. We had got into the real trade wind, and every day would carry us farther and farther out to sea. The only thing to do was to go ahead under full sail. After a week or so the sea grew calmer, and we noticed that it became blue instead of green. We began to go west-north-west instead of due north-west.

Thor Heyerdahl, "The Kon-Tiki Expedition," 1948

Trade-wind climates (Figure 8-12) are to some extent the equivalents of the tropical arid climates over the oceans, but the contrast in surface produces major differences. Köppen has no category for this type of climate, although it is the most extensive climatic environment of all. It occurs mainly over the tropical oceans, but also affects strips of land on the western margins of these oceans; in the Americas, the Caribbean and most of central America are included.

The five weather stations demonstrate several contrasts with the tropical arid stations shown in Figure 8-11. First, they generally have much smaller monthly and annual ranges of temperature and lower maximum temperatures. Most trade-wind stations have extreme temperature ranges of less than 20°C (36°F) compared to 40°C (70°F) in the arid areas. The only station with significant seasonal variations in temperature is in the Azores at the poleward limit of this region. Second, the relative humidity is generally in the range of 70 to 80 percent. Third, each station, apart from Ascension Island, has a precipitation of over 580 mm (23 inches). Precipitation totals and thunderstorm activity increase toward the western margins of the oceans. Precipitation at most places is distributed throughout the year, but there is a tendency to a summer maximum closer to the western margins and a winter maximum on the eastern margins.

The trade winds blow toward the equator around the eastern margins of the oceanic subtropical high-pressure zones; they are northeasterly in the northern hemisphere and southeasterly in the southern hemisphere. These winds are among the most constant features of the global atmospheric circulation, and the area they affect shifts little with the seasons. The opening excerpt gives an idea of the strength of the trade winds.

Variations Within the Trade-Wind Climatic Environment. The air on the eastern margins of the subtropical high-pressure zones is dry as a result of either descent in the subtropical high-pressure zones or blowing from the deserts to the east (Figure 8-13). When the dry air reaches the ocean surface, high rates of evaporation occur, raising the relative humidity. However, these are areas of cold ocean currents, and the cooling produces fog above the ocean. Clouds and rain are seldom produced because descending air above is warmer than this surface layer and the inversion traps the fog near the ocean surface.

In the western parts of the tropical oceans, the trade winds are less constant, the air has a higher humidity, and surface waters are warmer. The inversion layer rises and the greater instability in the atmosphere increases convection and creates weather systems that produce heavy summer rains. This is the zone where easterly waves, tropical storms, and hurricanes thrive. Although the pre-

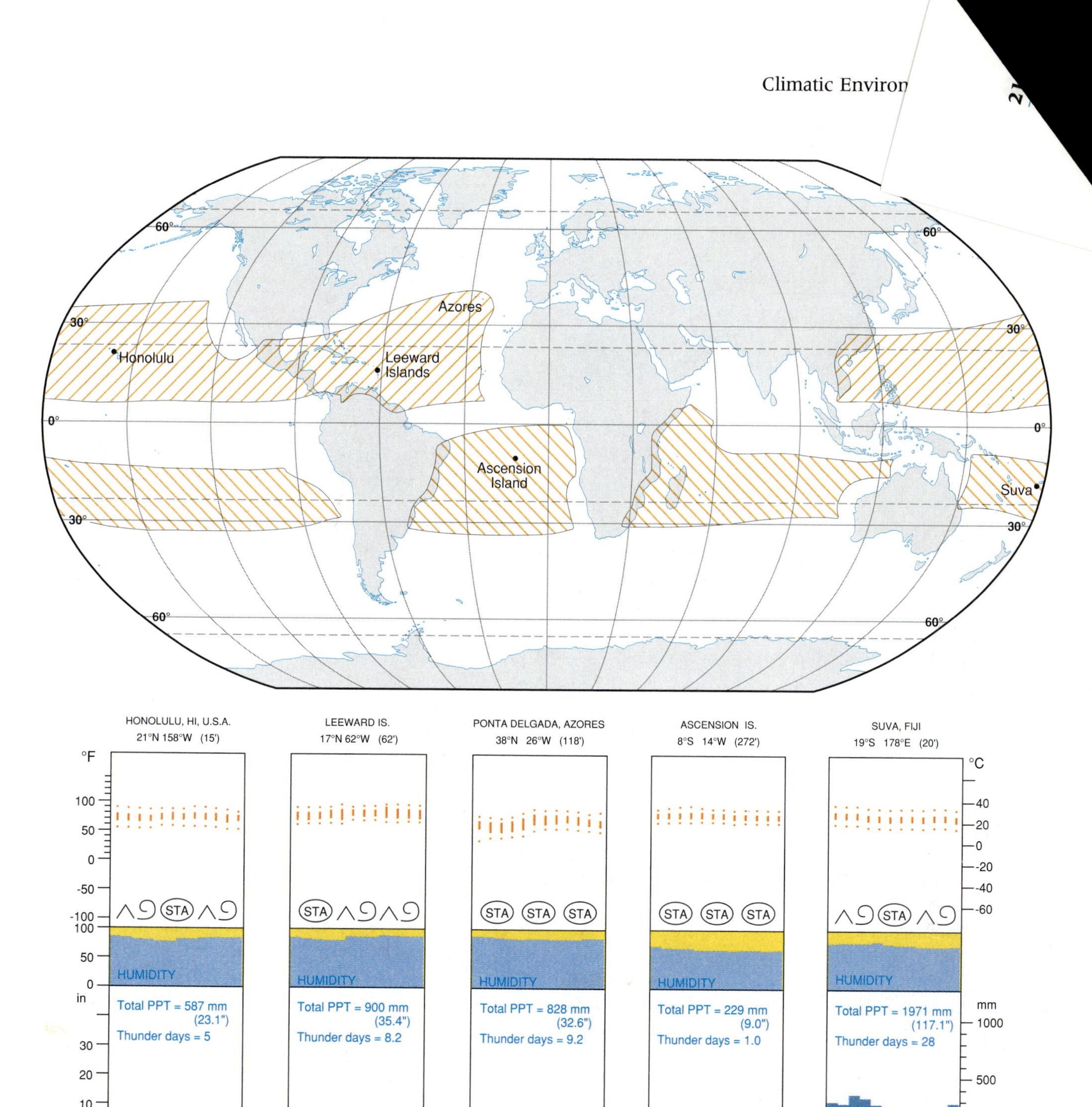

Figure 8-12 Trade-wind climates: world map, climographs.

cipitation totals of the eastern and western parts of the tropical ocean atmosphere differ greatly, the causal factors are part of the same dynamic atmospheric system lying between the intertropical convergence zone and the subtropical anticyclones.

The trade-wind climatic environment is largely oceanic, and its main effects on human activities are on islands and coasts. Islands often show contrasts in rainfall between the east-facing windward coasts and the leeward slopes. The island of Hawaii receives 3000 mm (120 inches) of rain on the southeastern side, while less than a tenth of this falls on the far side of the same island. Honolulu is on another island farther northwest, and its precipitation is also lower, since the moisture carried by the winds approaching the Hawaiian Islands tends to be "rained out" on the first island the winds encounter.

The trade-wind climatic environment also affects some coastal areas on the western sides of oceans, including northeast Brazil and eastern Africa.

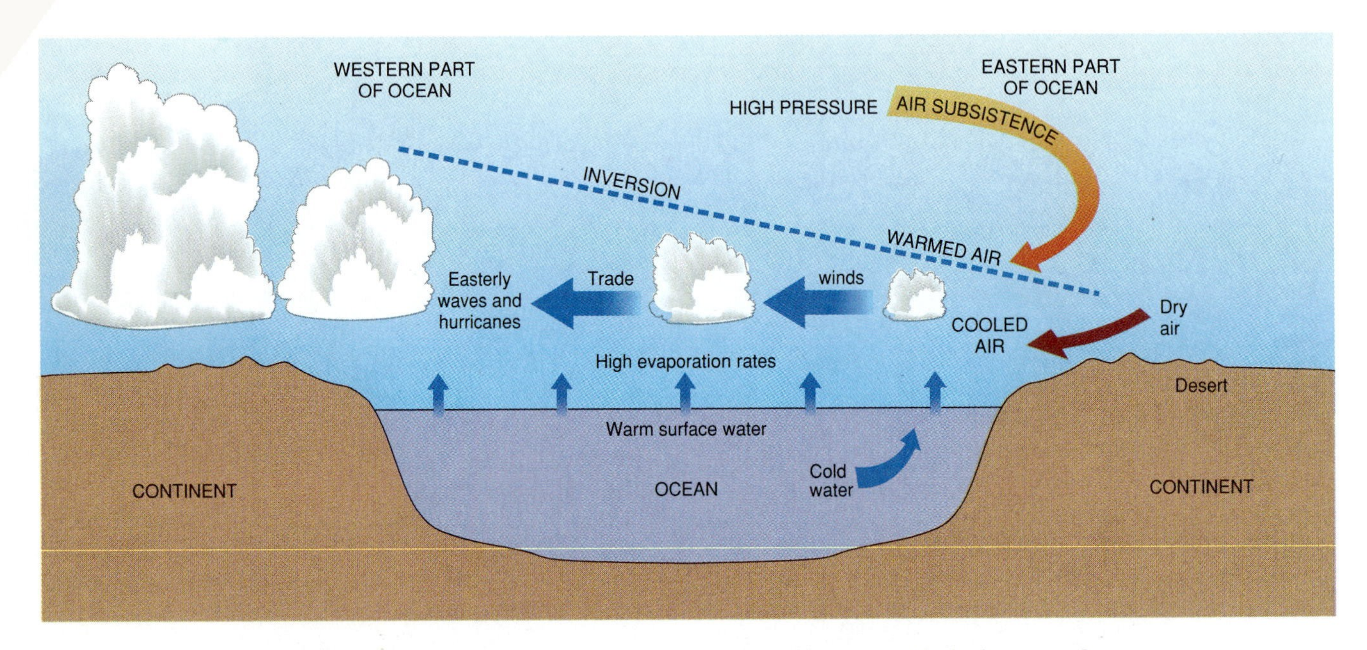

Figure 8-13 The trade-wind climatic environment: comparisons and links between the eastern and western parts of oceans.

Northeast Brazil experiences heavy coastal rains, but the inland plateau is a zone with great swings between arid and humid years, making life uncertain for the inhabitants. The poverty of people in this area led to political pressures for their resettlement in the Amazon basin, and that, in turn, to environmental pressure on the Amazon rainforest.

El Niño. The central significance of the tropical ocean climatic environments in the global climate system has been emphasized by studies of the El Niño phenomenon in the 1970s and 1980s. The **El Niño current** is a southward flow of warm water along the coast of northwestern South America from Colombia to Ecuador. Around Christmas ("El Niño" means "Christ child"), it pushes the cold Peru Current back along the Peruvian coast (Figure 5-2, *C*). In some years (e.g., 1953, 1957-1958, 1965, 1972-1973, 1976-1977, 1982-1983, 1986-1987, 1991-1992) the Peru Current retreats farther to the south and is replaced with warmer water for much longer. This may last for over a year and bring disaster to the Peruvian fishing industry, whose catch depends on the nutrients brought by the cold current.

The ultimate causes of this "enhanced El Niño" are unknown, but the sequence of associated events is becoming clearer. The normal situation is that the subtropical high-pressure zone over the eastern Pacific gives rise to trade winds that blow toward low pressure over Indonesia in the western Pacific (Figure 8-14, *A*) during the latter half of the year. The consistent trade winds blow warm surface water westward, reducing the depth of warm water off the South American coast. Cold water and nutrients rise from deeper water to the surface and maintain an ecosystem rich in plankton, fish, and sea birds. This keeps the shallow, warm El Niño current farther north off Colombia and Ecuador, except for the December to April period when it extends southward.

Every few years this normal situation breaks down. The low pressure over Indonesia moves east to the middle of the Pacific, the subtropical high-pressure zone weakens, and the trade winds are replaced by westerly winds over a large part of the tropical Pacific. These winds blow the surface waters eastward, so that warm water "piles up" against South America (Figure 8-14, *B*). The upwelling off the coast of Peru is now trapped under the deeper thickness of warm water. This longer-term stoppage of the upwelling is *enhanced* El Niño. The loss of cold, nutrient-rich water may devastate the local ecosystems. At the same time the low pressure brings clouds and rain to the coasts of Ecuador and Peru, and the replacement of the Indonesian low pressure by high pressure produces drought in Australia.

El Niño is part of a phenomenon in the atmosphere-ocean environment that affects the whole of the tropical Pacific Ocean, and thus covers over a fourth of the tropical atmosphere-ocean environment. As might be expected, events in an area of this size have wider repercussions, and it is thought that they may be linked to other global atmospheric

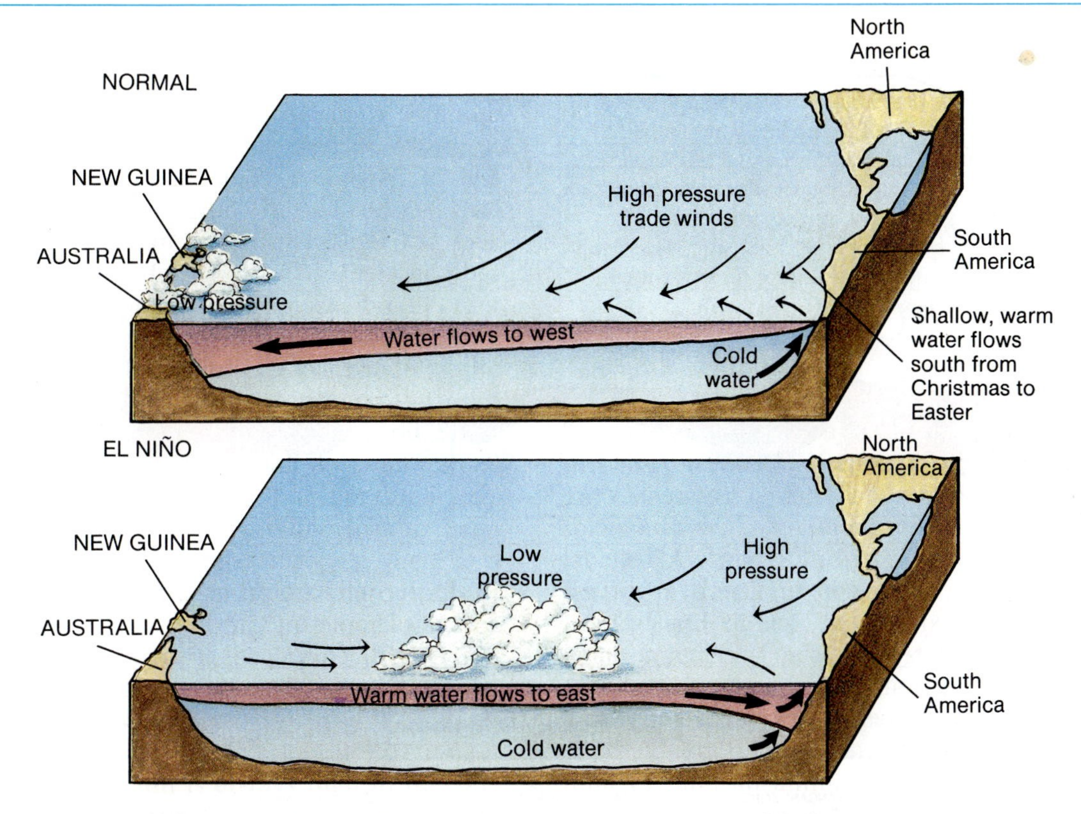

Figure 8-14 The El Niño effect: the shift in atmospheric and oceanic movements that results in the enhanced El Niño effect. The upper diagram shows how trade winds normally blow warm surface water westward and allow cold water to come to the surface along the South American coast. During the El Niño (lower diagram), winds blow eastward, deepening the warm water off South America.

conditions of similar scale, such as the Asian monsoons or equatorial convection centers over the Amazon and Zaire basins, but the nature of the connection is not clear. Rival explanations have suggested links with the events governing the occurrence of tropical cyclones in the western Pacific, or with volcanic eruptions. Research to establish wider connections to other climatic regions and shifts may help predict large-scale weather changes from year to year.

Equatorial Climatic Environments

. . . no matter what month, daily temperature variations are greater than monthly averages. This daily cycle of temperature plays a role in the scheduling of agricultural work. During my 1973-74 fieldwork in Vila Roxa, most colonists native to the Amazon began work around 06.30 hours and stopped by 11.00 hours. Between 11.00 and 15.00 to 16.00 hours they would go home, rest in the shade, and carry out moderate types of activities such as sharpening tools, feeding livestock, and visiting neighbours. This is the hottest time of day, when humidity levels go above the already high mean of 85 per cent. Cooling is difficult during strenuous work since humidity hovers between 86 and 90 per cent from March through October, the months of most intense work effort. In the course of strenuous work such as forest clearing and cutting, for instance, four men each consumed an average of a gallon of water in a half day . . .

The volume and relative constancy of precipitation in the humid tropics can blind one to the significant patterns of variation. Altamira has a marked dry season of four months, during which the rainfall is below 60 millimetres per month. The mean annual precipitation is 1705 mm, although variability from year to year can be great. Annual potential evaporation is 1595 millimetres. Variation is not only seasonal, but also daily. One third of the annual rainfall at Maraba in 1974 came down in a 24-hour period. In Altamira I noted that in March 1974 there were 24 consecutive rainy days, in which 522 mm of rain fell—an amount greater than any monthly total and 30 per cent of the annual mean.

Emilio Moran, "Developing the Amazon," 1981

Equatorial climatic environments occur over an area up to 10 degrees of latitude on either side of the equator (Figure 8-15). This is a narrow band, but it covers a large area, especially when the oceans are included. The equatorial climates are categorized as *Af* by Köppen signifying a coolest month with a temperature above 18°C (A) and rain all year (f).

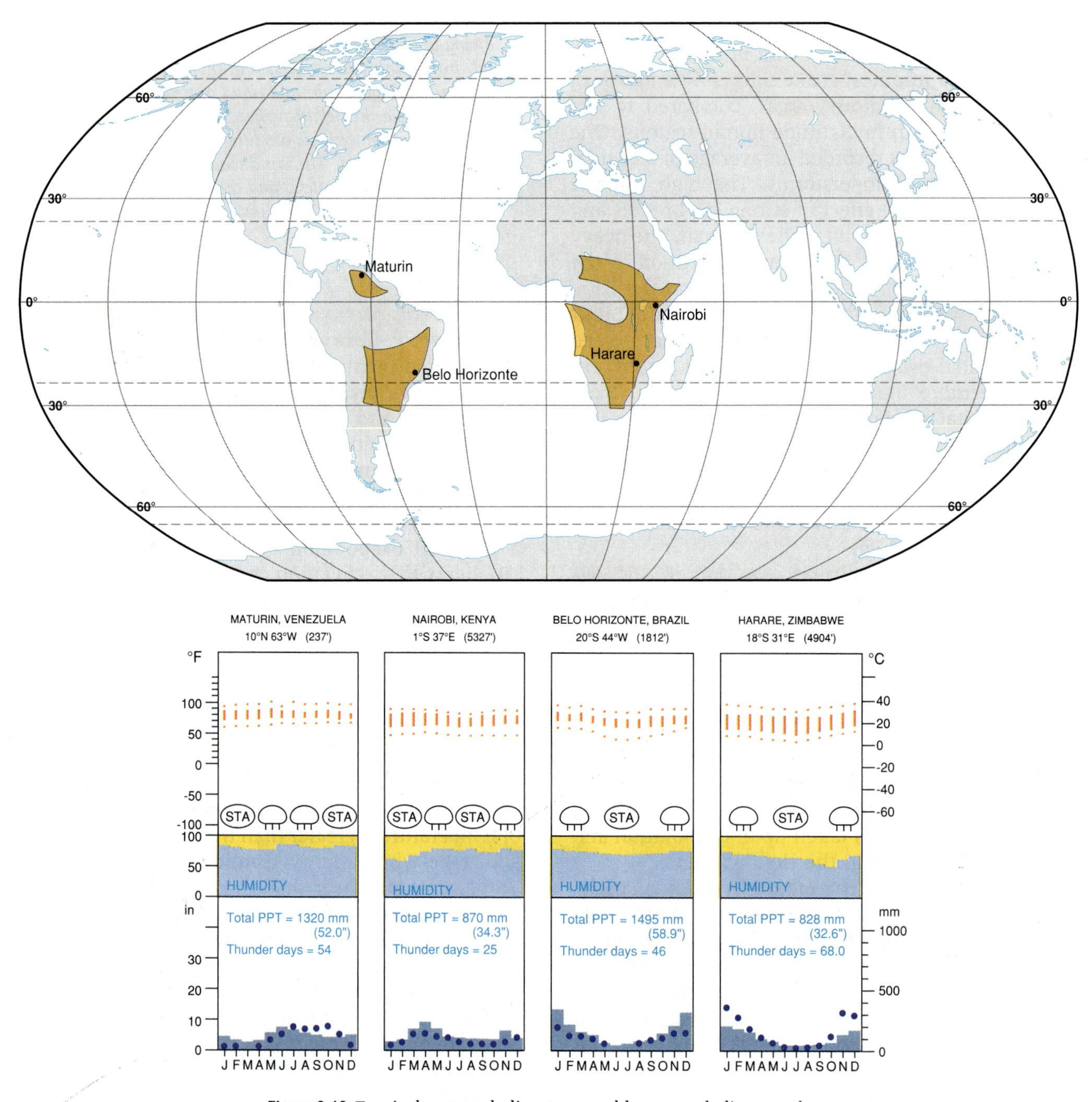

Figure 8-16 Tropical seasonal climates: world map and climographs.

winter season is basically an extension of desert conditions toward the equator as the intertropical convergence zone shifts to the opposite hemisphere and higher pressure systems move to cover the region. Figure 8-18 shows the vegetation typical of this seasonal climate—wooded savanna—which is adapted to the long dry season.

The Sahel Zone. The drier margins of the seasonal tropics immediately south of the Sahara in Africa (known as the "Sahel" zone) have been the subject of intense concern owing to the effect of prolonged drought on the lives of the inhabitants since the early 1970s (Figure 8-19). At first the debate was whether natural climatic change or human intervention was causing such "desertification." Human activities were blamed for plowing and overgrazing of the grassland vegetation by domestic animals, which increase the amount of bare soil and therefore the albedo. A higher albedo causes more insolation to be reflected back to space instead of being absorbed and heating the surface. Less heat energy is radiated into the atmosphere and this re-

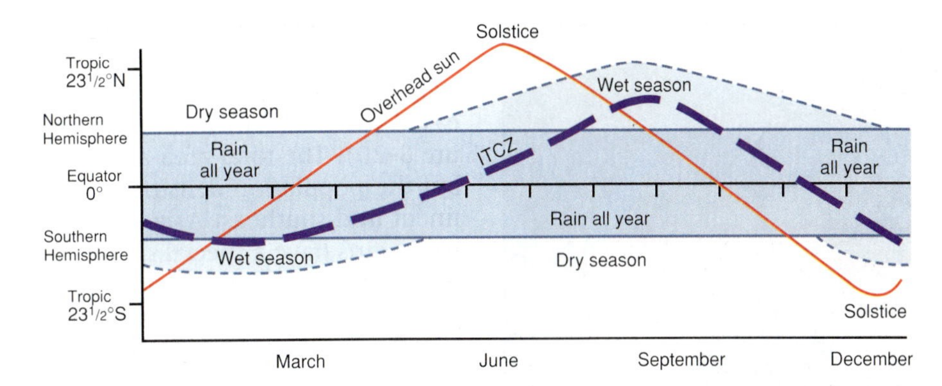

Figure 8-17 The relationship between the apparent seasonal movements of the overhead sun, the intertropical convergence zone, and the seasonal rains. Equatorial climates have rain all year, with slight increases in some seasons, especially at stations distant from the Equator. Seasonal climates have wet summers and dry winters.

Figure 8-18 A farmed area in Zimbabwe, southern Africa, near the end of the dry season. The hills in the background, and the vegetation covering them are typical of seasonal tropical areas.

duces atmospheric moisture, thereby decreasing rainfall. However, current scientific opinion favors the idea that climatic fluctuations in these areas are primarily natural, since the present period of drought is the third in this century. The human tragedies that result from drought are seen as resulting from a shift away from the traditional culture of these regions. The imported agricultural practices have less resilience and adaptation to the fluctuations in climate (see Box 21-1: Desertification).

Recent climatic research has demonstrated strong links between the dry conditions of the Sahel and higher temperatures in the ocean waters of the Gulf of Guinea off West Africa. This is another case where improved understanding of the atmosphere-ocean links may provide a better basis for future land use and water management decisions.

RECONSTRUCTING PAST CLIMATES

Since climatic change has continued from Earth's beginnings, a wide range of different types of evidence is available for reconstructing the past. Direct data—those collected by meteorologic instruments—are available only for the last 100 to 150 years. Historic writings provide evidence of changes over the last few thousand years. Beyond that, climatologists depend on evidence recorded by the natural environment during climatic changes. They learn about climates of the distant past from layers of glacial ice, ocean floor sediments, rocks, fossils, and pollen trapped in bogs or deposits on the bottoms of lakes. The study of ancient climates is known as **paleoclimatology.**

As the scale of past time increases, the precision of measurements decreases, and estimates of climatic conditions vary more widely. The best data are those of the last few years, while the tree rings or ocean floor sediments provide information that is based on phenomena indirectly connected with climate and requires careful interpretation. Figure 9-3 shows how graphs can be employed to study four time scales of climatic change and the evidence on which they are based. First, meteorologic evidence is used to reconstruct changes over the last 1000 years (Figure 9-3, *A*); second, pollen, archeologic, and historical records are used to reconstruct climates during the few thousand years since the last glacial phase (Figure 9-3, *A, B*); third, landforms and sediments are used to reconstruct the changes in climate over the last few hundred thousand years (Figure 9-3, *C*); and fourth, geologic evidence makes it possible to reconstruct Earth history, although in less detail, over tens and hundreds of millions of years (Figure 9-3, *D*).

LINKAGES

The measurement of changing climates highlights a number of linkages among aspects of Earth environments. The climate of a given time influences the nature of organisms in different parts of the world ***(living-organism environment).*** *It also produces distinctive soil types and landforms* ***(surface-relief environment).*** *When climate changes, it favors different organisms, soil types, and landforms. It is common to find the old and new together as a result of climatic change.*

The Period Covered by Meteorologic Records

Thermometers have been used to measure temperatures in enough locations around the world to provide a picture of climatic change since about 1860. Figure 9-4, *A*, illustrates the general warming trend, together with a series of fluctuations every 10 to 15 years within that pattern. The order of change is given as a total of 1°C during this period. The overall trend of rising temperature since the mid–nineteenth century, known as **global warming**, has caused widespread concern among scientists, which has been communicated to governments in the developed world (see Box 9-2: Political Impacts of Climatic Change). One fear is that global warming is caused by human activities and will cause shifts of climate that affect the geographic distribution of human economic activities. This may occur as melting ice sheets cause sea levels to rise, drowning ports and productive low-lying areas, or as drought occurs in major farming areas. The pattern of temperature for the United States since 1895 (Figure 9-4, *B*), however, suggests a slight general cooling over the last 50 years. This apparent conflict with the global pattern indicates that the effects of overall global changes are not the same for every area.

The pattern of precipitation over the United States since 1895 (Figure 9-4, *C*) shows that higher totals at the start of the twentieth century were followed by a period of lower precipitation from 1925 to 1940 (the main phase of Dust Bowl wind erosion in Oklahoma); this was followed by higher precipitation in the 1940s, and drier conditions in the mid-1950s and mid-1960s. Since then, precipitation has been higher, with some of the highest yearly totals on record.

Clearly, it is desirable to assemble as many elements as possible in reconstructing past climates. Temperature has been the main criterion used for this purpose, because the records of temperature measurements extend back farthest in time, and because temperature is the main weather element that can be reconstructed for more ancient history.

Postglacial Climatic Environments

Over the longer term history of Earth, a series of major ice ages has interrupted periods of greater warmth. **Ice ages** are periods of several millions of years when ice sheets form over the polar regions and spread out to cover extensive areas of the higher latitudes and even invade midlatitude regions. Within each ice age the climate fluctuates between

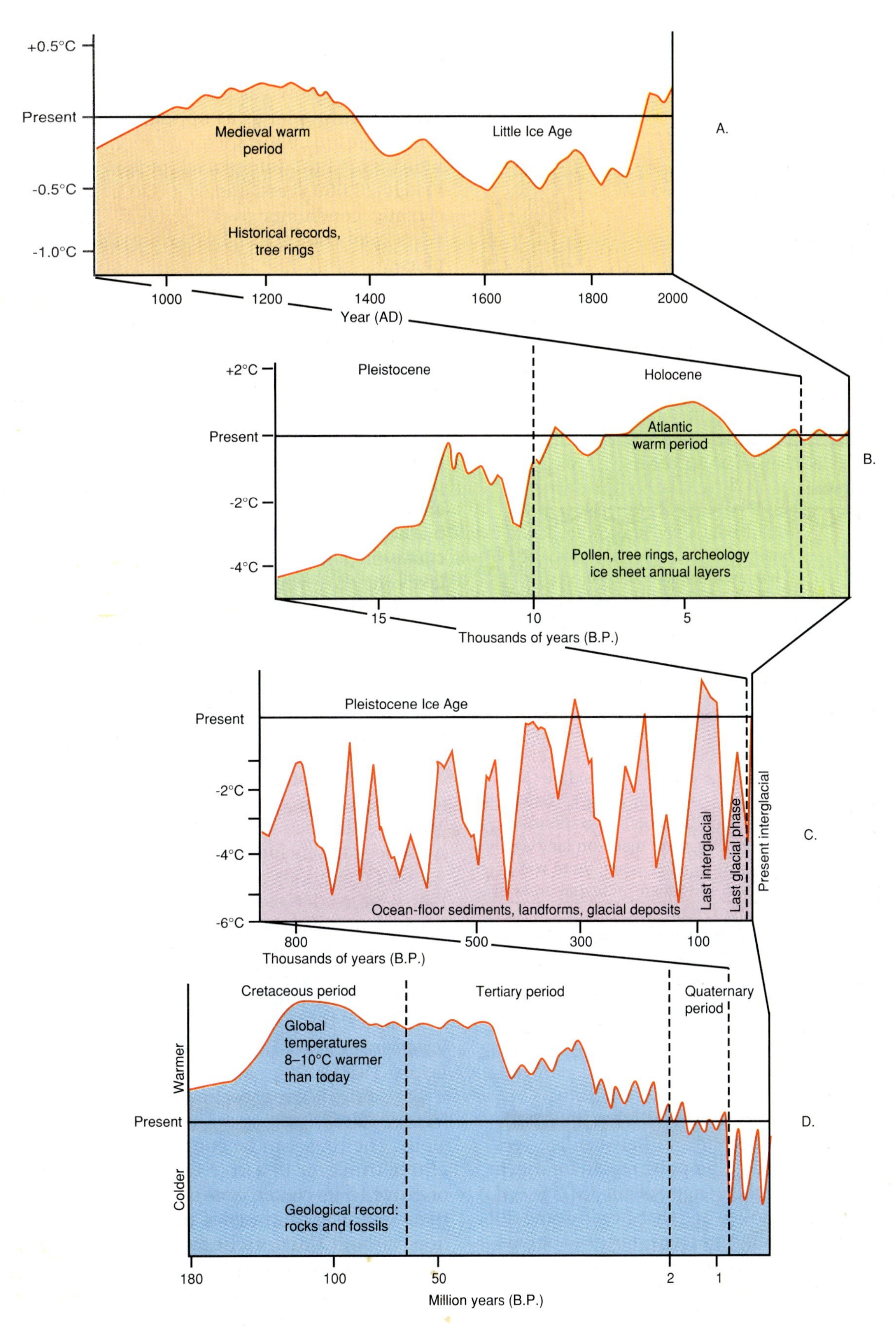

Figure 9-3 Time scale, varied types of evidence, and climatic change. These diagrams demonstrate how climatic variability and change work at different time scales. For the most recent time scale, paleoclimatologists must estimate the smaller fluctuations more precisely.

A full appreciation of the evolution of Earth's landforms rests on a study of the materials that compose the rocky layers of this planet and how they are arranged. Earth is composed of "building blocks" at different scales (Figure 10-1). They range from the largest, which is the solid planet as a whole, and move through internal layers of the planet, major relief features, smaller landforms, rocks and minerals, to atoms. Each scale contributes to an understanding of landforms. The excerpt describes how rock layers often provide a dramatic element in scenery.

The study of the "building blocks" is also significant for human interactions with this environment, since they provide basic resources used in human societies. The earliest human societies used stone as a resource for implements, such as axes, arrowheads, and a wide range of tools. New technologies developed with the use of metals and new solid-earth resources were extracted to make copper, bronze (copper with tin), iron, and eventually steel and aluminum goods. The use of timber and fossil fuels to produce the purified metals from rock-based resources increased alongside the metal technology. Both these trends stimulated interest in the rocks and minerals found at Earth's surface, and the science of geology developed. However, just as the materials extracted from the planet's outer layer have helped improve human life, they can also contaminate the atmosphere, oceans, and living organisms. This is a dilemma that increasingly faces humans as they continue to use Earth's resources for a rapidly growing population.

This chapter is about Earth materials, the ways in which they are arranged, and the surface features of the solid-earth environment. An examination of the basic structure of Earth—its internal layers—is followed by a consideration of the form of Earth's surface and the range of sizes at which landforms occur. The next section is about the main rock types that form the crustal layer of the solid-earth environment, and the contorted and fractured shapes in which they occur. The chapter concludes with an overview of the methods used to date the origin of rocks and a description of how surface rocks are depicted on geologic maps.

EARTH'S LAYERS

Earth is composed of a series of concentric "shells" of different materials in a variety of physical states, although most is solid (Figure 10-2). The atmosphere-ocean environment is the outermost of these layers and the others compose the solid-earth environment.

The internal structure of the solid-earth environment cannot be viewed directly from the surface. It has been worked out from the study of earthquake shock waves, or **seismic waves**. *Seismology* is the study of earthquakes or artificially produced shock

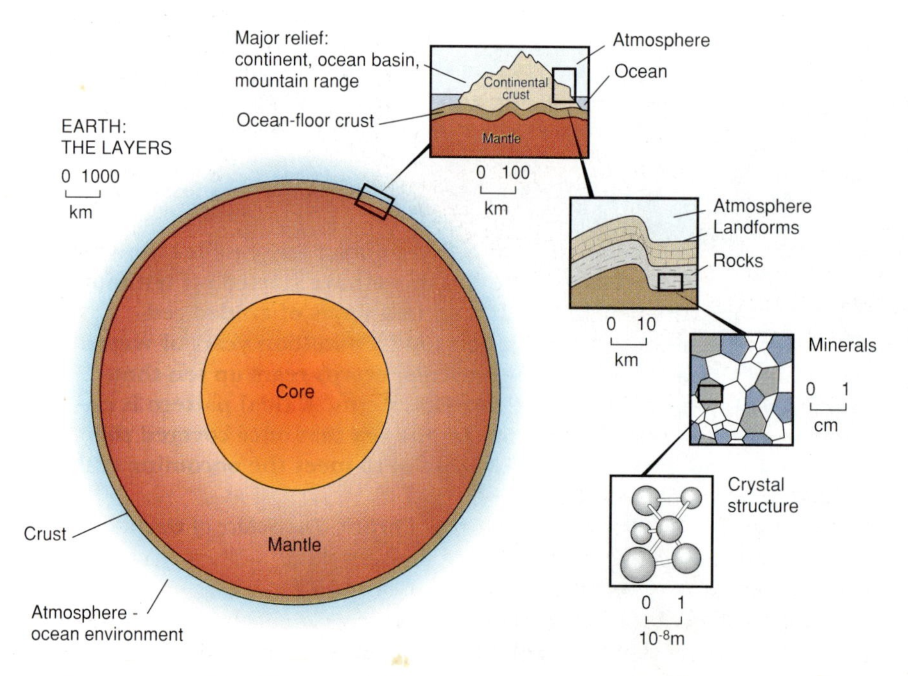

Figure 10-1 The building blocks of planet Earth. At the largest scale are the concentric layers of the atmosphere-ocean, crust, mantle, and core; then, largest-to-smallest, are the major relief features, the rocks, the minerals, and the atoms.

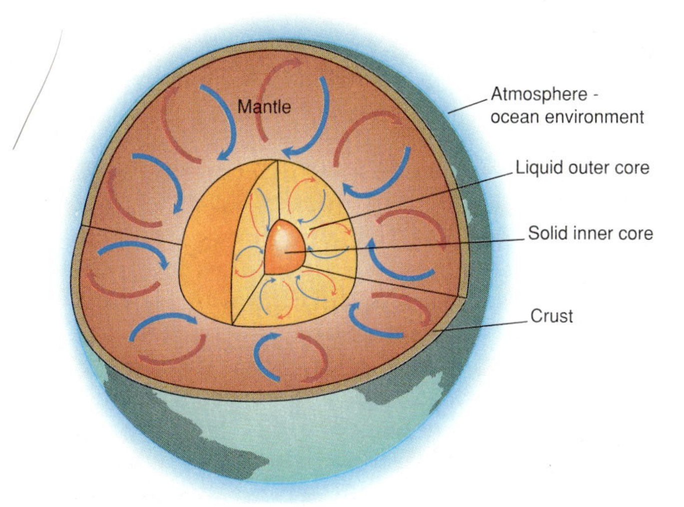

Figure 10-2 Earth's interior layers: a series of concentric shells. The core is the source of heat energy that is transferred through the mantle by conduction. The arrows show possible convection inside Earth.

waves. When an earthquake occurs, or an explosive charge is detonated, it sends seismic waves around the surface layer and through Earth's interior. These shock waves are measured at surface stations around the world. Artificial shock waves are produced to investigate the nature of rock structures too deep for firsthand observation. The techniques developed in seismology are used for detecting underground nuclear explosions.

The waves that travel through Earth change speed when they pass from a solid to a liquid, or through rocks of different densities inside the planet. Two types of waves pass through Earth. The slower of them, S waves, are blocked by liquid rocks inside the planet, while the faster P waves are not. The third type of wave (L) passes through the surface rocks around the outside of the planet. Speed of seismic wave transmission is faster in denser materials. By measuring changes in the speed and direction of the waves, seismologists have discovered that there are three main layers inside Earth—core, mantle, and crust. Each has its own properties and some internal variations.

Earth's **crust** is merely a veneer covering the planet's interior. Much more is known about the crust than about the mantle or core because the features of the latter two cannot be observed directly. The crust is composed of two layers with different compositions. The lower layer is called *sima* because it is composed of rocks containing large amounts of silicon and magnesium. This layer, which is about 10 km thick, consists of relatively dense, dark rocks. In the ocean basins, these form the ocean floor. The continental areas are composed of *sial* (rocks dominated by silicon and aluminum) and lie on top of sima. Sial is less dense than sima and includes many types of rocks. Earthquake seismic waves increase in speed as they pass from the sial into the denser sima rocks. Thus, the entire globe is covered by a layer of sima; in the oceans this is the only crustal layer, but the continents have the additional layer of sial. The continental masses stand higher because their rocks are less dense; the relationship has been likened to the way an iceberg (less dense) protrudes above water (more dense). The greater the thickness of continental rocks, the higher are their highest points. The total thickness of crust in continental areas averages 33 km (20 miles), but may be as much as 75 km (45 miles).

Earth's **mantle** is a mostly solid layer, nearly 2900 km (1800 miles) thick, in which both temperature and pressure rise toward the interior. Evidence for the rising temperatures comes from deep mines and the melted rock brought to the surface by volcanoes. Below the surface the temperature rises sharply, and then more gradually with greater depth. The rise in pressure results from the weight of overlying rocks, which increases steadily with depth. As pressure rises, so does the temperature at which rock melts, and for most of the mantle's depth the pressure is too high to allow rocks to melt. However, in a thin layer between 100 and 200 km (60 to 120 miles) below the surface, the temperature rise overcomes the effect of rising pressure. This causes the rocks to melt, at least partially, and to flow very slowly, like taffy. This zone where partial melting and flow occur is known as the **asthenosphere.**

The modern understanding of Earth's surface layers is that the two crustal layers, sial and sima, are bonded to each other and to the uppermost part of the mantle in a single rigid unit above the asthenosphere. This composite layer, called the **lithosphere,** averages 100 km thick, but is over 300 km thick beneath some continents and only 45 km thick under some ocean ridges. Figure 10-3 depicts the asthenosphere and lithosphere.

The outer surface of Earth's **core** is more than 2900 km below the surface. It appears to be composed of the densest materials of any layer of Earth. However, the S waves do not travel through this zone, so at least the outer part of the core must be in a liquid state. The outer core is liquid because the high temperatures there have melted the rock: hence, its state is known as *molten*. It is thought that this remains from a time when much more of Earth was molten. The innermost part of the core is probably solid; although it is very hot, with temperatures estimated at around 2500°C (4500°F), the high pressures there prevent it from melting.

The other planets in the solar system may also be layered internally. The smallest planetary bodies (Earth's moon, Mercury) have a rocky surface, but it is unlikely that much of their interior is still molten because they are too small to hold heat for billions of years, as Earth has. Mars appears to have had a longer phase in its early existence when internal heat produced surface features such as volcanoes, but this subsided as the internal layers solidified. Venus may be similar to Earth in its internal layered structure and level of internal heat; evidence from the Magellan spacecraft confirms that Venus has mountains and volcanic activity. It also shows that many of the features of the Venusian surface are quite different from those on Earth. The outer planets, such as Saturn and Jupiter, have relatively small rocky cores surrounded by much greater thicknesses of low-density fluids, often at very low temperatures. However, space probes have shown that some of the moons of outer planets continue to experience volcanic activity—a sign of internal heat energy.

THE FORM OF EARTH'S CRUSTAL SURFACE

The physical form of Earth's surface is defined by its relief. Formally, **relief** is the difference in height between two points on Earth's surface such as a hilltop and valley floor. It may be measured by comparison to a reference level such as sea level. The term *relief* is also used to describe the overall features of Earth's surface that result in changing elevation. In this sense, the relief of an area is made up of landforms of varying height and shape. Major landforms, such as mountains, are mainly formed by processes in the solid-earth environment, whereas smaller landforms (e.g., hills and stream channels) are mainly produced by surface processes. The science of **geomorphology** studies the nature and origins of landforms. *Topography* is another term that is sometimes used to describe the shape of the land surface. Strictly speaking, however, topography also includes other surface details such as vegetation, soils, and human constructions; all these features are included on topographic maps.

The most significant division of the surface relief is between continents and ocean basins. Twenty-nine percent of the outer surface of Earth's crust is

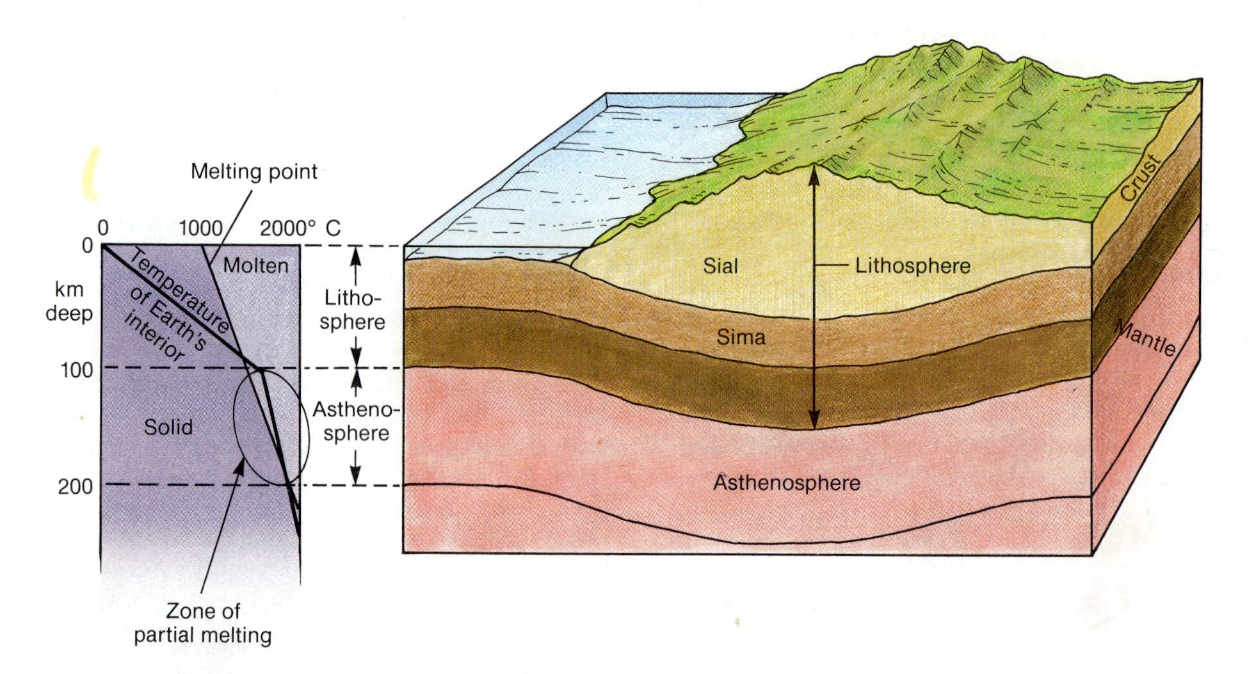

Figure 10-3 The crust and upper mantle. The crustal layers and uppermost mantle are bonded together in a rigid mass. Between 100 and 200 km deep, the rising temperature overcomes the effects of high pressure and rocks begin to melt. The partly melted rocks are mobile and movement within the asthenosphere causes movement in the overlying layers. The diagram on the left shows that the melting point of rock increases steadily with depth as the weight of overlying rocks increases the pressure. There is a band 100 km thick where the temperatures exceed the melting point.

covered by continental land masses and 71 percent by ocean water. The continents are characterized by major landforms such as narrow zones of high mountain systems and broader areas of high plateau and lowland plains. The ocean basins contain submarine ridges, together with extensive ocean floor plains and narrow, very deep trenches. The highest mountain, Everest in the Himalayas, is 8848 meters (29,030 ft) above sea level. The deepest point in the ocean is the Mariana Trench, in the western Pacific Ocean, 11,022 meters (36,163 feet) below sea level. The difference between these highest and lowest points is 19,870 meters (65,193 feet or 12.3 miles), only 0.15 percent of Earth's diameter. Just as the atmosphere-ocean environment was described as "the fuzz on a peach," so Earth's relief is merely a slight "roughness" on the surface.

Earth's major relief forms fit into a hierarchy of scales ranging from thousands of square kilometers to a few square millimeters in size. The five orders of landform unit listed in Table 10-1 range from continents and ocean basins through mountain ranges to small details on a rock face. The larger landforms are formed by internal Earth processes, while the smaller are caused by the interaction of internal processes with those in the atmosphere-ocean environment. Physical geographers must understand how the larger forms are produced by internal Earth forces to make sense of many of the smaller landforms shaped by surface processes. Chapters 10 to 12 focus on the former, while Chapters 13 to 18 focus on the latter.

ROCKS AND MINERALS OF EARTH'S CRUST

Rocks and minerals are the building blocks of Earth's crust. When exposed at the surface, their compositions cause them to react in a variety of ways with the atmosphere-ocean environment. **Rocks** are coherent masses of mineral matter; the mineral matter may be in the form of individual minerals interlocked together, or fragments of other rocks cemented together. Many of the rocks and minerals in Earth's crust are extracted for various uses (see Box 10-1: Environmental Impacts of Mineral and Rock Extraction).

Rock-Forming Minerals

The **minerals** making up rocks are naturally occurring combinations of chemical elements bonded together in orderly crystalline structures. A small number of elements and minerals dominate the crustal rocks. Although a great variety of elements are present in Earth's crust, oxygen (O, existing not as a gas, but with other elements in solid compounds) and silicon (Si) make up 74 percent of the total volume. Aluminum, iron, sodium, calcium, potassium, and magnesium make up a further 23 percent (Table 10-2). It is worth comparing the proportions of several of the economically valuable elements with those in the table: carbon (C) makes up 200 ppm (0.02%) of Earth's crust, copper (Cu) 35 ppm, and gold (Au) 0.004 ppm. The main ele-

Table 10-1 Scale of Landform Units

	ORDER	SIZE (sq km)	EXAMPLE	DOMINANT PROCESSES
MAJOR LANDFORMS	First	Billions	Continent, ocean basin	Internal processes
	Second	Millions to hundred thousands	Mountain system (Andes, Rockies, Appalachians), major plateau (Colorado), Mississippi lowlands, ocean trench	Internal processes
	Third	Ten thousands	Adirondacks, Blue Ridge Mountains, Mississippi delta	Internal processes plus surface processes
SMALLER LANDFORMS	Fourth	Hundreds to tens	Individual peaks: e.g., Mount St. Helens; valley; estuary	Mainly surface processes
	Fifth	Fractions (square meters)	Beach ripple, gully	Surface processes dominant

BOX 10-1

ENVIRONMENTAL IMPACTS OF MINERAL AND ROCK EXTRACTION

The minerals and rocks of Earth's crust provide essential raw materials that people have used widely since early in human history. The extraction of raw materials has had a range of environmental impacts on landforms, streams, lakes, and ecosystems.

In the Stone Age, hard, fine-grained rocks such as flint were chipped into tools such as ax heads and arrow points. The hollows excavated to obtain the flint left marks on the landscape. The early excavation of metals such as copper and tin also caused small-scale surface changes in Europe and the Middle East. On Dartmoor in southwest England, the shapes of the upper parts of valleys were altered by surface tin-ore extraction during a period between 900-1200 AD and the lower parts of these valleys became filled with the finer debris that was washed down the rivers from the mines—as shown on the diagram. After 1849, in the Gold Rush of California, larger companies used strong jets of water to flush out the precious metal ore from deposits along the valleys leading down from the Sierra Nevada, as shown on the map. In just over 30 years, from the discovery of the gold-bearing gravels north of Sacramento in 1853, until a court injunction in 1884 forbidding the discharge of mine debris into streams, more than 1.5 billion cubic meters of debris were flushed from the mining valleys. The sand and gravel moved down creeks and canyons into the Central Valley and San Francisco Bay areas. River currents deposited the coarser material at the canyon mouths, raising river levels and destroying farmland. Finer sand and mud washed into the bays at the mouth of the Sacramento River, filling channels and raising shoals. The navigable channels became too shallow and the Bay area diminished in size. As a result of the 1884 court injunction, the hydraulic mining industry closed down. The streams then gradually reworked their deposits and the channel of the Sacramento returned to its former state during the twentieth century.

After explosives became available in the seventeenth century, subsurface mining began. Coalfields in Europe and the United States were honeycombed by underground shafts and tunnels in the nineteenth and early twentieth centuries, and the ground often caved in following the collapse of the tunnels. Subsidence still damages houses and roads built over mine workings and can be the subject of expensive litigation. Subsidence is also common where rock salt, oil, and water have been extracted from underground rocks. Parts of the Los Angeles basin have sunk 3 meters (10 feet) as a result of pumping oil. Underground mining also contaminates subsurface and surface waters and produces waste that is dumped in surface heaps. In coalfields in uplands such as Appalachia, piles of waste that were not drained effectively became unstable as the water built up inside, and they often collapsed, forming floods of water and debris that destroyed property and killed people in their paths.

Surface extraction of rocks and minerals continues to be important and is modifying the land surface at increasing rates as the holes get larger. The expansion of coal mining in the United States in the late 1970s and 1980s was based on strip mining in Montana, Wyoming, and eastern Kentucky. Although laws now require mining companies to restore the land surface after

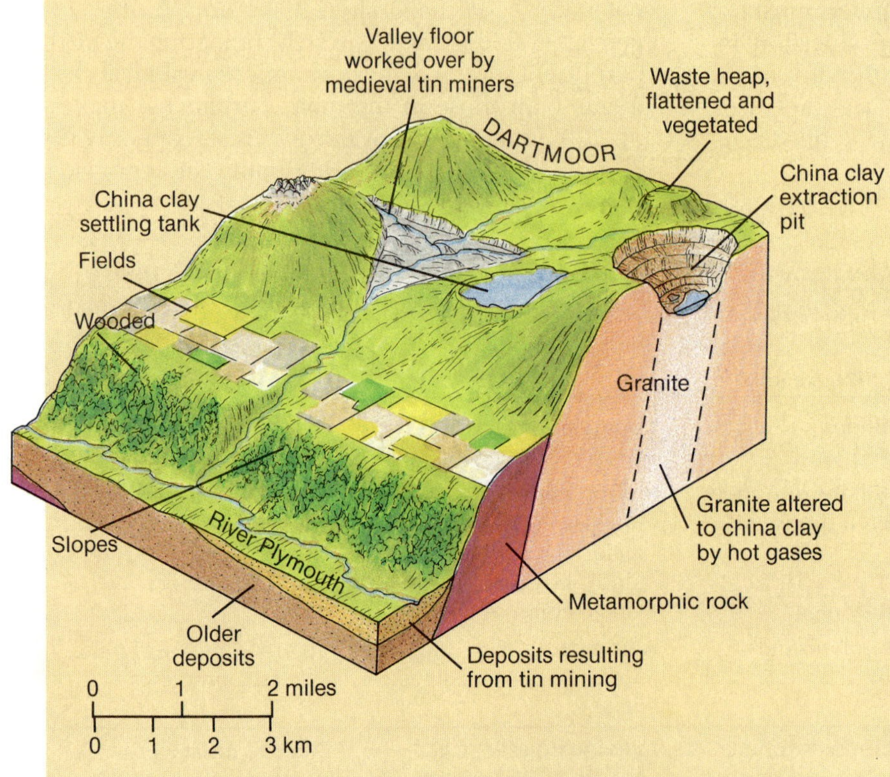

Impacts of surface mining, ancient and modern, on Dartmoor, southwest England. The medieval tin miners dug up the valley floors in the upland area. Modern china clay mining affects larger areas, digs deeper pits, and builds extensive waste mounds.

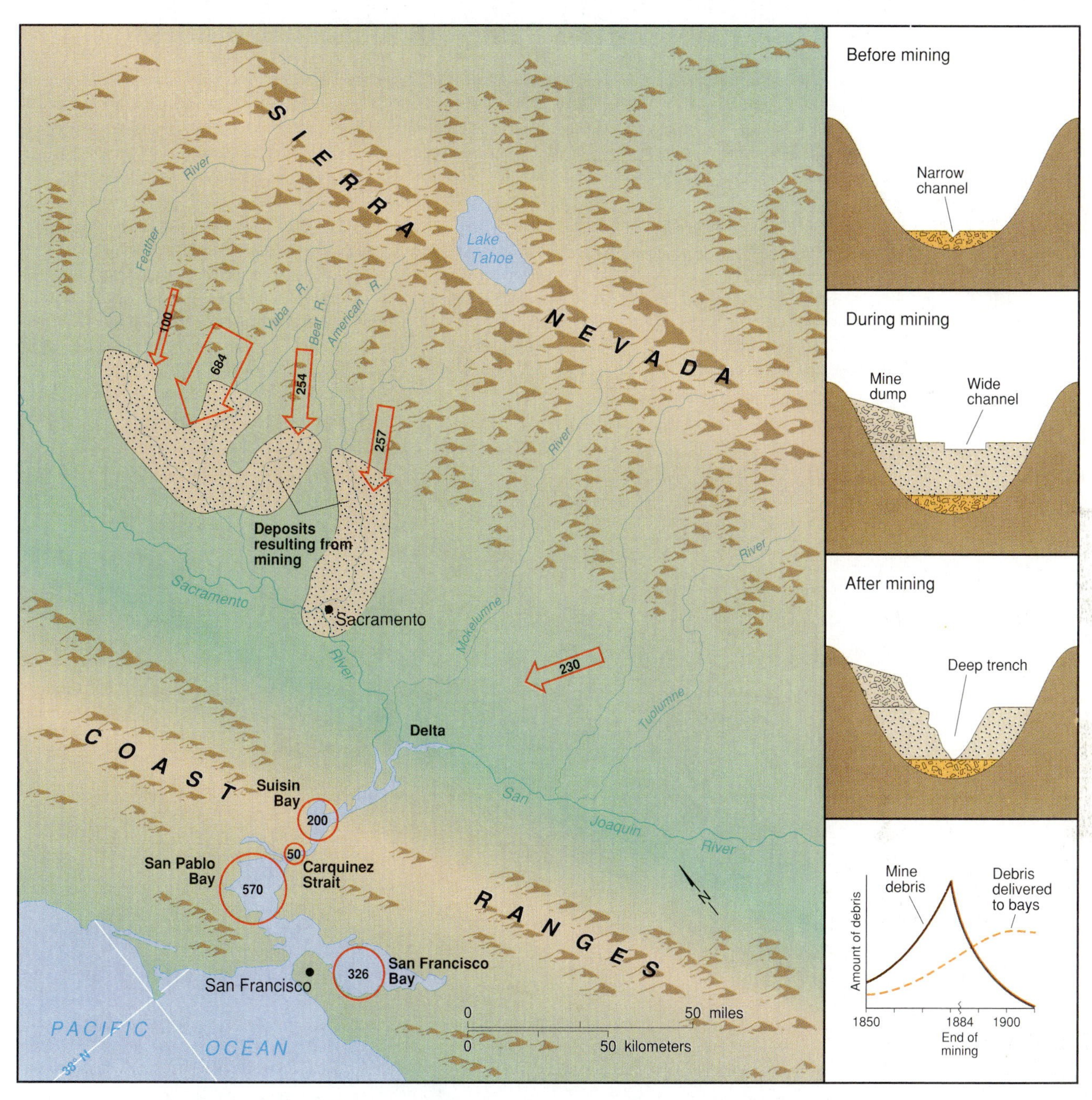

The impact of the California Gold Rush, 1849-1884, on the Sacramento River and the San Francisco Bay area. The map shows the areas affected by hydraulic mining that washed rock debris downstream; the red arrows show how much (million cubic meters) sediment was produced between 1849 and 1909 (total 1700 million cubic meters); the red circles in the bays show how much was deposited there between 1849 and 1914 (total 1146 million cubic meters). The inset valley cross sections show the impact on the mined creeks and canyons, and the graph gives an idea of the production of mine waste and its impact on the bays.

ENVIRONMENTAL IMPACTS OF MINERAL AND ROCK EXTRACTION—cont'd

mining, noise and dust create environmental impacts during the working period.

Quarrying for building stones, such as granite, slate, or marble, for road chippings, or for limestone to manufacture cement can produce giant open pits. In the twentieth century many have grown to be huge. To extract low-grade metal ores (often containing less than 0.5 percent of the metal), vast quantities of rock must be removed; underground mining is too costly except for valuable ores such as gold or platinum, and so surface pit working is necessary. The bottom photo shows one of the largest open pits in the United States.

Other significant excavations produced by surface extraction include the pits dug along the lower parts of river valleys to obtain sand and gravel for the construction industry. These frequently become filled with water and may eventually be used for water sports. Some quarries, especially those near urban areas, have been filled by the disposal of solid waste.

Mineral and rock extraction affects the immediate environment where the extraction takes place, but the impacts also extend to surrounding areas as sediment, dust, and chemical pollutants are carried farther afield by streams or the wind.

Strip mining for Appalachian coal in West Virginia.

Butte, Montana—open pit. This pit was dug by a copper mining company, but production ceased in the 1970s when it became uneconomical.

Table 10-2 Major Elements in Earth's Crust and Oceans

EARTH'S CRUST		OCEAN WATER	
ELEMENT	*PROPORTION (ppm)*	*ELEMENT*	*PROPORTION (ppm)*
Oxygen (O)	446,000	Chlorine (Cl)	19,000
Silicon (Si)	277,000	Sodium (Na)	10,500
Aluminum (Al)	81,000	Magnesium (Mg)	1,350
Iron (Fe)	50,000	Sulfur (S)	885
Calcium (Ca)	36,300	Calcium (Ca)	400
Sodium (Na)	28,300	Potassium (K)	380
Potassium (K)	25,900	Bromine (Br)	65
Magnesium (Mg)	20,900	Carbon (C)	28

Note: The quantities are given in parts per million (ppm), a measurement of volume: oxygen makes up 44.6% of the total volume of rock in Earth's crust, and chlorine 1.9% of the total volume of ocean water.

ments combine in a small number of minerals that make up nearly all the crustal rocks. Most of the rock-forming minerals are known as *silicates*, since they contain high proportions of silica (a compound of silicon and oxygen) in their formulas.

The internal structure of each mineral is based on the element or molecule combinations it contains, and determines physical properties, such as color, shape, density, and resistance to being worn away. The mineral olivine, for instance, forms at an early (high temperature) stage in the solidification of cooling molten rock deep in the crust. Its environment of origin is characterized by high temperature (over 1000°C) and pressure, and the mineral's crystal structure is very susceptible to change in the conditions of lower temperature and pressure at the surface. Quartz, on the other hand, does not solidify until the molten rock has reached a much lower temperature. When it is exposed to the atmosphere, quartz is among the most resistant minerals to chemical breakdown and physical disintegration. Most sand is made of quartz that has been exposed for thousands or millions of years without being changed. Figure 10-4 shows some rock-forming minerals; Table 10-3 lists the most common rock-forming minerals and some of their properties.

A further important distinction among the rock-forming minerals is based on the relative densities of minerals. The top three listed in the table have specific gravities of between 3.2 and 4.7, whereas quartz, the feldspars, and calcite have one of between 2.5 and 2.7. The density of the minerals in a rock determines its density.

These, then, are the building blocks of crustal rocks: elements combine to form a mineral, and minerals combine to form a rock. When a rock is exposed at the surface, it may be broken down into the constituent minerals or into fragments containing several minerals; some minerals are transformed into others by chemical reactions. The separated minerals and rock fragments may be recombined into a new rock.

Major Rock Groups

Rocks are classified into three major groups according to the way in which they are formed. **Igneous rocks** are formed by the solidification of molten rock, which often migrates upward from the place of melting to the place of cooling. The place of melting is often 50 to 100 km below Earth's surface; the place of cooling is higher in the crust or on the surface. **Metamorphic rocks** are formed by the alteration of minerals in other rocks under extreme heat or pressure, but without complete melting or migration. Metamorphic rocks can be formed from the alteration of older igneous, sedimentary, or metamorphic rocks. **Sedimentary rocks** are formed at Earth's surface from products of the breakup of other rocks—clay particles, sand grains, dissolved chemicals, and fragments up to boulder size—sometimes with the remains of organic matter such as shells and plant tissues. Sedimentary rocks can be formed from the materials that were part of older igneous, metamorphic, or sedimentary rocks.

Figure 10-4 Some rock-forming minerals—quartz, plagioclase feldspar, orthoclase feldspar, biotite mica, muscovite mica, augite, hornblende, olivine, and calcite. **A,** Hornblende: long, dark crystals in schist rock. **B,** Shiny mica crystals in schist rock. **C,** Quartz crystals. **D,** Quartz (gray) feldspar (white, pink) and mica crystals in granite rock.

Table 10-3 Major Rock-Forming Minerals

MINERAL	COMPOSITION	COLOR	DENSITY*	OCCURRENCE IN ROCKS
Olivine	$(Mg,Fe)SiO_4$	None	3.2-4.3	Dark mafic igneous
Augite	Ca,Mg,Fe,Al silicate	Black	3.2-3.5	Dark mafic igneous
Hornblende	Ca,Mg,Fe,Al silicate	Black	3.3-4.7	Igneous, metamorphic
Biotite mica	Mg,Al,K,Fe,H silicate	Brown, black	2.7-3.1	Igneous, metamorphic
Muscovite mica	Al,K,H silicate	White, none	2.76-3.0	Igneous, metamorphic, a few sedimentary
Orthoclase feldspar	$KAlSi_3O_8$	White, pink	2.57	Igneous, metamorphic, a few sedimentary
Plagioclase feldspar	$(Ca,Na)AlSi_3O_8$	White, gray	2.6-2.76	Dark mafic igneous
Quartz	SiO_2	None, white	2.65	Igneous, metamorphic, and sedimentary
Calcite	$CaCO_3$	White	2.71	Sedimentary, some meta-morphic
Garnet	Complex silicate	Green, red, pink, yellow	4.2	Metamorphic

*The measure of density used here is *specific gravity*—the ratio of the weight of a substance to an equal volume of water at a given temperature.

The formation of rocks is thus a continuing set of processes, and the materials are recycled over periods of millions of years. This "rock cycle" is depicted in Figure 10-5. Igneous rocks can be considered as "primary" rocks, because metamorphic and sedimentary rocks are produced by modifications of the materials *initially* supplied from Earth's mantle in igneous rocks. Some of the more common rock types found in Earth's crust are illustrated in Figure 10-6.

Igneous Rocks

Igneous rocks are formed by the solidification of molten rock material that is known as magma. **Magma** is a mobile mixture of liquids, solid rock, and gases that moves slowly and has the consistency of thick oatmeal. When the magma solidifies, the resultant igneous rock is composed of interlocking crystals.

There are two major types of magma. One forms where local heating melts the upper mantle rocks and is rich in dense, dark-colored minerals such as olivine, augite, hornblende, and plagioclase feldspar (Table 10-3). Igneous rocks forming from such magmas are known as *mafic*, and include fine-grained basalt and coarse-grained gabbro. The other type of magma forms where continental crust is dragged down to depths where it melts and mixes with upper mantle rock. This type of magma contains a wider range of minerals, including more of the less-dense and lighter-colored quartz and orthoclase feldspar. Igneous rocks formed largely of the lighter-colored minerals are known as *felsic*, and include fine-grained rhyolite and coarse-grained granite.

Magma typically flows toward Earth's surface, forced upward by the higher pressures below, and cools as it travels through cooler rocks. The rock-forming minerals in magma solidify into crystals at specific temperatures from as high as 1300°C (2400°F) down to around 500°C (950°F), by which time only a small fraction of the magma remains in the liquid or gaseous form. At each stage in cooling some minerals crystallize, leaving the remaining magma richer in other minerals that crystallize at lower temperatures. The dark minerals tend to crystallize at higher temperatures and the light-colored

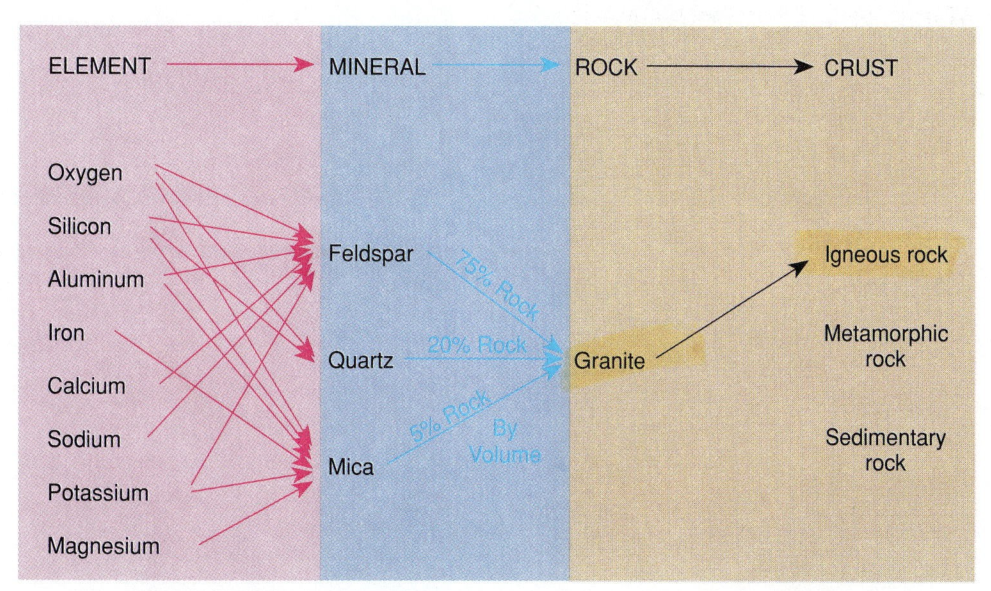

Figure 10-5 The rock cycle: relationships between different rocks and their origins. Igneous rock forms from magma generated in the mantle or lower crust. Igneous and sedimentary rock may be buried and metamorphosed. Igneous, sedimentary, or metamorphic rocks may be exposed at the surface and broken into fragments that form another sedimentary rock.

Figure 10-6 **A,** Shale. **B,** Limestone. **C,** Gneiss. **D,** Basalt. **E,** Sandstone. **F,** Marble. **G,** Slate. **H,** Quartzite.

at lower temperatures. When the magma that forms beneath continents rises through the crust, it produces a sequence of rocks dominated at first by dark mafic rocks and later at the higher levels by felsic rocks.

If the final solidification occurs in the crust before the magma reaches the surface, the resultant rock is said to be **intrusive,** because it occurs intruded between other rocks. Intrusive igneous rock is thus younger than the surrounding rocks. Intrusive rock may eventually be exposed at the surface when the rock that originally covered it is removed by surface erosion over periods of millions of years. When magma erupts at the surface, it is said to produce **extrusive** rock, including solidified lava flows and ashes produced by explosive disintegration of the lava. The processes of magma formation, flow, and solidification of igneous rock may take up to several million years.

The sizes of the crystals in an igneous rock reflect the time taken in cooling of the magma. Magma below the surface cools slowly and gives the minerals time to grow into large crystals. Igneous rocks that solidify several kilometers below the surface commonly have crystals from a few millimeters to a few centimeters in size. Rapid cooling in contact with the atmosphere or ocean produces fine-grained crystals, often too small to be seen with the naked eye. Grain size, or *texture,* and mineral composition are used to classify igneous rocks.

Intrusive Igneous Rocks. Intrusive igneous rocks produce forms, known as **plutons,** with a variety of shapes and sizes. The largest plutons are huge masses up to 1000 km across called **batholiths** (Figure 10-7). They commonly occur in the cores of mountain ranges and are often composed of granite. As the batholith rocks cool, they contract and cracks appear; these cracks, known as **joints,**

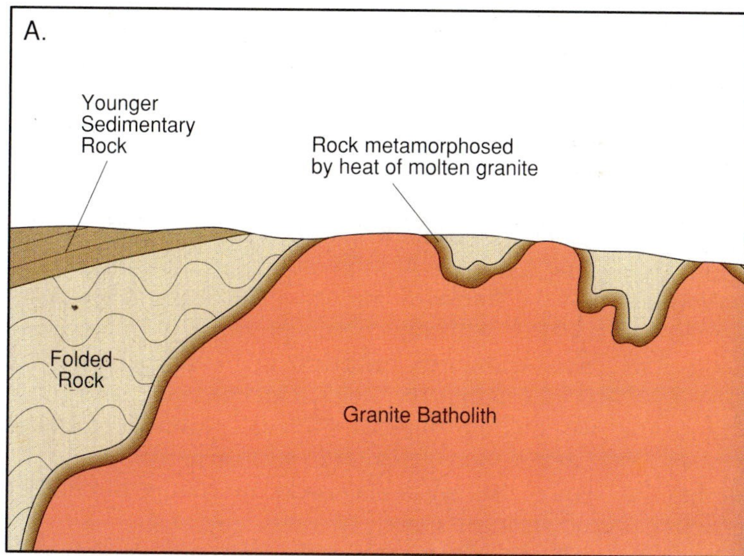

Figure 10-7 Granite batholiths. **A,** The diagram shows how the batholith affects older rocks in the process of intrusion—by pushing them apart, and by metamorphosing those that are adjacent to the hot magma. The batholith rocks will be exposed at the surface if the overlying rocks are worn away. Younger rock layers may form on top. **B,** Some batholiths exposed in western North America. Note their size and shape.

Figure 10-8 Vertical joints in granite in Yosemite National Park, California.

form at right angles to the directions of contraction. In coarse-grained granite there are usually two sets of such joints that produce rectangular blocks (Figure 10-8).

Smaller, sheetlike masses of intrusive igneous rock are normally between a few meters and several hundred meters thick, but may extend for hundreds of kilometers. When igneous rock is forced between layers of older rocks and generally runs parallel to them, it forms features known as **sills.** Sheets of intruded rock that cut across other layers are called **dikes.** Figure 10-9 shows some of these smaller intrusive forms. Exceptionally, some dikes are several kilometers wide, the best example being the "Great Dike" of Zimbabwe in southern Africa.

The rock types in igneous intrusions are determined by the origin of the magma and the site and rate of cooling (Figure 10-10). The most common rock occurring in intrusions is granite, a coarse-grained felsic rock that is produced from the intrusion of silica-rich magma. Both mafic and felsic rocks occur in the sheet intrusions. Medium-grained igneous rocks occur in the sheet intrusions where cooling was more rapid.

Extrusive Igneous Rocks. When extruded into contact with air or ocean waters at the surface, igneous magma erupts as **lava flows,** or explodes and is deposited as layers of volcanic fragments from smaller ashes to larger "bombs." Lava in the molten state may be either very fluid and flow several kilometers before solidifying if it is composed of mafic materials, or very thick and barely flow out of the volcano if it is felsic in composition. Figure 10-11 shows fluid lava erupted on Hawaii. Lava erupted on the ocean floor often occurs in rounded masses formed by the rapid solidification of the outer surface when it comes into contact with cold water; the rounded masses are known as *pillow lava.* Volcanic ash layers on land may be several tens of meters thick. The landforms associated with volcanic activity are studied in Chapter 11.

Lava flows may also be marked by vertical joints formed during cooling. These tend to be closer together in the finer-grained rocks than in the coarse-grained rocks of a batholith. When exposed at the surface, such joints cause the rock to break apart in six-sided columns, a structure that is known as *columnar jointing.*

Metamorphic Rocks

Rocks buried beneath the surface are subject to increased heating and pressure but do not always melt enough to become liquid magmas. Under these conditions the minerals and physical structures in the original solid rock are transformed, or metamorphosed. A mineral within a heated rock may melt and recrystallize when it cools, rearranging its atoms or molecules to produce a new mineral in the same location. A mineral subject to compression may be flattened and have its internal chemical structure rearranged. This produces a fine layering effect known as **foliation** in many metamorphic rocks (Figure 10-12). These processes take place over millions of years. Metamorphism often makes the rock more resistant to surface processes; for instance, soft, weak clay rocks are changed to harder and stronger slates.

There are several levels of metamorphism that result from combinations of heating and pressure conditions (Table 10-4). Low levels of metamorphism occur where high pressure is combined with low temperature, or low pressure with high temperature. High pressures and low temperature affect rocks where two sections of crust move against each other along a large crack, resulting in crushing of the existing rocks. The crushed zone may be several hundred meters wide, and the rocks in it are a mixture of fine and coarse fragments, sometimes smashed into a solid mass by the movement.

High temperature and low pressure conditions occur in the rocks surrounding an igneous batholith. The solidifying magma is at temperatures between 500 and 1000°C and it heats up the surrounding rocks by conduction. A zone from a few

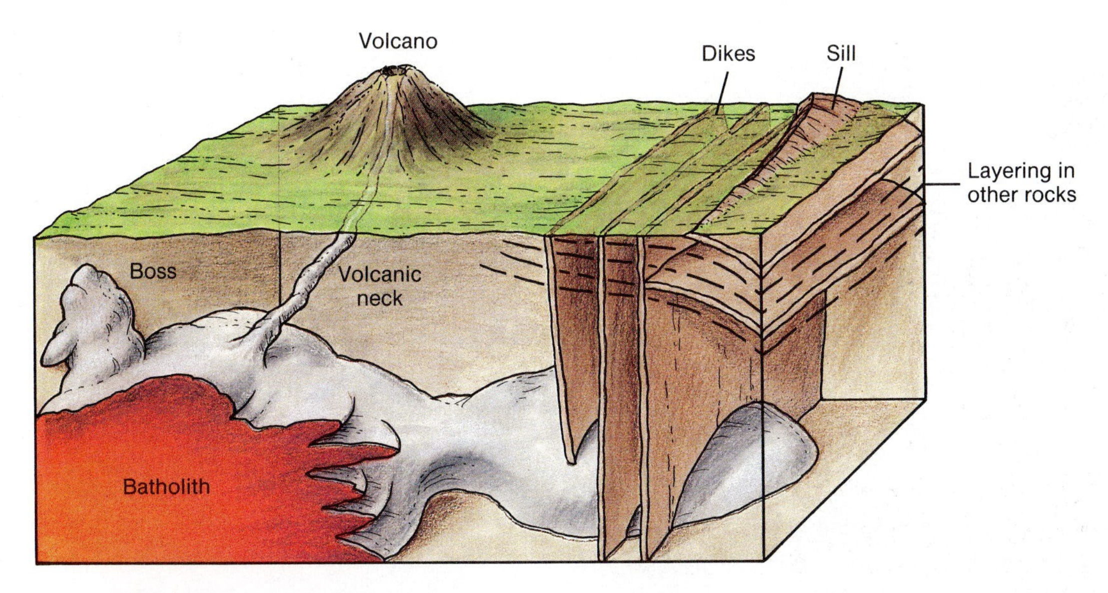

Figure 10-9 A range of plutons: the forms of igneous intrusions and their relationship to surface volcanic features. Batholiths are the largest forms. Sills are intruded parallel to the original rock layers; dikes cross those layers. A volcanic neck is a pipelike feature leading to a surface volcano.

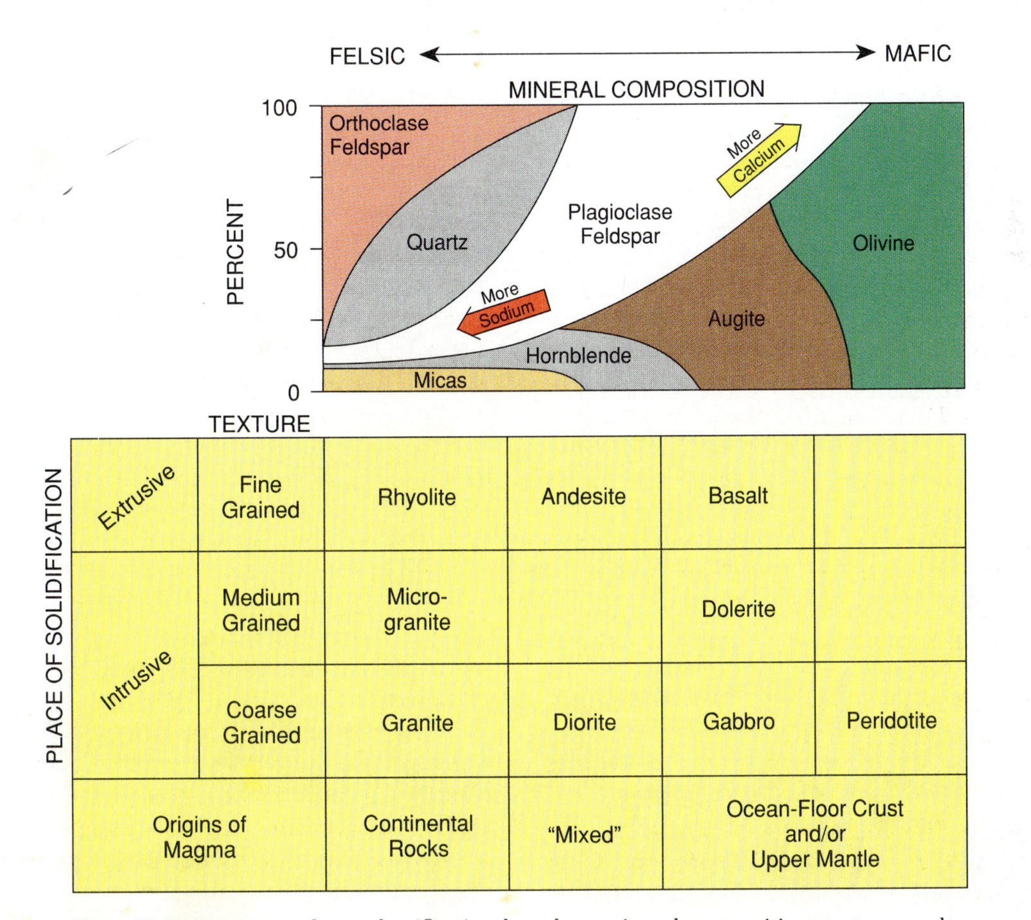

Figure 10-10 Igneous rocks: a classification based on mineral composition, texture, and place of solidification.

Figure 10-11 Molten lava on Hawaii. This lava covered the road in 1990.

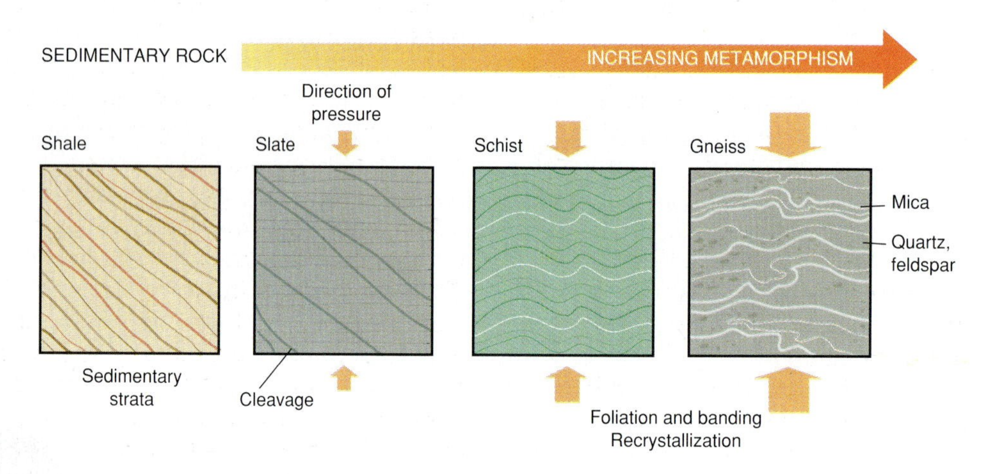

Figure 10-12 The formation of some metamorphic rocks from an original sedimentary shale. Moderate pressure and heat produce cleavage in a slate; further increases in pressure and heat give rise to different minerals and types of foliation in schist and gneiss.

meters up to a few kilometers wide is affected by this process, often termed **contact metamorphism** (see Figure 10-7). Such metamorphism causes minerals to recrystallize at higher or lower temperatures than those at which the original rock formed. A number of unusual minerals that have achieved the status of "precious" or "semiprecious" stones, including diamonds and garnets, are formed by this process.

Moderate levels of both heat and pressure occur when rocks are buried a kilometer or so deep. Clay minerals are susceptible to such pressures and are flattened in directions at right angles to the pressure. A rock with a high clay mineral content that is subjected to such pressures and some heating is changed to a *slate*. Slate is a rock that can be split apart (cleaved) into the thin sheets that are commonly used as roofing materials. The cleavage lines are perpendicular to the directional pressure that metamorphosed the rock.

At higher temperatures and pressures, which occur when the rocks are buried several tens of kil-

Table 10-4 Metamorphic Rocks

	HIGH TEMPERATURE	MODERATE TEMPERATURE	LOW TEMPERATURE
HIGH PRESSURE	REGIONAL METAMORPHISM: large-scale, deep burial forming schist, gneiss. Strong foliation and recrystallization of minerals	Transition: slate with foliation (cleavage)	Crush rocks, often in fault zone
MODERATE PRESSURE	Transition	Shallow burial: slate with poor cleavage	Some crushing
LOW PRESSURE	CONTACT METAMORPHISM: rock altered in contact with igneous intrusion. Minerals recrystallized to new forms, but no foliation	Some recrystallization	No metamorphism

ometers deep, the minerals are changed in chemical structure as well as physical orientation. Clay minerals are converted to micas, and rocks containing high proportions of such minerals are characterized by foliation layers with shiny surfaces: these rocks are known as *schists*. Metamorphism of granite under conditions of extreme temperature and pressure produces a *gneiss*. This is a strongly foliated rock with bands of quartz and feldspar (Figure 10-13). The environments that give rise to slates, schists, and gneisses often extend for thousands of square kilometers; when metamorphism occurs at this scale, the processes are combined under the term **regional metamorphism.**

Metamorphic rocks produced by regional metamorphism may occur over extensive areas covering several thousand square kilometers, commonly in the cores of mountain ranges adjoining granite batholiths. Examples are found in the Rockies and the Alps. The batholith and highly metamorphosed rocks formed together in conditions of great heat and pressure, which totally transformed older rock and caused some of it to melt.

Sedimentary Rocks

Sedimentary rocks form at Earth's surface. They result from the accumulation of broken rock material, the debris of living organisms (e.g., shells, plant remains), and chemical deposits. Conversion to hard layers of rock takes place by a combination of compaction and the addition of a cementing medium—a "glue"—over hundreds of thousands of years. Layering is a common feature of sedimentary rocks, with the individual layers known as beds or **strata** (Figure 10-14). Each bed is bounded by surfaces known as *bedding planes*, which mark the change from one phase of deposition to the next. A bedding plane often also marks a change in composition of the sedimentary rock.

Sediment is a mass of disconnected solid particles, grains, and fragments of rock that have been worn from older rocks, transported, and deposited. These products, in increasing order of size, are clay particles (less than 0.002 mm), silt particles (0.002 to 0.02 mm), sand grains (0.02 to 2 mm), and rock fragments from gravel to boulders (larger than 2 mm). Other materials in sedimentary rocks are fragments of organisms such as shells, bones, leaves, or tree trunks that have fallen into the accumulating sediment, or have been transported like the sand or gravel. Some sedimentary rock constituents, such as peat (decaying vegetation) layers or coral reefs, may have formed in the place where they are later incorporated in sedimentary rocks.

Sedimentary rocks also incorporate material that has been precipitated (solidified) out of a solution. When rock-forming minerals are dissolved following chemical reactions (Chapter 13), the dissolved material is transported in streams until a change in chemical environment (e.g., when fresh water reaches the salty ocean) causes it to be precipitated.

The processes that move sediments from their source to the point where they are deposited tend

Figure 10-13 A highly metamorphosed folded gneiss in the highlands of Scotland. The total width of the rock shown is approximately one meter.

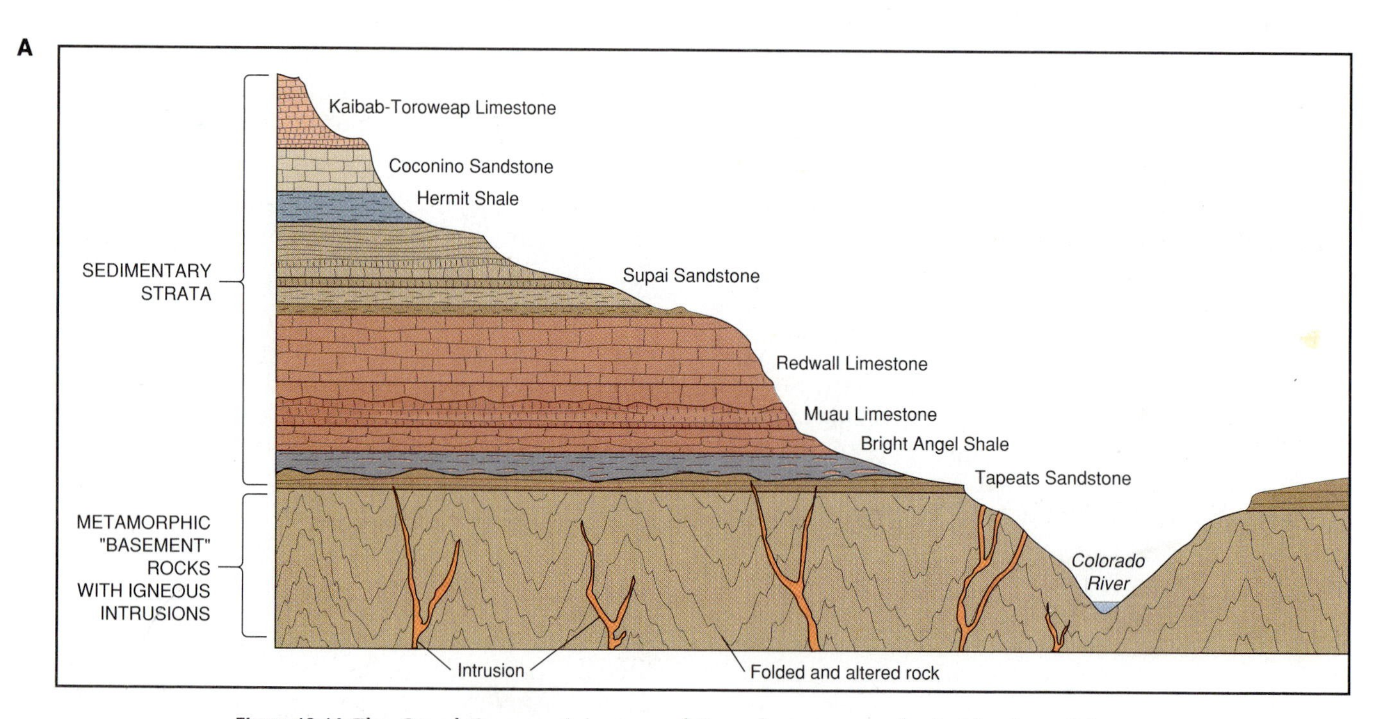

Figure 10-14 The Grand Canyon, Arizona, and its rock structures. **A,** A side view of the rocks as exposed on the sides of the Colorado river valley. The horizontal sedimentary strata were formed in layers in an ancient sea on top of metamorphic and igneous rocks that had been formed in a previous mountain-building phase and then worn down.

B

Figure 10-14 Cont'd B, The Grand Canyon, Arizona showing the horizontal strata at the tops of the canyon sides.

to separate different sizes of sediment, so one size generally predominates in a deposit. This process of differentiation by size is known as *sorting*. Larger rock fragments do not travel as far or as fast as sand grains, and clay particles are carried farthest.

The sediments, organic debris, and precipitated minerals are usually deposited in the ocean or in lakes and become buried as new layers form on top. New layers may be composed of similar or different materials; clay may be covered by sand, and the sand by boulders. The weight of the overlying materials compresses and pushes out water from the original deposit, reducing the thickness of the buried layer. The expelled water may leave behind a cement of calcium carbonate or silica that bonds the particles together. The combination of *compaction* and *cementation* changes the soft, disconnected sediment into a hard sedimentary rock.

The types of sedimentary rock are distinguished by their main components (Table 10-5). *Clastic* sedimentary rocks are formed mainly of inorganic particles, and are named according to their main constituent. A preponderance of clay minerals forms *shale*; of sand grains a *sandstone*; of larger fragments (gravel, boulders) a *conglomerate*. *Organic* sedimentary rocks are formed from the remains of living organisms: shells or coral reefs form the bulk of some *limestones*; peat layers form *coal*. *Chemical* sedimentary rocks are formed by the precipitation of minerals from water: *rock salt* forms in evaporating seas or lakes, and some limestones form when waters precipitate calcium carbonate. However, mixtures within and between the groups are also common: "boulder clay" is a common sediment resulting from the deposition together of large rock fragments and fine clay from melting ice; a limestone may be composed mainly of calcium carbonate mixed with smaller amounts of sand or clay.

LINKAGES

The formation of sedimentary rocks involves interactions between Earth environments, whereas igneous and metamorphic rocks are products of the solid-earth environment alone. The initial rock materials are brought to the surface by ***solid-earth environment*** *processes; the* ***atmosphere-ocean environment*** *interacts with these rocks to produce sediment and dissolved minerals that form the basis of sedimentary rocks; the* ***living-organism environment*** *gives rise to peat, coral reefs, and shell-based materials that may be incorporated in sedimentary rocks.*

The Distribution of Rock Types in North America

The major rock types are all well represented in the landscape of North America, but there are regional concentrations. Areas in which igneous rocks are predominant are mostly in the west, including the exposed batholiths of the mountain ranges (see Fig-

Table 10-5 Sedimentary Rocks

ORIGIN OF DOMINANT COMPONENTS	ROCK GROUP	ROCK TYPE
TRANSPORTED SEDIMENT	CLASTIC	Conglomerate Sandstone Clay, shale
PRECIPITATED MINERALS	CHEMICAL	Rock salt, potash Flint Some limestones
UNTRANSPORTED AND TRANSPORTED ORGANIC MATTER	ORGANIC	Most limestones: shelly, reef Coal

ure 10-10), the massive lava flows of the Columbia and Snake river basins in the northwestern United States, and the volcanic peaks such as Mount St. Helens topping the Cascade Mountains. Batholiths also occur in New England and the Maritime Provinces of Canada. Metamorphic rocks dominate northern Canada east of the Rocky Mountains and also occur in the Piedmont and Blue Ridge sections of the Appalachian Mountains. Such rocks also underlie much of the continent, as is shown where they are exposed in the bottom of the Grand Canyon (Figure 10-14). Sedimentary rocks form the surface of most of the interior of the United States, parts of the Appalachians (ridge-and-valley and plateau sections), and the south Atlantic and Gulf coastal plains. The arrangements of the different rock types, and the order in which they occur in each locality, give clues to the changes the crustal surface has undergone through geologic time.

CONTORTED AND BROKEN ROCKS

The rocks seen in cliffs or road cuttings (Figure 10-15) seldom occur in the original flat horizontal layers of sedimentary strata or lava flows. Stresses within Earth's crust result in deformation structures known as **folds** when the rock layers are deformed, but not broken, and as **faults** when the layers are broken and displaced on either side of the break. Joints are cracks where the rocks are broken but do not move along the crack.

Folds, faults, and joints occur at scales from a few centimeters to tens of kilometers. The internal Earth processes that give rise to stresses are known as *tectonic* processes. The sources of the tremendous pressures that cause such changes in rocks are discussed further in Chapter 11, and the distribution of the resulting structures in mountain systems and other major relief features is described in Chapter 12.

Folds

It is difficult to believe that a hardened and rigid rock layer can be folded in the same way that a tablecloth folds when it is pushed. In fact, the folding of rock can occur only when the rock characteristics and the forces acting on the rock combine to allow deformation. The rock may be pliable before final hardening. Forces applied in a particular direction may reduce a solid rock's rigidity. Rock subjected to heat and pressure deep underground is also susceptible to folding.

When pressure is not very intense, simple folds with open **anticlines** (upfolds) and **synclines** (downfolds) are produced (Figure 10-16, *A*). In describing the form of the rock layers in folds, geologists refer to the angle between the steepest surface of a bedding plane and the horizontal as the **dip,** and to the direction of a horizontal line on this surface as the **strike** (Figure 10-16, *B*). Such information is depicted on geologic maps, making it possible to work out the type and degree of folding from a two-dimensional representation.

The Appalachian ridge-and-valley section, extending from central Pennsylvania southeast into Tennessee, is formed of folded layers of sedimentary rocks (sandstones, shales, and limestones). As Figure 10-17 shows, the rocks have boat-shaped outcrops, reflecting the domelike shapes of the folds. The ridges are formed in limestones and sandstones. This example shows that the anticlines, which

A

B

Figure 10-15 **A,** An anticline fold in sandstone rock, Wales. **B,** A normal fault in sandstone rock, near Liverpool, England.

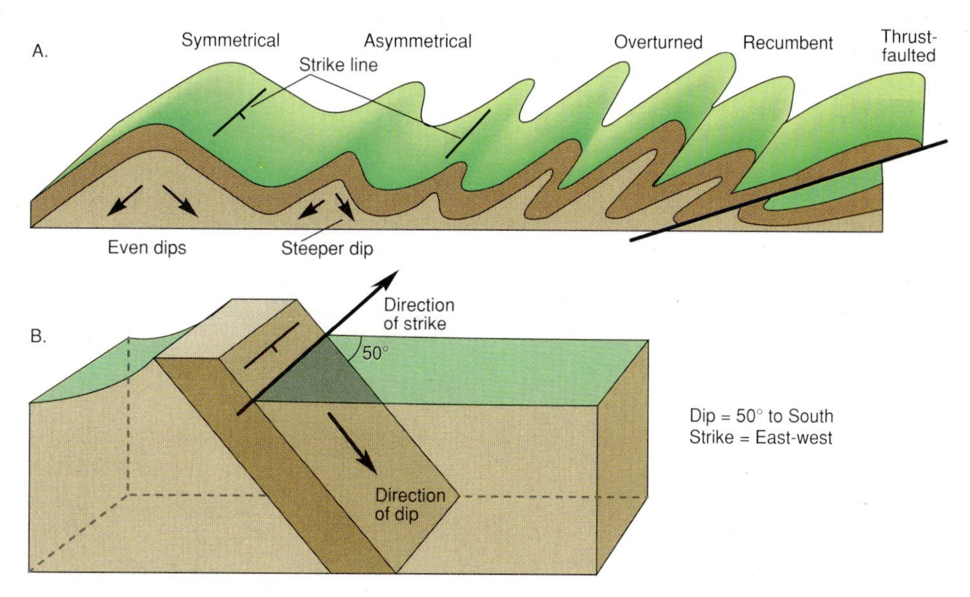

Figure 10-16 Folded rocks. **A,** Types of fold, with some dip directions and strike lines added. **B,** Measuring strike and dip: a horizontal strike line is marked on a rock surface and its direction determined by a compass; the dip angle can then be calculated by drawing a line at right angles to the strike and measuring the angle between this line and a horizontal plane. This is normally carried out with a clinometer, which is basically made up of a protractor combined with a compass and level.

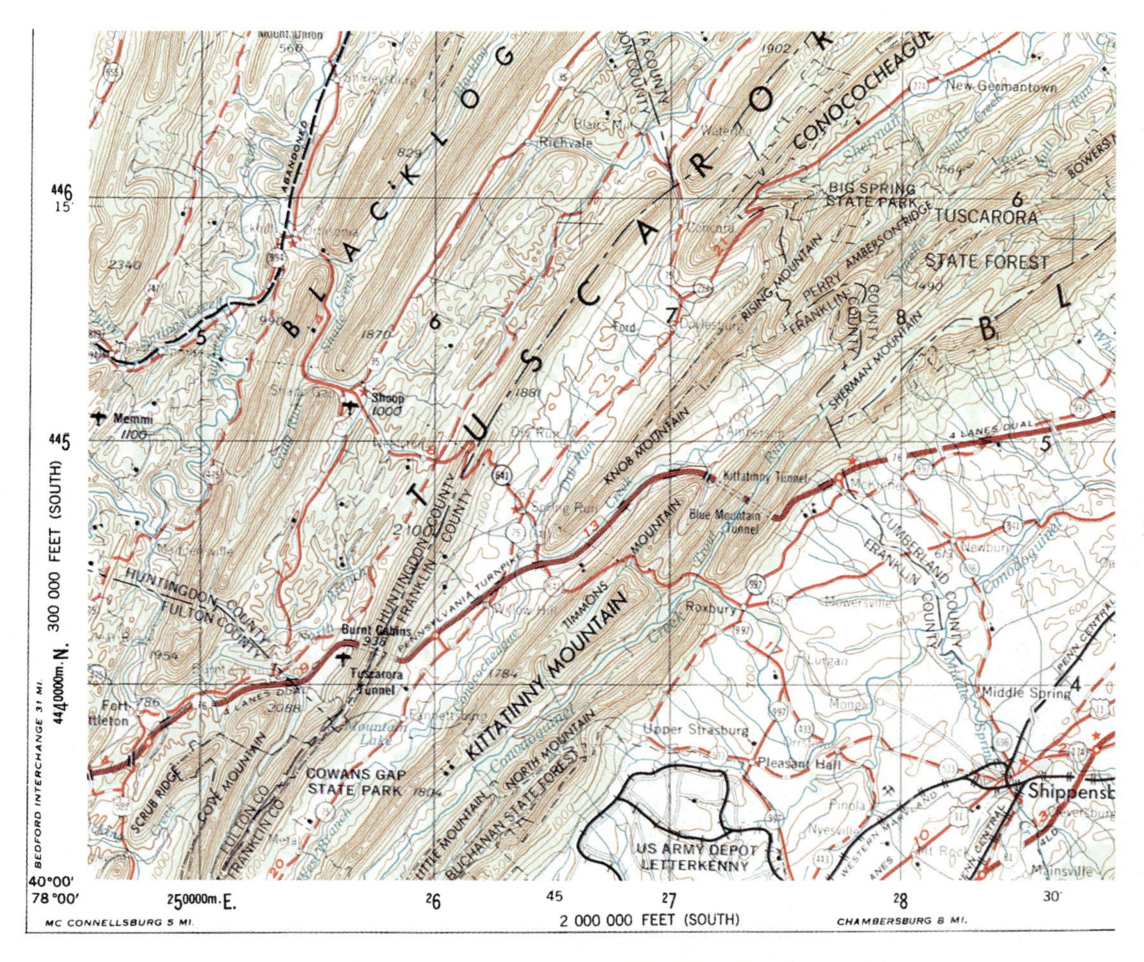

Figure 10-17 Folded rocks in the Appalachian Mountains of Pennsylvania.

might be expected to form hills if surface relief reflected the internal structure, have been worn away in their centers to form valleys, and that the synclines sometimes form hills. Surface processes have been dominant in the development of this landscape, and differences in rock resistance to wearing away result in limestones and sandstones forming ridges while the shales form lower land.

As pressure from a particular direction becomes more intense, the folds may become *asymmetrical* in cross section, with one limb dipping more steeply than the other; at the most intense pressures, the folds are pushed right over and are known as *overturned.* Such folds are characteristic of high mountain ranges, such as the Alps, Himalayas, and Andes.

Faults

Rigid rock fractures under stress; when the rock also moves along either side of a fracture it is said to be faulted. Faults often occur in groups along a *fault zone.* One of the best-known fault zones is the San Andreas in California, where rock on one side of the fault has periodically moved parallel to that on the other side, but in opposite directions. Over several million years some rock has been displaced over 100 km (Figure 10-18).

Faults are grouped according to the ways in which rock is displaced, as defined by the dip and strike of the fault surface. Faults that involve horizontal displacement along the fault plane are termed *strike-slip faults* (Figure 10-19, *A*). Horizontal movements along the San Andreas fault occur in localized short, sharp jerks, with displacements of a few meters at a time.

Faults that involve primarily vertical displacement up or down the fault plane are known as *dip-slip faults* (Figure 10-19, *B, C*). Total vertical movements of up to several thousand meters can occur. When rock on each side of the fault is pulled away in opposite directions by tensional forces, the rock above a sloping fault plane moves down the plane relative to the rock below it, producing a **normal fault;** when the crust is compressed, the upper block is forced over the lower block and a **reverse fault** forms.

Intense pressures may shear off overturned fold structures (see Figure 10-16, *A*), with the top part moving forward over the lower. The resulting almost horizontal fault plane is known as a **thrust fault.**

A

B

Figure 10-18 The San Andreas fault. **A,** A large slice of land on the coastal side of the fault is moving in fits and starts to the northwest. The inset map shows how rocks are broken and displaced on either side: the numbers refer to the age of the rock in millions of years. **B,** The photo shows how the fault has affected surface landforms. It crosses halfway up the picture: the foreground has moved to the left and taken the rivers with it. Carrizo, California.

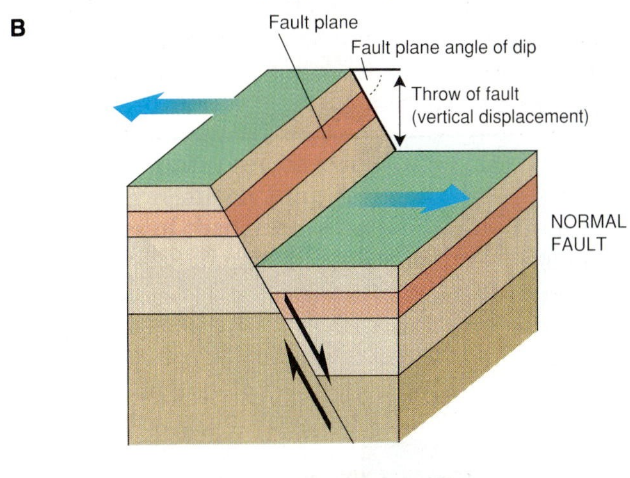

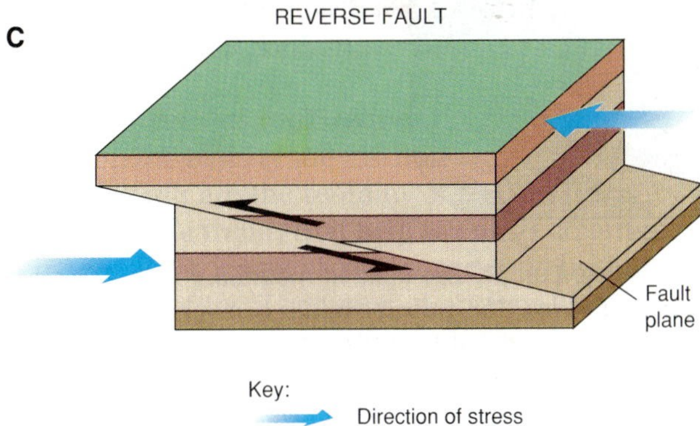

Figure 10-19 Types of fault. The thick arrows show direction of pressure that causes the rock to fracture; the half arrows indicate the relative displacement of rock on opposite sides of the fault. **A,** A strike-slip fault, where the movement is horizontal. **B,** A normal fault, where tension causes fracture as rocks are pulled apart, and the rock above the fault plane moves down relative to the rock below. **C,** A reverse fault, where compression causes the upper block of rock to be pushed over the lower.

The uplift of large sections of Earth's crust produces higher areas, such as the plateaus of the western United States, while subsidence often leads to the drowning of continental margins, as in the North Sea basin of Europe. When combined with faulting, uplift may produce *fault-block mountains,* such as the Sierra Nevada of California. The Rhine valley area features fault-block mountains uplifted between parallel faults; this landform is known as a **horst** (Figure 10-20). The sinking of a "keystone" valley between two horsts produces a **graben,** as in the upper Rio Grande valley at Albuquerque.

Joints

Cracks in rock that is not displaced on either side of the crack are known as joints. In igneous rock, joints form by cooling and contraction, producing four sets at right angles in coarse-grained rock and six sets in fine-grained rock. In sedimentary rock, joints form as the rock contracts following the expulsion of water. Joints in igneous, metamorphic, and sedimentary rock are also caused by other stresses, such as the flexing and stretching of rocks over the top of an anticline fold.

> *LINKAGES*
>
> *Rock type and rock structures have an important influence on the shape of surface relief features. Rock type also affects the characteristics of the soils produced from their broken debris, and the type of plants that grow in the soil. Such linkages reflect interactions between the* ***solid-earth environment,*** *the* ***surface-relief environment,*** *and the* ***living-organism environment.***

ROCKS AND TIME

The length of time over which rocks are formed, often known as *geologic time,* makes the human occupation of Earth appear insignificant. Consider the age of the earliest "modern human" (approximately 30,000 years ago), the age of the oldest civilizations (approximately 5000 years ago), and the length of the period during which human technology has begun to have major impacts on the natural environment (the last 200 years). These are all very short compared to the billions of years of geologic time. One implication of this difference beween human history and geologic time is that natural environmental change occurs over much longer periods than humans have existed. Virtually all human history, for instance, occurred after the rocks of the Grand Canyon (Figure 10-14) and the valley itself

The literary excerpt that opens this chapter attempts to portray the experience of descending into an active volcano. Earth's internal heat does much more than produce volcanic eruptions. On a global scale, the internal energy of the planet gives rise to forces that move huge sections of the lithosphere and create major landforms. The slow but momentous processes examined in this chapter are powered by these forces.

The 1960s heralded the modern era of understanding how the solid-earth environment works. The fundamental changes in thinking that occurred then have been called a revolution in the earth sciences. The changes arose from a buildup of new observations that added to the existing geologic data of continental rocks and fossils. Studies of rock magnetism reopened the debate over whether continents were fixed or mobile on Earth's surface. Mapping of the ocean-floor relief and of magnetism led first to the idea that material is added to the crust under the ocean (sea-floor spreading), and eventually to the more comprehensive theory of plate tectonics, which explains the long-term migration of the continents as a balance between the addition and removal of material at the edges of large sections of lithosphere. Further studies since the late 1960s have extended these ideas and have looked to Earth's deeper interior for an explanation of the processes at work. Previously, phenomena such as earthquakes, volcanic activity, ocean trenches, and continental mountain ranges were often regarded as unconnected to each other; now they are all considered parts of a single grand scheme.

This chapter focuses on the theory of plate tectonics and its implications. It begins by describing the operation of plate tectonics and then investigates the solid-earth processes that cause them. This leads to a discussion of some of the shorter-term (earthquakes, volcanoes) and longer-term effects (movements of continents, formation of mountain ranges) of plate tectonics. Other solar-system planets are then examined to see whether plate tectonics occur more widely.

PLATE TECTONICS

Preparing the Ground

Although much evidence of solid-earth movements was available to earth scientists from their study of the continental rocks, it was evidence from a previously hidden zone—the ocean floor—that led to a comprehensive theory of Earth crustal evolution in the late 1960s. Technology developed during World War II was used in the late 1950s and early 1960s to map the topography of sections of the ocean floor from surface ships. A pattern of ocean ridges, flat plains, and deep trenches emerged (see Figure 12-1).

Patterns of magnetic "stripes" were also discovered on the ocean floors. Studies of magnetism in rocks in the 1950s were based on the knowledge that iron-rich volcanic rocks, such as basalt, contain a record of Earth's magnetic field at the time of their cooling. As the rock cools from 500°C to 450°C (930°F to 850°F), the atoms in the iron minerals become aligned parallel to the magnetic lines of force that encircle the globe. When the temperature falls below 450°C (850°F), these lines become "frozen," and only change if the rock is heated again above 450°C. Lavas erupted within the last few thousand years have atoms aligned to the current magnetic field, while older lavas do not. The study of the magnetism of ancient rocks is known as **paleomagnetism.**

Studies of paleomagnetism led to two results. First, the studies indicated that Earth's magnetic poles had moved over the last 500 million years. This movement was initially termed **polar wandering,** but it was soon found that rocks in Europe and North America demonstrated different polar-wandering patterns. This implied that the continents, rather than the magnetic poles, had done the wandering, a theory that renewed interest in continental drift.

The second result from paleomagnetic studies arose from mapping during the 1950s of ocean-floor stripes hundreds of kilometers long and 20 to 30 km (12 to 18 miles) wide. The magnetic minerals in the rocks within each stripe were identically aligned, but the alignment differed from stripe to stripe. In the early 1960s, researchers demonstrated that the pattern of magnetic stripes across the Carlsberg Ridge in the Indian Ocean and the mid-Atlantic ridge southwest of Iceland formed mirror images on either side of the center (Figure 11-1). This observation and evidence from the ocean-floor relief led to the explanation that the mid-ocean ridge was a source of new ocean-floor crust. The rigid crust spreads out on either side of the ridge as new molten rock is forced up along the center. When such ocean-floor crust reached a trench on the ocean margin thousands of miles away, it descended back to the mantle. This "conveyor-belt" idea was termed **sea-floor spreading.**

The magnetic stripes also suggested that Earth's magnetism reverses approximately four times every million years, although the timing is very irregular. The north magnetic pole becomes the south mag-

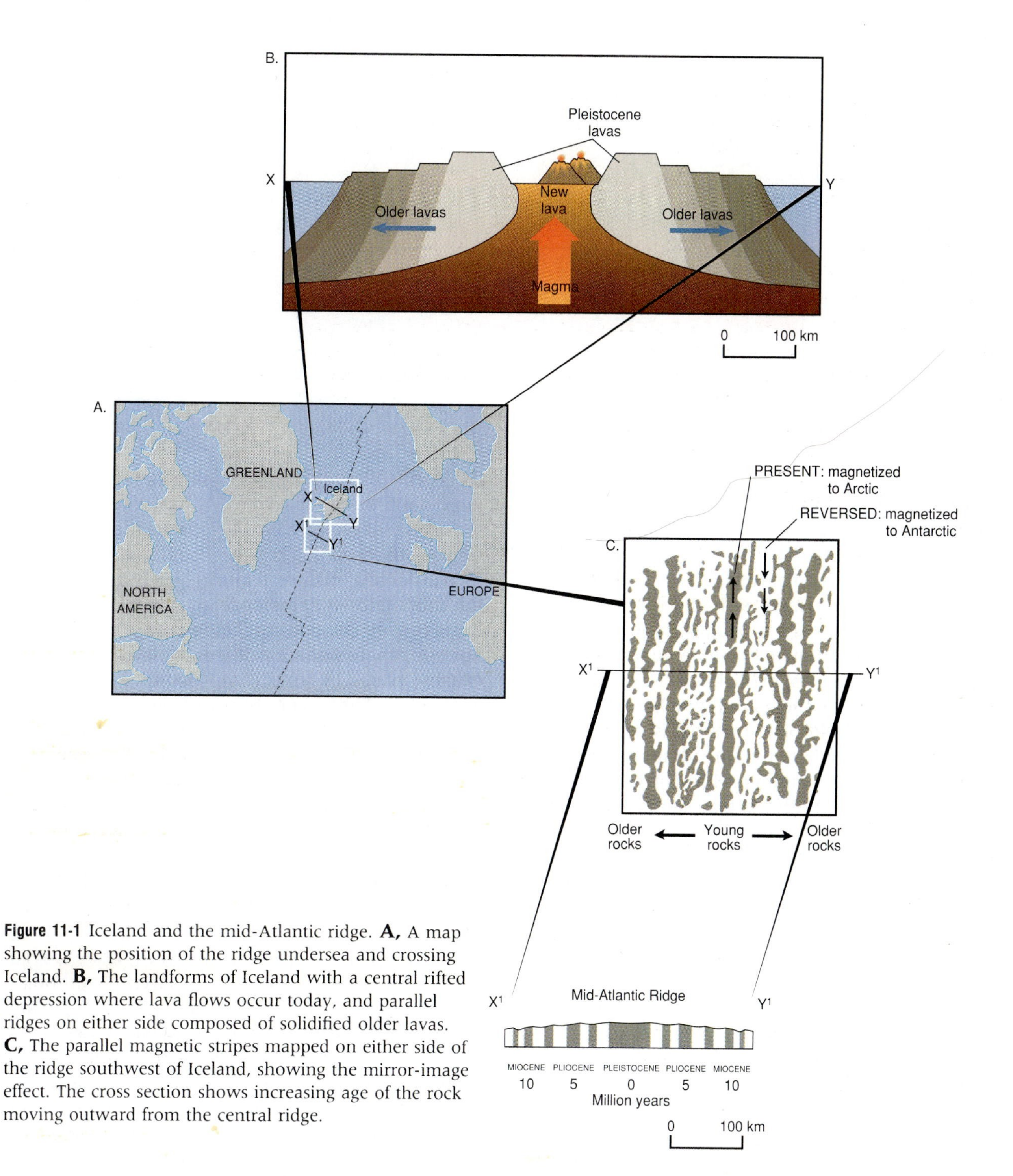

Figure 11-1 Iceland and the mid-Atlantic ridge. **A,** A map showing the position of the ridge undersea and crossing Iceland. **B,** The landforms of Iceland with a central rifted depression where lava flows occur today, and parallel ridges on either side composed of solidified older lavas. **C,** The parallel magnetic stripes mapped on either side of the ridge southwest of Iceland, showing the mirror-image effect. The cross section shows increasing age of the rock moving outward from the central ridge.

netic pole, and vice versa, as the direction of magnetization is reversed. The Deep Sea Drilling Project of the late 1960s and early 1970s brought rock samples to the surface so that the paleomagnetism of the ocean-floor rocks could be worked out and their dates of formation established by radiometric dating. The results linked the ocean-floor stripes to the geologic timescale. Only a small area of ocean floor proved to be more than 150 million years old.

Tectonic Plates and Margins

In the late 1960s the various sources of new evidence from studies of paleomagnetism, ocean-floor relief mapping, extended seismic surveys, and the radiometric dating of ocean-floor rock samples led from the idea of sea-floor spreading to the new theory of **plate tectonics.** This theory of plate tectonics relates the movements of continental masses to the movements identified in the ocean basins,

focused on the damage caused. Today earthquake magnitude is measured by the increasing number of stations around the world that record seismic waves using instruments known as *seismographs.* In 1935 Charles F. Richter established a scale for use with seismograph records in which each value on the scale is a tenfold increase in magnitude over the value one unit lower (Table 11-1). A magnitude of 2 is 10 times greater than that of 1, and a magnitude of 6 is a million times greater than 1. The vibrations from earthquakes of magnitudes 1 and 2 cannot be distinguished, even by seismographs, from the vibrations caused when a truck drives along a local road. It is estimated that some 9000 natural earthquakes of magnitude 1 or greater occur around the world every day. The earthquakes of highest magnitude are those with the deepest foci; they involve the greatest *stresses* (forces acting on the rocks) and produce the greatest *strains* (resulting deformation and displacement).

Earthquakes such as the one of magnitude 7.1 at San Francisco in 1989, or those over magnitude 8 in Armenia and Iran, cause extensive damage and loss of life (see Box 11-1: Earthquake Hazards). The main surface damage occurs within seconds of the initial earthquake shock, although smaller aftershocks may continue for several days as further slippage occurs along the fault. Even in the most stressed parts of the crust, earthquakes may occur only once every century at a particular place.

The world distribution of earthquakes since 1960 is plotted on Figure 11-6. This shows that the highest proportion of earthquakes, and virtually all the major ones, occur along plate margins. The greatest concentrations occur around the margins of the Pacific Ocean and in the mountainous areas of Europe and southern Asia through to Indonesia. These are convergent plate margins, where the collision and subduction of plates give rise to strong and deep-seated earthquakes (see Figure 11-4). Clear linear zones of less intense earthquake activity occur along the divergent plate margins such as the mid-Atlantic ridge, which runs down the center of the Atlantic Ocean and continues into the Indian and Pacific oceans; another such line occurs in eastern Africa along the rift valley region. The San Andreas fault is a strike-slip fault in which the western side moves northward relative to the eastern side (see Figure 10-18). Earthquakes do take place outside of the plate margin zones, but they are less frequent and their causes are less easy to link to specific zones of movement in Earth's crust.

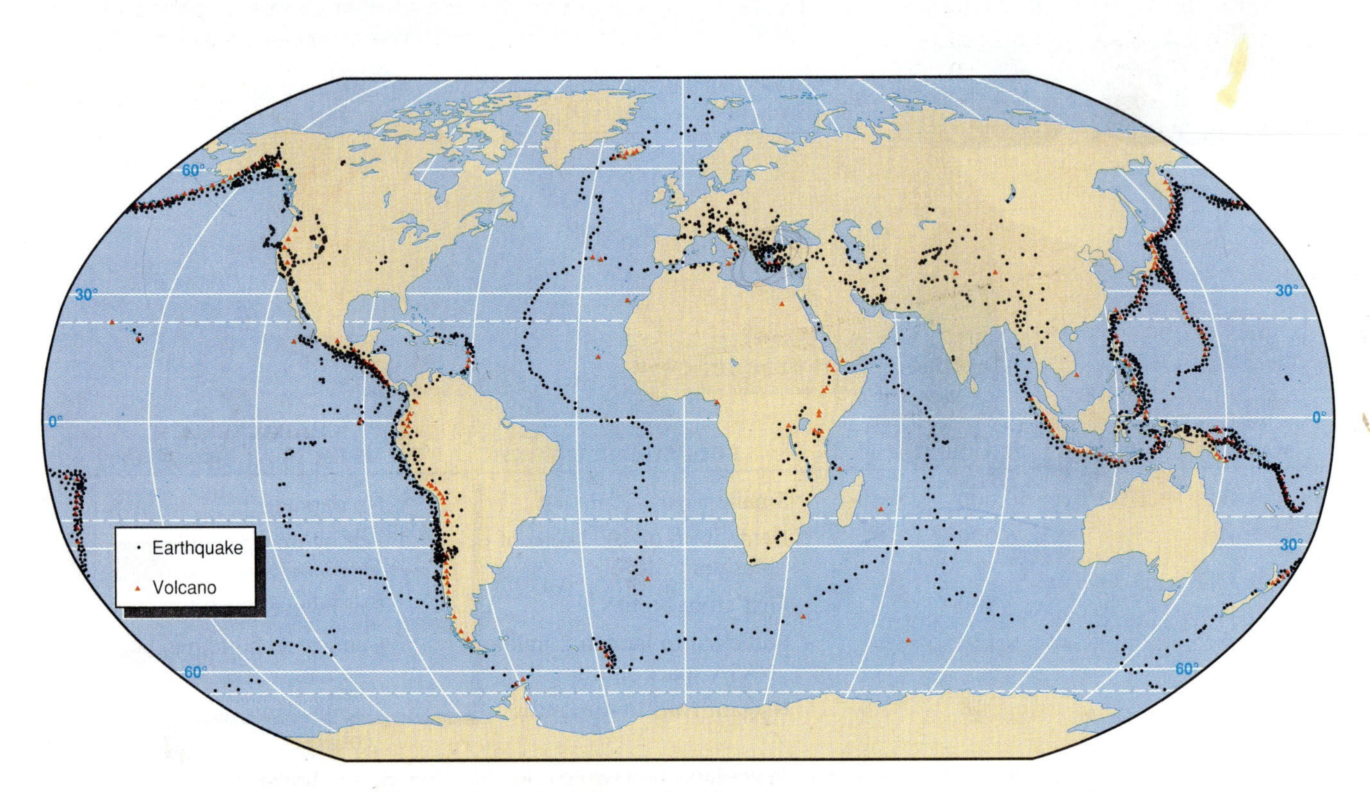

Figure 11-6 World map of the distribution of earthquakes and volcanoes. This should be compared to Figure 11-3 so that the occurrence of earthquakes and volcanoes can be linked to plate boundaries.

Volcanic Activity

Volcanic activity includes all the ways in which igneous magma erupts at Earth's surface. This occurs principally through pipelike *vents* or elongated *fissures* in Earth's crust. It produces a variety of surface rocks and relief features that can be linked to the type of eruption, and to the movement of plates. The world distribution of volcanic activity in historic time resembles that of earthquakes (see Figure 11-6). It is concentrated along plate margins, especially around the Pacific Ocean.

A **volcano** is a landform that is produced above a pipelike vent. Most volcanoes occur in ocean basins because of the formation and destruction of plates in and around ocean basins, and it has been estimated that there are 50,000 on the floor of the Pacific Ocean alone. Approximately 600 continental and island volcanoes around the world are recognized as *active,* since they have erupted in historic time. Sixty percent of these occur in and surrounding the Pacific Ocean, which is often called the "Pacific Ring of Fire" because so many volcanoes have erupted there. Several thousand volcanoes that have not erupted in historic time are regarded as *extinct* (unlikely to erupt again) or as *dormant* (may erupt in the future).

In addition to the volcanoes along the margins of plates, many occur in the interior portions of plates. Mid-plate volcanoes are most common on the Pacific Ocean floor and on the African plate, where they include the offshore peaks of the Azores and the Canary Islands and the inland Mounts Kenya and Kilimanjaro. They form over zones where magma rises toward the surface, known as **hot spots.** Hot spots form where there is localized heating that produces melting and expansion of upper-mantle rock. This causes thinning and cracking of the overlying crust, letting the magma through to the surface. Since the plates often move over a hot spot, the volcanoes built by the eruption of magma will occupy different surface locations at different times. Some volcanoes occur in clusters over a hot spot, as in the Canary Islands. Others are strung out in lines, especially in the Pacific Ocean. The Hawaiian Islands, for example, are part of a "conveyor belt" system in which the hot spot continues erupting volcanic rock (at present under the island of Hawaii at the southeastern end of the group) while the spreading plate movement of the Pacific Ocean floor carries each pile of volcanic rock to the northwest, away from the hot spot (see Figure 11-7). Earlier-formed volcanic islands eventually sink beneath the ocean surface in the course of this movement; their submarine remnants are known as *seamounts.* The seamounts that were formed over the Hawaiian Island hot spot extend for over 3000 km (2000 miles) northwest from the present islands, which lie over the hot spot today.

It is instructive to compare the Hawaiian Islands with Iceland in the North Atlantic (see Figure 11-1). Both are located over hot spots, but their plate situations are very different. Iceland has been built on a divergent plate margin in an ocean where volcanic activity is less intensive than in the Pacific. Hawaii annually produces six times as much lava as Iceland, but because Iceland has stayed in the

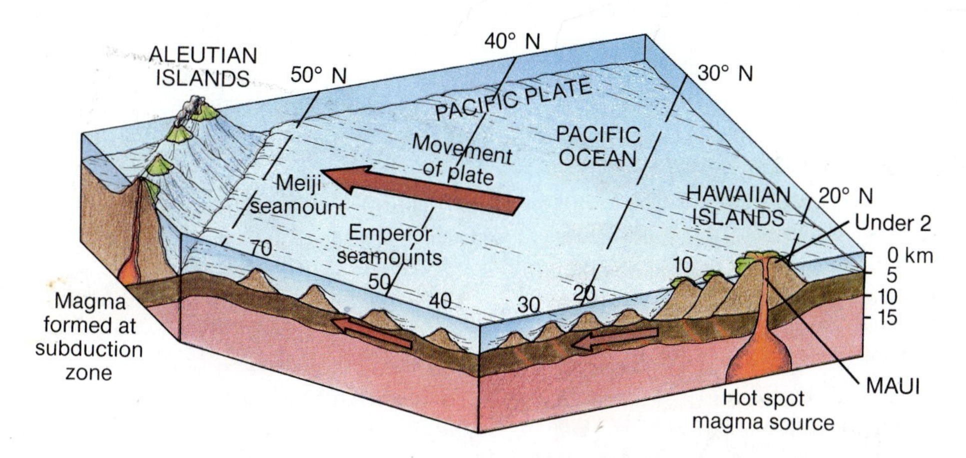

Figure 11-7 The Hawaiian Islands: a hot spot and its relation to an overlying plate. The hot spot continues to supply lava, as it has done for 70 million years. Former peaks move northwest on the Pacific plate. The numbers above the peaks on the near side of the diagram indicate the ages of the oldest lavas in millions of years. The Aleutian Islands are shown to compare the modes of formation of volcanic islands formed over a hot spot, and those formed along a convergent plate margin above a subduction zone. Note: The vertical scale is greatly exaggerated.

Figure 11-8 Lava advancing over a field in Hawaii.

same place above the divergent margin for more than 50 million years, the total volume of volcanic rock is five times that which forms the Hawaiian Islands.

Volcanic eruptions vary in their explosiveness. The greatest explosions blow off the top of any previous buildup of volcanic rock and leave a huge crater, known as a **caldera.** An explosion on the island of Krakatoa in Indonesia erupted 10 cubic km of rock into the atmosphere in 1883, and some 75,000 years ago an explosion on Toba in the same area erupted an estimated 20,000 cubic km of material. In the least explosive volcanoes, liquid lava wells up steadily and there is little explosive activity apart from an occasional "fireworks" display of glowing clots of lava (Figure 11-8). The Hawaiian volcanoes are typically quieter volcanoes, in which the lava wells up and erupts as molten flows. When the lava draws back, the central part of such a volcano may collapse, forming a *collapse caldera.*

Whether an eruption is explosive or nonexplosive is determined by the fluidity of the lava. Very fluid lavas flow easily and spread out over wide areas. Silica-rich lava is very thick and moves slowly: it is said to be highly *viscous,* producing short flows or explosive eruptions. The fluidity of the lava influences how it traps the gases, which separate from rising magma as pressure diminishes at or near the surface. In the more fluid types of lava, these gases bubble up and escape into the atmosphere. In viscous lava, the gas cannot escape; pressure builds up until eventually it bursts the lava apart in an explosive eruption. The clots of ejected lava solidify in the cooler atmosphere and fall to earth. These solidified particles and lumps of igneous material are known as **tephra.** Tephra range in size from *fine ash* (less than 2 mm across) to *lapilli* (2 to 64 mm) and *volcanic bombs* (over 64 mm). One volcanic bomb weighing 200 tons was hurled a distance of 14 km in an eruption of the Andean volcano Cotopaxi.

Whether a magma is viscous or fluid depends on its chemical composition. Magmas formed largely of felsic minerals are highly viscous, whereas mafic magmas are fluid. Magmas intermediate between the two have moderate viscosity. Granitic and intermediate magmas are common along high mountain systems above subduction zones, as for instance in the Andes and the western margins of the United States. They produce explosive eruptions, such as that of Mount St. Helens, or alternately explosive and fluid eruptions. The rocks produced from granitic and intermediate lavas are typically rhyolites or andesites (see Chapter 10). Basaltic magmas result from the melting of mantle rock at hot spots (e.g., Hawaii, Iceland) and produce lavas with a high proportion of mafic minerals and low percentages of silica.

The explosiveness of volcanic eruptions also determines the form of the volcanic landforms that result. A detailed range of eruptive types has been classified and related to landforms (Table 11-2). The 1980 eruption of Mount St. Helens was of the Vulcanian type. Where explosive eruptions center on a single pipelike vent, viscous granitic lavas will produce steep-sided, dome-shaped forms, or piles of explosive tephra forming the *cinder cone* type of volcano, which may be several hundred meters high and up to a kilometer or so across (Figure 11-9, *A*). The most extreme volcanic explosion causes any previous cone and surrounding rocks to disintegrate in a mass of hot gas and fluid lava droplets known as a *nuée ardente* ("glowing cloud"). On the Caribbean island of Martinique in 1902 the explosion of Mont Pelée produced a *nuée ardente* that destroyed the local town and its inhabitants (Figure 11-9, *B*).

Fluid basalt erupted at a vent builds up shallow-sloped layers, eventually forming huge *shield volcanoes* such as the Hawaiian Islands (Figure 11-9, *C*). The Hawaiian Islands rise over 10 km from the ocean floor to their highest points, and cover several thousand square kilometers of the ocean floor. Most continental and island volcanoes combine layers of tephra and lava and are known as *composite volcanoes* (Figure 11-9, *D*). The peaks surmounting the Andes, such as Cotopaxi, are of this type, and typically have slopes that become less steep with increasing

Table 11-2 A Classification of Volcanic Eruptions

VIOLENCE OF ERUPTION	ERUPTION TYPE (named after a volcano or place)	MAGMA TYPE	EXPLOSIVE ACTIVITY, LAVA ERUPTION	LANDFORMS AROUND VENT
Least violent, mainly lava flows	Icelandic, Hawaiian	Mafic, flows easily	Very weak explosive activity. Extensive lava flows from fissures and vents	Broad lava cones, and shield volcanoes; plains and plateaus along fissures
Moderately violent with mixture of lava and ashes	Strombolian, Vulcanian, Vesuvian	Mixed mafic/felsic to mainly felsic; flows slowly	Few flows; thick, moderately extensive	Cinder cones and lava flows; large composite cones; explosive craters
Extremely violent, with ashes, nuées ardentes	Plinian, Peléan, Krakatoan	Felsic, viscous	Very few flows; small domes	Widespread lapilli; small cones, domes; large caldera

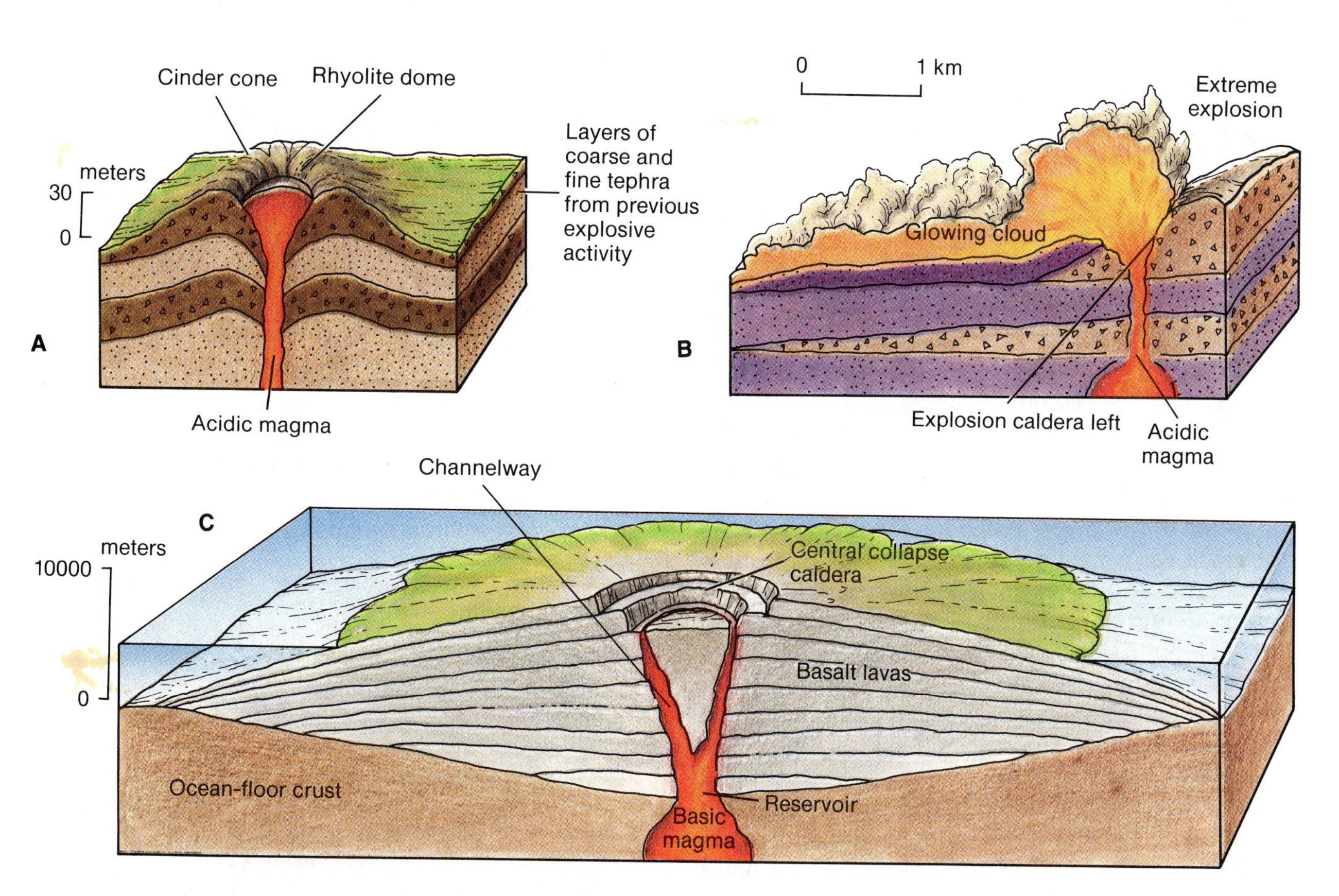

Figure 11-9 Forms produced by volcanic activity. **A,** A cinder cone with a rhyolite dome inside, formed by an explosive, acidic (felsic) magma. **B,** A nuée ardente formed by an extreme explosion. **C,** A shield volcano formed of many layers of fluid basalt lava.

tually breaks apart (Figure 11-11, *A*). Volcanic activity and shallow earthquakes are common above upward currents. The two sections on either side of the rift are carried away from each other as the plates diverge (Figure 11-11, *B* and *C*). Over millions of years an ocean forms as the rift widens, and then continues to expand in area until its margins are above a downward current in the upper mantle convection cell (Figure 11-11, *D*). When that occurs, subduction begins. Deep-seated earthquake shocks and explosive volcanic activity are characteristic of areas above subduction zones because of the stresses caused when one large mass of rock pushes down into another. The ocean may eventually begin to close if the ocean crust is consumed faster at the subduction zone than it is produced at the divergent plate margin. Closure of an ocean may bring together two masses of continental crust that combine to form a massive mountain system (Figure 11-11, *E*).

Ocean-floor plate motion at divergent margins is slower in the early phase during which the ocean basin widens (0 to 20 mm per year, as in the Atlantic Ocean) than it is in the later, ocean-closing stages (up to 100 mm per year, as in the Pacific Ocean). It is thought that in the earlier phase upward convection at the divergent margin provides the only driving force and so has to push continental crust apart by itself. In the later stages, the subduction zones add a pulling force to the push of the divergent margins, and so the rate increases (see Figure 11-11, *C* and *D*).

Tectonic processes recycle a portion of the materials of Earth's surface through the renewal of ocean-floor crust every 200 million years or so. The continents accumulate rocks and minerals over a much longer period, and contain Earth's oldest rocks (3.8 billion years old). It appears that the continents grow when new crustal material is added at their margins by converging plates. Continents are also involved in a different type of recycling of rock materials as the mountains uplifted by plate

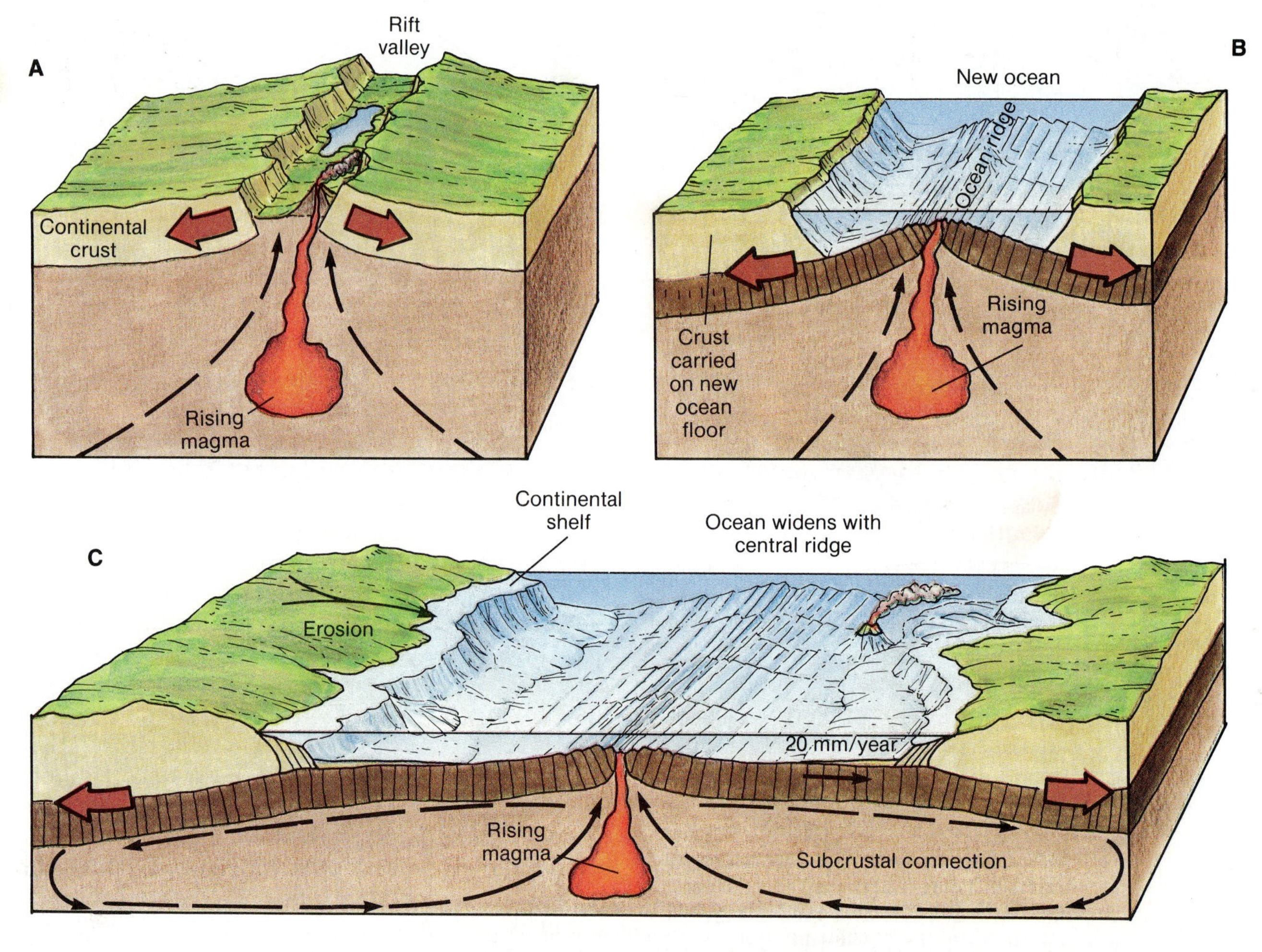

Figure 11-11 The opening and closing of an ocean, and the formation of continental mountain ranges. The stages of this process, **A-E** are known as the Wilson Cycle, are described in the text.

tectonics are worn down by streams, glaciers, and the wind. The broken rock debris is transported to the oceans or deposited on land, where it forms new sedimentary rocks.

Mountain Building

Distinctive relief features of the continents include mountain systems, such as the Rockies of North America, the Alps of Europe, the Andes of South America, and the Himalayas of Asia. These mountain systems occur along convergent margins of plates. Most of the highest ranges within the mountain systems are still in the process of uplift and deformation, signalled by major earthquakes and volcanic eruptions. They are also characterized by highly folded rocks and batholith intrusions. The process of mountain formation is known as **orogenesis** ("mountain forming"), and its range of outcomes is discussed in Chapter 12.

Epeirogenesis

While the most dramatic effects of plate movements result from horizontal compression or tension at the convergent and divergent margins of the plates, broad, vertical movements may affect large areas in plate interiors. These produce uplifted plateaus such as most of Canada, Africa, and Australia, or subsiding basins such as that beneath the North Sea in Europe. The processes involved in vertical movements are poorly understood, but may result from internal thermal expansion or contraction in the upper mantle. Because of variations in temperature in the weak asthenosphere the lithosphere may sag in some places but may be pushed up in others. The resulting crustal movements are known as **epeirogenesis** ("continent forming") and commonly tend to occur over larger areas than orogenesis. Movements such as these may involve some shallow earthquake activity, folding, and faulting.

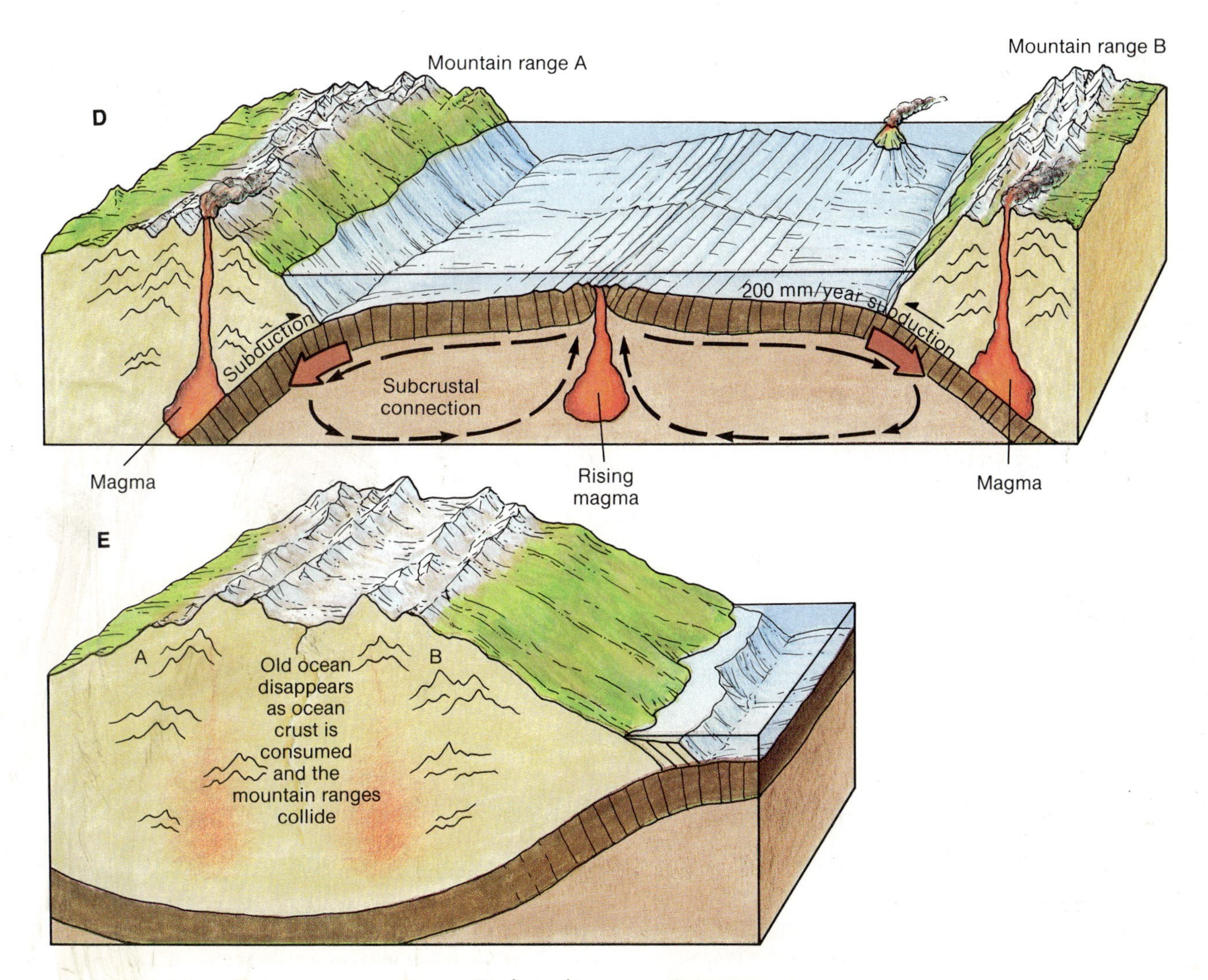

For legend, see opposite page.

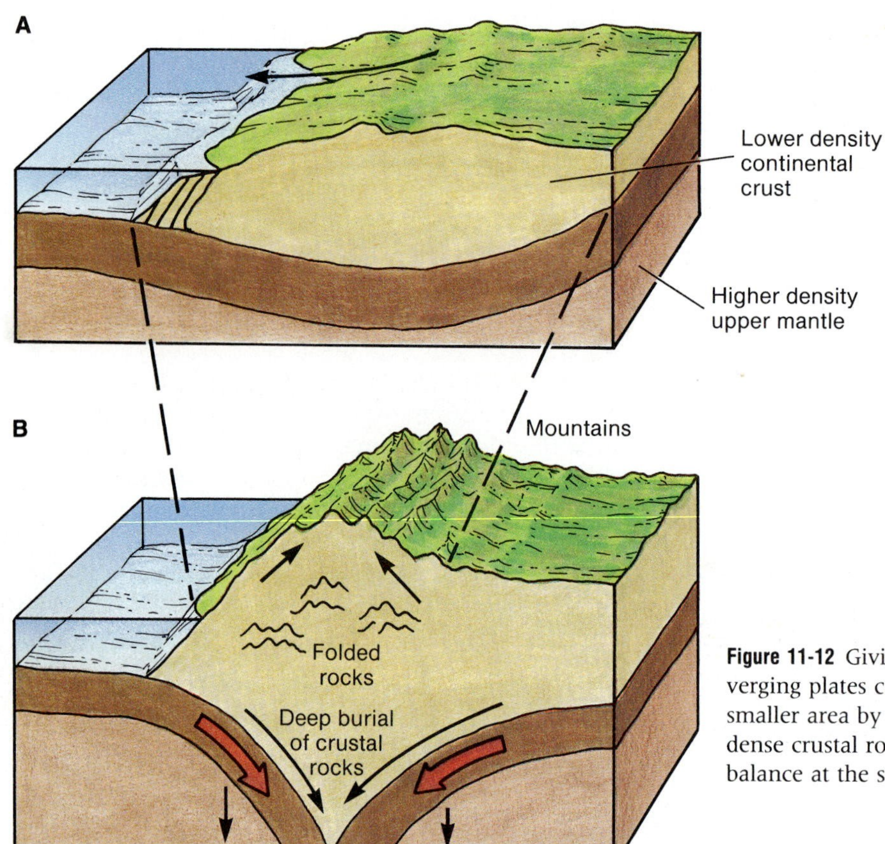

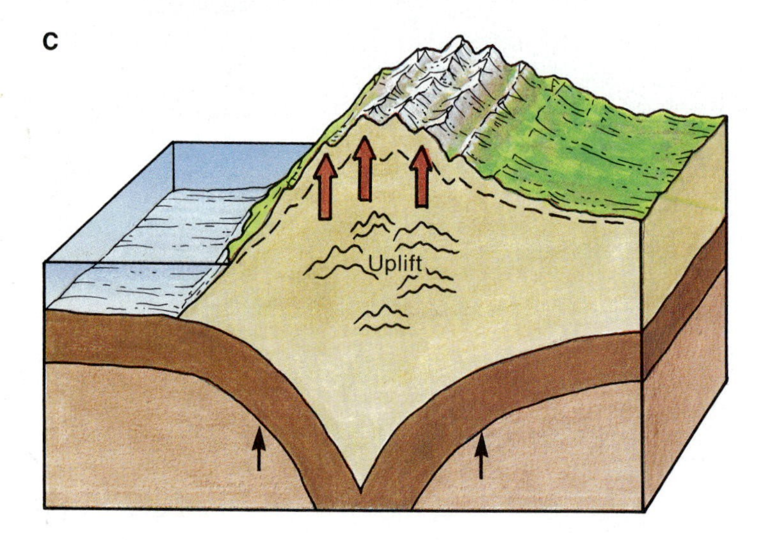

Figure 11-12 Giving extra height to a mountain range. Converging plates cause crustal rock to be concentrated in a smaller area by compression, and this thickens the less-dense crustal rocks. After a time they rise to maintain a balance at the surface.

A form of epeirogenesis may occur near the end of the orogenic process. Before orogenesis, a continent composed of lower-density sial lying on top of denser sima sags to a depth that depends on the combined thickness and density of the two layers (Figure 11-12, *A*). The sag is compensated by outflow in the asthenosphere. During orogenesis, horizontal compression at convergent plate margins results in thickening of the crust along the mountain belt and in its being carried deeper at the subduction zone (Figure 11-12, *B*). This increases the total thickness of the lower-density layer of sial. An imbalance in density occurs as the proportion of sial increases in depth. Eventually this causes the thickened mountain zone to rise and even out the distribution of density at Earth's surface (Figure 11-12, *C*). This process of vertical adjustment of Earth's surface in response to variations in the depth of lighter sialic rocks is known as **isostasy.**

The Himalayas are the world's highest mountains because they have the greatest thickness of low-density continental crust beneath them. The situation is often likened to an iceberg floating in the ocean, where the lower density of ice makes the top of the iceberg rise above the surface of the denser water. Crustal rocks cannot spring upward as a depressed iceberg might do in water, and they take millions of years to redress the imbalance. Thus, mountain systems continue to rise long after the main plate collision and thickening of the crust take place. While the mountain system is rising, surface processes wear down and remove rock, which reduces the weight of the mountains and encourages further uplift. This process also explains how surface processes can wear down deeply into crustal rock, since wearing down processes result in further uplift and further erosion.

Another type of isostasy takes place when material accumulates on top of the continental or ocean crust in other ways. One example was during the Pleistocene Ice Age, when large ice sheets formed over northern North America and northern Europe. The extra weight of the ice caused the crust beneath it to sink (Figure 11-13, *A, B*). When the ice sheets melted, the land surface rose slowly toward its earlier level (Figure 11-13, *C, D*). For instance, the weight of the ice sheet that covered Scandinavia in northern Europe caused the land beneath to sink. When the ice melted some 15,000 years ago, the central section of crust did not rise immediately and was drowned to form the Baltic Sea. The crust in this area is rising at 8 to 10 mm per year and will take several thousand years to return to its preglacial level.

A further example of isostasy is where the deposition of sediment is concentrated in relatively small areas, imposing an additional load on the crust. The world's largest deltas are one example of such areas. The Mississippi River deposits large quantities of sediment along its lower course and on its delta in Louisiana. The deposited sediment forms additional land, but its weight adds to the load on the crust so that the surface is sinking, and the Gulf of Mexico waters may eventually drown the area. The subsidence resulting from the load of sediment on the crust is enhanced by compaction of the sediments. An excess of depositional buildup over subsidence is vital for the continued existence of this land area, on which the city of New Orleans rests. The features of the Mississippi delta as a coastal landform are discussed in Chapter 17. The Nile delta in Egypt is in a similar state of balance between deposition and subsidence. For a long

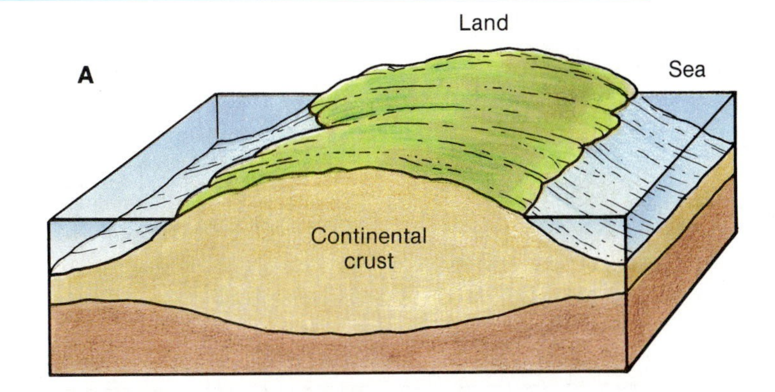

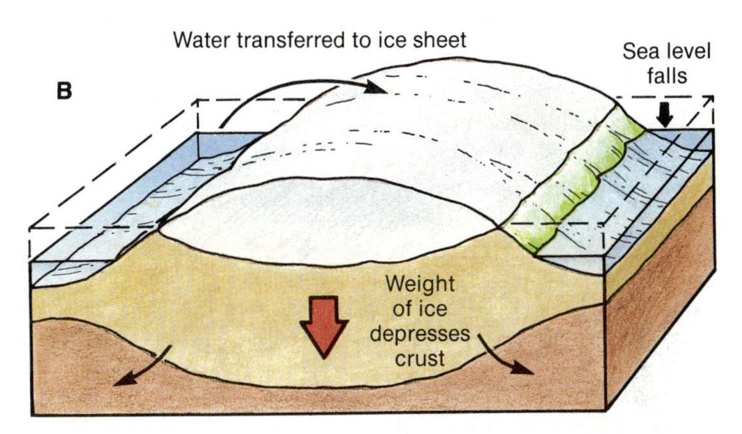

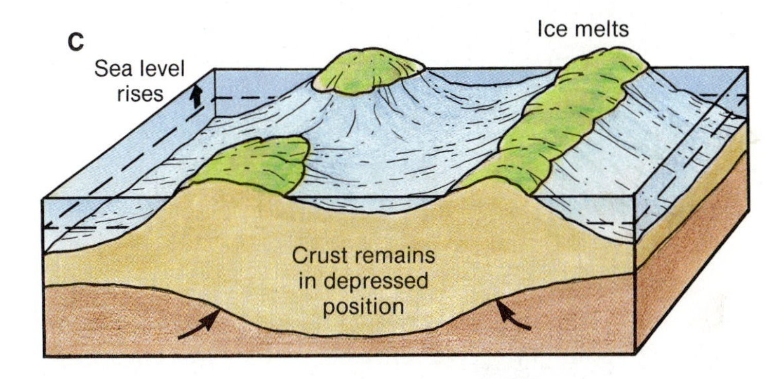

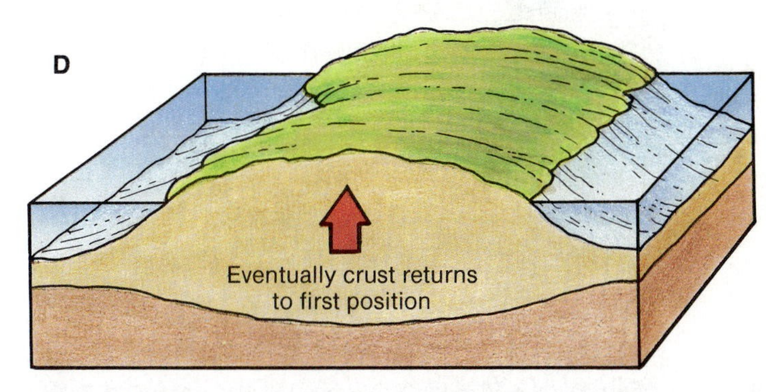

Figure 11-13 The impact of loading a continent with an ice sheet. The land is depressed by this weight *(B)*, and adjustment takes place in the asthenosphere. When the ice melts *(C)*, it takes some time for the land to recoil to its former position *(D)*: the area drowned by the Baltic Sea was depressed during the Pleistocene Ice Age and is rising at 8 mm per year at present.

time, the deposition during floods maintained the delta surface just above sea level. However, the trapping of sediment behind the Aswan Dam since it was built in the 1960s has reduced the amounts deposited on the delta. Subsidence resulting from past deposition is no longer balanced by current deposition and is beginning to result in drowning and waterlogging of the outer delta—a problem that is likely to get worse.

To summarize, changes in levels of land surfaces that result from additions or removals of crustal material are known as *isostatic* changes. The crustal loading or unloading is caused by either internal solid-earth forces or external atmosphere-ocean environment activity (e.g., ice or sediment accumulation, or erosion). These adjustments take place in the weak asthenosphere layer beneath the lithosphere, when hot, flowable rock moves to accommodate the changes.

LINKAGES

*The plate movements of the **solid-earth environment** give rise to new distributions of continents and oceans and new ranges of high mountains. This influences global air and water circulations **(atmosphere-ocean environment)** and the dispersal of plants and animals **(living-organism environment).***

PLATE TECTONICS ON OTHER PLANETS?

The phenomena examined in this chapter testify to the links between changes in major surface features and internal activity in the solid-earth environment. Some parts of the globe are much more active than others in earthquakes, volcanic activity, and mountain building. These tectonic activities are principally concentrated in narrow zones along plate margins. Other parts of Earth's land surface appear to be inactive, but are in the process of slow, continuous horizontal or vertical movement, and may have been involved in more dramatic changes in the past. Earth has few surface features such as meteorite craters that indicate the intervention of matter from outside the planet. The meteor crater at Winslow, Arizona, is one of a small number that have been identified. Any that formed in the past have been obliterated by plate tectonics and surface processes, which are continually changing Earth's surface relief.

The major surface features of other planets also reveal secrets about their interiors, and may provide insights to geologists studying Earth. The surfaces of Earth's moon and Mercury are dominated by meteorite impact craters. Although these bodies experienced volcanic activity in their early histories, it was often started by the impacts of large meteorites that punched holes in the crust when their interiors were largely molten between 3 and 4 billion years ago. Such impacts were responsible for virtually all the surface relief features on these planets. It appears that the internal layers of both bodies have cooled and solidified since that time. Earth must have been subject to a similar meteor bombardment, which in early Earth history would have been slowed only partially by the atmosphere. The recycling activities of plate tectonics and atmospheric processes, however, have formed new surface features that have obscured almost all the meteorite craters from that period. The presence of so many meteorite craters on Earth's moon and Mercury makes it clear that plate tectonics has not dominated there.

The surface of Mars is marked by distinctive large shield volcanoes and huge canyon systems (see Figure 1-9). They suggest that internal processes almost led to something akin to plate tectonics. Lava poured out of single vents over many millions of years to form Olympus Mons and other huge volcanic edifices several hundred kilometers in diameter, but plate movements are not evident. Apparently, a rift began to form in the Martian crust (the Valles Mannensis), but the activity ceased some 3 billion years ago.

Venus is approximately the same size as Earth, and has a similar density. It might be thought that its surface relief would be affected by plate tectonics.

Until the Magellan probe in 1990, the total cloud cover of Venus made it almost impossible to determine whether this is so, although a few clues were obtained. Russian spacecraft sent back some images that showed masses of small broken rock fragments, probably of basaltic composition. Radar images taken from outside the planet's clouds showed that much of the planet is fairly smooth and the difference in height between the highest and deepest points is about 13 km (compared with nearly 20 km on Earth). The surface appeared to be divided into continentlike and ocean-basin-like areas. This division, and the basaltic lavas, suggested similarities with Earth. However, there are more circular features resembling impact craters on Venus than are preserved on Earth, and this suggests that the lower range of relief on Venus may not be due to more active surface processes than those on Earth.

The early results of the Magellan observations confirm that Venus has features such as mountains and volcanoes that may be caused by plate tectonics (Figure 11-14), but it has other features such as small-scale fracturing of the surface that are quite different from anything that occurs on Earth. It is possible that Venus may have its own unique internal mechanisms for generating relief.

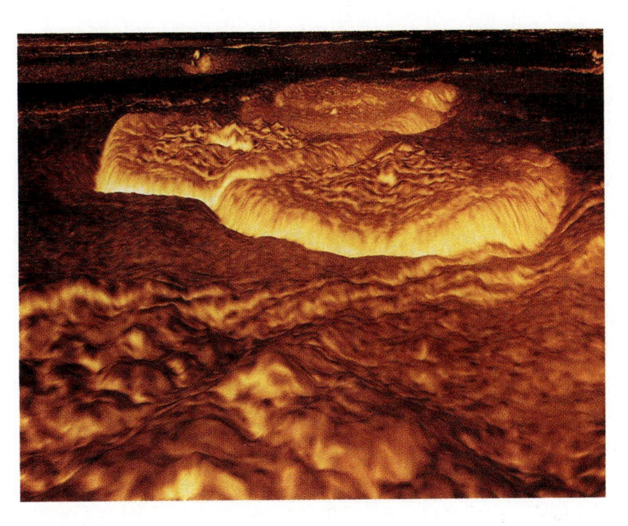

Figure 11-14 Part of the surface of Venus. The three flat-topped hills are 25 km (15 miles) across and up to 750 m (2500 feet) high. They may be formed of viscous lava eruptions or by uplift over shallow intrusions. Some surface fractures can also be seen. This picture was produced from radar images and the simulated colors are based on those sensed by Russian spacecraft.

FRONTIERS IN KNOWLEDGE

Plate Tectonics

Although geologists' understanding of solid-earth processes has grown tremendously since the 1960s, much remains to be learned before scientists can forecast earthquakes and volcanic eruptions precisely and answer other important questions. The answers to some may not be known for a long time, if ever. For example, a complete understanding of the nature and functioning of Earth's core is unlikely in the forseeable future. Scientists believe Earth's magnetism is generated in the core, but do not know how and do not have the means of finding out. Magnetic reversals are a mystery, subject to speculative ideas. The way heat energy is transferred through the mantle is basic to understanding what drives plate tectonics, but present investigations are limited to the study of seismic waves and to laboratory models that attempt to simulate some of the conditions in the mantle. The mechanism by which mantle convection drives plates is still a matter of debate. The process of continent growth and continent splitting will be better understood as geologists continue to assemble and correlate a worldwide knowledge of rocks and plate structures.

SUMMARY

1. Earth's surface features change continuously as a result of internal solid-earth processes.
2. Plate tectonics is the set of processes that produces large-scale changes in Earth's surface features. Plates are almost rigid sections of Earth's lithosphere, which converge and diverge, colliding and pulling apart.
3. Plate motion is driven by convection in the upper mantle.
4. The solid lithosphere is produced by the cooling of rising magma along diverging plate margins and is destroyed by subduction beneath lower-density (mainly continental) crust.
5. Plate tectonics gives rise to short-lived phenomena such as earthquakes and volcanoes along plate margins.
6. Plate tectonics produces the broad features of major landforms, such as the shapes and distribution of ocean basins, continents, and mountains. Areas of broader warping—uplift or sinking—of sections of the crust occur in inner sections of plates.
7. Earth's moon and Mercury probably did not have plate tectonics processes, but Mars shows some signs of them, and Venus displays a range of features that may have been caused by processes akin to plate tectonics.

KEY TERMS

paleomagnetism, p. 294
polar wandering, p. 294
sea-floor spreading, p. 294
plate tectonics, p. 295
plate, p. 296
divergent margin, p. 296
convergent margin, p. 296
subduction, p. 296
transform margin, p. 296
earthquake, p. 298
focus, p. 298
epicenter, p. 298
earthquake magnitude, p. 298
volcano, p. 303
hot spot, p. 303
caldera, p. 304
tephra, p. 304
orogenesis, p. 309
epeirogenesis, p. 309
isostasy, p. 311

QUESTIONS FOR REVIEW AND EXPLORATION

1. List the evidence you would use to defend the theory of plate tectonics to someone who believes that Earth's major relief features have not changed over time.
2. What evidence do earthquakes and volcanoes provide about the causes of plate movements?
3. Note from newspapers and TV where earthquakes and volcanoes take place. Relate their positions and strengths to location on the world map and test whether they occur along plate boundaries. Should people be prevented from living in places subject to such natural disasters?
4. What are the links between plate tectonics and continent and ocean basin changes, orogenesis, and epeirogenesis?
5. Draw a series of labelled diagrams to explain isostatic relationships between crustal areas as a result of orogenesis, the formation and melting of ice sheets, and the deposition of sediment in a delta.
6. This is a chapter that deals with ideas covering the entire solid-earth environment. Identify aspects where further research might be concentrated.

FURTHER READING

Boraiko AA, McDowell B: When the Earth moves, *National Geographic,* May 1986, 638-675. Accounts of the volcano Nevado del Ruiz erupting in Colombia, and the Mexican earthquake of 1985 are used as examples of short-lived extreme hazard events and have linked them to plate tectonics.

Burchfiel BC: The continental crust, *Scientific American,* September 1984, 130-142. The author reviews cycles of tectonics, erosion, and sedimentation.

Francheteau J: The oceanic crust, *Scientific American,* September 1984, 114-129. This entire issue is devoted to Earth's environments; this article summarizes knowledge about the ocean floor and its origins.

Gore R: Our restless planet, *National Geographic,* August 1985, 142-181. This article emphasizes dynamic change in the solid-earth environment.

Mackenzie DP: The Earth's mantle, *Scientific American,* September 1984, 67-78. Emphasizes the convection cycle in the upper mantle that drives plate tectonics.

Tilling RI: *Eruptions of Mount St. Helens,* U.S. Geological Survey (1987). This is a well-illustrated account of the eruptions through the mid-1980s.

Tilling RI, Heliker C, Wright TL: *Eruptions of Hawaiian volcanoes,* U.S. Geological Survey (1987). This booklet considers the Hawaiian islands in the context of plate tectonics and examines the details of earthquake and volcanic activity and their impacts on human lives.

Vink GE, Morgan WJ, Vogt PR: The Earth's hot spots, *Scientific American,* April 1985, 50-57. Explains hotspot "plumes" as a key part of plate tectonics.

Wallace RE, editor: *The San Andreas fault system, California,* U.S. Geological Survey Professional Paper 1515 (1991). This publication is an "overview of the history, geology, geomorphology, geophysics and seismology of the most well known plate tectonic boundary in the world."

Ward PL: *The next big earthquake in the Bay Area may come sooner than you think,* U.S. Geological Survey (1991). This publication was designed to provide education about living in an earthquake zone following the 1989 Loma Prieta earthquake. It can be obtained from: Earthquakes, U.S. Geological Survey, 345 Middlefield Road, Menlo Park, CA 94025.

Wegener A: *The origin of continents and oceans,* Dover (1966-reprint of third edition of 1924). The fullest account of Wegener's ideas of continental drift.

The final stage of ocean-basin history begins when the basin is reduced to a small area occupied by an inland sea. The Mediterranean Sea is the remnant of a former ocean basin that extended between Europe and Asia to the north and Africa and India to the south.

LINKAGES

The shapes of ocean basins are the result of plate tectonics ***(solid-earth environment).*** *They influence the circulation of ocean water and the heating of the atmosphere* ***(atmosphere-ocean environment).*** *They also affect the distributions of living organisms, both on land and in the sea* ***(living-organism environment),*** *and the remains of marine plants and animals provide a large proportion of ocean floor sediments.*

MAJOR LANDFORMS OF CONTINENTAL ENVIRONMENTS

Continents are large land masses that rise sharply from the floor of ocean basins. As landform environments, they show many contrasts with the ocean basins. Continents are mostly above sea level, although their margins may be covered by shallow seas. Continents stand high in Earth's relief because they are composed of thicker masses of crustal rock having a felsic composition and low density. The low density prevents these rocks from being subducted, and so, unlike that of the ocean floor, continental rock is not recycled to the asthenosphere. Thus, continents preserve the oldest rocks on Earth. Continental rock accumulates and large masses are added to continental areas, a process known as *accretion*. However, the landforms created by this buildup, unlike those of the ocean basin, are exposed to disintegration by running water, glaciers, and wind.

The major landforms of the continents are shown in Figure 12-4, and are described in the following sections. The highest points of continental relief occur in relatively narrow mountain systems, and there are more extensive areas of lowland and plateau. A *plateau* is an area of higher land with a relatively flat top. Some of the largest plateaus are formed of the oldest Precambrian rock. A significant feature of the continental surface is that most of it was originally formed as mountain ranges. Older mountain ranges have been worn down by the processes of the atmosphere-ocean environment, but can still be identified through geologic studies.

Young Folded Mountains

The highest mountain systems of the present day are the Himalayas of northern India, the Andes of South America, the Rockies of western North America, and the Alps of southern Europe. They generally exceed heights of 3000 meters (10,000 feet) above sea level and rise to some 8000 meters (25,000 feet) at their highest points but occupy narrow zones and account for a relatively small part of the total continental area. These mountain systems are often termed the **young folded mountains** because their rocks have been folded by compression, and they are the most recent mountains to have been formed. They consist of rocks formed over the last 250 million years that have been crushed at convergent plate margins and uplifted over the last 50 million years. There are two major belts of these mountains on Earth. One extends along the west coast of the Americas, where ocean plates collide with plates carrying continents. The other consists of the ranges from the Alps to the Himalayas and beyond, where plates carrying continents have collided with other continent-carrying plates.

The folded mountains have greater thicknesses of low-density continental granites, gneisses, and sedimentary rocks than the nonmountainous areas adjacent to them. For instance, in the Alps the continental crust is 10 km (6 miles) thick, but rock of similar type formed in the same period in northern Europe is 2 km (1.2 miles) thick. Another feature of these folded mountains is the occurrence of huge granite batholiths aligned with the trend of the ranges, as, for example, in western North America (see Figure 10-7). The combination of thick low-density sediments and low-density granite intrusions causes the ranges to stand high in relief because of isostasy, as was explained in Chapter 11. The complex structures into which the rocks have been contorted, including large overturned folds and huge thrust-fault fractures (Figure 12-5, *A*), show that they have been subject to intense compression.

The strongest evidence of the stresses involved in the formation of fold mountains is found in the rocks exposed after erosion of the inner parts of the ranges. These rocks were once mainly clays and sandstone but have been deeply buried, folded, metamorphosed, and intruded by granites during the mountain-building processes. The temperatures and pressures that caused such changes only occur several tens of kilometers below the surface. Much overlying rock must have been worn away to expose these rocks at the surface today (Figure 12-5,

B). In the outer zones of young folded mountains, the rocks are less deformed and consist of a greater variety of recognizable sedimentary rocks, including limestones, shales, and sandstones.

Fault-Block Mountains

A **fault-block mountain** area is typically an uplifted section of crust with one or more faulted boundaries. Recall that a fault is a crack in Earth's crust where the rocks on one side are displaced relative to the other. Examples of fault-block mountains include the Sierra Nevada of California, which rises to 3500 meters (14,000 feet), the basin-and-range area of Nevada and Utah, and the horsts of central and northern Europe (see Figures 10-19 and 10-20) which rise to just over 1000 meters (3000 feet).

Normal faults most commonly form the margins of fault-block mountains, indicating that these mountains are produced when tensional stresses pull the rocks apart. This tension may result from broad upwarping that leads to rifting of the crust.

The rocks of the fault-block mountains of central Europe were formed mainly between 300 and 600 million years ago and closely resemble the rocks observed in young folded mountains. The overturned fold and thrust-fault structures also resemble those found in the highest mountains of today. It is clear that fault-block mountains were folded mountains some 250 to 300 million years ago and were worn down over hundreds of millions of years before being uplifted again by faulting in the last 50 million years. Evidence from these areas is used to reconstruct ancient plate margins, since it may be assumed that they were folded mountains formed by the collision of ancient plates.

Precambrian Shields

A **shield** area (or craton) is a large section of the continental crust that has been unaffected by mountain-building since Precambrian times. The name *shield* originates from the fact that such areas often produce plateaus, rather like a shield lying on the floor with its front side up. Exposed shield areas constitute about 25 percent of continental surfaces. Shields also underlie large areas of younger rocks, forming the structural cores of the continents; hence they are often termed the *basement*. The surface relief of exposed shield areas shows little relation to the underlying fold or fault structures that remain from long-gone folded mountain ranges. The knobbly low relief of northern Canada, the plateaus of Africa, and the submerged rocks beneath the Baltic Sea are all parts of shield areas that evidently were at the core of mountains in Earth's distant past.

Internally, shields are made up of a series of elongated zones of highly altered rocks resembling those of the inner sections of young folded mountains. Each shield area commonly contains evidence of several phases of orogenesis between 3.8 billion and 600 million years ago (Figure 12-6). This repetition of rocks and structures in continental crust of different ages demonstrates that cycles of mountain formation have molded the continental surfaces since the earliest crustal rocks were formed.

Some shield areas have been covered by thin layers of sedimentary rock. Following the erosion of older mountains, these areas were invaded by shallow seas during the last 600 million years. The Gulf and Atlantic coastal plains, the Midwest of the United States, the Paris Basin in France, and large parts of Africa are examples of such areas. The sedimentary rocks have sometimes been tilted and simply folded as a result of epeirogenic warping of the shield areas on which they lie. In the Deccan plateau of southern India, the Precambrian shield rocks have been covered by little-disturbed layers of basalt lava.

The shield areas bear testimony to the dynamic nature of the solid-earth environment. The parts that were formed long ago in the heart of folded mountains and are now exposed have had several kilometers of overlying rock removed by the grinding action of running water and moving ice. Other parts have been covered by more recent sedimentary rocks or lava flows after the wearing down phase. Continental shields demonstrate the effects of both internal processes lifting up the crustal rock and surface processes wearing it down and providing the materials for new layers of sedimentary rock.

Continental Margins

Today most continental margins are covered by ocean water but are considered parts of the continents because they are composed of continental crust. Continental margins are designated active or passive according to their relationship to plate margins. An *active* continental margin is one, such as the western coast of North America, where a convergent plate boundary is producing earthquake activity, volcanic activity, uplift, and the rapid wearing down of rocks exposed at the surface. A *passive* con-

SUMMARY

1. Plate tectonics produces two major environments—ocean basins and continents—in which distinctive groups of major landforms and rocks are formed.
2. The ocean basin environment gives rise to ocean ridges, abyssal plains, ocean trenches, island arcs, and isolated volcanic peaks.
3. Ocean basins evolve from rifted sections of continental crust that widen by sea-floor spreading. Eventually subduction at ocean trenches consumes ocean-floor crust, and the ocean basins are reduced in size and may close altogether.
4. Continental environments are dominated by folded mountains—both recently formed "young" mountains and worn down remnants in continental interiors. Thick sediments formed along continental margins may become the basic material of future folded mountains.
5. Folded mountains are formed when plates collide. The main results are the cordilleran and Himalayan types of folded mountains.
6. Tectonic processes give rise to conditions of heat and pressure under which igneous and metamorphic rock forms underground; surface processes resulting from contact with the atmosphere wear down the rocks of the mountains produced by tectonic processes and provide the deposits that make up sedimentary rocks.
7. A pattern of continental accretion and rifting repeats through geologic time. This cycle influences sea level, long-term climatic change, and the distribution of living organisms.

KEY TERMS

ocean basin, p. 318
ocean ridge, p. 318
abyssal plain, p. 319
continental rise, p. 319
ooze, p. 319
ocean trench, p. 320
island arc, p. 320
young folded mountain, p. 322
fault-block mountain, p. 323
shield, p. 323
continental shelf, p. 327
continental slope, p. 327
cordillera-type folded mountain, p. 330
Himalaya-type folded mountain, p. 330
metallic ore, p. 333
Wilson cycle, p. 333
supercontinent cycle, p. 333

QUESTIONS FOR REVIEW AND EXPLORATION

1. What are the special features of the ocean basin or continental "solid-earth environment"?
2. How do plate tectonic processes create folded mountains?
3. What distinctive major landforms and groups of rocks are formed when (a) an ocean is opening, and (b) an ocean is closing?
4. What evidence suggests that the fault-block mountains and shields of today began as high ranges of folded mountains?
5. How does the "supercontinent cycle" influence climatic change?
6. Reconstruct an outline geologic history of North America in terms of plate tectonics and orogenesis; refer to shields, the Appalachian Mountains, and the western mountains.
7. Compare Figure 12-14 and Figure 11-10, which show the changing positions of the continents in recent Earth history. Match the stages shown in the two diagrams.
8. To what extent can human activities influence the features of ocean basins and continents?
9. Given that Earth is approximately 4 billion years old, about how many supercontinent cycles could have occurred? Is it likely that that many have occurred?

FURTHER READINGS

Bonatti E: The rifting of continents, *Scientific American,* March 1987, 96-103. Focuses on the opening of the Red Sea area, which it sees as a fundamental part of continent and ocean evolution.

Macdonald F: The mid-ocean ridge, *Scientific American,* June 1990, 72-79.

McGregor BA, Lockwood M: *Mapping and research in the exclusive economic zone,* U.S. Geological Survey and National Oceanic and Atmospheric Agency (1985). Goes into more detail than the booklet produced on the topic, and includes a range of photos and diagrams.

McGregor BA, Oldfield TW: *The exclusive economic zone: an exciting new frontier,* U.S. Geological Survey (1987). This free booklet assesses the potential economic effects of extending U.S. sovereignty to 200-nautical-mile zones around the coasts. It refers to plate tectonics.

Molnar P: The structure of mountain ranges, *Scientific American,* July 1986, 70-79. Relates the different formats of fold mountains to plate tectonics.

Nance RD, Worsley TR, Moody JB: The supercontinent cycle, *Scientific American,* July 1988, 72-79. This wide-ranging, stimulating article investigates and links the major features of Earth's relief.

Summerfield MA: *Global geomorphology,* 1991, Longman/Wiley. Chapters 3 and 4 cover the landforms of plate margins and interiors.

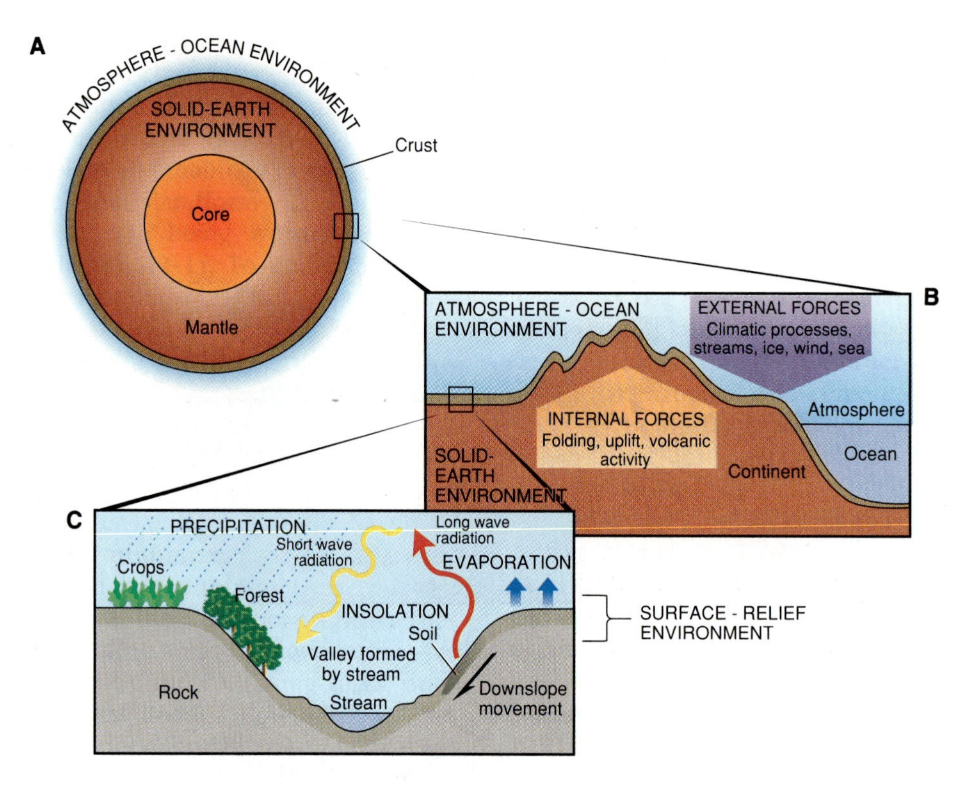

Figure 13-3 The surface-relief environment at various scales. **A,** The global scale: relationships to other major environments. **B,** The continental scale: the zone between internal and external forces. **C,** The local scale: landforms produced by interactions between the atmosphere-ocean environment and the solid-earth environment.

ters 11 and 12). Most of the processes that result from internal forces lift the continental surface or produce volcanoes or lava plateaus that raise the surface level. The same processes can also produce subsidence of large areas or rifting that lowers the surface.

The processes that result from the action of external forces are the main focus of Chapters 13 through 18. They result from atmosphere-ocean conditions acting on the surface provided by internal forces and mainly act to lower the land surface over tens of thousands to millions of years. This overall wearing down of the landscape is known as **denudation.** Over shorter periods, the breakup and removal of rock produce valleys, plane off the surface layers of wider areas, and create cliffs. While denudation is wearing down some sections of the surface relief, other parts are raised by materials deposited by running water, glaciers, winds, and the sea. This building-up process is known as **aggradation.**

The major processes in the surface-relief environment that result in denudation and aggradation are weathering, erosion, transport, and deposition. **Weathering** is the disintegration and decomposition of solid rock material. It is the initial stage of denudation, but does not involve movement of the broken rock. The products of weathering range from large blocks several meters across, through sand grains (0.02 to 2 mm across) and fine clay particles, to dissolved matter. **Erosion** is the wearing away and removal of rock by an agent such as running water, flowing ice, wind, or waves. **Transport** is the movement of rocks, rock debris, and dissolved rock materials by agents. **Deposition** is the dropping of rock debris, which often produces sediments having distinctive composition and form.

The wearing-down processes work toward an elevation known as the **base level** of denudation. For most continental areas, sea level acts as base level, since rivers flow downhill to sea level and no lower. In basins of inland drainage, however, the rivers do not reach the sea and the local base level may be higher (e.g., Great Salt Lake, Utah) or lower (e.g., Dead Sea, Israel) than the sea level. Inland drainage basins are common in arid or semiarid regions, including much of northern Africa.

Table 13-1 Interacting Processes of Landform Change in the Surface-Relief Environment

	RAISING OR BUILDING UP	LOWERING OR WEARING DOWN
INTERNAL PROCESSES Solid-Earth Environment	Mountain building Crustal uplift by warping and faulting Volcanoes, lava plateaus	Crustal subsidence by warping, faulting, and rifting
EXTERNAL PROCESSES Atmosphere-Ocean Environment	Deposition by rivers, glaciers, wind, and sea	Rock breakup and decomposition Erosion by rivers, glaciers, wind, and sea

WEATHERING AND MASS MOVEMENT

As soon as rock is exposed to the atmosphere-ocean environment, it is subjected to forces that break it up and then remove the pieces produced. Once rock has been broken into pieces by weathering, it can be transported. **Mass movement** is the downhill transfer of sections of solid rock or of masses of weathered debris under the influence of gravity. It is sometimes known as mass wasting.

Some of the processes involved in weathering and mass movements are familiar. Frost action breaks up sidewalks and roads, and acidic water dissolves the surface minerals in building stones and monuments. Landslides and debris avalanches are dramatic forms of mass movement. The slow movement of soil particles down a slope is less noticeable and changes landforms slowly. The passage that opens this chapter demonstrates that people in the early nineteenth century were unaware of previous landslides in the White Mountains; they did not understand the evidence that showed the area was prone to landslide hazards. Knowledge of this sort can be valuable in better management of the environment. The passage also demonstrates that processes create landforms over time scales longer than human lives and communal memory.

WEATHERING PROCESSES

The term weathering refers to a complex set of physical, chemical, and biologic processes that lead to rocks being progressively broken down. The resultant rock fragments may disintegrate further into their constituent minerals, decompose into chemical products such as clay minerals, or dissolve. Although weathering occurs mainly when rock is exposed to the atmosphere, weathering processes can take place up to several meters below the surface.

The following measurements reflect worldwide variations in the types and rates of weathering:

- *By 1928 several centimeters of crude soil had formed on new volcanic ashes and lavas erupted on the island of Krakatoa (Indonesia) in 1883.*
- *A 45 centimeter layer of decomposed granite formed beneath building foundations in Rio de Janeiro (Brazil) within 20 years of construction.*
- *No soil or weathered layer has developed on 1200-year-old mudflows around Mount Shasta in northern California.*
- *Rock surfaces bared by the action of glaciers during the last glacial advance in Britain show little evidence of 10,000 years of postglacial weathering.*

These examples suggest that weathering intensity varies from one part of the world to another, and in some places can be very significant in rock destruction (Figure 13-4). In general, weathering is more rapid in the humid tropical areas such as Indonesia and Brazil than in midlatitude areas such as northern California and Britain.

Weathering can be separated into physical (mechanical), chemical, and biologic processes (Figure 13-5), although these usually act in concert. Biologic weathering processes can be considered as subgroups of physical and chemical weathering processes, since plants and animals cause physical or chemical effects on rocks. The physical processes cause disintegration without significant chemical alteration, and may be assisted by the growth of plant roots or animal burrowing. The chemical processes create chemical changes and frequently involve living organisms. When considering weathering processes, it is important to remember that the chemical processes are by far the most important, and that many of the physical processes involve some chemical action as well.

THE SURFACE-RELIEF ENVIRONMENT

Little Pigeon River—Great Smoky Mountains National Park, Tennessee
Kenneth Murray—Photo Researchers, Inc.

Running Water and Fluvial Landforms

14 CHAPTER

In Concord it (the Concord River) is in summer from four to fifteen feet deep, and from one hundred to three hundred feet wide, but in the spring freshets, when it overflows its banks, it is in some places nearly a mile wide. Between Sudbury and Wayland the meadows acquire their greatest breadth, and when covered with water, they form a handsome chain of shallow vernal lakes, resorted to by numerous gulls and ducks. Just below Sherman's Bridge, between these towns, is the largest expanse; and when the wind blows freshly in a raw March day, heaving up the surface into dark and sober billows or regular swells, skirted as it is in the distance with alder swamps and smokelike maples, it looks like a smaller Lake Huron, and is very pleasant and exciting for a landsman to row or sail over. . . . The shore is more flat on the Wayland side, and this town is the greatest loser by the flood. Its farmers tell me that thousands of acres are flooded now, since the dams have been erected, where they remember to have seen the white honeysuckle or clover growing once, and they could go dry with shoes only in summer. Now there is nothing but blue-joint and sedge and cutgrass there, standing in water all the year round.

Henry David Thoreau,
"A Week on the Concord and Merrimack Rivers," 1849

Running water and flowing ice are two closely related agents of denudation. When rain and snow are precipitated out of the atmosphere, the resulting rainwater, snowmelt, or solid ice flows across the surface of the continents, moving downslope under the pull of gravity. In the course of flow they act as agents of denudation by picking up loose rock material and using it to erode the surface; they transport the rock debris in their flow and eventually drop it to form deposits.

Both streams and glaciers are part of the hydrologic cycle. Streams are a rapid means of returning water precipitated on the continents back to the oceans. Not all the water returns immediately to the oceans, for rain and snow precipitated on the continents are often stored in ice sheets, lakes, the soil, or rocks. Such storage retards the flow for periods varying from days to thousands of years and provides water that maintains streamflow in periods between rain or snowfalls.

A comparison of the roles of streams and glaciers as agents of denudation in the surface-relief environment shows that stream and glacier flow is generally channelled along valleys. Water can also cover wider areas as local unconfined sheets of running water, while ice sheets may cover whole continents. Streams are the dominant agent of denudation everywhere outside of glaciers and ice sheets. They even flow from time to time in deserts. This makes them more common than glaciers and ice sheets, which are confined to smaller regions. The two are often linked, however, with glaciers in the upper parts of high mountain regions providing meltwater to nourish the streams in the lower parts. Streams are by far the most important of all the agents of denudation acting on Earth's land surfaces: they deliver 85 to 90 percent of sediment worn from the continents to the oceans.

This chapter examines how fluvial processes—those associated with streamflow—carve and construct landforms. Chapter 15 focuses on the work of ice as an agent of denudation.

WATERSHEDS

Fluvial processes take place in an area drained by a river and its tributaries known as a **watershed,** or drainage basin (Figure 14-1). A watershed provides a distinctive surface-relief environment in which water is supplied by precipitation, flows across and through slopes toward stream channels, and is then channelled in a flow of water and sediment that shapes the channel. Other landforms are produced where the sediment is dropped. A watershed is bounded by a **drainage divide**—the boundary along a ridge top that determines whether runoff will flow toward one river or another.

Running water is organized into branching stream systems within watersheds. A watershed is composed of a set of slopes, which are connected to a network of stream channels. The channels carry water from the base of each slope, channelling it along the valley floors to form the main river and its tributaries. The network of streams begins in the upper parts of the basin with small streams that have no tributaries leading into them and are known as the *headwaters*. Headwater tributaries begin at a source and join other streams at a *confluence*. Tributaries add water along the length of the stream until it flows into a lake or sea at its *mouth*.

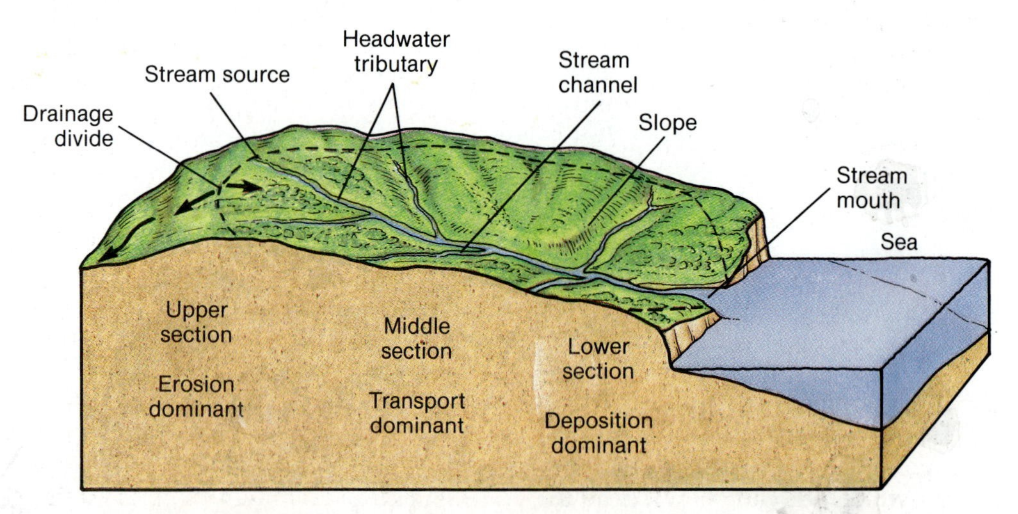

Figure 14-1 Some of the features of a watershed. It is drained by a network of permanent stream channels in humid regions.

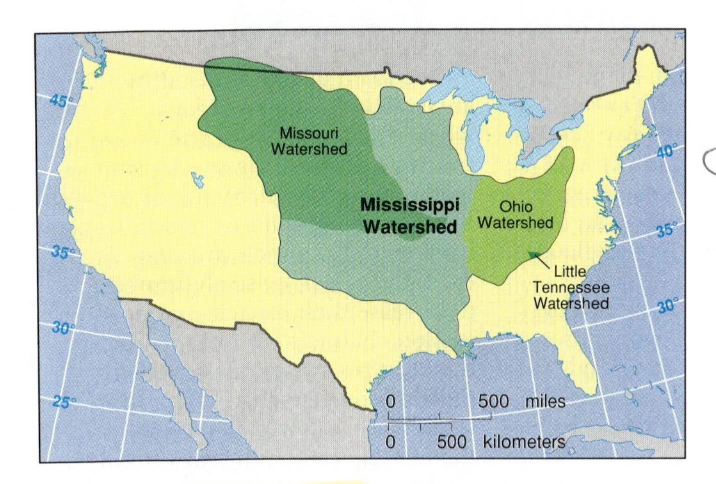

Figure 14-2 The Mississippi River watershed is one of the world's largest. Within it are the watersheds of the tributary Missouri and Ohio Rivers. The Little Tennessee watershed is one of the drainage areas contributing to the Ohio and Mississippi flow and has headwater tributaries in the Great Smoky Mountains.

A watershed may be divided in general terms into an upper section, in which water and sediment are concentrated into streams, a middle section where the sediment is transported, and a lower section where it is deposited. This division is merely a convenient way to describe aspects of the dynamic activity of streams in their watersheds; in different proportions, all three elements are present in every section of a stream.

Watersheds can be studied at a variety of geographic scales (Figure 14-2). For instance, the Mississippi watershed covers over 40 percent of the continental United States (3,222,000 km^2, or 1,244,000 miles^2). It is made up of several major tributary watersheds, such as those of the Missouri and Ohio, which in turn are composed of smaller watersheds. The smallest watersheds contain only a single tributary stream and commonly have drainage areas of 1 or 2 km^2. Small watersheds are contained within larger; the scale of a watershed is directly related to the size of the stream network it contains.

SURFACE FLOW AND GROUNDWATER

The term **hydrology** refers to the study of the distribution and properties of water within the atmosphere-ocean and surface-relief environments. The hydrology of a watershed can be described as a flow sequence that begins with a supply of water from precipitation, continues with the movement of water through soil and rock and downslopes, and ends with streamflow. Figure 14-3 summarizes the stages of the hydrologic flow sequence that is considered in this section. A basic understanding of *hydraulics*—the mechanics of streamflow—is also necessary to link streamflow to sediment movement and the production of landforms.

Sources of Water

The sources of water include not only rain, but snow melt and glacier-ice melt; each of these sources has a different pattern of supply throughout the year that is reflected in fluctuations of streamflow. Rainfall varies in magnitude (amount), intensity (volume of rain per unit time), and frequency (how often it occurs). If intense rain lasts for a sufficiently long period, and the ground is saturated with water, local flooding is likely since the amount of runoff will exceed the capacity of the stream channel (see Box 14-1: Flood Hazards). Many parts of the world have rainy and dry seasons for part of each year, but some have rain throughout the year, and others have hardly any (Chapter 8). All these factors affect the supply of water to rivers and fluctuations in flow.

Snow melt is usually concentrated in spring or early summer (Figure 14-4). During the winter, several falls of snow may accumulate without melting. In spring the snow melts rapidly and the extra volume of water may produce flooding during the melt season. Glacier-ice melt occurs throughout the summer, increases to a peak in mid or late summer, and then diminishes as winter freezing resumes.

Flow Through the Watershed

Rain or snow falling on the watershed (Figure 14-3) is first intercepted by the vegetation cover of trees, grass, or crops—if they are present. Grass and trees hold raindrops on their leaves. **Interception** is the percent of rain or snow that plants prevent from directly striking the ground; some of the water then evaporates from the plants' surfaces. Any water that is intercepted by vegetation, but not evaporated, drips off the leaves or runs down branches and trunks to the ground. Snow may remain on trees and plants until it melts.

Different plant covers intercept varying proportions of rainfall. Deciduous trees intercept some 15 percent of falling water, conifers 20 to 25 percent, and the tropical rainforest up to 30 percent. A grass cover intercepts about 20 percent of rain. Interception prevents or slows the arrival of the water or

where a cliff blocks the valley, and there is often a swallow hole at the foot of the cliff. Larger *polje* depressions up to several kilometers across are created by wider rock removal over an impermeable layer. The flat floors of poljes are often subject to seasonal flooding as the water table in the limestone rises. This flooding was more common during the Ice Age, as is shown by the fact that the underground streams draining into them are often clogged by debris produced long ago by frost action.

The underground features of karst include **caverns** and their connecting tunnels. The chambers and tunnels run vertically or horizontally according to rock structure. Their size and shape are determined by whether they were formed above or below the water table. In the saturated zone beneath the water table, groundwater movement produces tunnels that are almost circular bore tubes. Deposits of dripstones, rock piles from collapsed cavern roofs, and mud brought in by streams fill caverns and tunnels to varying extents. The dripstones are formed as calcium carbonate is precipitated from water dripping from a cavern roof; some grow down from the roof *(stalactites)*, and some grow up from the floor *(stalagmites)*. Large caverns in the United States occur particularly in Mississippian limestones around the margins of the Appalachians, including western Virginia and Kentucky.

Karst landforms occur in different varieties in different climatic environments. Tropical karst is marked by higher rates of chemical decomposition and greater quantities of surface water. The forms of the tropical cone karst, with its wider depressions, or *cockpits*, are caused by the greater influence of surface water flow following heavier rain. *Tower karst* in southeast Asia has high, steep-sided hills that are honeycombed by massive caverns. Karst areas that were covered by ice during the Pleistocene, such as the Niagara escarpment of Canada, have been scraped bare and exhibit widened joints

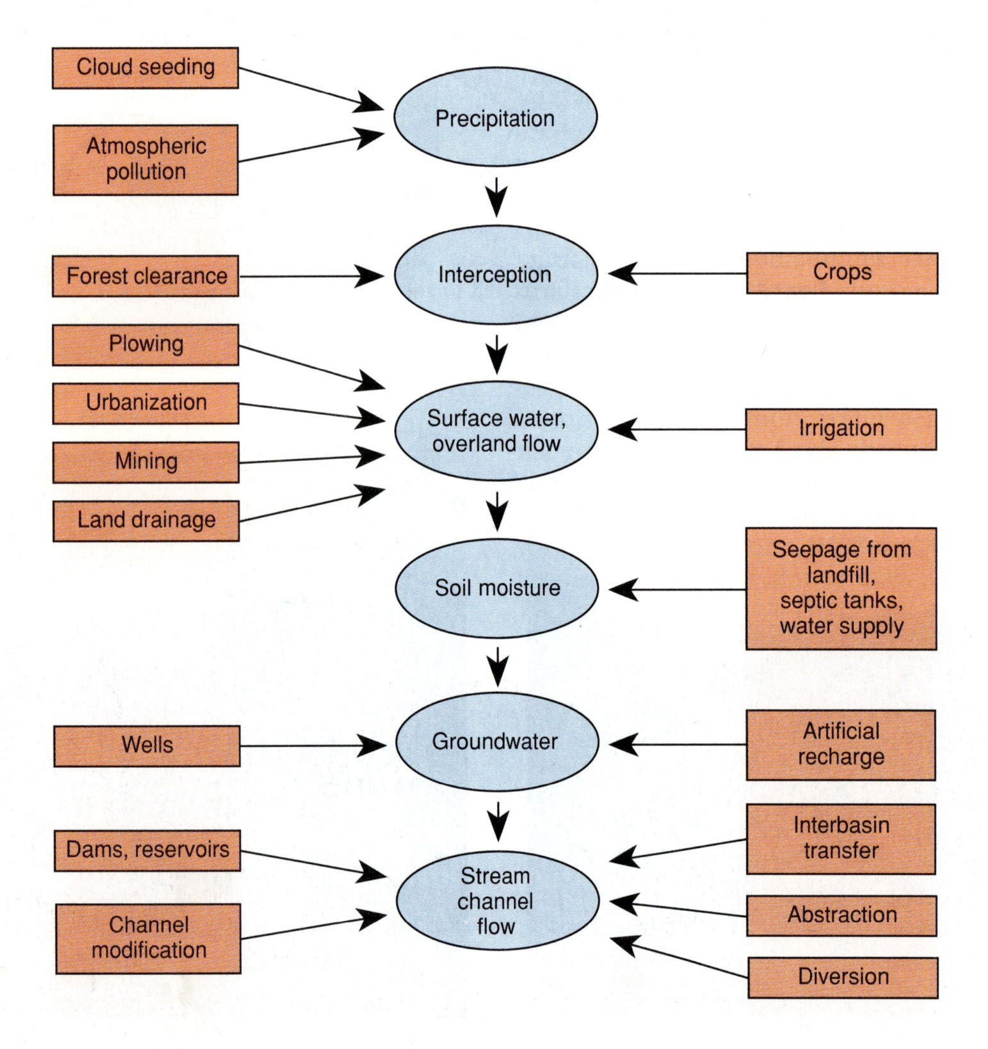

Figure 14-31 Human modifications of stream processes. In some places such modification has been so great that it has become the main influence on landform change.

that form *limestone pavement.* Karst occurs in arid areas such as northern Egypt and Arabia, but the forms there were produced under earlier, more humid conditions.

HUMAN IMPACTS ON FLUVIAL PROCESSES AND LANDFORMS

The watersheds of the world have long been occupied by human beings who have produced many changes in the natural processes that affect landforms. This is true even in arid areas, where the rivers have been used for irrigation since the earliest days of civilization. The impacts of human activities are often evident in the landscape, even centuries after a particular activity has ceased. In humid midlatitudes, natural landform change is relatively slow, and intense human development has become the main agent of landform change (Figure 14-31).

Some human activity has affected streamflow discharge and the potential of stream channels to modify the landscape. In many areas stream channels have been made into canals by lining them with concrete, which renders the streams inactive as agents for landform change. Streamflow has been affected directly by the building of dams and reservoirs that divert and reduce flow, as along the Colorado River. The natural flow of water through the Florida Everglades has been altered by the canal building, and extraction of water for metropolitan Miami (see Box 19-1: Living on Water). This has influenced the rest of the natural environment by reducing the flow south of Lake Okeechobee (Figure 14-32).

The hydrologic cycle in the wider sense has been affected by all types of changes in land use. Farming has had a major impact by removing natural forest. In Europe this began with the Neolithic and Bronze Age clearance of upland areas. Clearing of forest reduces interception and increases overland flow.

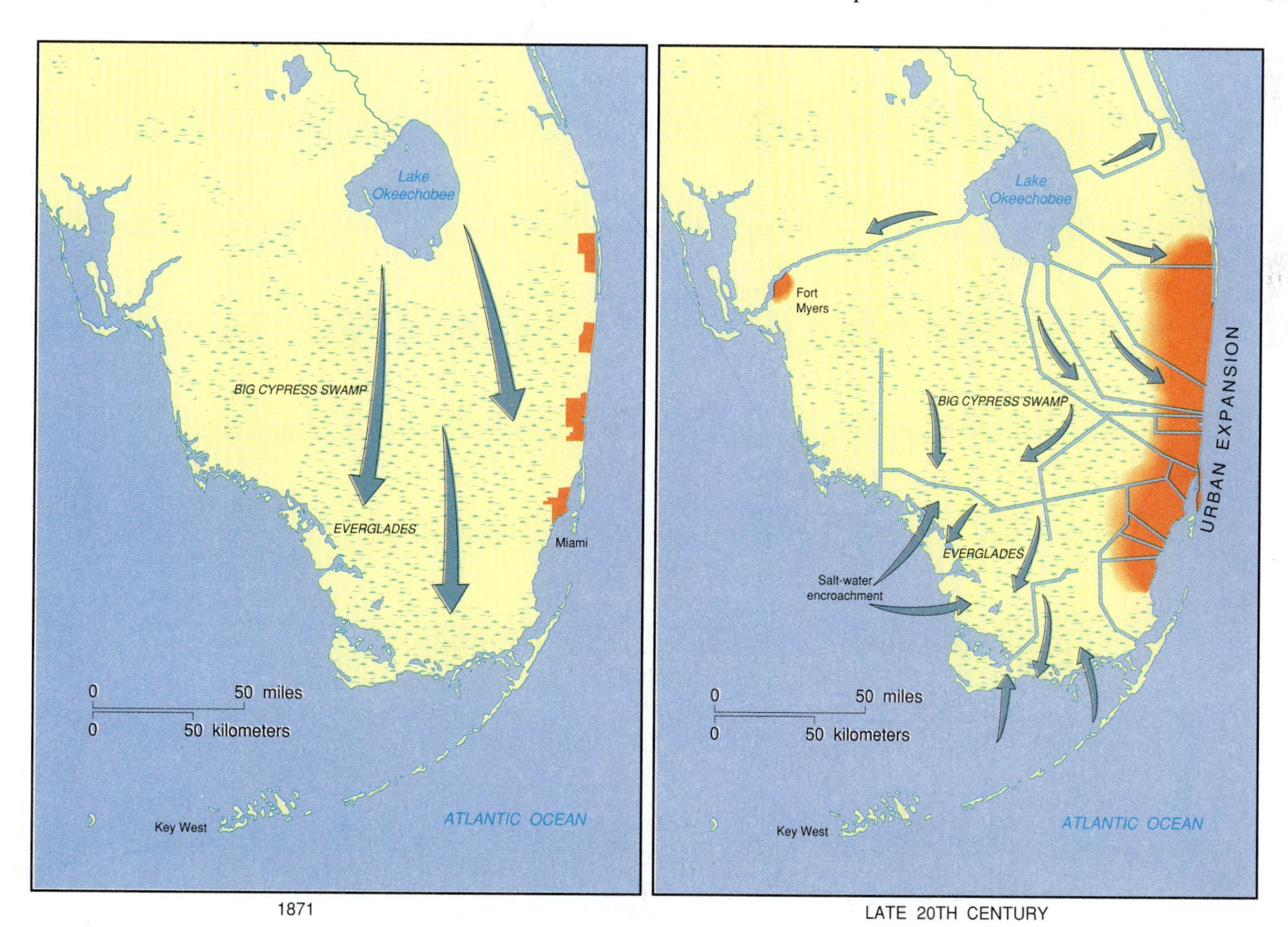

Figure 14-32 The modification of natural drainage in Florida. Up to 1871 the natural drainage was maintained by seasonal flow southward from Lake Okeechobee through the Everglades. Then canals were dug, diverting water to the east coast and lowering the water table.

Soil modification and erosion follow, leading to the deposition of sediment in the downstream valley floors. By medieval times, much of the Italian peninsula had been reduced to bare rocky slopes upstream, while swamps developed downstream on the deposits of eroded soil. The southeastern United States suffered similar effects in the nineteenth century from tobacco and cotton cultivation: upstream erosion and downstream deposition destroyed good land. In Wisconsin, some valley floors have been buried by up to 4 meters (12 feet) of sediment since farming began there in the 1830s. Farming continues to accelerate soil erosion. The effect is particularly disastrous in tropical regions because of the high rainfall (Figure 14-33).

Logging also removes natural vegetation cover and exposes hillsides to erosion. Experiments have been carried out in Oregon to determine which method of timber cutting results in least erosion. It appears that clear cutting produces smaller increases in sediment erosion than patch cutting, which necessitates the building of longer, highly erodible logging roads between the patches. The road building enhances soil erosion and adds to stream sediment loads, which are deposited downstream.

Mining can also have a drastic impact on natural streams. The California gold miners of the mid-nineteenth century had a major influence on the landforms of the state (see Box 10-1: Environmental Impacts of Mineral and Rock Extraction).

Urbanization affects the functioning of drainage basins in several ways. Construction often entails removal of the vegetation cover and topsoil. This exposes the land to erosion by storms, and local streams may be clogged by the resulting debris. Once the urban area has been built, the cover of roads, houses, and other impermeable surfaces drastically increases stormflow. This is especially true when storm sewers are constructed. Storm sewers rapidly remove rain from high-density, urbanized areas. Since the rain has little chance to soak into the ground, stormflow to the stream increases. This often leads to flooding and the widening and deepening of stream channels (see Box 1-1: Four-Mile Run). Figure 14-34 compares runoff in built-up Moscow with that in the surrounding rural areas.

Human-caused changes affect areas downstream of urban areas. The rapid erosion during construction moves sediment into streams and deposits it on flood plains lower in the watershed. Rapid runoff after storms from a large urban area can add to the stream's power below the city and cause more erosion of stream banks or stream beds.

An example of how human activities modify hydrology in a midlatitude humid climatic environment can be found in the piedmont area of the eastern United States. This area has a history of changing stream response to human activities since the beginning of the nineteenth century (Figure 14-35). In the early 1800s the spread of farming through this area led to soil erosion, gullying, and increased sediment load. Later in the 1800s and in the early 1900s, farming was abandoned over much of this land, soil erosion fell, and stream load decreased. In the mid-twentieth century the suburbs of Washington, D.C., spread into the area. At first,

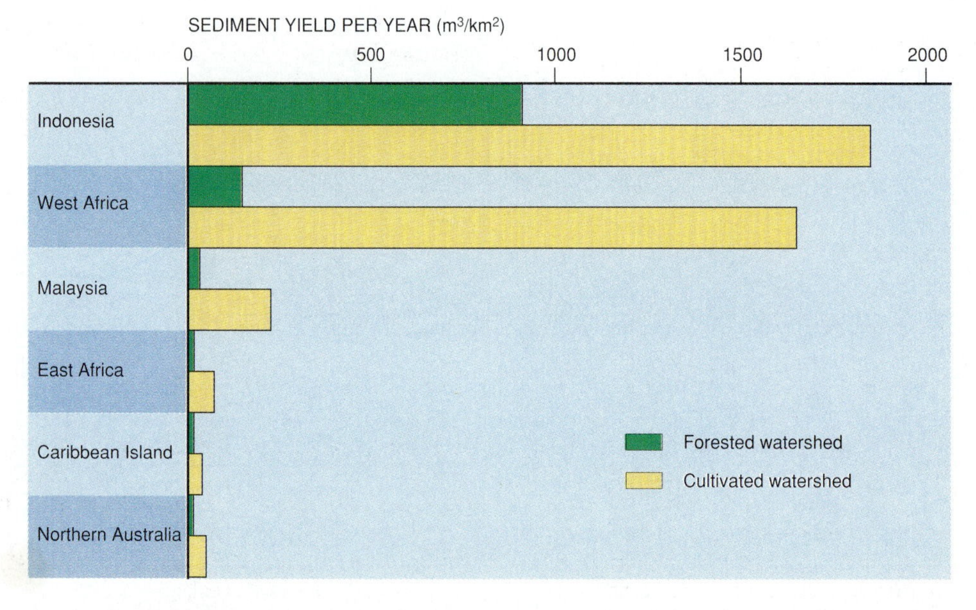

Figure 14-33 Stream erosion of forested and cultivated watersheds in tropical areas.

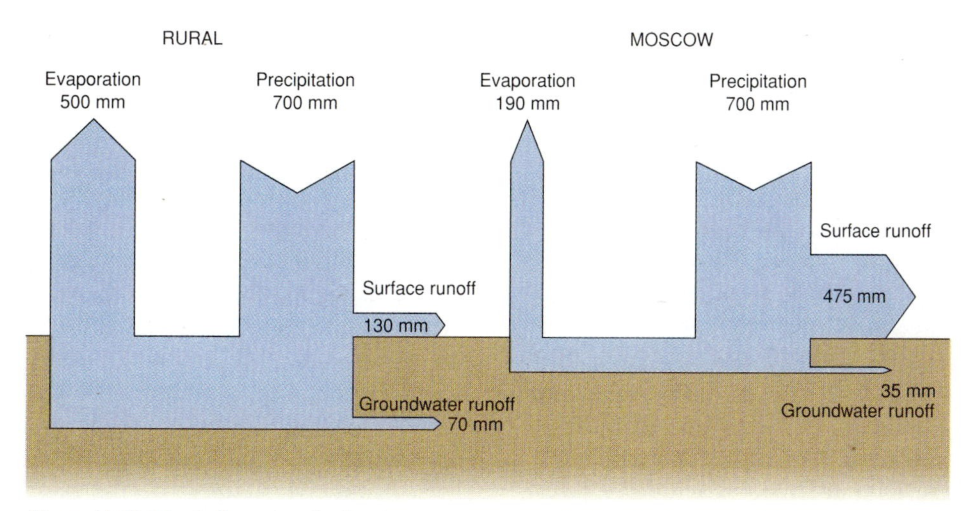

Figure 14-34 The influence of urbanization on watershed hydrology: the higher proportion of surface stream flow results in erosion of stream channels.

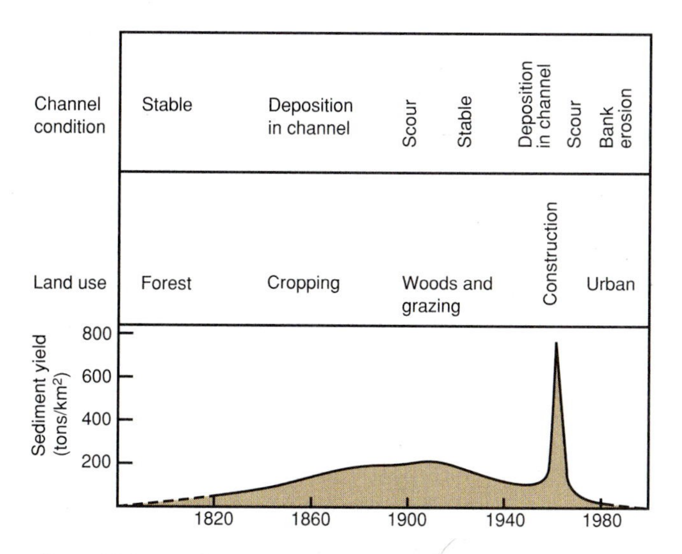

Figure 14-35 Land-use changes and their influence on stream erosion and sediment loads in streams draining the piedmont just west of Washington, D.C.

construction caused massive soil erosion and the lower parts of valleys were clogged with debris. Later, the soil was protected or built over, and the stream load fell to almost nothing.

Human activities thus affect stream processes and landforms both directly and indirectly. The direct influences include the modification of stream channel sizes and shapes, the reduction of streamflow by extraction, and adding to streamflow by diversions. They result in the channel becoming artificially protected from erosion, or in the channel becoming narrower or wider in response to changing discharge. The indirect influences result from land-use changes such as the removal of forest or the expansion of urban areas. Such changes mainly affect the humid midlatitudes, but are spreading to other parts of the world. Virtually every land-use change has implications for landforms.

FRONTIERS IN KNOWLEDGE
Fluvial Geomorphology

Stream activity is studied by physical geographers, hydrologists, and engineers, who have learned much about how water flows and interacts with its channel and watershed. Knowledge is still inadequate, however, to predict what will happen when streamflow increases after a storm—whether the extra water will flow faster through the same channel, widen or deepen the channel, or flood over its banks. Many studies focus on the relationship between the processes and the landforms produced, and on the ways in which rivers adjust to change. Stream processes such as bed load movement and the formation of stream channels and alluvial fans continue to be a focus of research in this area.

A further challenge is to extend the understanding of how rivers act in the surface-relief environment to explaining the features of larger areas of landscape over longer periods of time. Much current research is based on small areas over short time periods.

SUMMARY

1. Running water is confined mainly to stream channels, but may also flow in a thin layer of sheetflow over the surface.
2. Water comes to watersheds from precipitation (rain or snow) and glacier meltwater.
3. Water falling on a watershed passes through, but its passage is slowed if it is intercepted by vegetation, enters the soil or rock, or is stored in lakes or ice masses.
4. A stream hydrograph depicts the relationships between rain falling on a watershed and the flow of water in the stream over time.
5. The sediment load of streams includes solutes, suspended load, and bed load.
6. The movement of water and sediment in streams acts as an agent of denudation, carrying out erosion, transport, and deposition of rock debris.
7. The shapes and sizes of stream channels are controlled by the flows of water and sediment that pass through them. The shape of a stream channel also affects how water flows through it.
8. Flood plains and alluvial fans are the major depositional landforms in watersheds.
9. Stream action varies in different climatic environments.
10. Different rock types provide different degrees of resistance to stream action. Water acting on limestone rocks produces distinctive karst landforms.
11. Humans have greatest impact on the relief of humid-region watersheds. Farming, logging, mining, and urbanization all have a major influence.

KEY TERMS

watershed, p. 372
drainage divide, p. 372
hydrology, p. 373
interception, p. 373
infiltration, p. 376
overland flow, p. 376
permeability, p. 376
throughflow, p. 377
percolation, p. 377
water table, p. 377
groundwater, p. 377
aquifer, p. 377
aquiclude, p. 378
spring, p. 379
baseflow, p. 379
discharge, p. 380
stream hydrograph, p. 380
stream load, p. 382
dissolved load, p. 383
suspended load, p. 384
stream capacity, p. 384
bed load, p. 384
stream competence, p. 384
corrasion, p. 385
corrosion, p. 385
alluvium, p. 388
stream channel, p. 388
knick point, p. 389
meandering channel, p. 390
braided channel, p. 390
anastomosing channel, p. 391
flood plain, p. 392
natural levee, p. 393
backswamp, p. 393
point bar, p. 393
alluvial fan, p. 393
karst landscape, p. 398
doline, p. 398
cavern, p. 400

QUESTIONS FOR REVIEW AND EXPLORATION

1. Summarize how the features of a drainage basin can modify the passage of water from inputs of rainfall to outputs of streamflow.
2. Compare the annual patterns of input to a specific stream from rainfall, snowmelt, and ice melt.
3. Obtain data on rainfall and river flow and construct a flood hydrograph for a local catchment. Relate its shape to the features of the drainage basin.
4. Find out how groundwater is used in your area.
5. How is an understanding of streamflow important to engineers designing an irrigation project to deliver water to fields several kilometers from a river?
6. Devise a sediment budget diagram for a river in your area.
7. See Box 14-2: Stream Networks and Streamflow. Trace a stream network from a topographic map and analyze its interior and exterior links and its generations. If you have streamflow data for the area, compare the results with those of a typical hydrograph.
8. Examine a local stream channel (in the field or on a topographic map) and divide its course into straight, meandering, braided, and anastomosing sections. What are the proportions of each? Do they relate to slope or bed materials?
9. What are the distinctive features of a karst landscape?
10. How do climatic and geologic variations influence streams' effects on landforms? For systems mentioned in this chapter, such as the changing lower Mississippi, devise your own diagrams to summarize (a) changes in stream water and sediment flows, or (b) impacts of human activities.

FURTHER READING

Carries J: Water and the West—The Colorado River, *National Geographic,* June 1991, 2-35. An up-to-date account of competition for the Colorado water.

Cooke RU, Doornkamp JC; *Geomorphology in environmental management,* Oxford University Press (1990). Four chapters are relevant to the study of fluvial processes and landforms: Chapter 4, Soil Erosion by Water; Chapter 6, Rivers and Floodplains; Chapter 7, Drainage Basins and Sediment Transfer; and Chapter 14, Problems in Limestone Terrain.

Hamilton L: The recent Bangladesh flood disaster was not caused by deforestation alone, *Environmental Conservation* 15(4) (1988), 369-370.

Summerfield MA: *Global geomorphology,* Longman/Wiley (1991). Chapter 8, Fluvial Processes, and Chapter 9, Fluvial Landforms, provide a more advanced discussion of the material covered in this chapter.

United States Geological Survey, *Ground water and the rural home owner,* Government Printing Office (1988). This free booklet explains many of the basic concepts with profuse illustrations.

United States Geological Survey, *National Water Survey of 1987,* Government Printing Office (1990). This annual publication summarizes national and state water budgets and provides accounts of specific hydrologic events in 1987.

THE SURFACE-RELIEF ENVIRONMENT

East Front Rocky Mountains Montana, U.S.A. Pine Butte Preserve

Stephen J. Krasemann/Nature Conservancy

Ice, Glacial, and Periglacial Landforms

15 CHAPTER

For two hours we battled into the gale, seeing enough of the peaks around to steer by. A slight moderation of the wind happily coincided with our arrival in an area of crevasses. They were clearly defined on the hard surface and were rarely more than two or three feet wide. It was impossible to dodge them so we kept going, winding our way along and crossing them where they were narrowest. Sometimes the bridges held under the weight of the tractors, but usually there was a violent lurch and we'd pass over safely but leave a gaping hole behind.

Ahead of us stretched the steepest part of the glacier. The main body of ice tumbled down in a huge icefall to our left, but by following a devious route to the right it was possible to climb on much more reasonable gradients. The glacier rose in two great sweeps, called by us the Lower and Upper Staircase, with a flattish area in the middle, which we called the Landing. If the snow was very soft on these steep areas we knew it would delay us considerably, although we were determined, if necessary, to winch our way to the top.

Sir Vivian Fuchs and Sir Edmund Hillary,
"The Crossing of Antarctica," 1958

Glaciers in mountain valleys and the ice sheets on Antarctica and Greenland are among the most striking features of Earth's surface-relief environment (Figure 15-1). Ice is the one agent of denudation that occurs in forms that rival the landforms it produces in size; where it occurs, it covers most or all of the landscape, and huge domes of ice several kilometers thick can dominate a continent. Another unique feature of ice as an agent of denudation is that human activities have hardly affected its working to date—although that may change if global warming occurs. The account by Fuchs and Hillary that opened this chapter describes some surface features of ice sheets and glaciers that have impressed explorers.

The presence of ice on a large scale produces surface-relief environments in which ice sheets, glaciers, and frozen ground create distinctive groups of landforms. Today ice cover is restricted to high latitudes and high altitudes, where precipitation falls as snow and temperatures seldom rise above 0°C (Figure 15-2). The zone that includes these areas is often termed the *cryosphere* owing to the predominance of frost and ice (Greek *kryos* = icy cold). In these regions ice occurs both beneath the surface in permanently frozen ground known as *permafrost* and also in flowing surface ice sheets and glaciers that mold the underlying land. Areas close to ice sheets in polar environments do not have an ice cover but are influenced by permafrost and often have ground ice: they are called **periglacial** areas.

Ice had a much wider impact during the glacial phases of the Pleistocene than it has today (Chapter 9). Only 15,000 years ago—a short period in terms of the generation of relief features—much of northern Europe and North America was covered by ice sheets. That ice has retreated in the present interglacial phase, but it produced many landforms in the areas it once covered.

Figure 15-1 The edge of the Greenland ice sheet.

This chapter begins by considering surface-relief processes involving ice—how ice forms from snow, the different forms of ground ice and surface ice, and how surface ice flows. The study of ice processes leads to an examination of how landforms are produced by surface and ground ice and the ways in which groups of landforms are associated with each other, in particular glacial or periglacial environments.

SURFACE ICE AND GROUND ICE

From Snow to Ice

Snow forms in the atmosphere as tiny, crystalline flakes (see Chapter 6). In subfreezing temperatures, freshly fallen snow accumulates as a downy powder. Wetter snow falls over large areas, particularly where high mountains in middle latitudes intercept air flows coming from the oceans. Air temperatures there are close to the melting point of snow. When snow piles up faster than it melts or sublimates (changes directly from ice to water vapor), it gradually changes to ice, a process that takes several years. The increasing weight of accumulating snow compacts the lower layers (Figure 15-3), and the melting of surface snow forms water that percolates downward, refreezing to fill pore spaces between the snow crystals. The density of this material rises with successive accumulations. Packed snow has a density of 0.1 gram per cubic centimeter. *Firn* is an intermediate stage of dull, white frozen material that still has much air trapped in its pores; its density is 0.5 g/cm^3. True ice has a density of 0.8 to 0.9 g/cm^3, is crystalline in structure, and contains virtually no air. Ice is less dense than liquid water (1.0 g/cm^3), which is why ice floats in water.

The change from snow to ice may take 3 to 5 years in places such as the Yukon (central Alaska), where temperatures fluctuate just above and below freezing for most of the year. However, these changes may take up to 200 years in central Greenland, where temperatures seldom permit surface melting. Ice masses often accumulate more rapidly in slightly warmer conditions, especially where wet snowfall occurs.

Permafrost and Ground Ice

Permafrost is ground that is permanently frozen. Strictly, it is ground in which the temperature re-

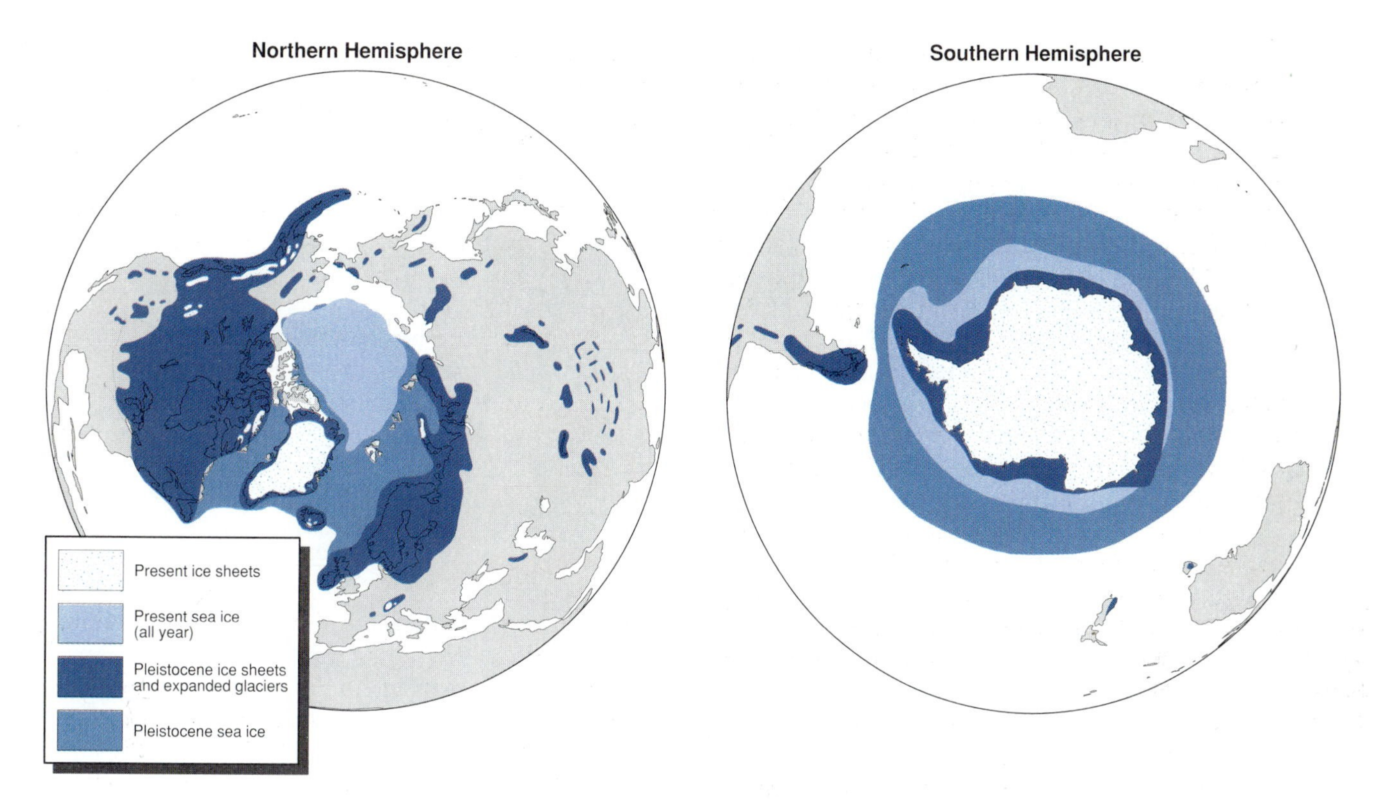

Figure 15-2 Hemisphere maps of ice cover.

Figure 15-3 Annual layers of ice shown in a crevasse in an Alaskan glacier. Some of the layers are marked by dirt bands formed by material washed over the surface or deposited from the atmosphere during the summer.

mains below 0°C for more than 2 years, but in many parts of northern North America and Asia it has been present for thousands of years. Water in permafrost occurs mostly as ice, which cements together loose particles, but not all permafrost contains water. The ground ice associated with permafrost is either dispersed through soil or rock in pores or concentrated in wedge-shaped bodies or mounds that may influence surface relief.

The features of permafrost are shown in Figure 15-4. It affects soil and rock. A surface *active layer,* 50 to 80 cm deep, lies above the permafrost and is subject to alternate melting in the summer and freezing in the winter. The *permafrost table* marks the upper limit of permafrost. The lower limit of permafrost is determined by the balance of heating from Earth's interior and the effect of cooling by the atmosphere: it is up to 400 meters (1250 feet) deep in northern Canada and 1000 meters (3000 feet) deep in Siberia. The deepest permafrost was formed during the last glacial phase and has not melted since.

Continuous permafrost occurs mainly in the northernmost parts of North America and Siberia, and in total it covers some 21 million square km (8.2 million square miles). *Discontinuous permafrost* is thinner and divided by unfrozen ground and covers 17

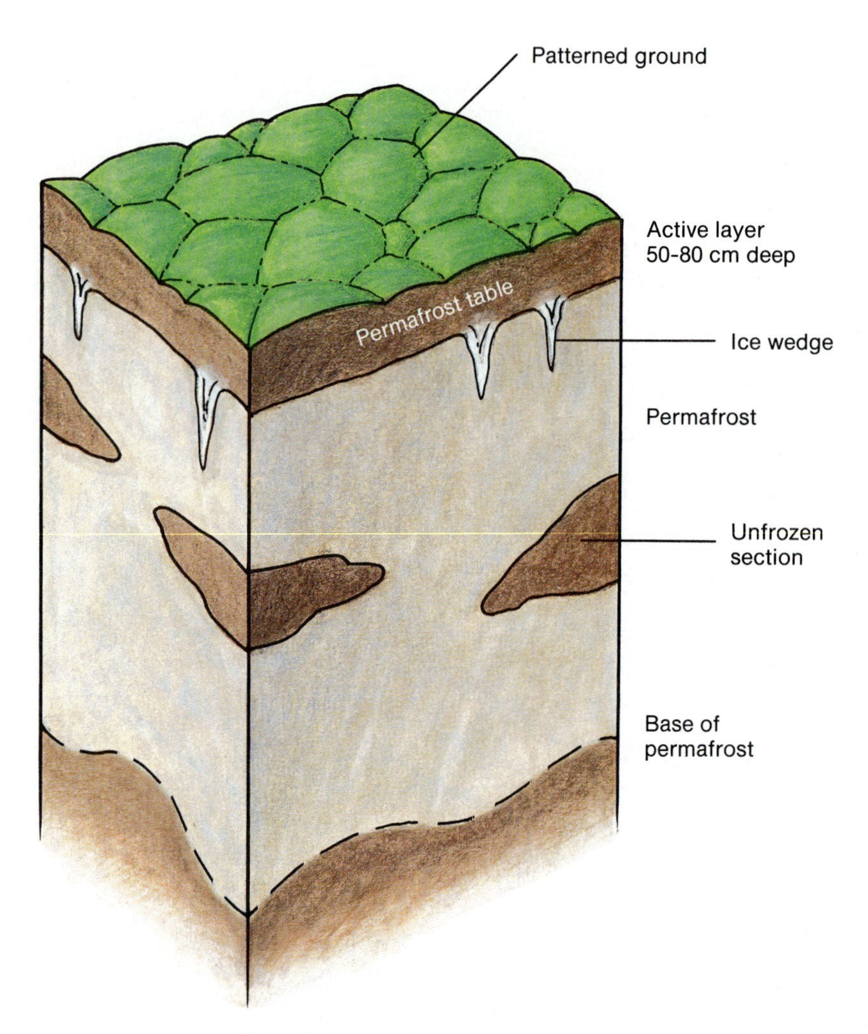

Figure 15-4 Permafrost: major features.

million km² (6.6 million miles²). Discontinuous permafrost is being reduced in extent and thickness by current warming, but continuous permafrost shows fewer signs of change. Permafrost provides a distinctive environment that settlers need to treat with care (Box 15-1: Problems of Living on Permafrost).

Forms of Surface Ice

Surface ice forms have varying shapes and sizes. They become large enough for the ice to flow when the weight of accumulated ice is sufficient to overcome its internal strength.

The largest surface ice forms are **ice sheets,** sometimes known as *continental glaciers* since they cover such huge areas. Antarctica and Greenland are Earth's two ice sheets, and they submerge almost the entire landscape in broad domes that thin toward the margins (Figure 15-5). The two ice sheets make up 96 percent of the continental area covered by ice. Antarctic ice covers over 12.5 million km² (4.9 million miles²), and the Greenland ice covers 1.7 million km² (0.7 million miles²); all the remaining ice covers just 0.6 million km² (0.2 million miles²). Areas of ice of less than 50,000 km² (20,000 miles²), such as Vatnojokull on Iceland, are called **ice caps** and resemble ice sheets in many respects.

The weight of the ice in ice sheets and ice caps is so great that it causes the crust beneath to subside by isostasy (Figure 11-13). The rocks of the central parts of Greenland and Antarctica are buried by several kilometers of ice and lie below sea level.

A **glacier** is a smaller ice mass that is confined by surface features. Most glaciers form in high mountains and take a variety of forms. They usually begin with **cirque glaciers,** small bodies of ice occupying amphitheater-shaped hollows just below mountain crests. Cirque glaciers supply **valley glaciers,** which fill the main part of a valley trough and may be joined by a network of tributary glaciers (Figure 15-6). Some valley glaciers do not extend

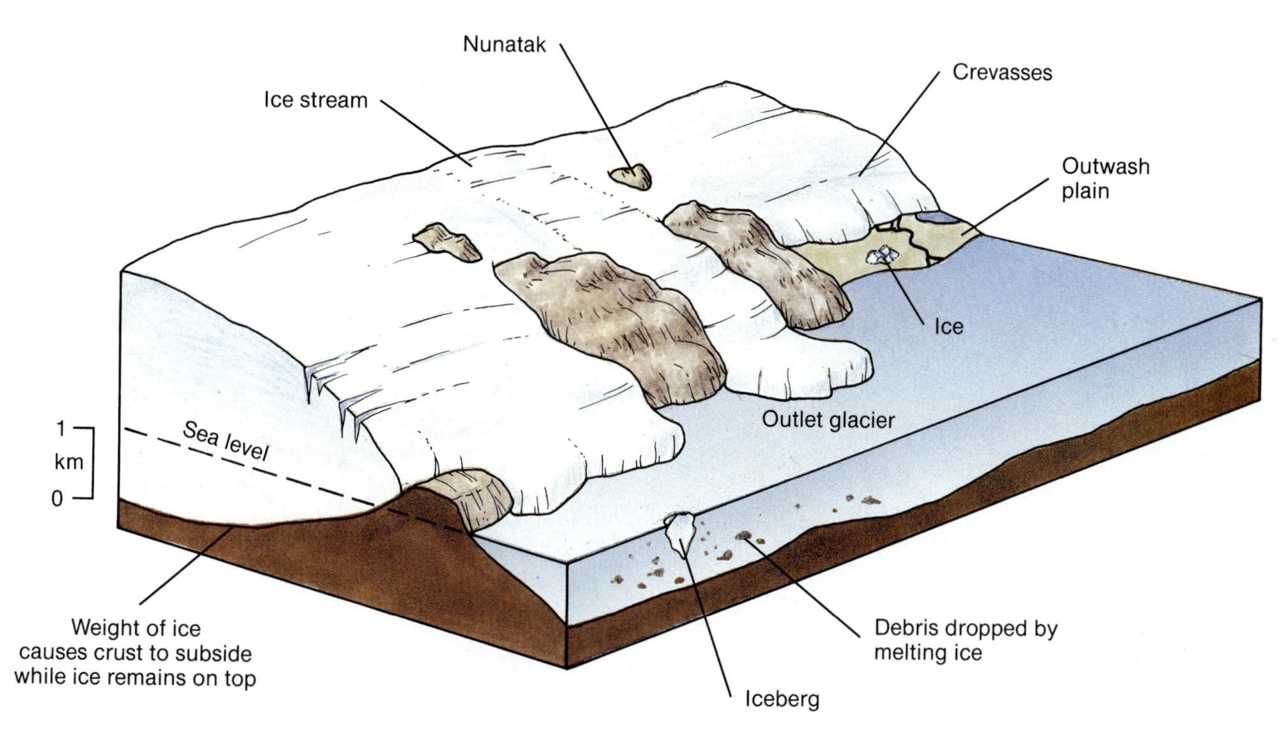

Figure 15-5 Features of an ice sheet and its margins on land and in the ocean.

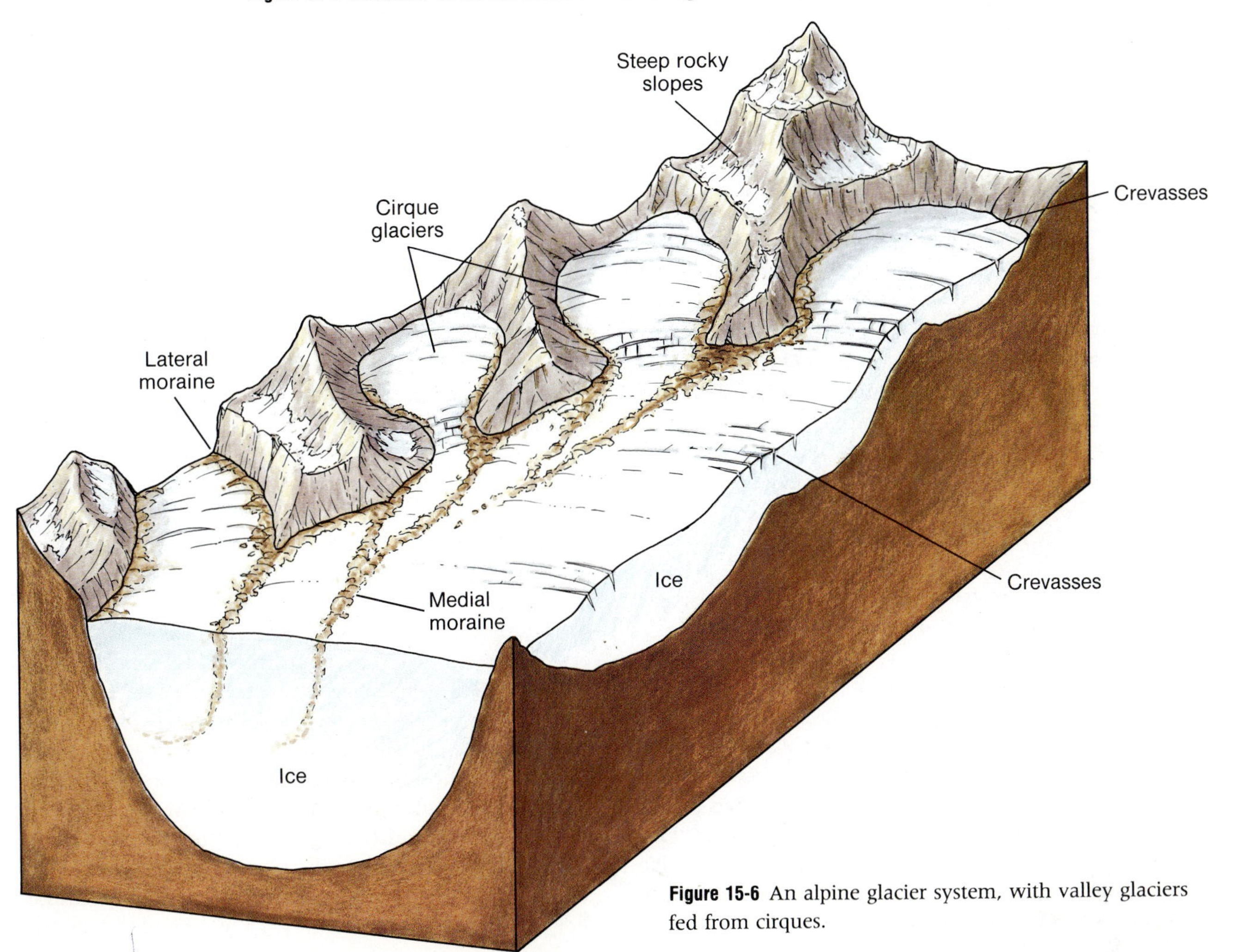

Figure 15-6 An alpine glacier system, with valley glaciers fed from cirques.

Figure 15-7 Space Shuttle view of the Malaspina piedmont glacier, southern Alaska, together with valley glaciers and the mountain ice field. Some of the valley glaciers have zigzag patterns on their surface; others enter the sea before melting. Meltwater rivers bring clouds of sediment to the sea.

beyond the deep valley in which they flow, but others flow out on a lowland area. There they may join several others to form a wide, shallow area of stagnant ice, known as a **piedmont glacier.** The Space Shuttle view of southern Alaska (Figure 15-7) shows the Malaspina (piedmont) glacier and several related glacier types. Cirques, valley glaciers, and piedmont glaciers occur commonly in mountains of many latitudes, and are entirely different types of ice forms from the ice sheets and ice caps. The two major groups of surface ice forms are often termed *alpine glaciation* and *continental glaciation.*

Some glaciers occur on the margins of ice sheets, where mountains pierce the ice cover and divide the flow into *outlet glaciers.* Outlet glaciers often extend back into *ice streams* within the ice sheet that are characterized by concentrated, more rapid flow and show some of the characteristics of glaciers, but are not bounded by rocky valley sides.

Ice also occurs floating on the oceans in polar regions. Around Antarctica the floating ice is partly ice that has flowed off the land. **Sea ice** forms when the sea surface freezes. The largest areas of sea ice occur around Antarctica and in the Arctic Ocean, which is completely covered by sea ice in winter. The ice breaks off (or "calves") into **icebergs** that later melt as they float into warmer areas. Both icebergs and sea ice make navigation difficult in polar waters.

Ice Flow

Ice in ice sheets and glaciers flows from source areas, where snow **accumulation** and conversion to ice exceed losses by melting (Figure 15-8). It flows to zones where ice is lost by **ablation**—melting and evaporation—or breaks up into icebergs if it reaches the ocean. The transition from accumulation to ablation zones takes place at what is known as the *equilibrium line.* The position of this line is determined by altitude, latitude, and snow supply. The lower margin of a glacier or ice sheet is known as its *snout.*

Ice begins to move when stress builds up to a critical level as the weight of accumulating snow and ice increases. The pressure imposed distorts individual crystals, and this **internal deformation** is the basic cause of solid ice flowing as a plastic substance—that is, changing its shape without fracturing. Ice flow may be assisted by meltwater lubricating the base of the glacier, a process known as **basal sliding** (Figure 15-9). It has been shown that glaciers flow faster over loose rock debris saturated with water than over dry rock.

A surface layer of ice up to 60 meters (180 feet) deep remains rigid since it requires that depth of ice to build the stresses causing flow. This layer is carried along by the flowing ice and cracks when subjected to stresses, such as bending over a small hill or drop in the valley floor. The cracks are called *crevasses* and extend to the base of the rigid surface ice layer. Surface crevasses reveal some of the patterns of glacier movement, such as the tension cracks that open because of the faster movement in the center of a valley glacier compared with the margins, where the ice movement is slowed by friction with the valley walls. Crevasses also play an important role in allowing meltwater and rock debris to penetrate the glacier ice: this water is important in basal sliding.

Ice sheets and glaciers flow very slowly compared with rivers. Rates vary but are generally in the range of 3 to 300 meters per year. In valley glaciers, movement rates vary internally. Flow is quickest near the top because friction with the sides and base slows ice at the margins (Figure 15-9). Flow rates also vary along the length of the glacier, rates being fastest near the equilibrium line where the cumulative volume of ice reaches a maximum. Increasing ablation downstream of this line reduces the volume of ice and its velocity.

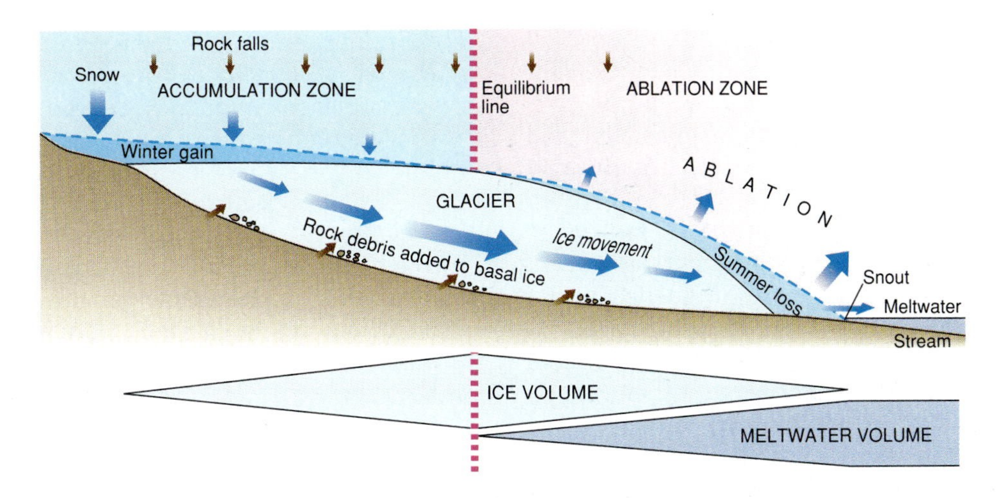

Figure 15-8 A glacier system. Snow accumulates above the equilibrium line and changes to ice; ice flows downhill and melts in the ablation zone. The blue arrows are proportional to the volumes of snow and ablation, and to the speed of flow.

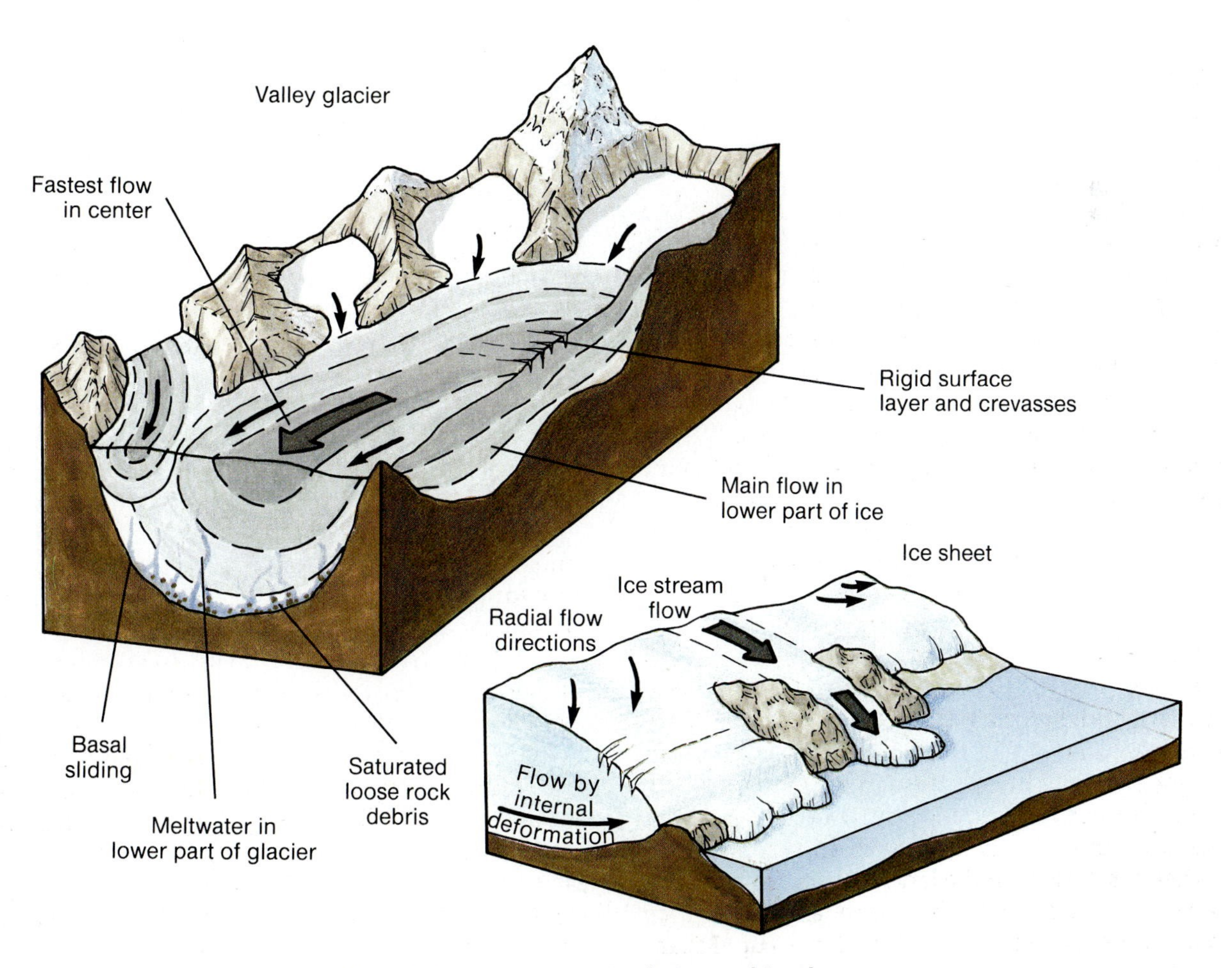

Figure 15-9 Ice movement in glaciers and ice sheets.

Ice sheet flow is outward from the center in all directions because there are no confining rock walls as there are in valley glaciers. The weight of ice is greatest at the point of maximum thickness and causes the outward movement, which is most rapid a little away from the center of the sheet. Within ice sheets, ice streams and outlet glaciers provide local surface zones of faster movement.

Temperature has an important influence on the movement of ice masses. There are **warm glaciers** where the ice is just below 0°C and **cold ice sheets and glaciers,** in which temperatures are much lower. Warm glaciers occur in the mountains of humid midlatitude and tropical regions, such as southern Alaska, the Swiss Alps, the Andes, and the Himalayas. The summer rise in temperature causes surface melting, rapid formation of firn, and percolation of water to the base of the glacier. Movement is caused in such glaciers by the combination of internal deformation and basal sliding; the latter may account for 90 percent of the movement. Cold ice sheets and glaciers occur in polar regions. The major ice sheets of Greenland and Antarctica are examples. Movement in cold ice sheets and glaciers is totally by internal deformation.

Meltwater is important in the movement of warm glaciers (Figure 15-10). Meltwater flowing on the ice surface may wear away shallow channels in the ice or rock material. When the water meets a crevasse, it plunges in. Water moving freely through the ice tends to cut deep passages due to vertical erosion. Where water flows beneath the water table in the glacier, it may be squeezed through the ice under pressure, and hollow out circular, tubelike passageways.

At times the presence of large quantities of meltwater can cause a glacier to *surge*—a rapid forward movement of part of the glacier at speeds 10 to 100 times the normal flow. The Tweedsmuir glacier on the British Columbia/Yukon border surged several kilometers in a few days in 1973, and others in Alaska do so regularly. Over 200 surging glaciers have been identified in North America. Surging is not fully understood. Some glaciers surge every few decades, others are unpredictable. Earthquakes may act as a trigger in some cases.

Figure 15-10 Meltwater issuing from the Glacier Blanc, French Alps.

LINKAGES

The character of ice masses results from the interactions of major earth environments. The ***atmosphere-ocean environment*** *determines the low temperatures and level of precipitation; the* ***solid-earth environment*** *determines the configurations of the continents and oceans and the positions of high mountain ranges.*

LANDFORMS OF ICE EROSION

Glaciers and ice sheets are important agents of denudation. The landforms of high mountains in particular demonstrate their effects. It has not been easy to identify the precise mechanism that is at work when flowing ice detaches and moves rock debris, since scientists can penetrate only the outer margins of ice sheets and glaciers.

One mechanism of ice erosion is **abrasion** (or corrasion), in which the debris held in the base of the ice scratches and scrapes underlying rocks—rather like sandpaper on wood. In an experiment in Iceland, blocks of rock were bolted down beneath moving ice. In 3 months, during which the ice moved forward 9.5 meters (30 feet), the surface of a marble block was lowered by 3 millimeters (0.2 inches) and a basalt block by 1 millimeter.

Moving ice also removes rock by the process of **plucking.** On the downstream side of protruding rocks, joints and fractures fill with ice joined to the surrounding glacier ice. As the glacier moves forward, it carries with it the ice and attached blocks of rock (Figure 15-11). Plucking can also excavate basins in well-jointed rock.

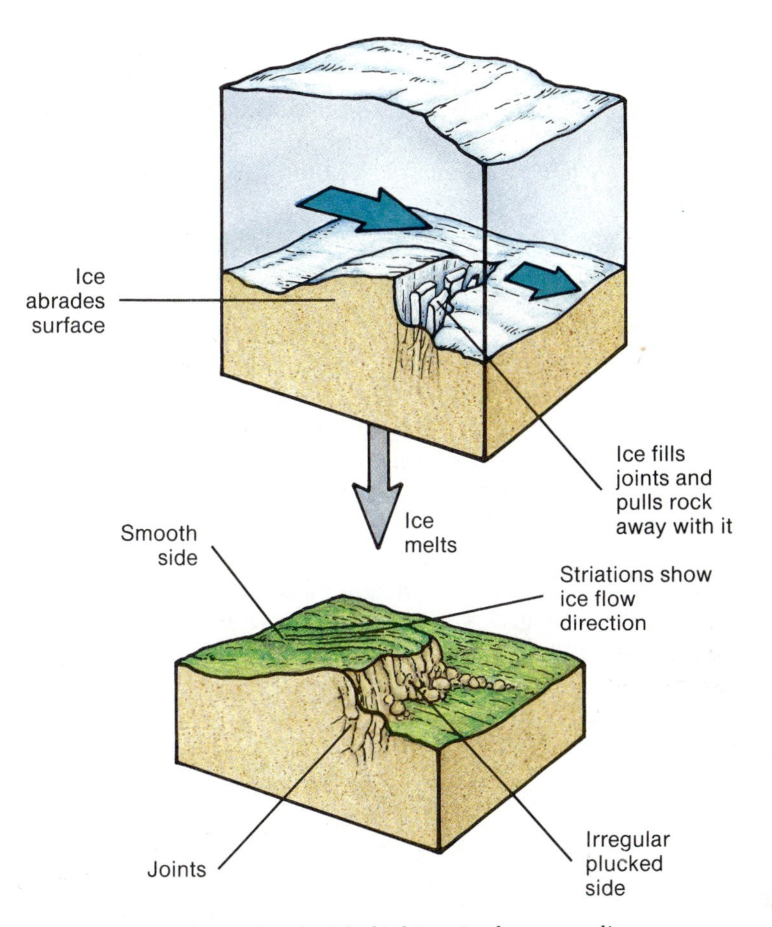

Figure 15-11 Erosion by glacial plucking. In the upper diagram the ice melts and refreezes around the blocks of a jointed rock. When it moves forward, the blocks are pulled out. After melting, the protruding rock has a smoothed upstream side and an irregular downstream side.

Glaciologists debate whether abrasion or plucking is more important in glacial erosion. Ice at the bottom of a glacier carries large blocks of rock rather than finer debris that has been ground down, suggesting that plucking may be more significant. However, plucking is most effective in well-jointed rocks. It may be that abrasion is more important for some rocks and plucking for others.

Although the mechanisms of glacial erosion are not fully understood, their impact is clear in formerly glaciated regions such as northern Canada, where the shield area has been scraped bare in many places, and in the Rockies, where distinctive deep valleys occur. Some of these landscapes are smoother, suggesting abrasion is more important; others have more angular features, suggesting that plucking is of greater significance.

In addition to the abrasion and plucking by which ice sheets and glaciers erode solid rock, much of the rock debris removed by them is loose regolith that becomes covered by ice as it advances. As in streams, glacier erosion is aided by previous and current weathering that provides rock debris for use in abrasion.

Glacial erosion may occur across extensive regions beneath a thick and mobile ice sheet, or it may be concentrated along valleys by valley glaciers in upland regions. Following removal of regolith and scouring by ice, erosional landforms are often characterized by bare rock. In regions of lower relief, ice erosion may scour the entire landscape, scraping up loose soil and regolith, freezing it into the ice base, and using it to abrade solid rock underneath to a smooth surface. This smoothing can be seen in **whaleback hills** or **basins.** The hills are up to 200 meters (600 feet) high and several kilometers long (Figure 15-12). Smaller forms a few tens of meters high are known as *roches moutonnées* and often have a smooth scoured slope upstream

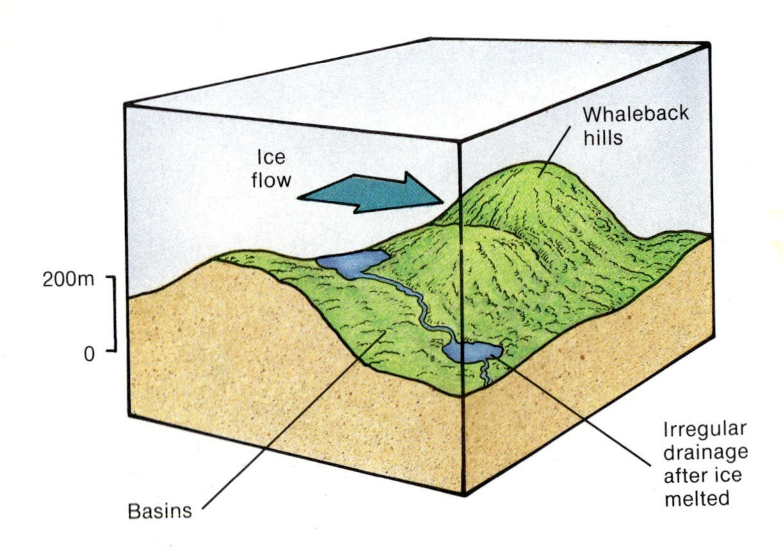

Figure 15-12 Whaleback hills and basins, formed by ice abrasion.

and an irregular plucked face on the downstream side. Some rock surfaces are marked by grooves, or *striations,* cut by the harder fragments in the ice base. Striations have been used to establish ice-flow direction in former ice sheets.

Valley glaciers deepen and straighten their valleys through a combination of ice movement, freeze-thaw weathering, and meltwater activity. The resultant troughlike valleys often have an open "U" shape in cross section, and are called **U-shaped valleys** (Figure 15-13). Along its length, a U-shaped valley often has an uneven floor with flatter sections alternating with steps formed where more resistant rocks come to the surface. In Antarctica, U-shaped valleys up to 50 km (30 miles) across and 3.5 km (2.5 miles) deep are found. Those in the Rockies and Alaska are seldom more than 1 km across and deep. The difference reflects the

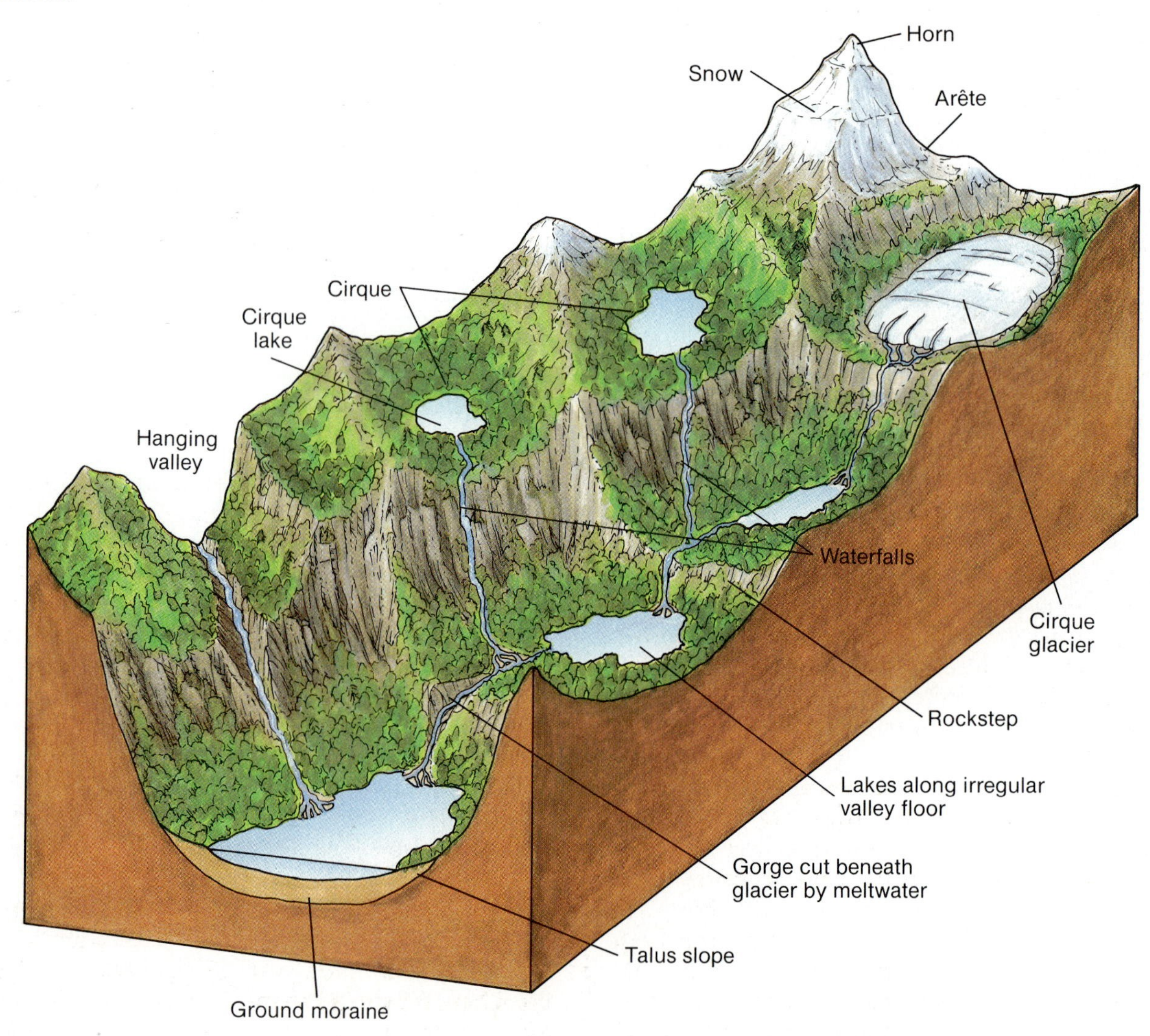

Figure 15-13 The landforms of a glaciated valley after most of the ice has melted. The U-shaped valley, hanging valleys, cirques, and sharp ridges are typical features.

Figure 15-14 A hanging valley with a waterfall in Yosemite National Park, California.

Figure 15-15 The Breithorn, near Grindelwald, Switzerland. The sharp, angular peak rises behind a cirque and a valley glacier that has been retreating since the early 1800s.

greater thickness of ice and length of time that Antarctica has been glaciated. Glacial troughs may be drowned by the sea at their mouth to form fjords (Chapter 17).

Tributary glaciers are usually much smaller than the main glacier and deepen their valleys in accordance with their size and debris load. When the ice melts, the shallower tributary valleys are perched above the main valley and are called **hanging valleys** (Figure 15-14). Rivers flowing out of hanging valleys into the main valley often do so in waterfalls.

At the heads of valley glaciers are amphitheater-shaped hollows known as **cirques.** Most cirque ice flows into valley glaciers, but in some cirques the ice is restricted to the hollow and forms a cirque glacier. Cirques are usually 0.5 to 1 km across and may be up to 1 km deep. Cirques form by a combination of weathering, mass movement, and ice flow. Hollows where snow gathers and is converted to ice are enlarged by frost action and mass movement. These processes continue on the slopes above the ice and maintain steep back and side walls. When there is sufficient ice in the hollow for flow to occur, a cirque glacier forms; further growth of the ice results in its flowing down a valley below the cirque. Rock debris is incorporated in the ice as meltwater washes it into crevasses and down to the base of the glacier. Abrasion of the bottom of the cirque produces the rounded shape. Cirques are often oriented so that their outlets face away from the direction of the sun's rays at midday, reflecting the conditions necessary for the accumulation of snow and ice and the effective operation of weathering and mass movements. A lip at the mouth of a rounded cirque basin sometimes traps water to form a lake after the ice has melted.

If a series of cirques form back-to-back, the walls between may be reduced to knife-edge ridges known as *arêtes.* Sharp peaks rising from such ridges often have a pyramid shape, like the Matterhorn in Switzerland, and are known as *horn* peaks (Figure 15-15). Peaks that rise above the margins of a continental ice sheet are known as **nunataks.**

LANDFORMS OF DEPOSITION FROM ICE

Ice sheets and glaciers can transport large quantities of rock debris. They build up impressive deposits with distinctive shapes, especially near their margins where ice flow is slower and melting becomes increasingly significant.

The size and rigidity of ice sheets and glaciers enable them to move blocks of rock as heavy as several tons that could not have been shifted by streams or the wind. They act as giant conveyor belts, with the uppermost ice and debris borne on the plastic flowing ice beneath. The rock material is dropped when the conveyor belt slows or stops flowing at the ice margin.

An assortment of rock debris is transported by ice—in terms of both origin and composition. Some comes from falling on the glacier edges following frost weathering on the exposed steep slopes above. On the surface it forms piles of angular fragments along the sides of the glacier. Other debris comes from erosion of the rocks, soil, and regolith over which the ice advances (Figure 15-16). The rock debris transported inside the ice mass gets there by falling or being washed into a crevasse or by being swept up into the base of the ice. The basal load may be more rounded by abrasion and may also contain more finer materials from abrasion and soil that has been eroded. Most rock fragments are not altered significantly by ice transport, although frost action may continue to break down blocks on the ice surface.

Deposition by ice takes place during ice movement and melting. An important distinction is made between deposits formed by ice movement and ice melting, and those deposited by the meltwater. The former are generally poorly sorted, a jumble of particles of varied sizes; the latter are better sorted and often form layers. The ice deposits are called **till.** Till and stratified meltwater deposits often occur together in glacial environments.

Tills form where local melting occurs, either beneath the ice in the process of basal sliding when rock debris is "plastered" on the ground beneath, or where ablation at the snout of a glacier or margin of an ice sheet causes rock debris to be dropped. Ablation is widespread during the rapid disintegration of an ice sheet at the end of a glacial phase (Figure 15-17).

Most landforms formed of till are called **moraines,** which are piles of unsorted rock debris having a variety of shapes. The piles of rock on the surfaces of valley glaciers are also termed moraines, and often end up as recognizable piles along the valley sides when the ice melts. *Lateral moraines* form along the sides of valley glaciers where rock fragments fall from higher slopes, and *medial moraines* form where two glaciers meet and their lateral moraines join.

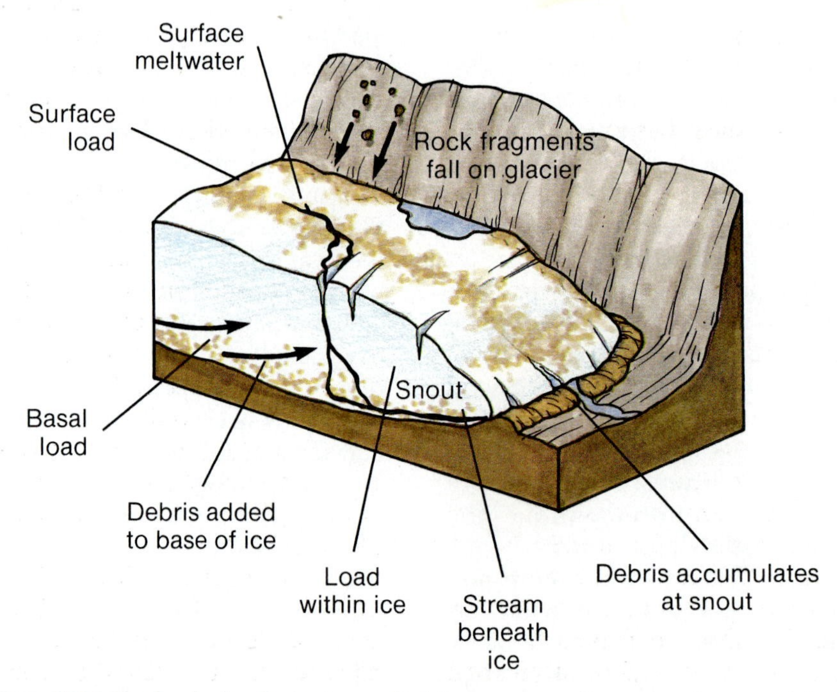

Figure 15-16 The load of a glacier is supplied from surrounding steep slopes and the valley floor; it is carried on and in the glacier ice and by meltwater.

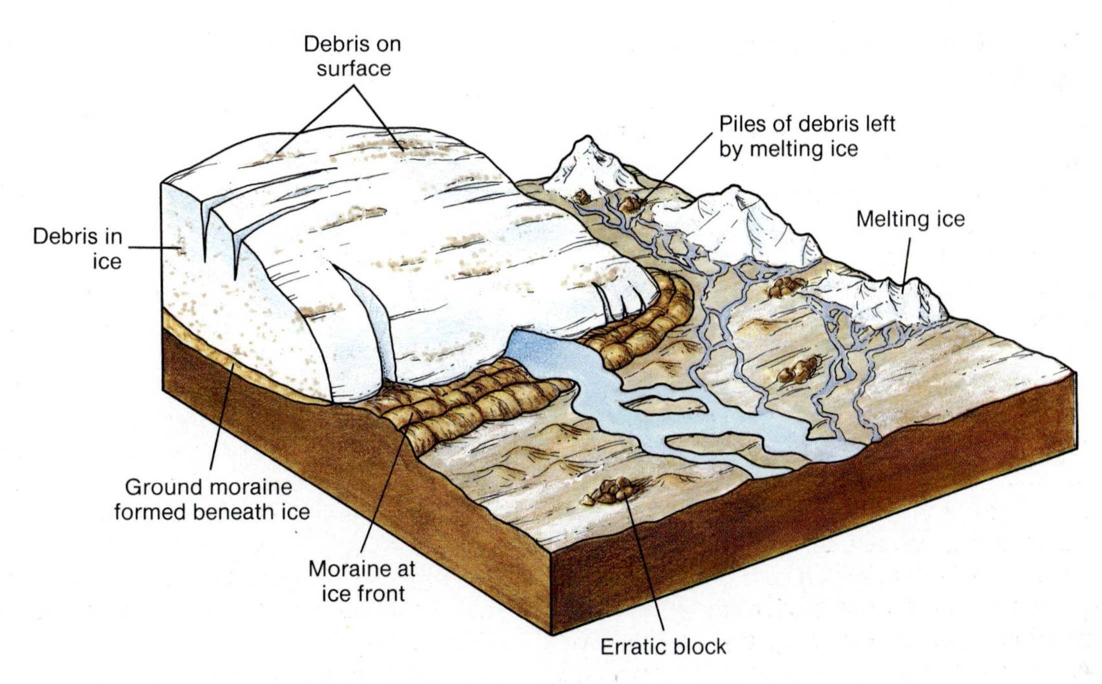

Figure 15-17 Deposition by ice takes place from the base of active, advancing warm ice and when ice melts.

Terminal moraines form at the snouts of valley glaciers and the margins of ice sheets, where ablation causes the rock debris carried down the glacier conveyor belt to accumulate. Strictly, the term is used for moraines that mark the farthest advance of a glacier or ice sheet but is often used for any cross-valley morainic ridge that formed at an ice front. The glacier or ice sheet has to have an unchanging front for some time to build up sufficient debris that produces a low ridge across the valley floor along the line of the glacier snout or one stretching across the ground along the line of the ice sheet margin. Glaciers and ice sheets are marked by alternating advances and retreats of their margins. If the glacier is expanding, such terminal moraine deposits show internal evidence of having been bulldozed by the ice. As the glacier retreats up-valley, a series of *recessional moraines* is formed at each interruption of the retreat process as the climate fluctuates. Most "terminal moraines" are recessional in origin, since those formed by advancing ice are soon overridden and removed.

Ground moraine covers ground beneath the outer section of an ice sheet to depths of up to several tens of meters. It is formed of mixed rock debris, some of which was plastered on the ground beneath the moving ice, and some that formed as the ice disintegrated and dumped the debris it had carried on its surface and within the mass. The surface of ground moraine is often irregular, with hollows occupied by lakes and a deranged drainage pattern following melting of the ice sheet. Such landforms are characteristic of northern Wisconsin and Minnesota. If the deposits are mainly clay, they form a featureless *till plain,* characteristic of central Illinois and Indiana.

The irregularity or monotony of ground moraine is often broken by smoothed, oval hills known as **drumlins.** These are 1 to 2 km long and up to 50 meters high, and are formed of the same material as the ground moraine (Figure 15-18). They often occur in large numbers, or "swarms." The smooth outline indicates that they were formed by moving ice rather than ice disintegration. It has been suggested that drumlins are formed by ice advancing over and reshaping ground that was moraine deposited at the end of a previous glacial phase.

Areas covered by ice in the past often contain large **erratics,** blocks of rock quite different from those of the local area. These blocks may have travelled several hundred kilometers and indicate that the area was once covered by ice and the direction of ice flow.

SUMMARY

1. Ice is a dominant feature of Earth's surface landscapes in high latitudes and the highest parts of mountains. The presence of ice has a major impact on landforms.
2. Ice accumulates where it is too cold to melt all the snow that falls. The conversion from snow to ice is caused by compaction, which is enhanced by some surface melting and refreezing lower in the snow mass.
3. Ground ice occurs in areas of permafrost.
4. Warm ice (middle latitudes) and cold ice (high latitudes) have distinctive features that affect how they flow and create landforms.
5. Ice sheets and glaciers advance and retreat with climatic fluctuations.
6. Ice sheets and glaciers may erode or protect the underlying land; they transport large quantities of rock debris and produce distinctive depositional forms.
7. Some typical landforms of glacial erosion are roches moutonnées, U-shaped valleys, and cirques.
8. Some typical landforms produced by glacial ice and meltwater deposition are moraines, drumlins, eskers, kames, kettles, and outwash plains.
9. Ground ice produces small landforms, including ice-wedge polygons and pingos.
10. In polar glacial environments, large ice sheets composed of cold ice influence the landforms of continental-size areas.
11. In alpine glacial environments the warm-ice valley glaciers and cirque glaciers are confined to valleys and valley-head zones of accumulation. These glaciers show signs of retreat over the last 150 years.
12. Ice-front glacial environments are dominated by depositional forms.
13. Former continental glacial environments extend into many midlatitude areas.
14. Periglacial environments have landscapes molded by ground ice, rivers, and wind.

KEY TERMS

periglacial environment, p. 408
ice sheet, p. 410
ice cap, p. 410
glacier, p. 410
cirque glacier, p. 410
valley glacier, p. 410
piedmont glacier, p. 412
sea ice, p. 412
iceberg, p. 412
accumulation zone, p. 412
ablation zone, p. 412
internal deformation, p. 412
basal sliding, p. 412
warm glacier, p. 414
cold ice sheet, glacier, p. 414
ice abrasion, p. 415
ice plucking, p. 415
whaleback forms, p. 415
U-shaped valley, p. 416
hanging valley, p. 417
cirque, p. 417
nunatak, p. 417
till, p. 418
moraine, p. 418
ground moraine, p. 419
drumlin, p. 419
erratic block, p. 419
esker, p. 420
kame, p. 420
kettle, p. 421
valley train, p. 421
outwash plain, p. 421

QUESTIONS FOR REVIEW AND EXPLORATION

1. What evidence is there that glacier ice erodes?
2. Compare channels, flow mechanisms, and deposits of glaciers with those of streams.
3. Compare warm and cold glaciers in terms of flow mechanisms and landforms produced.
4. Assess the significance of glacial activity from maps or photos of erosional and depositional landscapes.
5. How far is the active layer above permafrost a key to understanding the origins of periglacial landscapes?
6. Why is human influence less obvious in glaciated areas than in connection with drainage basins?

FURTHER READING

Eliot JL: Glaciers on the move, *National Geographic,* January 1987, pp. 104-119. A study of glaciers with special reference to Alaska.

Hodgson B: Antarctica: a land of isolation no more, *National Geographic,* April 1990, pp. 2-51. This article focuses on Antarctica as an environment for living organisms, but includes information on the ice mass, landforms, and political issues. Illustrated.

Matthews SW: Ice on the world, *National Geographic,* January 1987, pp. 79-103. Examines the nature and extent of ice and its past and present influences on the landscape

Sharp RP: *Living ice,* Cambridge University Press (1988). A well-written and illustrated account of glaciation based on a lifetime of research by one of the world's leading authorities.

Summerfield MA: *Global geomorphology.* Longman/Wiley (1991). Chapter 11, Glacial processes and landforms, and Chapter 12, Periglacial processes and landforms, provide detailed accounts of the landforms discussed in this chapter.

US Geological Survey: *Glaciers: clues to future climate?* Government Printing Office (1983). This free booklet provides examples of glaciers and puts the changing extent of ice masses in the context of climatic change.

THE SURFACE-RELIEF ENVIRONMENT

Dust storm, Springfield, Colorado

OMIKRON/Photo Researchers, Inc.

Wind and Aeolian Landforms

16 CHAPTER

In the last part of May the sky grew pale and the clouds that had hung in high puffs for so long in the spring were dissipated. The sun flared down on the growing corn day after day until a line of brown spread along the edge of each green bayonet. . . . The surface of the earth crusted, a thin hard crust, . . .

In the water-cut gullies the earth dusted down in dry little streams. Gophers and ant lions started small avalanches. . . . Every moving thing lifted the dust into the air; a walking man lifted a thin layer as high as his waist, and a wagon lifted the dust as high as the fence tops, and an automobile boiled a cloud behind it. The dust was long in settling back again. A day went by and the wind increased, steady, unbroken by gusts. The dust from the roads fluffed up and spread out and fell on the weeds beside the fields, and fell into the fields a little way. Now the wind grew strong and hard as it worked at the rain crust in the cornfields.

The wind grew stronger, whisked under stones, carried up straws and old leaves, and even little clods, marking its course as it sailed across the fields. The air and sky darkened and through them the sun shone redly, and there was a raw sting in the air. . . . The dawn came, but no day. In the grey sky a red sun appeared, a dim red circle that gave a little light, like dusk; . . .

In the middle of that night the wind passed on and left the land quiet. The dust-filled air muffled the sound more completely than fog does. All day the dust sifted down from the sky, and the next day it sifted down. An even blanket covered the earth. It settled on the corn, piled up on the tops of fence posts, piled up on the wires; it settled on roofs, blanketed the weeds and trees.

John Steinbeck, **Grapes of Wrath, 1939**

Wind processes and the landforms they produce are often termed **aeolian,** after the Greek god of wind, Aeolus. Winds are most effective as agents of transport and deposition in areas where there is little vegetation, such as deserts, glacial ice-front areas, sandy beaches, and plowed fields. Compared to running water or moving ice, wind is less powerful and more intermittent in its effects. Winds also whip up waves at sea, and therefore influence coastal landforms, which will be studied in Chapter 17.

The excerpt from Chapter 1 of John Steinbeck's *Grapes of Wrath* dramatically portrays wind erosion in the 1930s Dust Bowl area of Oklahoma. Dust storms such as those that disrupted the lives of the characters in *Grapes of Wrath* have occurred in many regions. Much evidence suggests that arid areas are growing (see Box 21-1: Desertification), adding to the importance of studies of wind processes.

Figure 16-1 Wind erosion has destroyed this field by removing the fine clay particles and blowing the sand into piles.

This chapter is the shortest in the book, which reflects the relatively small influence of wind processes on landforms. It begins by examining how wind acts as an agent of denudation and then considers the conditions required before wind action can erode, transport, and deposit sediment. Landforms produced by wind erosion and deposition are examined in the context of the various environments in which wind action is a major influence in producing groups of landforms.

WIND PROCESSES

Wind as an Agent of Erosion

The differences in potency between wind and running water are due partly to differences in the density of the two media. Owing to its greater density, water can move boulders and hold sand in suspension, whereas wind seldom moves anything larger than sand and holds only dust in suspension. Wind must be faster than water to move even fine sediment.

Another feature of wind is that its strength fluctuates from moment to moment. Strong gusts get particles airborne, but a lull in wind strength returns the particles to the ground. Wind turbulence is very important in providing vertical lift and increases with wind speed and the unevenness of the ground.

Wind erosion is strongly reduced by the presence of a vegetation cover and by water in the surface soil (Figure 16-1). Plants lower wind speeds near the ground and thus protect the surface. Water binds soil particles together, thereby increasing surface resistance to wind action. Wind is limited in effectiveness as an agent of erosion because of the small particle sizes it can transport (clay, silt, and sand) and the fact that most of the world's land surfaces are covered by vegetation.

Wind erodes the ground by two main processes. Wind uses the particles it carries to **abrade** (or corrade) rock or sediments, wearing them away by bombardment—rather like sandblasting is used to clean building stones. The process is restricted to the lowest 1 to 2 meters above the ground. Coarser sand is most effective in this process but only rises a few centimeters. Finer silt grains may produce some smoothing of rocks. Abrasion is greatest where the wind can pick up sand and blast it against a weak rock.

Deflation is the removal of loose particles from the ground to produce a depression or lowered surface level. It mainly affects clay and silt particles that can be lifted in large numbers by wind turbulence and carried away in suspension. An estimated 130 million to 900 million tons of material is deflated from continents each year, with 60 to 200 million tons coming from the Sahara alone. Deflation events require loose deposits and high wind speeds and produce dust storms lasting several days.

LINKAGES

Wind action depends on air movement in the ***atmosphere-ocean environment*** *and acts on rocks and minerals provided by the* ***solid-earth environment.*** *Plants* ***(living-organism environment)*** *provide resistance to wind action.*

Wind Transport

Wind's main geomorphic role is to transport and deposit loose clay, silt, and sand. In dry and unvegetated areas, the main factors affecting wind transport are wind speed and particle size (Figure 16-2). Wind moves particles by a combination of lift and drag. *Lift* forces are dependent on pressure differences within the air—rather like those over an airplane wing—that cause particles to rise from the surface. *Drag* is caused by differences in pressure on the upwind and downwind sides of a particle and starts movement at lower wind speeds than lift. A threshold wind speed is necessary before any sediment movement takes place, and the turbulence of gusty winds is important in initiating movement.

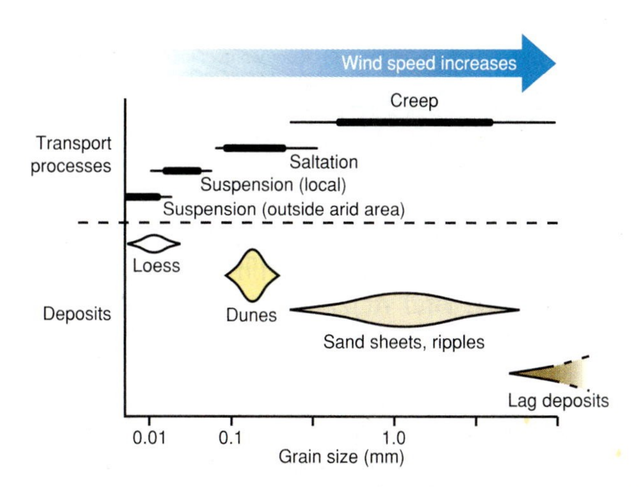

Figure 16-2 Wind transport and deposition. As the wind velocity increases, it is able to move larger particles (top graph). The other parts of this diagram show how the type of movement and resulting deposits are related to grain size and movement velocity.

Very fine particles (less than 0.1 mm diameter) include dry clay, silt, and salt crystals. The finest particles resist the forces that would make them airborne because they present little surface roughness to encourage turbulence and because they stick to other particles of the same size. They are more difficult to lift off the ground than slightly larger particles (up to 1 mm diameter), but once in the air they remain in suspension and may be moved long distances. Large quantities are blown from the Sahara over the Atlantic Ocean, particularly during the winter when the dry monsoon wind is dominant. Space Shuttle photos (Figure 16-3) show clouds of fine particles over the Atlantic. They are brought down by rain and sink to the ocean floor.

Fine sand grains up to 0.4 mm in diameter are lifted into the air for a short time and then fall to the ground—a bouncing motion called **saltation** (Figure 16-4). Most saltation takes place within a few centimeters of the ground, but individual grains may rise to 1.5 meters (5 feet). Once turbulence raises a few grains off the ground, the higher wind speeds a few centimeters above the surface carry them downwind. The forward motion causes particles to disturb any grains they land on, setting other particles in motion.

Surface creep, rolling and sliding action, affects larger grains, and is caused by wind drag and the bombardment of saltating grains. Even in coarse sand, surface creep is seldom responsible for moving more than 25 percent of the total material in motion.

The rate of sand transport is proportional to the third power of the wind speed: if wind speed doubles, sand transport may increase by a factor of eight. Therefore, most sand is moved at moderate to high wind velocities. Particles larger than 2 millimeters in diameter are not moved by the wind.

Wind Deposition

Deposition occurs when wind speeds fall below the threshold necessary to transport the different sizes of particles. Clay and silt particles are often deposited outside the source area, but sand usually stays in the area of origin because it is not lifted high into the air or carried a long time. Particles larger than 2 mm remain when finer materials have been blown away, forming **lag deposits,** which may be influenced by other agents, such as running water.

SUMMARY

1. Wind is a less effective agent of denudation and of more localized influence than running water or flowing ice.
2. Wind transports fine dust easily and can move it long distances. Wind moves fine or medium sand only a few kilometers. Larger particles are not moved by wind.
3. The landforms produced by wind erosion vary in scale from small to large. They include ventifacts, yardangs, and deflation hollows.
4. The range of wind depositional landforms includes various dune forms and loess depositional plains and hills.
5. Wind has its greatest effects in certain environments, including deserts, coasts, cold landscapes with little vegetation cover, and plowed fields in which the surface soil has dried.

KEY TERMS

aeolian, p. 434
wind abrasion, p. 434
deflation, p. 435
saltation, p. 435
surface creep, p. 435
lag deposit, p. 435
hammada, p. 436
ventifact, p. 436
yardang, p. 437
deflation hollow, p. 437
desert pavement, p. 438
erg (sand sea), p. 438
dune, p. 438
ripple, p. 439
barchan, p. 439
transverse dune, p. 439
star-shaped dune, p. 439
linear dune, p. 439
parabolic dune, p. 440
loess, p. 441

QUESTIONS FOR REVIEW AND EXPLORATION

1. How are dune forms determined by sand supply, wind strength, and wind direction?
2. What are the differences between loess and erg deposits in terms of materials, conditions of formation, and location?
3. Compare and contrast the transport of clay and silt, sand, and pebbles by wind, ice, and running water.
4. Assess the relative impacts of running water, ice, and wind in weathering, mass movement, erosion, and transport.
5. Investigate examples of wind action in an area near you.

FURTHER READINGS

Cooke RU, Doornkamp JC: *Geomorphology in environmental management*, Oxford University Press (1990). Chapter 9, Aeolian Processes and Hazards, covers a number of practical uses for a knowledge of the geomorphology of aeolian environments.

McKee ED, editor: *A study of global sand seas*, U.S. Geological Survey Professional Paper 1052 (1979). A classic text on dune forms, based on a satellite image survey.

Summerfield MA: *Global geomorphology*, Longman/Wiley (1991). Chapter 10, Aeolian Processes and Landforms, discusses in greater depth the topics covered in this chapter.

Thomas DSG, editor: *Arid zone geomorphology*, Halsted Press (1989). Discussions of recent studies in the geomorphology of arid environments.

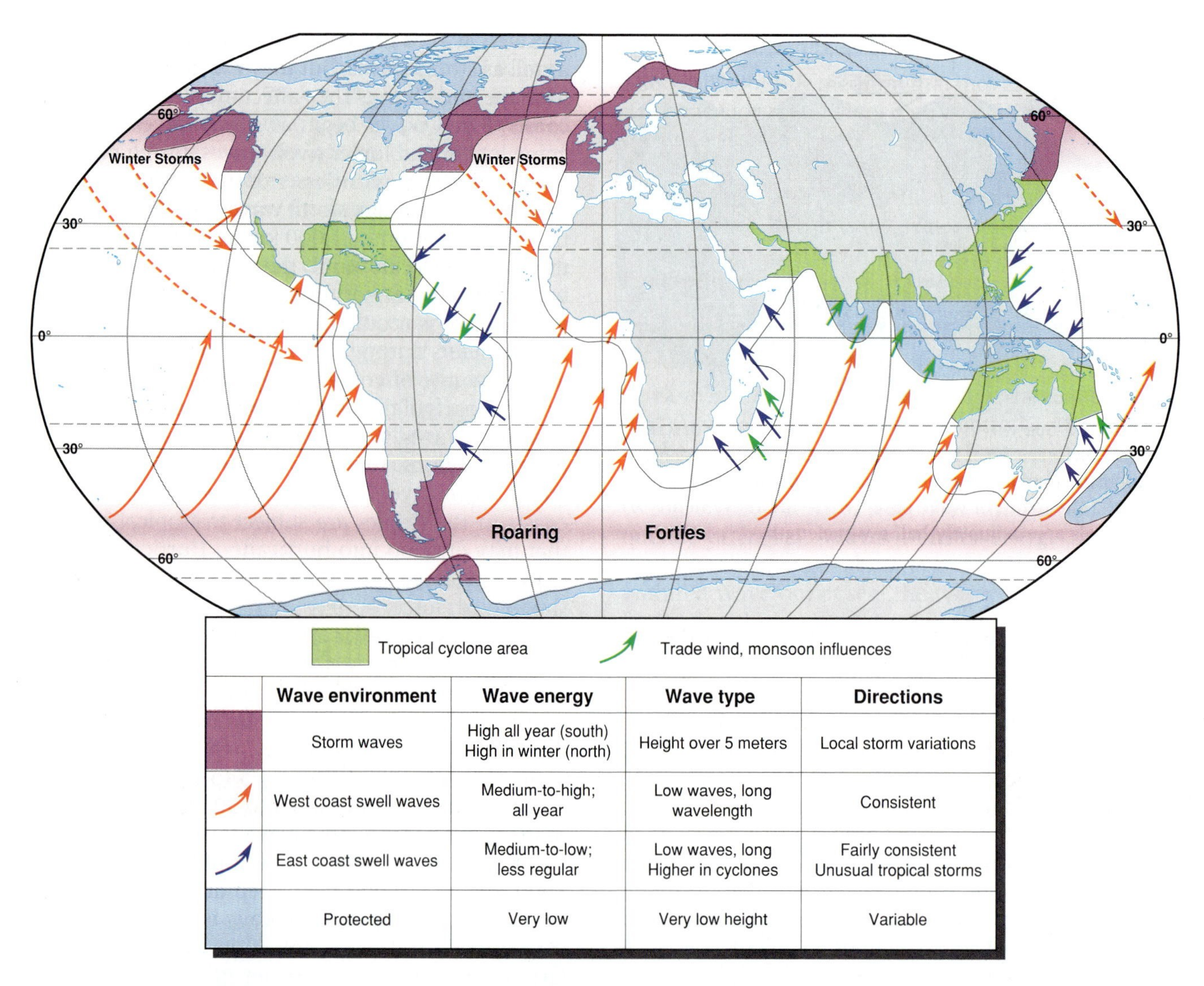

Figure 17-6 World wave environments. The main types of wave are storm waves and swell waves. Some coasts are protected from both and experience little wave action.

depth of water beneath them and topple over in the **breaker zone** (Figure 17-8). Waves break in a variety of ways that are determined by the slope of the shore and the wave size. *Spilling breakers* form offshore, where water cascades down the wave front as high waves reach gently sloping beaches. *Plunging breakers,* where the crest curls over and crashes downward, occur more frequently on steeply sloping shores. *Collapsing breakers* have less of a curl and plunge. *Surging breakers* form close inshore where low waves reach a steeply sloping shore and break in a sliding movement.

After wave break the water rushes shoreward in the **surf zone,** and this grades into a shallow sheet known as the **swash** that flows up the beach until it is slowed and stopped by friction. The water then returns down the slope of the beach as **backwash,** gathering momentum as it goes.

When the waves are high and plunge steeply, they bring more energy to remove material from the beach, and tend to remove sediment out to sea in strong backwash. When they are low in height, the energy is directed horizontally up the beach. These waves deposit sand on the upper part of the beach because much of the water drains away through the beach instead of forming a strong backwash down the beach surface.

For much of the time, wave crests arrive at the shore at a slight angle after refraction. This causes a flow of water along the shore in the breaker and surf zone, known as a **longshore current.**

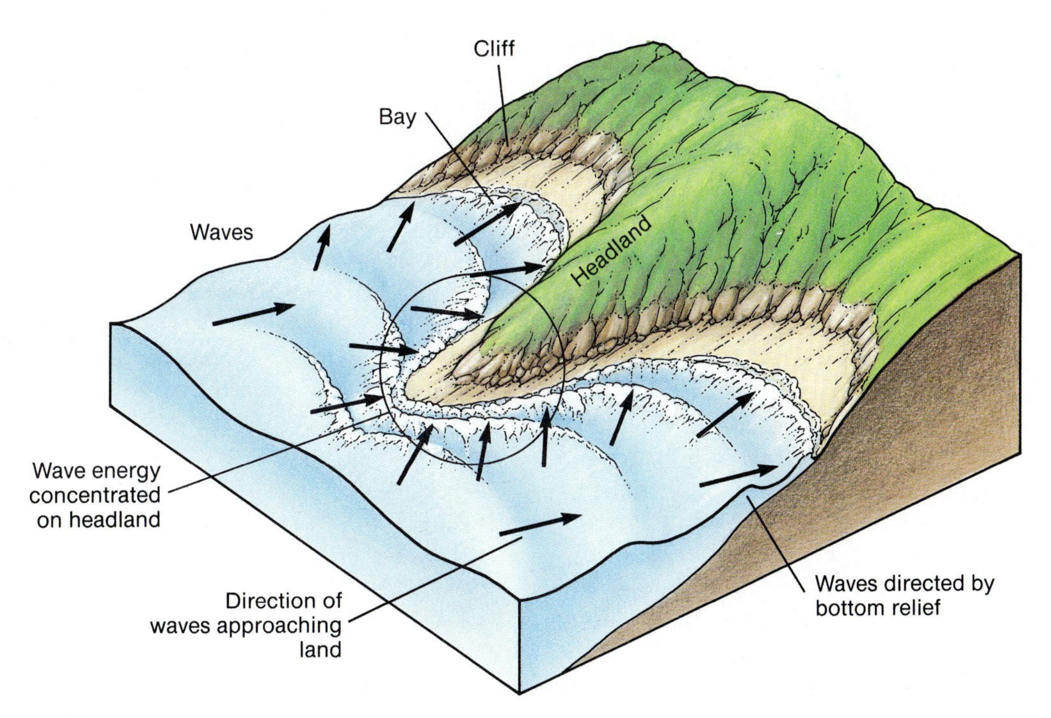

Figure 17-7 Wave refraction. The water slows as it enters shallow water. Waves turn to approach a headland, but spread out in bays: this focuses or dissipates the wave energy.

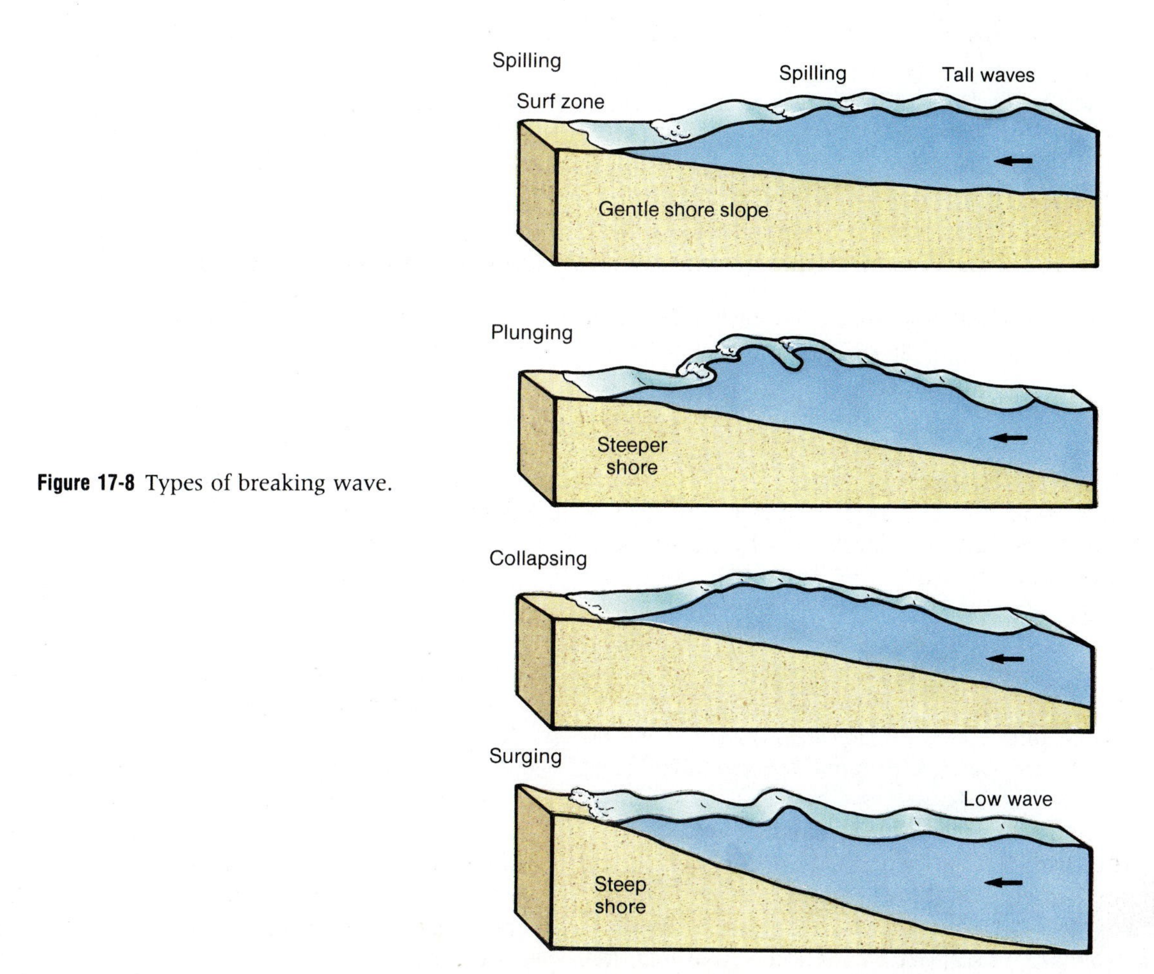

Figure 17-8 Types of breaking wave.

The nearshore zone of many beaches has a circulating system of currents. There are narrow corridors in which currents moving water toward and away from the shore are concentrated (Figure 17-9). Offshore water movements known as *rip currents* move water seaward across breaking waves and then spread out, losing impetus. The full circulation with strong rip currents occurs when high-energy waves approach the shore.

Tides

Tides are regular daily rises and falls of the ocean surface caused by the gravitational pull of the sun and Earth's moon. The effect varies in a monthly cycle depending on whether the sun and moon are pulling together (high range, or *spring tides*), or opposing each other (low range, or *neap tides*).

Tidal Environments. Tides vary in height along coasts according to the arrangement of continents and oceans and local relief. The **tidal range** is the vertical difference between the highest and lowest water levels and defines the tidal environment. Coasts with tidal ranges of over 4 meters are said to be *macrotidal;* those with ranges under 2 meters are *microtidal;* those with ranges between 2 and 4 meters are *mesotidal.* Tidal range is as much as 17 meters in the Bay of Fundy in eastern Canada (Figure 17-10), but scarcely 1 meter in enclosed seas such as the Mediterranean. The narrowing Bay of Fundy increases the tidal range; changes in the volume of water in the enclosed space cause large variations in sea level. Water entering the Mediterranean from the Atlantic, on the other hand, is able to spread out over a wide area.

The tidal range affects the width, as well as the height, of the tidal environment. The zone affected by such processes as wave action and the alternation of wetting and drying (a significant process along coasts protected from strong wave action) is generally wider along macrotidal coasts than microtidal coasts. The tidal zone is narrow (a few meters wide) around the microtidal Mediterranean, but may be over 1 km wide in macrotidal environments, where wider expanses of mud flats or sandy beaches are exposed at low tide. Local relief and the slope of the shore also affect the width of the tidal zone. Steeply sloping shores have a narrower tidal area than more horizontal shores extending between the same heights. Some of the most extensive tidal environments occur in very flat areas on the surfaces of deltas.

Tidal Currents. A greater tidal range produces stronger **tidal currents,** which are horizontal movements of seawater in response to the rise and fall of the tides. Tidal currents are the second major energy input to coastal zones after waves. One type of tidal current is that which flows through Hell's Gate on East River between Manhattan and Long Island, New York. This current exists because the high tide arrives from the east along the north side of Long Island while the low tide is ebbing out to the west. Flow is rapid and turbulent as the high tidal flow tumbles into the low-tide area. Such currents can prevent sediment from accumulating in narrow channels and can enhance erosion of the channel margins. Other tidal currents are responsible for flows into and out of estuaries (where rivers meet the sea) and bays. During a falling tide, water and sediment flow outward; on rising tides they flow inward. Where an estuary narrows inland, the tidal flow upstream at spring tide may form a wave a meter or so high, known as a *tidal bore.*

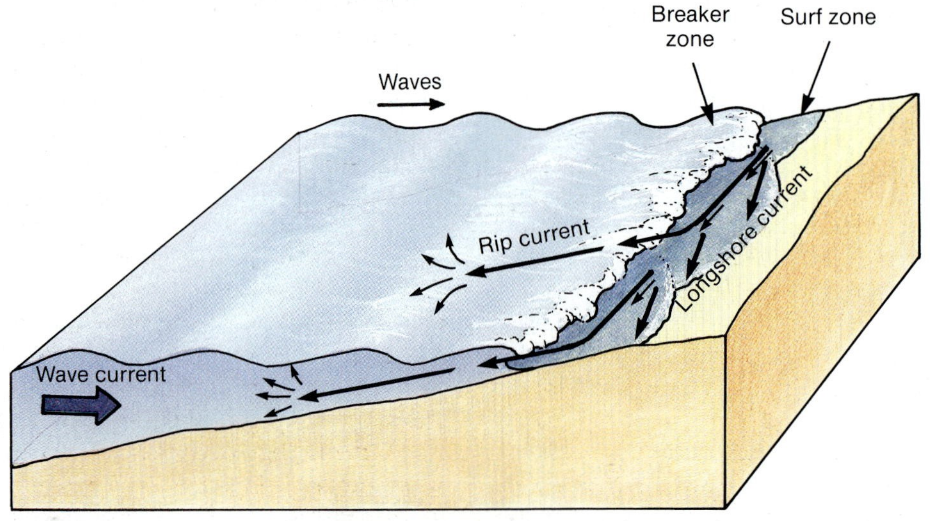

Figure 17-9 Inshore water circulation. The onshore wave current produces longshore currents and these turn offshore in a rip current.

Figure 17-10 High and low tide in a Bay of Fundy cove, showing the extreme tidal range.

COASTAL LANDFORM ENVIRONMENTS

Coastal landforms are the outcome of interactions among waves, tides, currents, freshwater, coastal rocks, sediment, and living organism structures. Marine processes form the basis of the coastal landform environments (Figure 17-11). Coastal landform environments involving interactions between the sea and the land produce landforms such as *cliffs* and *beaches*. Coastal environments where fresh water interacts with marine processes include *deltas* and *estuaries*. *Coral reefs* and *mangroves* are distinctive tropical coastal environments produced by organic structures.

Cliff Environments

Cliffs are steep rock faces that are common coastal landforms produced largely by marine erosion. Their vertical or sloping rock faces are most prominent in storm-wave environments. On more protected coasts, cliffs have only short vertical sections. In humid regions, upper slopes are covered by vegetation. Cliff forms are more common along the convergent margins of plates where hills and mountain ranges form areas of high relief along the coast. They also occur along some coasts where oceans have opened by splitting former mountain ranges: the ranges often run at right angles to the coast and end in spectacular cliffs, as in southwest Ireland.

Coastal cliffs form by a combination of processes that begin with the erosion of rock at the foot of a cliff below high-tide level when it is attacked by high-energy waves. The impact of waves breaking against a cliff base is a powerful force, and it compresses air between the wave front and the shore and also in joints in the rock. This combination dislodges blocks of rock and loose particles. Rock is worn away by *abrasion* (corrasion) when waves fling pebbles only at the cliff base. Large boulders can be moved by the largest waves and are less important in cliff erosion than pebbles, which can be moved by waves of moderate energy.

A notch is often formed at the cliff base below high-tide level and demonstrates the effectiveness of wave erosion. The formation of the notch and the excavation of rock fragments undermine the rocks above, and they fall to the shore. This maintains the steepness of the cliff. The roles of the force of water and abrasion are debated by scientists. The proportions depend on the type of rock being attacked and particularly on the presence of joints and other fractures that facilitate the rocks being dislodged by wave attack alone.

Since cliffs are usually steep rocky slopes, they are also subject to weathering and mass movements; rock falls and landslips are common events (Figure 13-13). Some cliffs are formed in unconsolidated materials such as glacial drift. These cliffs are often subject to rapid undercutting by wave erosion, accompanied by erosion of their surface by running water (Figure 17-12).

A **shore platform** is a horizontal, or nearly horizontal, surface formed by rock at the base of a cliff. There are several types. Shore platforms that result from cliff erosion are known as *wave-cut platforms*, although other processes such as weathering also mold them. They occur below high-tide levels and slope seaward. As the wave-cut platform gets wider, marine erosion of the cliffs is slowed because the water shallows as the platform rises toward the cliffs. A second type of shore platform occurs in low-

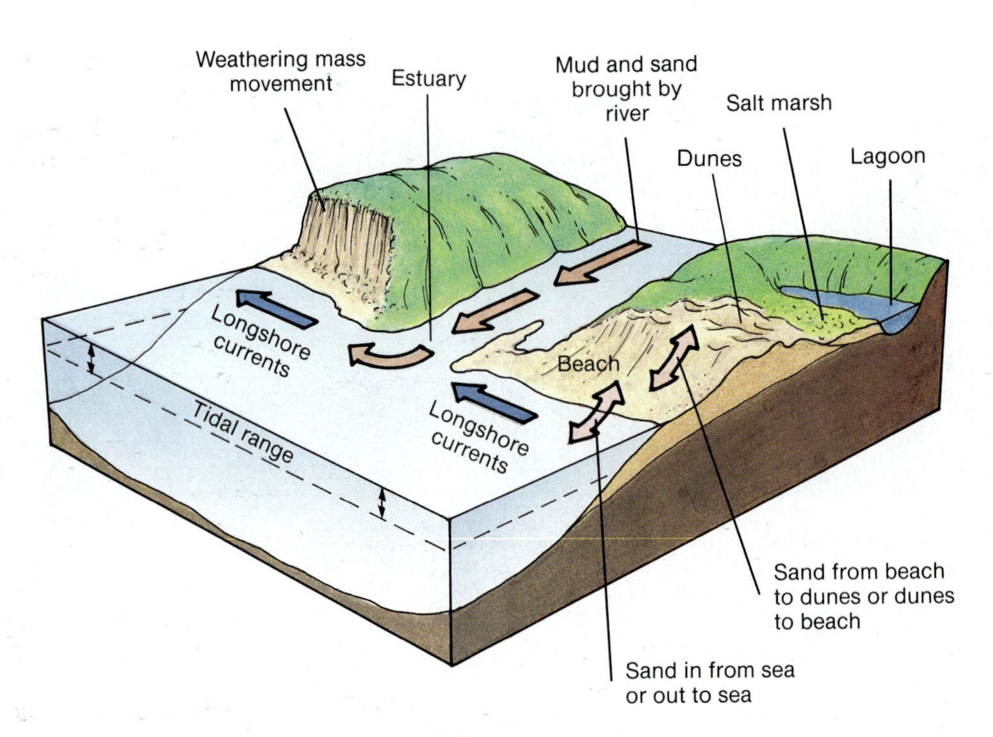

Figure 17-11 Marine processes and the movement of sediment in coastal environments: some of the links between different parts of the environment.

Figure 17-12 Cliffs in glacial deposits, New England. Waves undercut the base of the till, dislodging pebbles. The cliff face is at an angle determined by the angle of repose of the fragments that compose it. Small streams carve gullies in the weak material.

Figure 17-13 Cliffs on the north coast of France, showing an arch and stack.

energy wave environments. Horizontal shore platforms form where wave erosion is slow as a result of the wetting and drying of rock surfaces between tides, which causes some rocks to flake, together with other processes such as solution, cementation, and biologic activity. Horizontal shore platforms are most common in the tropics.

The detailed shapes of cliffs are influenced by rock type. In areas of resistant or recently uplifted rocks, cliffs tend to be higher. Where weaker and more resistant rocks alternate and where strong rocks are highly fractured, features such as caves, stacks, and arches form (Figure 17-13). A sea cave is eroded by waves along lines of weakness at the base of a cliff. A stack is a rock pillar that has been isolated by waves removing the rock between the stack and a backing cliff. An arch is a stack joined at the top to the backing cliff; if the upper section of an arch collapses, a stack will be formed.

Beach Environments

Beaches are coastal depositional landforms, composed mainly of loose sand or pebbles. They occur mainly along coasts of low relief but also along coasts of higher relief in places where sediment accumulates in protected zones. Sand produces wide, flat beaches, but pebbles produce narrower, steeper features. Most pebble beaches occur in the middle and high latitudes, where storm-wave environments coincide with regions that were glaciated in the Pleistocene Ice Age when larger rock fragments were deposited. The Grand Banks of Newfoundland and the floor of the North Sea have pebble covers for this reason. The central Atlantic states of the United States, on the other hand, have sandy beaches and offshore sand deposits supplied by streams from inland areas. Tropical beaches tend to be formed of fine sand with much calcareous shelly material. The shells dissolve and the calcium carbonate cements the sand to form *beach rock*.

The landforms of beaches are often small, measuring up to a few meters in size (Figure 17-14). At the highest point of the beach there may be a more steeply sloping *storm beach* several meters high, composed of pebbles that are moved only by high waves. On the sandy part of a beach the vertical relief is of the order of a few centimeters: *runnels* are small, flat-floored channels a centimeter or so deep that are formed by low-tide runoff; they alternate with *rippled* surfaces and low sandy *bars* formed by breaking waves. The ripples and bars cover most of the beach, and the runnels are more localized. *Beach cusps* are crescent-shaped sand or pebble accumulations around semicircular depressions that may be caused by powerful swash and backwash water movements. Most minor beach features exposed at low tide are normally removed during the following high tide (Figure 17-15).

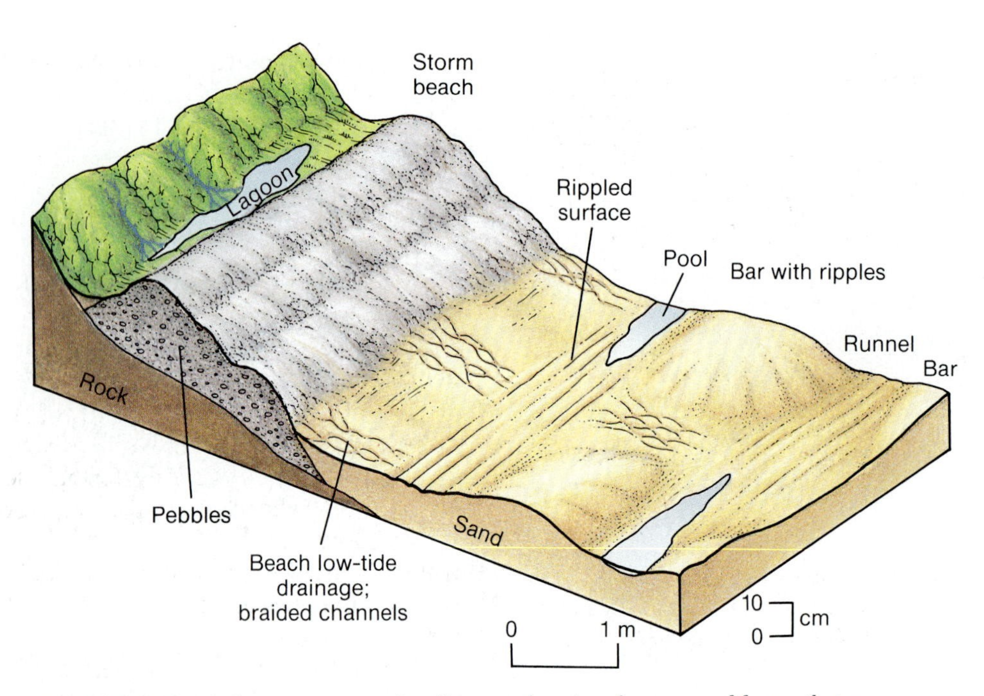

Figure 17-14 Beach forms: a composite diagram showing the range of forms that may occur. The actual occurrence is determined by the wave and tidal environments and the supply of sand or gravel.

Figure 17-15 Beach features at Cape Hatteras, North Carolina. The exposed beach has a series of low-amplitude bars and runnels.

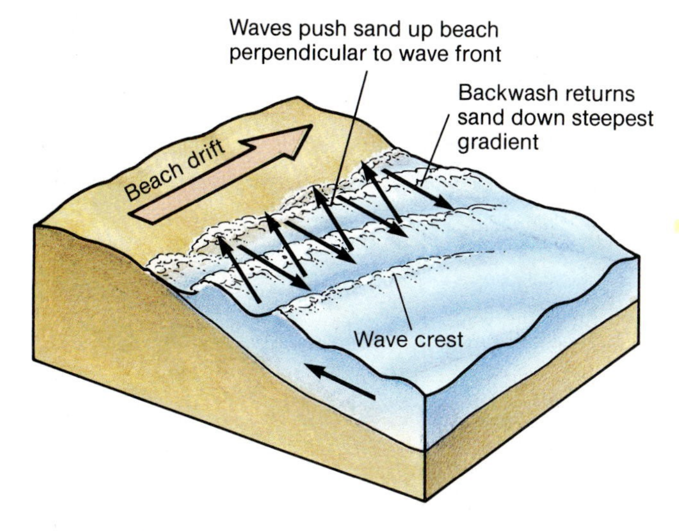

Figure 17-16 Longshore drift. Waves push sand up the beach in swash; backwash takes it down the beach's steepest slope. This moves the sand along the beach.

Beaches change form more frequently than cliffs, often over a yearly (or shorter) cycle. For example, stronger winter waves may pull sand out to sea, with gentler waves returning the sand the following summer. Strong and consistent onshore winds drive high waves that produce wide sandy beaches with several sequences of bar-runnel alternations in macrotidal environments where there is an abundant supply of sand. The high-energy waves take sand far up the beaches, and the strong backwash currents take much back down the beach slope, spreading it out to form a wide feature. Weaker winds, microtidal environments, and a small sand supply give rise to narrower beaches with a single bar-runnel sequence.

Longshore currents move large quantities of sediment parallel to the coast, a process known as **longshore drifting.** It occurs in the breaker and surf zones and is most active with steep, high-energy waves. Longshore drift causes sediment to be pushed up the beach at the angle of the incoming waves. Backwash returns it down the steepest slope of the beach. The overall result is movement of sediment along the beach (Figure 17-16). The movement of beach material along the coast is responsible for several important beach forms. If the coast consists of alternating bays and rocky headlands, the sandy sediment will accumulate in the heads of the bays. On low coasts, on the other hand, it is moved along the shore and the resulting deposits may dominate the coastal landforms.

On the coasts of the eastern United States and the Gulf, barrier beaches extend for over 4300 km (2700 miles) (Figure 17-17). **Barrier beaches** are elongated sand or pebble banks lying parallel to the coast. They are formed by several different processes. First, where a copious supply of sand is moved by longshore drift, changes in coastline direction cause the deposition of sand carried forward into deeper water. Second, dune ridges may be drowned by a rise in sea level, such as occurred following the last glacial melting. This is how the Outer Banks of North Carolina were formed. Third, deposition from longshore drift may take place on *offshore bars,* semi-submerged deposits of sand just outside the breaker zone. Offshore bars may be built above sea level, as occurred to form some Gulf islands. Barrier beaches enclose areas of protected water known as *lagoons,* which often have freshwater rivers draining into them, and develop into wetland marshes as vegetation grows on their surface.

Barrier beaches vary in their degree of attachment to the land, from bay barriers to barrier spits and barrier islands (Figure 17-18), but the last group is most common. *Bay barriers* are connected at both ends to headlands. They are usually small, and may have temporary inlets through them. Bayhead barriers occur toward the back of a bay with saltmarsh behind. Baymouth barriers connect the headlands and are common in microtidal conditions, which inhibit inlet formation. Brackish water lagoons develop behind them. A *tombolo* is a type of bay barrier that connects an island to the mainland.

Barrier spits are attached at one end to a source of sediment and extend into open water. They are beaches formed by the longshore drifting of sediment into the mouths of bays or estuaries where the coast changes direction. The example of Sandy Hook on the northern New Jersey coast occurs where a high tidal range coincides with longshore drifting into deeper water. The tip of Sandy Hook is curved by waves approaching from several directions. Water ponded behind the beach caused marsh accretion, and the exposure of sand above high-tide level has led to formation of dunes on top of the spit feature.

Barrier islands are not attached to the mainland. They are formed by the emergence of dune ridges and offshore bars and their features depend on their tidal environment and sand supply. Microtidal conditions with a low sand supply produce long, narrow, and low-lying islands that are vulnerable to storm action. Where there is a greater supply of sand, several parallel dune ridges may form on top of the island, giving the island greater stability. The

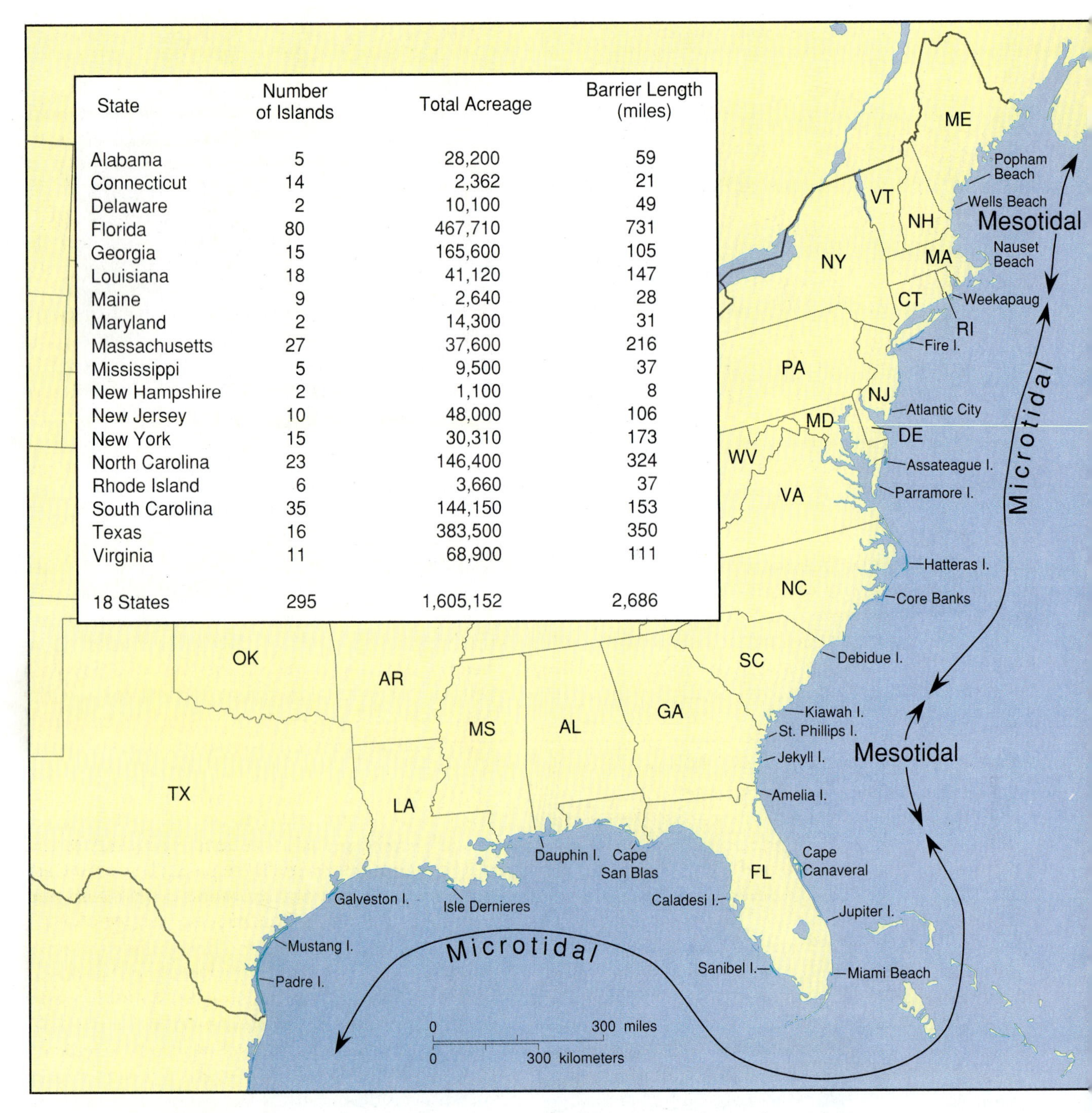

State	Number of Islands	Total Acreage	Barrier Length (miles)
Alabama	5	28,200	59
Connecticut	14	2,362	21
Delaware	2	10,100	49
Florida	80	467,710	731
Georgia	15	165,600	105
Louisiana	18	41,120	147
Maine	9	2,640	28
Maryland	2	14,300	31
Massachusetts	27	37,600	216
Mississippi	5	9,500	37
New Hampshire	2	1,100	8
New Jersey	10	48,000	106
New York	15	30,310	173
North Carolina	23	146,400	324
Rhode Island	6	3,660	37
South Carolina	35	144,150	153
Texas	16	383,500	350
Virginia	11	68,900	111
18 States	295	1,605,152	2,686

Figure 17-17 The distribution of barrier beaches and tidal environments in the southeastern United States.

Outer Banks of North Carolina, Fire Island, New York, and Galveston Island, Texas, are typical examples (Figure 17-19). Where the tidal range is mesotidal, the islands are shorter because the greater rise and fall of water creates more tidal inlets that are subject to daily low-tide scour. The smaller islands are more stable because the water is channelled around them. The bays and lagoons behind are normally filled by marshes. Along the eastern United States coastline, mesotidal barriers are less common than microtidal ones, but mesotidal barriers occur in Maine, South Carolina, and Georgia.

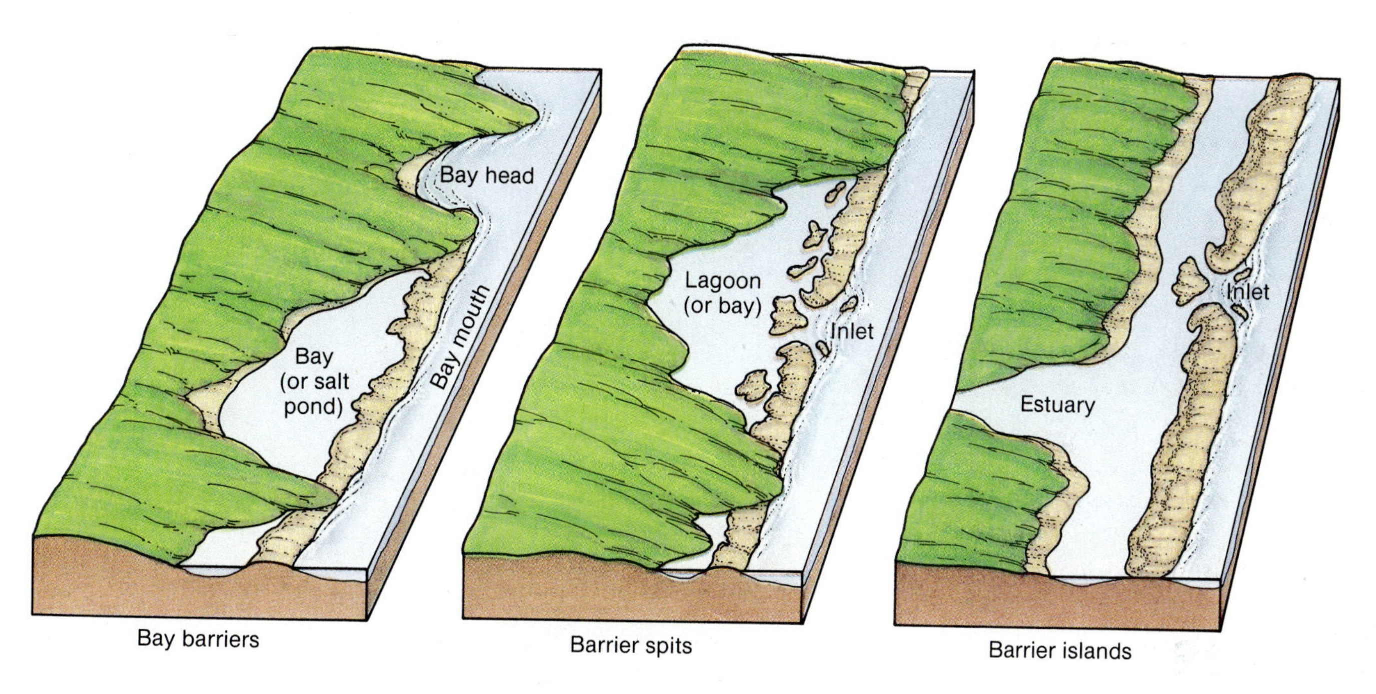

Figure 17-18 Barrier beaches: the main forms.

Figure 17-19 A barrier island backed by tidal marshes on the coast of Mozambique, Africa, near Vilanculos.

Deltaic and Estuarine Environments

River mouths form breaks in the cliffs and beaches, and produce mixed environments where fluvial and marine processes interact. Deltas and estuaries are the main features produced in these conditions and provide environments that contain a range of landforms and deposits.

Deltas. A **delta** is a fan-shaped deposit at a river mouth. Deltas occur on both upland and lowland coasts, and also in lakes. The essential condition for their formation is that the rate of deposition on the delta surface surpasses the ability of marine processes to remove the deposits to deeper water. The accumulation of sediment produces new land.

The features of particular deltas are determined by whether most sediment is dropped on the surface of the delta or on the seaward margin and by how marine erosion and tides modify the outer features. Deltas may be dominated by river, wave, or tide (Figure 17-20). Those dominated by river action flow into seas without strong currents or tidal ranges, and extend their distributary channels into the sea. The Mississippi delta is of this type, and its shape is often described as "birdsfoot." When wave

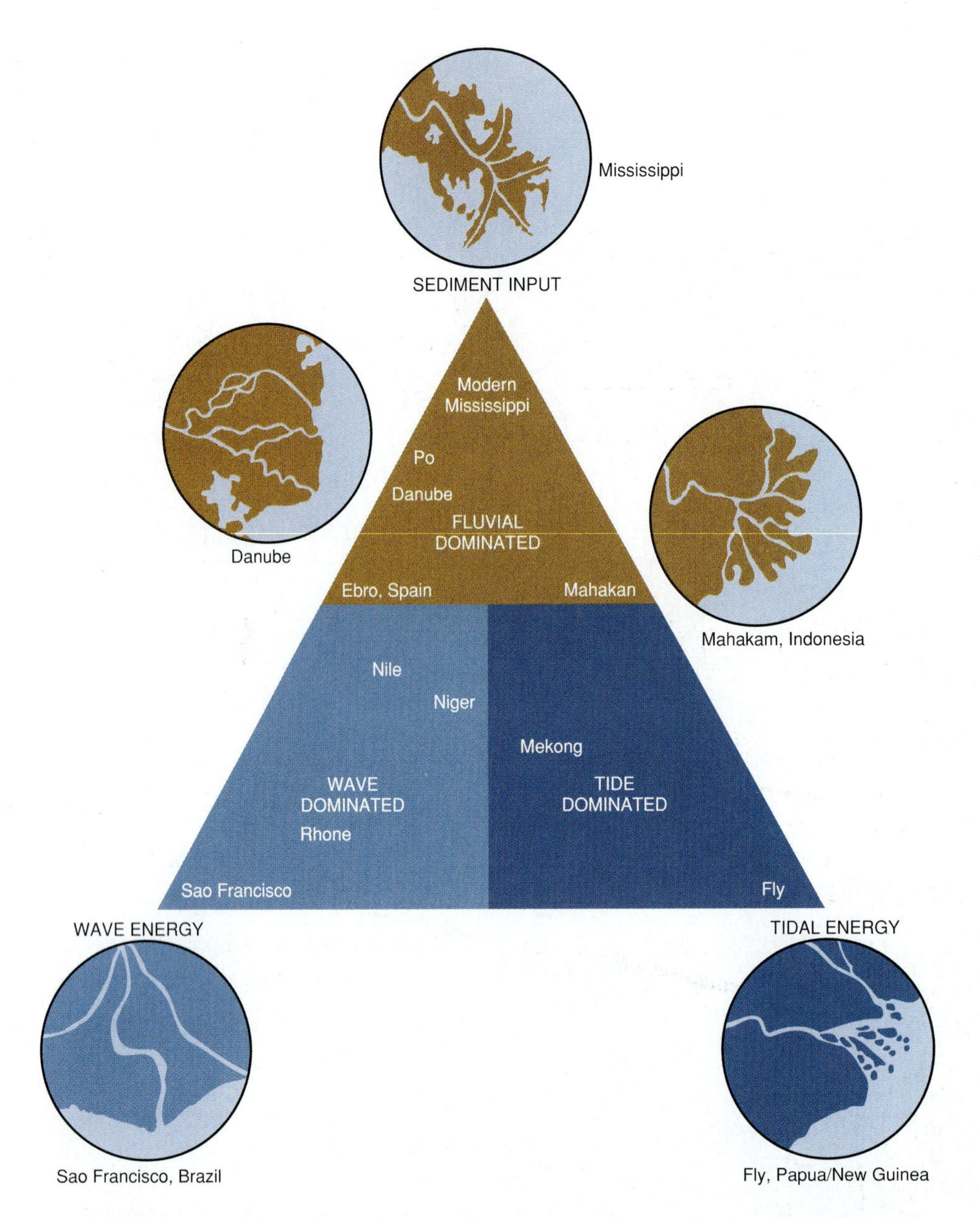

Figure 17-20 Types of delta form related to dominant conditions affecting the origin of the landform-high sediment input, wave attack, and tidal range. Where more sediment is carried by the river than the sea can remove, and when tidal currents are weak, the river builds new land along its distributaries, forming a birdsfoot delta like that of the Mississippi. Where wave action is strong, a smooth coastal outline forms. Where there is a high tidal range, the delta deposits are separated by deep channels.

action is significant, the delta front is smoothly curved by beach deposition, as is the Nile delta or the Sao Francisco delta in Brazil. Tidal environments influence the forms of deltas in macrotidal environments, where low-tide flows form deep channels between long islands. Some examples of deltas as seen from Space Shuttles are shown in Figure 17-21.

The weight of deposited sediment and its compaction make delta surfaces liable to subsidence. If deposition does not keep pace with subsidence, some or all of the delta is drowned. Deltas are thus unstable features where balances must be maintained between inputs and losses of sediment, between erosion and deposition by the sea, and between subsidence and sediment accumulation.

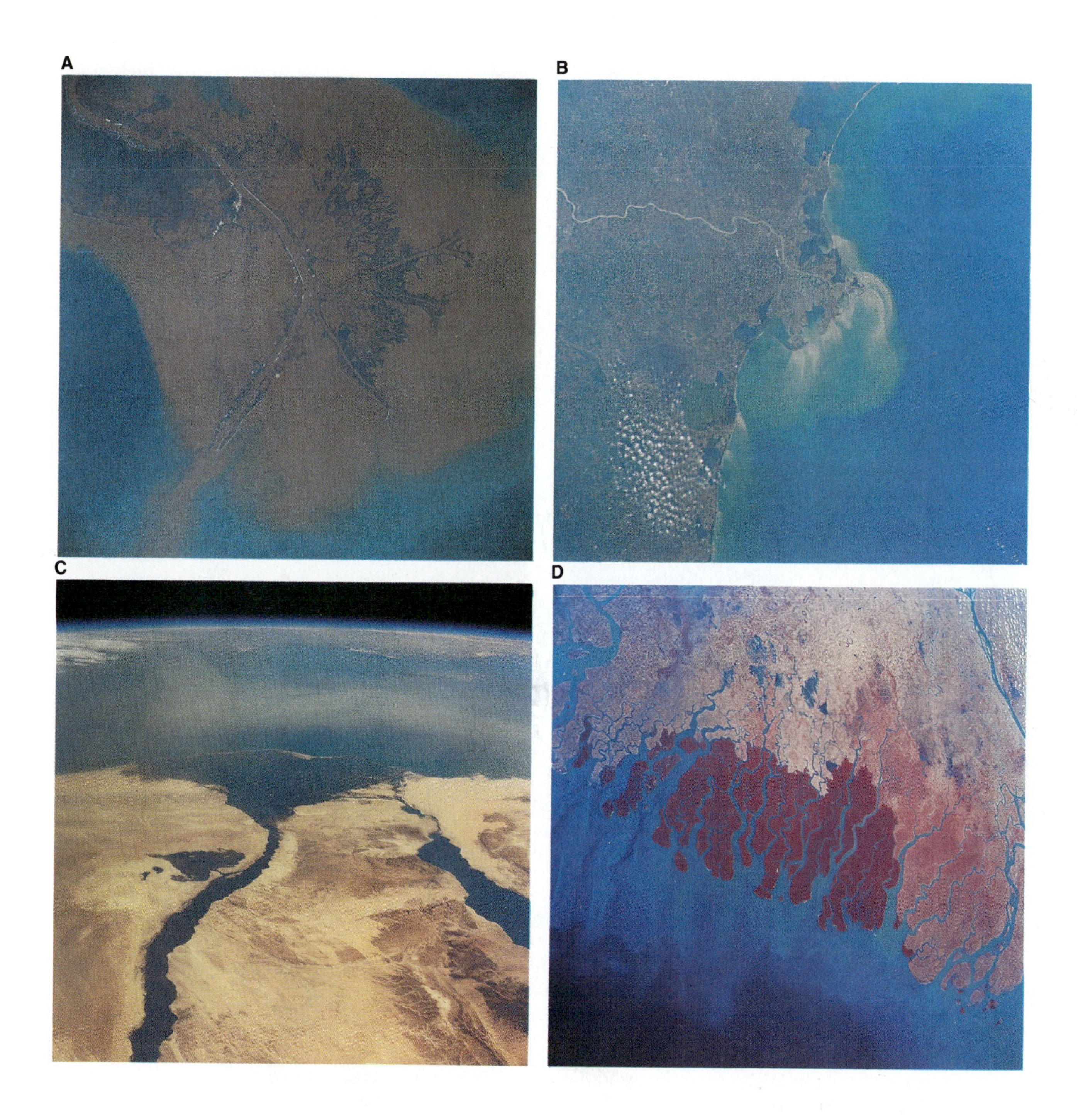

Figure 17-21 Types of delta photographed from Space Shuttles. **A,** Mississippi delta, Louisiana. **B,** Po delta, northern Italy. **C,** Nile delta, Egypt. **D,** Ganges delta (Sundarbans), Bangladesh.

Estuaries. An **estuary** is the mouth of a river that broadens into the sea and is affected by tidal processes. Estuaries are common on coasts where the rivers carry small quantities of sediment, but most river mouths, including delta distributaries, have estuarine sections. Most estuaries are alternately filled by seawater at high tide and then flushed by fresh water on the falling tide; some, however, particularly those in microtidal environments, remain brackish with little change in salinity. *Brackish water* is a mixture of seawater and fresh water, with a salinity of 15 to 30 parts per thousand. In macrotidal environments, the stream will carve its own channel in the estuary during falling-tide flow out to sea. At high tide, the fresh water flow is ponded back by incoming seawater and the reactions between salt water and clay particles in the stream load cause the clay particles to stick together and fall to the bottom.

Muddy sediment, composed of clay and silt particles with a high water content, is a characteristic of estuaries and also of lagoons formed on delta surfaces and behind barrier beaches. It is deposited to form **mud flats**—sections of the shoreline covered by silt or clay and submerged at high tide. When the mud flats rise above low tide, they are colonized by salt marsh plants and become stabilized (Figure 17-22). Salt marsh plants are those that will grow in water of changing salinity and will tolerate submergence at high tide.

Salt Marshes and Mangrove Environments

Salt marshes and (in the tropics) mangrove colonies develop in lagoons behind barrier beaches and in estuaries and deltas protected from wave activity. **Salt marsh plants** colonize exposed mud flats at

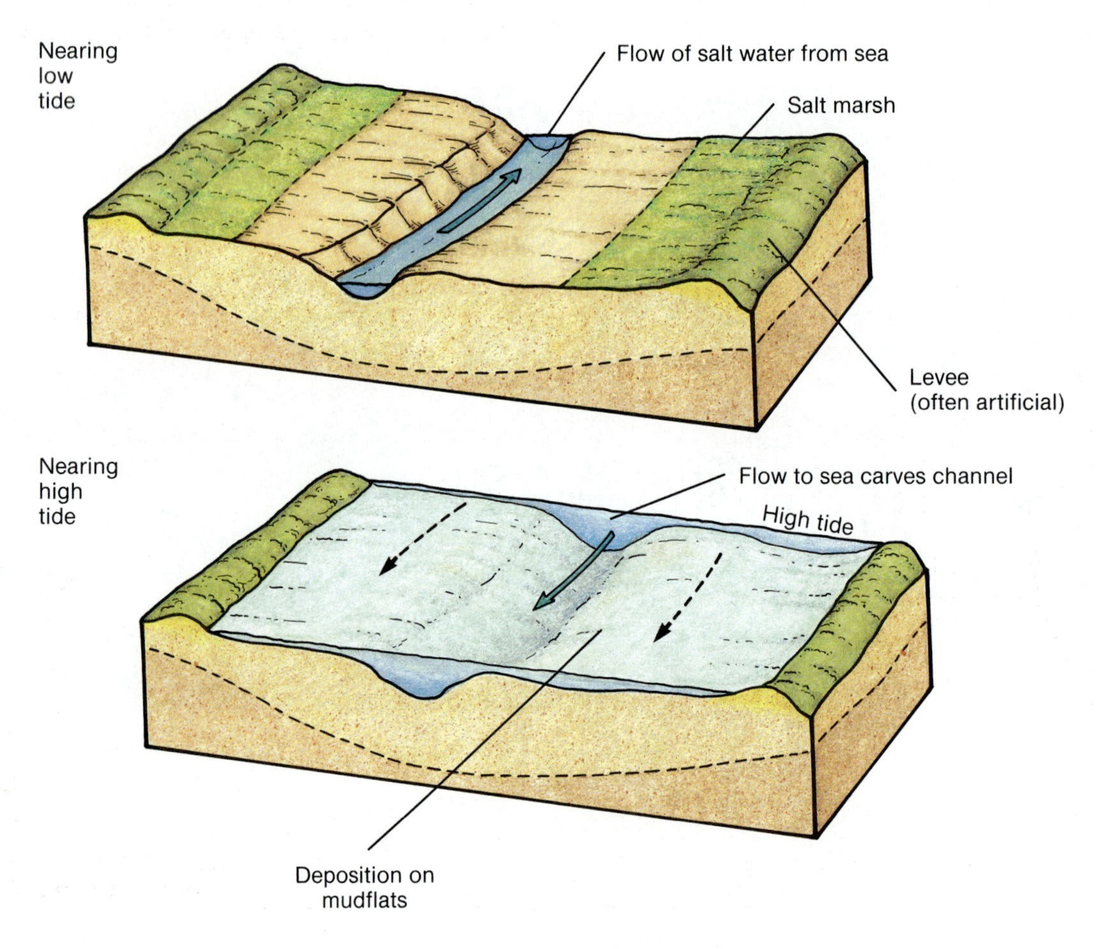

Figure 17-22 Features of an estuarine tidal environment. Mud flats form by deposition at high tide, and are colonized by salt marsh plants when they rise above the low-tide level.

low tide and trap muddy sediment, raising the level to just below high tide. **Mangroves** colonize low-lying shorelines and estuaries. The roots extend through the water to the bottom sediment, and the leaves and branches rise several meters above high tide. The roots act as traps for sediment in the brackish water environment and such deposition often removes solutes from the still water in a filtering effect. The area beneath mangroves is built toward high-tide level. Mangroves protect a coast from erosion and longshore drifting, but they are occasionally ripped out by hurricane surges.

Reef Environments

The high temperatures of subtropical and tropical coasts encourage the development of coral reef structures immediately offshore. **Coral reefs** are structures built by animals that live in colonies and secrete skeletons of calcium carbonate. The skeletons of many coral colonies accumulate, building on dead colonies to form reefs. Coral reefs may be up to several thousand meters deep and several kilometers wide, and so form important constructional features along coasts.

The growing part of a coral reef is just below sea level. The distribution of modern coral reefs (Figure 17-23) shows that greatest coral growth occurs where the sea-surface temperatures are between 22°C and 29°C (72 to 84°F), where the photosynthesizing algae associated with coral can flourish. Coral reefs grow most rapidly on the windward sides of islands, where breaking waves increase levels of dissolved oxygen. Reefs do grow in quieter conditions, but more slowly. Reef debris produced by storms is sometimes piled up to form an island. Some islands, such as Barbados, have been formed by the uplift of a coral reef.

Many coral reefs have formed around volcanic islands in the Pacific and Indian oceans. Charles Darwin noted that a sequence of three main types of reef—fringing reef, barrier reef, and atoll (Figure 17-24)—formed as a volcanic island was created but then became extinct and subsided. *Fringing reefs* are attached to land; *barrier reefs* occur where there is a wide lagoon between the reef and land; *atolls* occur where there is no land. Figure 17-25 is a view of some of these forms from the Space Shuttle. Not all coral reefs are attached to volcanic islands: those around Florida, for example, and northeastern Aus-

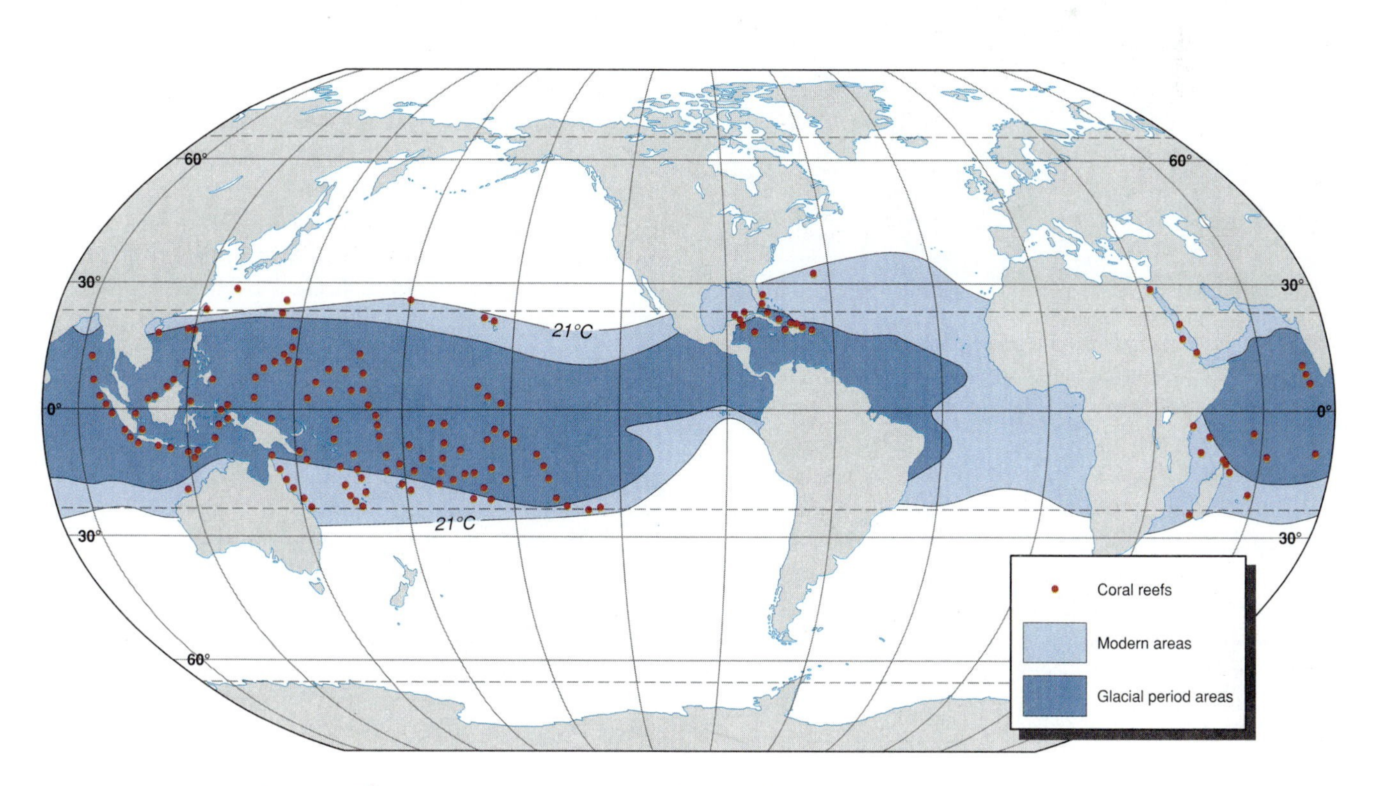

Figure 17-23 A world map of coral reefs, showing how active reefs are confined within water of relatively high temperatures. Not all the 425 atolls could be shown.

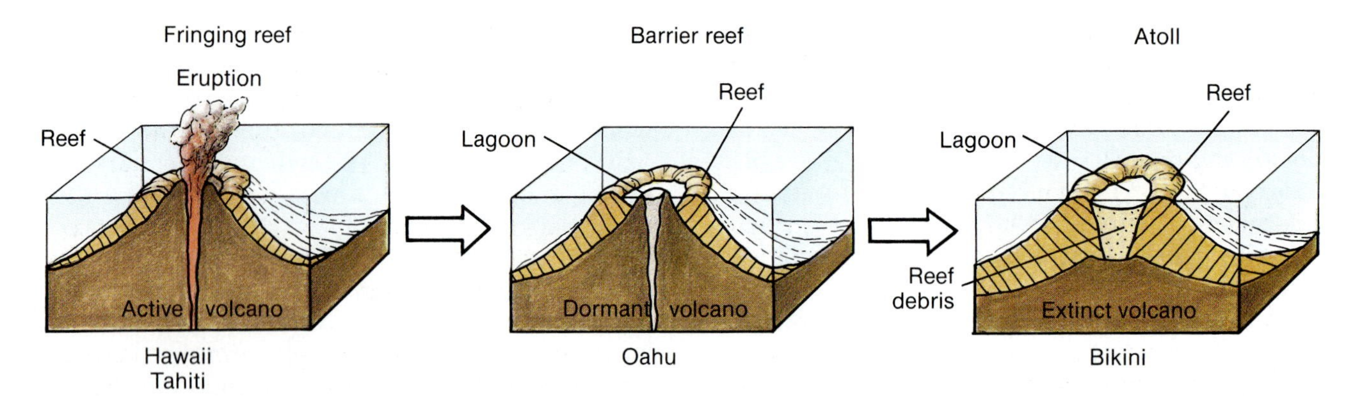

Figure 17-24 A developmental series of coral reefs. The volcanic peak provided an initial foundation, with the reef attached to the island as a fringing reef. The barrier reef is separated from the subsiding island by a wide lagoon. The atoll has no central island.

Figure 17-25 Coral reefs in the Society Islands in the Pacific Ocean, as seen from the Space Shuttle. There are barrier reefs around Tahaa-Raiatea (southeast), and the atoll in the northeast is Motu Iti.

tralia have formed along coasts that have provided a stable anchor for the reefs.

Although Darwin could not know, his sequence can be linked to plate tectonic processes, as illustrated by the Hawaiian Islands and the line of volcanic peaks and coral reefs extending to the northwest (Figure 11-17). These volcanic peaks were formed by eruptions above a hot spot. As the lithosphere plate moved northwestward, it carried with it the series of volcanic peaks formed over the magma source; the volcanoes became extinct as they moved away from the hot spot, were worn down by surface-relief processes, and gradually subsided by isostatic adjustment of the crust to their weight. The youngest islands, those nearest the hot spot at the southeastern end, have a few coral reefs that are all close inshore—the beginnings of fringing reefs. The other Hawaiian Islands have only extinct volcanoes but have well-developed fringing reefs. Northwest of the Hawaiian islands, barrier reefs surround the smaller islands, and farther along the line the reefs are all atolls. Cores drilled through Bikini Atoll indicate that the reefs have been built up since Eocene times, 50 million years ago, but the reef materials have always been formed in shallow water. Even major changes of sea level, such as the falls that occurred during Pleistocene glacial phases, appear to have had little effect on the reefs apart from the formation of karst-like features when the coral limestone was exposed during periods of low sea level. It is clear that the forms of many coral reefs are related to the sequence of events in the life of a volcanic peak on a moving plate.

COASTAL ENVIRONMENTS AND CLIMATIC CHANGE

The influence of climatic differences on marine processes is less than on other surface-relief processes. Steep cliffs and pebbly beaches are most common in middle latitudes where high-energy waves attack the coasts. Vegetation-covered cliffs, coral reefs, and mangroves are features of tropical coasts. Sandy beaches and estuarine mud flats, however, occur in all climatic conditions, where wave and tidal influences override the influence of climatic factors.

Changes in climate, however, have a major influence on coastal landforms by changing the level of the sea. Since wave action is concentrated in a narrow zone between high and low tide, changes in sea level cause wave action to take place at a higher or lower zone. Old cliffs, wave-cut platforms, beaches, and reefs may be left high and dry if the sea level falls; rises in sea level drown older coastal landforms and the mouths of river valleys.

When the ocean level as a whole changes, the change is said to be **eustatic** in nature. Eustatic change occurs, for example, when a climatic change causes ice sheet formation or melting, or when the shapes of the ocean basins change (see Chapter 12). This is different from isostatic change (see Figure 11-13), which causes movements of part of the crust relative to sea level. Ice sheet growth and decay during the Pleistocene caused both isostatic and eustatic changes of sea level. Unravelling the local distinctions between eustatic and isostatic changes can be difficult, but the movements of sea and land during the Pleistocene had an important influence on coastlines. The present interglacial sea level has only been in place for some 7000 years, a short time in the workings of surface-relief processes. Coastlines show relatively little modification by marine processes and often preserve remnants of past coastlines.

Because of the importance of changing sea levels, coastal landforms are sometimes classified on the basis of whether they indicate an emergent or submergent coast. **Emergent coasts** form where sea level falls relative to the land, and are marked by raised beaches and marine terraces. A raised beach has beach landforms—wave-cut platform, sand deposits, shells, or rounded pebbles—and is backed by a cliff but occurs above the level of present wave action (Figure 17-26). A marine terrace is a former wave-cut platform that occurs above the level of present wave action, and may occur in a staircase of terraces, as along the coast of southern California.

Submergent coasts are formed where sea level rises relative to the land and are characterized by drowned valleys and deeper water offshore. Features of submergent coasts are common around the world at present because of the rise in sea level when the ice sheets melted 15,000 to 10,000 years ago. San Francisco Bay and Puget Sound are examples of drowned valleys on the west coast of the United States; the Hudson estuary and Delaware and Chesapeake Bays are examples on the east coast. The drowned valley mouths in the hilly region of Spanish Cantabria are known as *rias,* and this name has been applied widely to such features. **Fjords** are deep ocean inlets in areas of high relief.

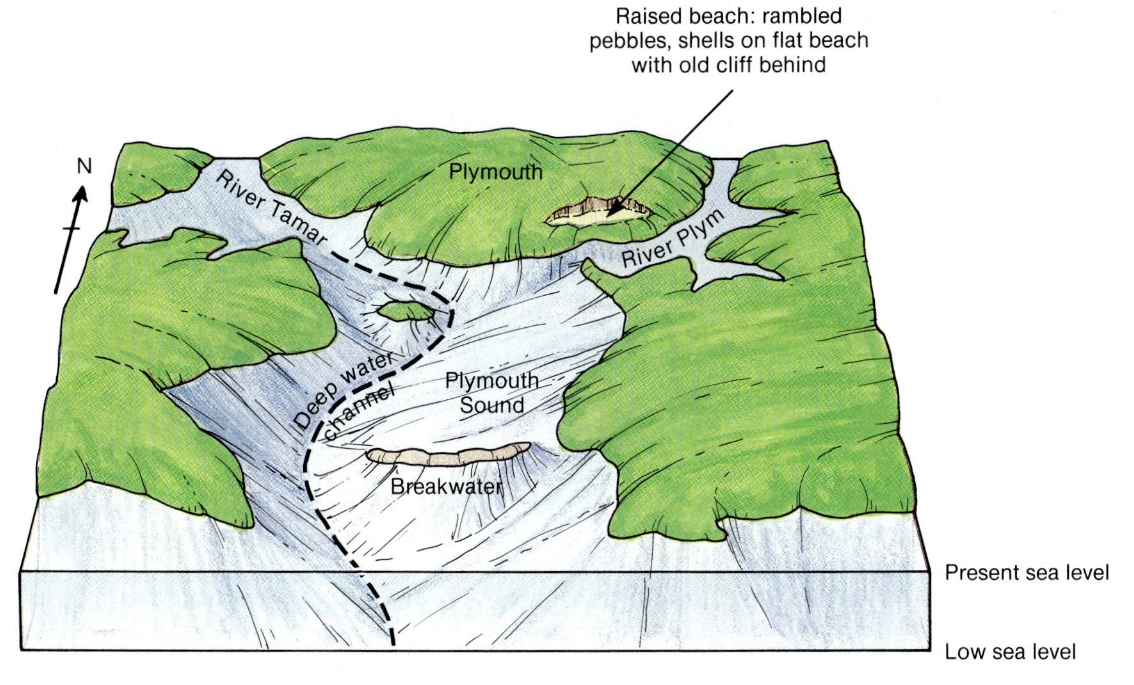

Figure 17-26 Coastal landforms at Plymouth, England. Plymouth Sound is a drowned river mouth, or ria. A raised beach on Plymouth Hoe shows that the sea level was also once at a high level.

Figure 17-27 A fjord in Alaska: Endicott Arm in the Coast Range of southeastern Alaska. The Dawes glacier terminates at the head of the fjord.

Greenland and Antarctica ice sheets and raise sea level (see Chapters 9 and 15), the causes and impacts of changing sea levels in relation to climatic change are an important topic. A rise in sea level of a meter or so would mostly affect lower coasts such as deltas, estuaries, reefs, barrier beaches, salt marshes, or mangrove colonies. It would also affect the many major cities in coastal locations. Sea level changes may result from factors other than climatic or geologic changes, as explained in Box 17-1: The Shrinking Aral Sea.

HUMAN IMPACTS ON COASTAL LANDFORM ENVIRONMENTS

Human beings make intensive use of the coastal zone. Harbors are built to link land and ocean transport systems and to service naval ships. Many forms of recreation are located in the coastal zone, encouraging the construction of highways and housing. People expect the coastline to stay the same under the pressures resulting from such activities, but the coastal environment is subject to complex interactions and continuing changes. The increasing value of coastal land has led to attempts to reduce erosion, but success in one section of a coast often causes losses in other sections.

Many local problems arise because breakwaters, landing wharves, and groins interfere with the natural longshore movement of sediment. *Breakwaters* are substantial walls built out into the sea to protect harbor entrances. *Landing wharves* are built at right angles to the coast and either are built of rock or are planked on wooden or metal support structures. *Groins* are low fences or walls built at right angles across a beach to trap sediment and protect the local coast. Groins may protect one area, but they prevent sediment from moving to replenish another area that it formerly protected, so the erosion is merely shifted along the coast.

Sea walls are built to insulate a section of cliff or dune from marine action and protect property from erosion. They prevent waves moving up the beach, however, and so increase backwash, lowering the level of the beach, and may necessitate the building of groins to trap sediment and provide protection for the sea wall. It is common for one attempt to preserve a section of coast from change to lead to a need for further action.

The effects of human activity on coastlines also influence larger regions. They are illustrated by five case studies of seaside areas—the California coast in San Mateo County, the coast of the Carolinas, the Mississippi delta, Cape Cod, and reclaimed tidal lands.

The California Coast

The coast of San Mateo County, south of San Francisco, is formed by the headland of Pillar Point and Half Moon Bay to the south (Figure 17-28). Much of the coast consists of softer sedimentary rocks, landslide debris, and unconsolidated stream and beach deposits. Sections of this coast have been retreating at rates of 3 meters per 10 years. A breakwater built in 1959 at the northern end of Half

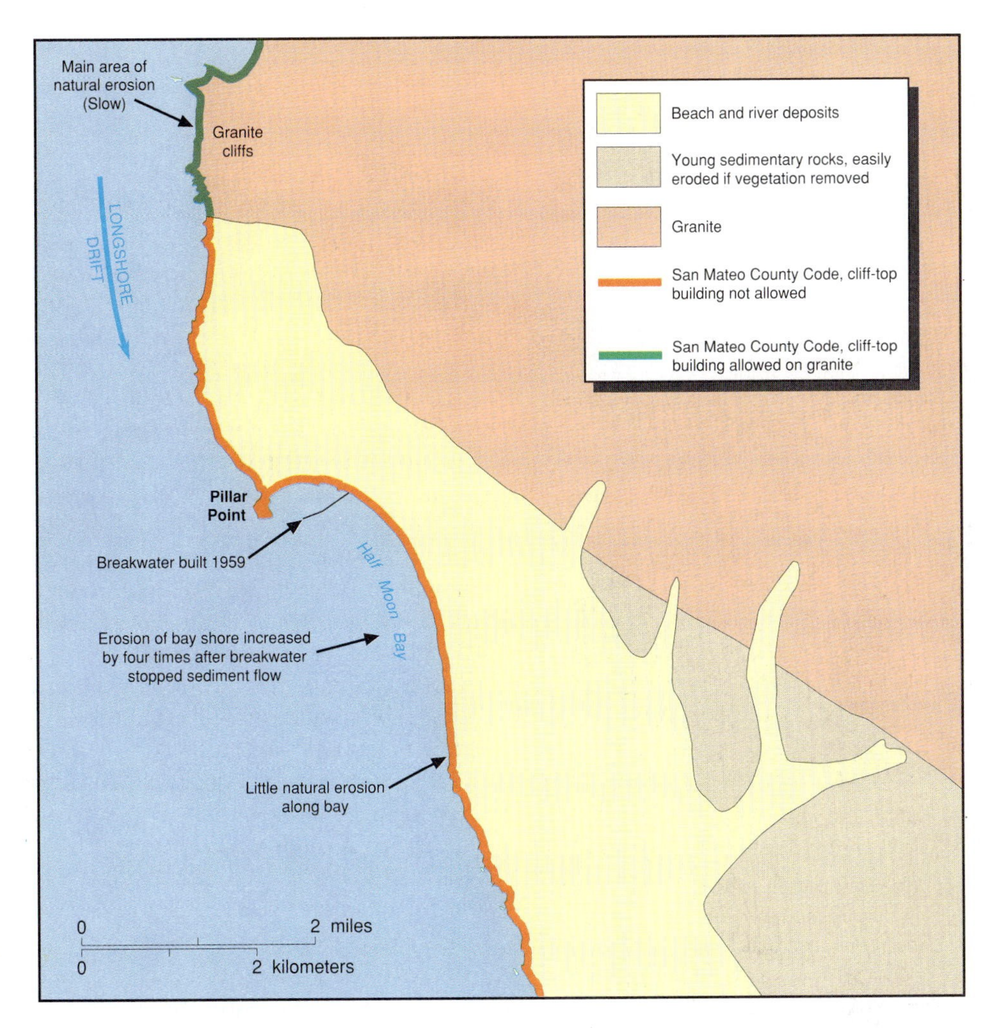

Figure 17-28 Half Moon Bay, San Mateo County, California. The natural situation involved slow erosion of the granite cliffs and deposition in the sheltered bay. Human activities altered this.

Moon Bay caused the most intense marine erosion to shift to a point just to the south, where maximum erosion rates rose to 13 meters per 10 years. New building regulations have been imposed on the eroded section in attempts to prevent further property losses.

Cape Hatteras, North Carolina

Cape Hatteras is formed of mobile sandy deposits that form a microtidal barrier island. Dry sand on top of the bar has been piled into dunes by the wind. In the natural course of events, hurricane storm surges wash over the beach and gouge gaps through to the marshy area inland of the dunes. As people came to use this area for homebuilding and recreation, they attempted to prevent the storm-surge hazard by building local sea walls to protect coastal towns and by stabilizing the dune area with grasses and tree plantations. The walls encouraged erosion along the beach front, since they increased the water depth offshore. At times, the dune line has been breached. The stabilization policy adopted in the past worked against natural processes. Authorities now recognize that overwash should be allowed.

The Mississippi Delta

The history of the Mississippi delta illustrates the constant battle between deposition, erosion, and subsidence, and the problems facing human at-

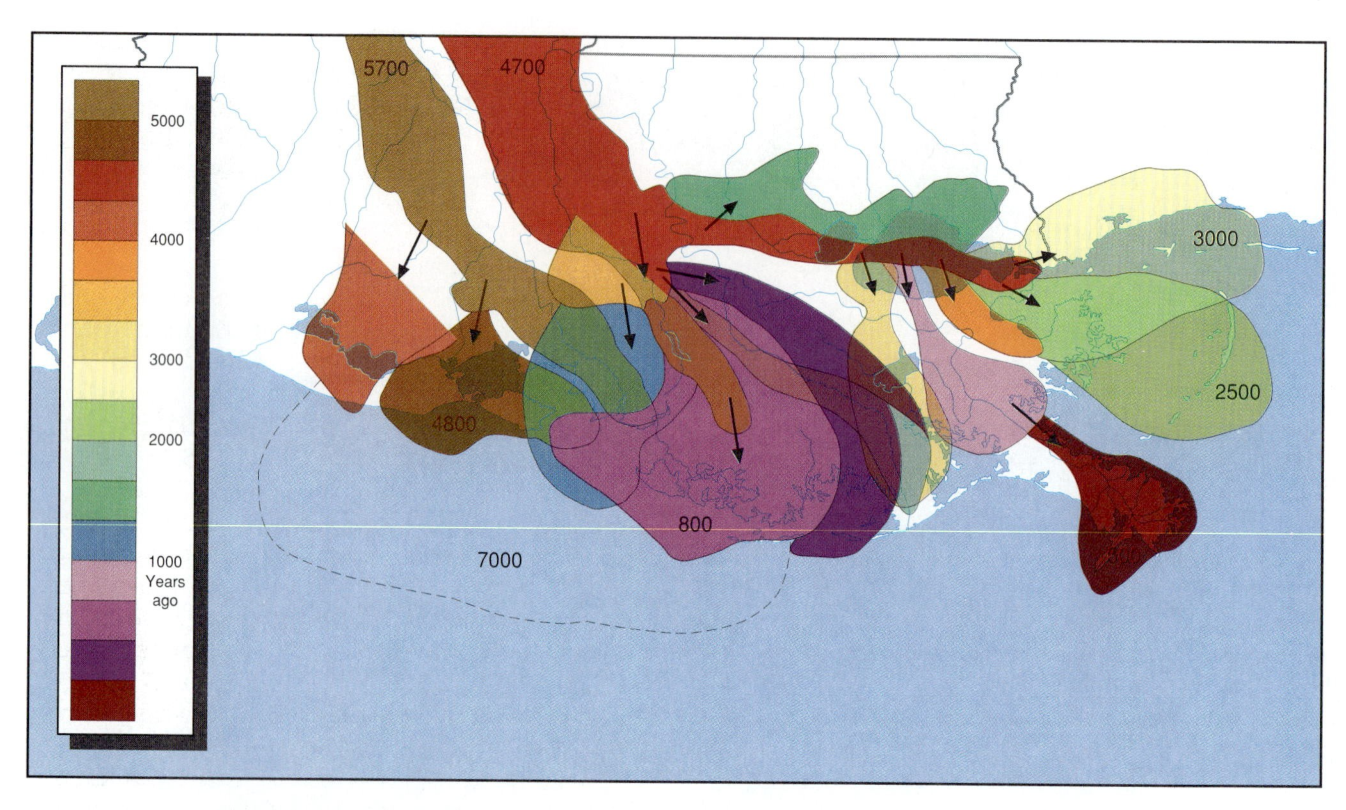

Figure 17-29 The evolution of the Mississippi delta over the last 7000 years. Engineering works preserve delta 16 and cause it to be extended farther south as other sections subside for lack of deposition.

tempts to manage such an environment. The part of the delta above sea level is almost flat and only a few meters at most above high tide. The river crosses its delta in a series of distributaries, which have a history of shifting courses (Figure 17-29). The main part of the delta has been built out into the Gulf of Mexico by distributaries extending their channels (Figure 17-21, *A*). Waves and tides are less influential in molding the delta form and operate mainly on sections of delta where river deposition is no longer occurring.

In the natural state, high flow in the Mississippi caused flooding. In the past, overbank flow and dropping of stream load formed natural levees, which grew higher than the backswamp areas between distributaries. The river sometimes broke through a natural levee to deposit sediment on adjacent wetlands, at times forming a new outlet to the sea.

The first French settlers in the area built small artificial levees to protect their settlements. Since the beginning of the twentieth century, the U.S. Army Corps of Engineers has constructed a series of artificial levees and flood-water outlets to protect New Orleans from river floods and to maintain the course of the river across its present delta. The Atchafalaya River, south of Baton Rouge, provides the shortest and steepest alternative route for the river to the sea, and it is likely that the river would take this route permanently if it once overflowed through it.

The waters of the engineered channel now take most of the sediment into the Gulf of Mexico through the long distributaries of the present delta. Overbank floods no longer occur, and sediment is

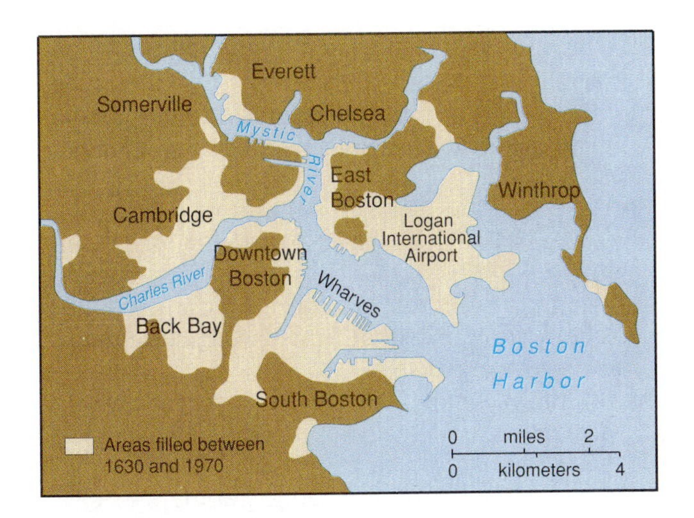

Figure 17-30 Boston Harbor. Reclamation of land has reduced the area of the harbor by half since AD 1800.

no longer deposited on wetlands. The supply of sediment has been restricted further by the building of dams in the upper portions of the Mississippi-Missouri system, since the reservoirs act as sediment traps. During the 1980s, loss of coastal land in Louisiana became a concern because subsidence is greater than deposition on the delta surface.

Reclaiming Tidal Lands

The pressure on coastal lands has led to the reclamation of tidal flats, a process that began on a large scale during medieval times in the Netherlands of northern Europe. Dikes were built around the area to be reclaimed, and then the water was pumped out. The reclaimed land was below sea level, so pumping had to continue to remove rainwater. More recently, land has been reclaimed by filling areas close to sea level with rubble and trash in order to raise it above sea level. Such fills are vulnerable to earthquake shaking (see Box 11-1: Earthquake Hazards).

In the United States, reclamation of tidal lands has been extensive. Most large cities on the Great Lakes have extended their area by building out into the lake. Boston filled in much of its bay (Figure 17-30), creating the Back Bay residential area and Logan International Airport.

The Hackensack Meadows of New Jersey were once a tidal area covering 72 square km (28 square miles) and stretching for 25 km along the Hackensack River. The area was used to dump solid waste, and it smelled from the sewage works. In the 1960s, several local jurisdictions planned its reclamation in a grandiose urban project that involved the building of Meadowlands Stadium where the New York Giants now play.

To build Miami and Miami Beach, miles of mangrove swamp were filled between the outer beach and the mainland. Artificial islands were constructed in this zone. More recently, extensive tracts of single-family homes with waterfront access have been built along the coast to the south. Since the mangroves no longer perform their filtering function, places south of Miami are experiencing problems with stagnant water and sewage disposal.

FRONTIERS IN KNOWLEDGE
Coastal Landforms

An understanding of coastal areas is crucial because such areas face intense pressure from human activity. In many cases, coasts have been developed before the natural systems of the area were understood.

Coastal geomorphology studies have continued to focus on the mechanisms of erosion and deposition and on the formation of particular landforms. Precisely how the sea abrades cliffs and undercuts them and how sediment moves along coasts are topics of particular significance. Engineers continue to search for ways to build coastal structures that will not produce extra deposition or erosion in areas where it is not wanted. Physical geographers also devote much attention to the possible rise of sea level caused by global warming.

SUMMARY

1. The sea plays an important part in producing coastal landforms in interaction with other surface-relief and living-organism environment processes.
2. The action of waves, tides, and currents is among the marine processes that affect coastal landforms.
3. The intensity of wave action varies around the world in relation to the strength and continuity of storm-force winds.
4. Variations in the heights of tides cause marine processes to affect different vertical ranges on coastal landforms.
5. Following marine action, cliffs typically have steep rocky surfaces.
6. Most beach environments are sandy, but some are muddy or pebbly. Beaches change rapidly because their unconsolidated materials are easily moved.
7. Estuarine and deltaic landform environments occur where fresh water and stream load meet the sea.
8. Mangrove and reef landform environments occur in tropical areas and involve structures built by living organisms.
9. Climatic change can cause sea level to vary, leaving coastal features at levels not now affected by marine action.
10. Human impact on coastal landforms is locally increasing rates of marine erosion, delta subsidence, and destruction of mangroves, wetlands, and reefs.

KEY TERMS

coastal zone, p. 450
wave, p. 452
fetch, p. 453
storm wave, p. 453
swell wave, p. 453
seismic sea wave, p. 453
wave refraction, p. 453
breaker zone, p. 454
surf zone, p. 454
swash, p. 454
backwash, p. 454
longshore current, p. 454
tide, p. 456
tidal range, p. 456
tidal current, p. 456
cliff, p. 457
shore platform, p. 457
beach, p. 459
longshore drift, p. 461
barrier beach, p. 461
delta, p. 463
estuary, p. 466
mud flats, p. 466
salt marsh, p. 466
mangrove colony, p. 467
coral reef, p. 467
eustatic sea-level change, p. 469
emergent coast, p. 469
submergent coast, p. 469
fjord, p. 469

QUESTIONS FOR REVIEW AND EXPLORATION

1. To what extent do nonmarine processes mold landforms in the coastal zone?
2. Observe coasts that you visit—lake shores as well as the ocean. Note the landforms present and try to account for their origins. Also notice if human activities are affecting the coastal landforms.
3. How do processes that affect storm-wave coasts, swell-wave coasts, and protected coasts differ? How do the landforms differ?
4. How does tidal range affect the landforms produced along a coast?
5. Why do some rivers produce estuaries, and others deltas?
6. Describe the development of the Mississippi delta over the last 5000 years. How does its history exemplify general features of deltas?
7. Describe the origins of different types of coral reef features.

FURTHER READING

Cobb C: Awash in change: North Carolina's outer banks, *National Geographic,* October 1987. Emphasizes the changeability of the coastal environment.

Ellis W: The Aral: a Soviet sea lies dying, *National Geographic,* February 1990. Highlights the human and environmental tragedies.

Kohl K: Man against the sea: the Oosterschelde Barrier, *National Geographic,* October 1986. Focuses on efforts by the Dutch to keep the sea off their land.

Leatherman S: *Barrier island handbook,* University of Maryland (1988). Provides a detailed study of the landforms that dominate nearly one half of coasts around the United States.

Lee D: Mississippi Delta: the land of the river, *National Geographic,* August 1983. Focuses on the human implications of changes in the physical environment.

Pethick J: *An introduction to coastal geomorphology,* Edward Arnold, 1984. A text that summarizes modern research.

Precoda N: Requiem for the Aral Sea, *Ambio,* May 1991. A detailed account of the environmental changes involved in the shrinking of the Aral Sea.

U.S. Geological Survey: *Geologic history of Cape Cod, Massachusetts,* Government Printing Office (1981). A well-illustrated free booklet that provides a good case study of coastal processes.

Williams SJ, Dodd K, Gohn KK: *Coasts in crisis,* U.S. Geological Survey Circular 1075 (1990). A publication in the free series, *Public Issues in Earth Science.* Focuses on changes in the Great Lakes and ocean shorelines and wetlands of the United States.

Earth's surface is subject to so many interacting forces that understanding the origins of landforms is a complex undertaking. But developing that understanding becomes more important as human beings increasingly influence the operation of natural processes. Physical geographers must try to understand whether human processes work with natural forces or against them. There is also the reverse side of this interaction: natural landforms and processes affect human activities through both catastrophic events and also what appears on a human time scale to be the long-term stability of the land surface.

So far, our examination of the surface-relief environment has concentrated on individual landforms and the specific processes that fashion them. This chapter examines how landforms are produced by interacting forces on the local and regional scales. In attempting to understand these issues, a study is made of both the small-scale individual slope elements and also large-scale groups of landforms occurring over wider geographic areas. The hills, valleys, and surface deposits in a location are not a haphazard collection, but reflect the action of a particular group of processes and a particular history of environmental change.

Figure 18-1 shows a group of landforms that have been created by one such sequence of environmental conditions. Western Scotland is a region of resistant rocks that was part of the mountain chains that were linked to the Appalachians some 250 million years ago (see Chapter 12). Erosion wore down the mountains. As the Atlantic Ocean opened, Scotland was close to the eastern edge and its western area was uplifted by the crustal movements. Precipitation on the high land increased and it became the center of one of the Pleistocene ice sheets. The deep U-shaped valley was eroded by glaciers and then filled with water as the ice sheets melted and formed a fjord. Surface runoff from steep slopes produced the delta in the center of the photo as stream load was deposited in deep water. Waves have created beach forms around the edges of the delta. The humid midlatitude climate has enabled trees to cover the area, slowing the rate of stream erosion and deposition. Human activities have assisted stabilizing influences, with the most recent development of recreational facilities.

THE COMPLEXITY OF LANDFORM ORIGINS

To understand the evolution of groups of landforms, one must understand the interactions among the processes that create them in particular surface-relief environments. No surface-relief environment is subject to only one sort of process. Chapter 17 emphasized that coasts are distinctive in being molded

Figure 18-1 A landscape in western Scotland that is the product of a complex set of interacting forces.

by marine forces, but that they are also influenced by internal earth activity and other surface processes—weathering, mass movements, running water, ice, and wind. In each of the world's environments, a unique combination of external surface processes acts on the rocks—the raw material produced by internal earth activity.

One approach to this complexity is to make linkages between surface-relief environments and the processes at work. Table 18-1 illustrates how the groups of processes studied in Chapters 13 through 17 operate together in four types of surface-relief environment. For example, in humid areas streams are the most obvious agents that produce landforms, but weathering and the slower forms of mass movements provide a large proportion of the stream load, while human activities often influence these processes. Such an approach takes a broad view that contrasts with the detailed analysis of flow in a stream channel or the workings of a particular weathering process. Local exceptions may be found for most broad generalizations. The broad approach, however, is useful in attempting to see the "wood" as well as the "trees." At the broader level, generalizations and speculative ideas must be checked against the observations made at the detailed level. If our knowledge were more complete, we would be able to use a broad understanding of landscape evolution as a basis for identifying crucial local and specific studies. As yet, however, broader systems of large-scale landform intepretation are some way from being able to provide such a framework for detailed landform or process studies.

Another approach is to begin with the detailed study of individual slopes in the landscape, or of the workings of small watersheds, and to measure and analyze the processes acting on them. Such investigations make it possible to deal with questions raised by specific tests that are difficult to apply in the broader context. They also have the potential for building up an understanding of the overall picture by piecing together local studies. It has proved difficult, however, to extend the results of studies at smaller geographic scales to interpretations at the larger scale. At present the two approaches are difficult to reconcile. This chapter approaches the issue by considering one example of each type of approach.

Table 18-1 Landform Environments and Processes

PROCESSES	HUMID TROPICAL OR MIDLATITUDE	ARID TROPICAL OR MIDLATITUDE	GLACIAL, PERIGLACIAL HIGH LATITUDE OR HIGH ALTITUDE	COASTAL
WEATHERING Chapter 13	Chemical	Physical	Physical	Important
MASS MOVEMENT Chapter 13	Slow forms	Rapid forms	Rapid forms	Important
RUNNING WATER Chapter 14	Important	Occasional	Meltwater	Some
SURFACE ICE Chapter 15			Important	
GROUND ICE Chapter 15			Important in periglacial	Polar ocean margins
WIND Chapter 16	(Some)	Important	Ice margins, periglacial	Arid regions, beach dunes
OCEAN Chapter 17				Important
HUMAN BEINGS	Important	Margins	Not very important	Important

LANDFORM SLOPES

Each landform is bounded by surface **slopes** of varied sizes, angles, and orientations. They are the basic landform unit of landscapes and reflect the interactions between the slope materials and surface processes acting on them. Slopes are local landforms, the study of which may lead to broader conclusions. A hillside can be mapped to show the arrangement of slopes (Figure 18-2) and it may be possible to link the processes at work on the slopes to a pattern of evolution for the hillside.

Slope Materials and Processes

If a slope is perceived as a location where the ground materials are in a constant battle with the processes acting on them, the crucial factors in the battle are the strength of the materials and the forces available to break down the materials. Stable slopes occur where the strength of the materials is greater than the effect of the processes. Unstable slopes occur where the reverse is true and there is failure of the slope. Some slopes are initially stable, but are liable to become unstable rapidly if their strength is reduced or the processes become more active.

Rocks and regolith are the two materials composing slopes. Rock is hard and coherent, but often crossed by joints and other fractures that are potential lines of weakness. Rocks are not greatly weakened by saturation with water. Regolith (including soil) is weak because it is formed of unconsolidated particles and is further weakened when saturated with water.

The processes that act on slope rock and regolith are varied and have been studied in Chapters 13 to 17. Weathering weakens rock resistance by break-

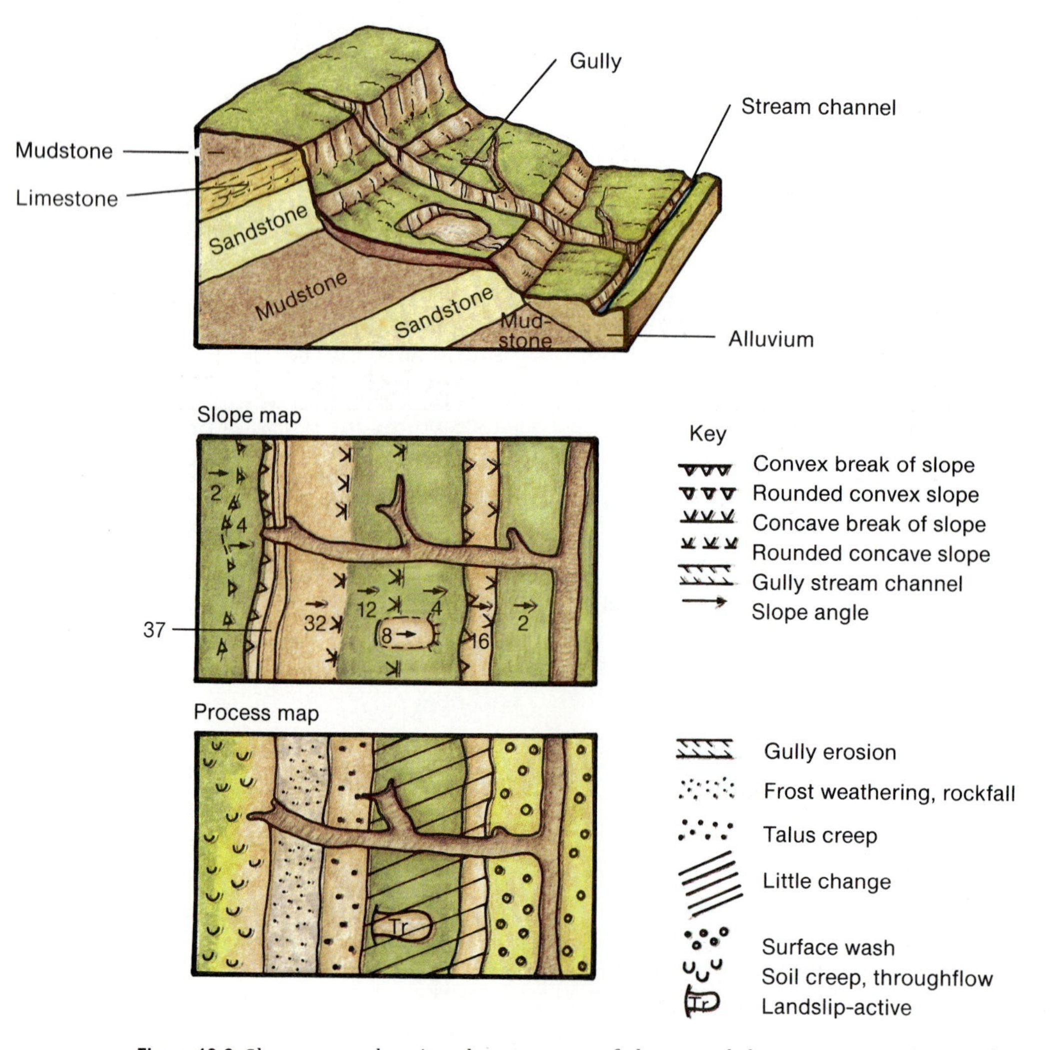

Figure 18-2 Slope maps showing the geometry of slopes and the processes acting on them.

ing the rock into particles and by dissolving minerals, and also reduces the size of regolith particles. Mass movements transport rock and regolith down slopes. Water movement moves regolith by the impacts of raindrops dislodging surface particles, by overland flow in sheetwash, rills, and gullies, and by throughflow. Surface water flow moves particles up to sand size; throughflow moves solutes out of the soil. Other processes, such as surface iceflow, ground ice, wind, and the sea, also affect the formation of slopes.

The rates at which slope processes work vary according to the slope angle, slope material, and climate (Figure 18-3). Faster rates occur on steeper slopes, in clay and loose rock materials, and where freeze-thaw fluctuations are frequent.

Slope Forms

Slopes are essentially three-dimensional forms. A simple profile of a hillside shows three basic units—a *convex element* (getting steeper downslope) at the top, a *straight segment* in the middle, and a *concave element* (getting less steep downslope) at the base (Figure 18-4, *A*). Several of these units can often be identified on individual hillsides. The three-dimensional form can be described in terms of these units (Figure 18-4, *B*) and is an important consideration in tracing routes taken by water, sediment, and solutes.

A straight segment with rock at the surface is known as a *free face*, and its slope will be controlled by the rock strength. Free-face slopes are most common in glacial, arid, or coastal environments, and

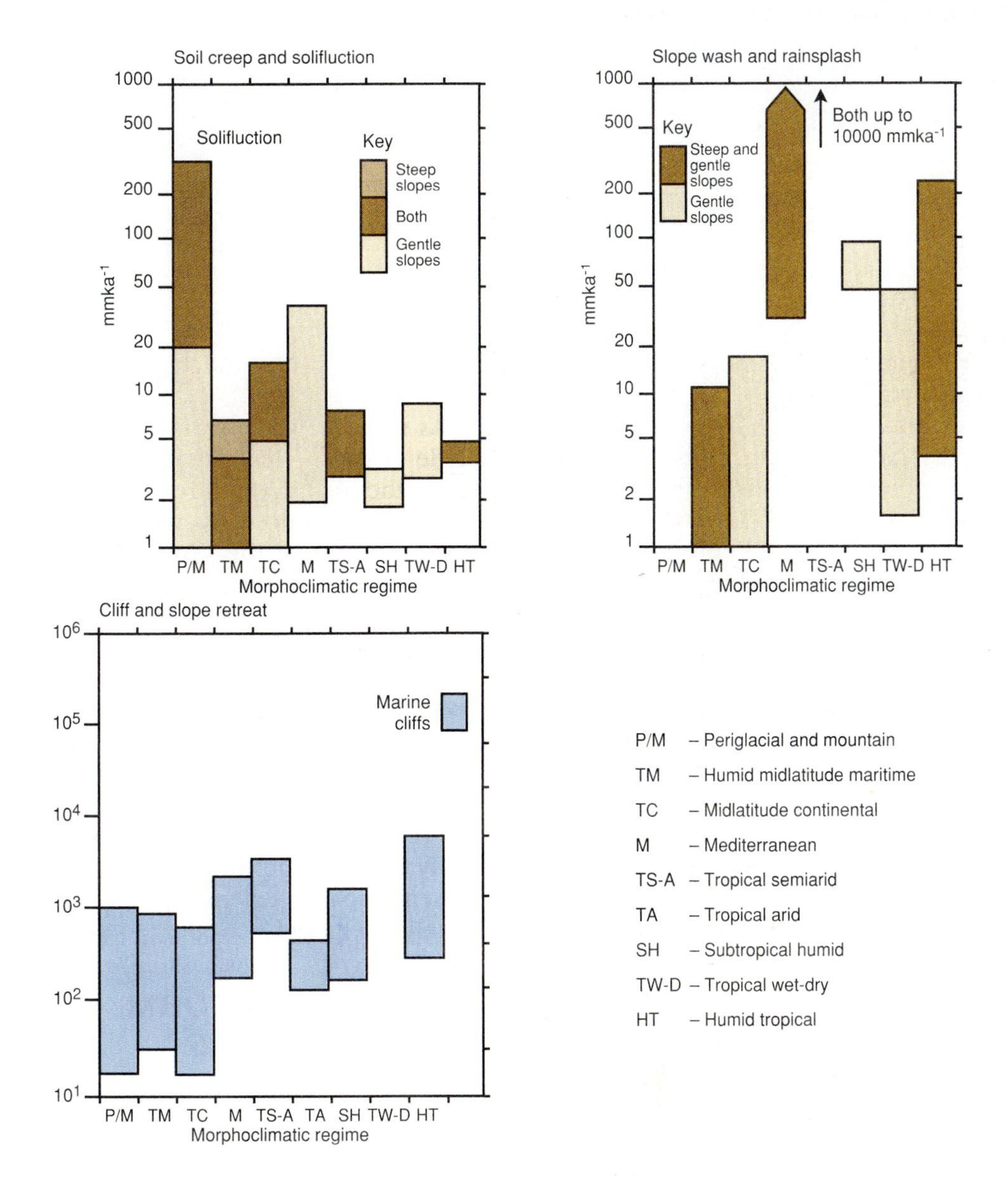

Figure 18-3 Rates of slope processes measured in millimeters per thousand years.

processes acting on them. In Chapter 12 the continental areas were divided on the basis of geologic processes into young fold mountains, fault-block mountains, broad plains, plateaus, shields, and continental margins. The study of the surface-relief environment, however, demonstrates the significance of climatic factors in the generation of distinctive assemblages of landforms. An analysis of world landforms on a global scale must take both geologic and climatic factors into account.

Early in the twentieth century, William Morris Davis devised an approach to understanding landform evolution based on the threefold themes: geologic ("structure"), climatic ("process"), and change over time ("stage"). The ideas of Davis dominated geomorphology until the 1950s, when there was a shift to an emphasis on how surface processes created landforms. In the late 1960s came the better understanding of internal Earth processes.

A more recent scheme has been proposed by the German geomorphologist, Julius Büdel. Although the subject of debate among geomorphologists, it provides an approach to modeling reality at the broadest scale that is worth consideration. It features several advances over the Davisian scheme. Büdel's analysis of landforms is based on differences among groups of landforms produced in varied climatic environments, and how they are caused by the processes at work. Each study is based on field observation, and allows for geologic influences and changing climate. Davis paid little attention to the measurement of process, to internal processes, or to the significance of climatic change.

Büdel divided the world into major landform regions (Figure 18-6). Not content with merely relating these regions to current climatic conditions, he also distinguished between regions on the basis of the rates at which the surface relief is being molded. In this section, we consider what he regarded as regions of most rapid change and least rapid change. Although Büdel recognized the importance of solid-earth forces, he argued that surface-relief processes have, for example, lowered the Alps to half their greatest height and that the present surface is thus essentially erosional. In taking this position, he did not sufficiently emphasize that the most rapid erosion occurs in high mountains in any of the climatic zones he recognized. For example, erosion rates are often greater in mountains of arid or humid midlatitude environments than in the lowlands of his regions of rapid change.

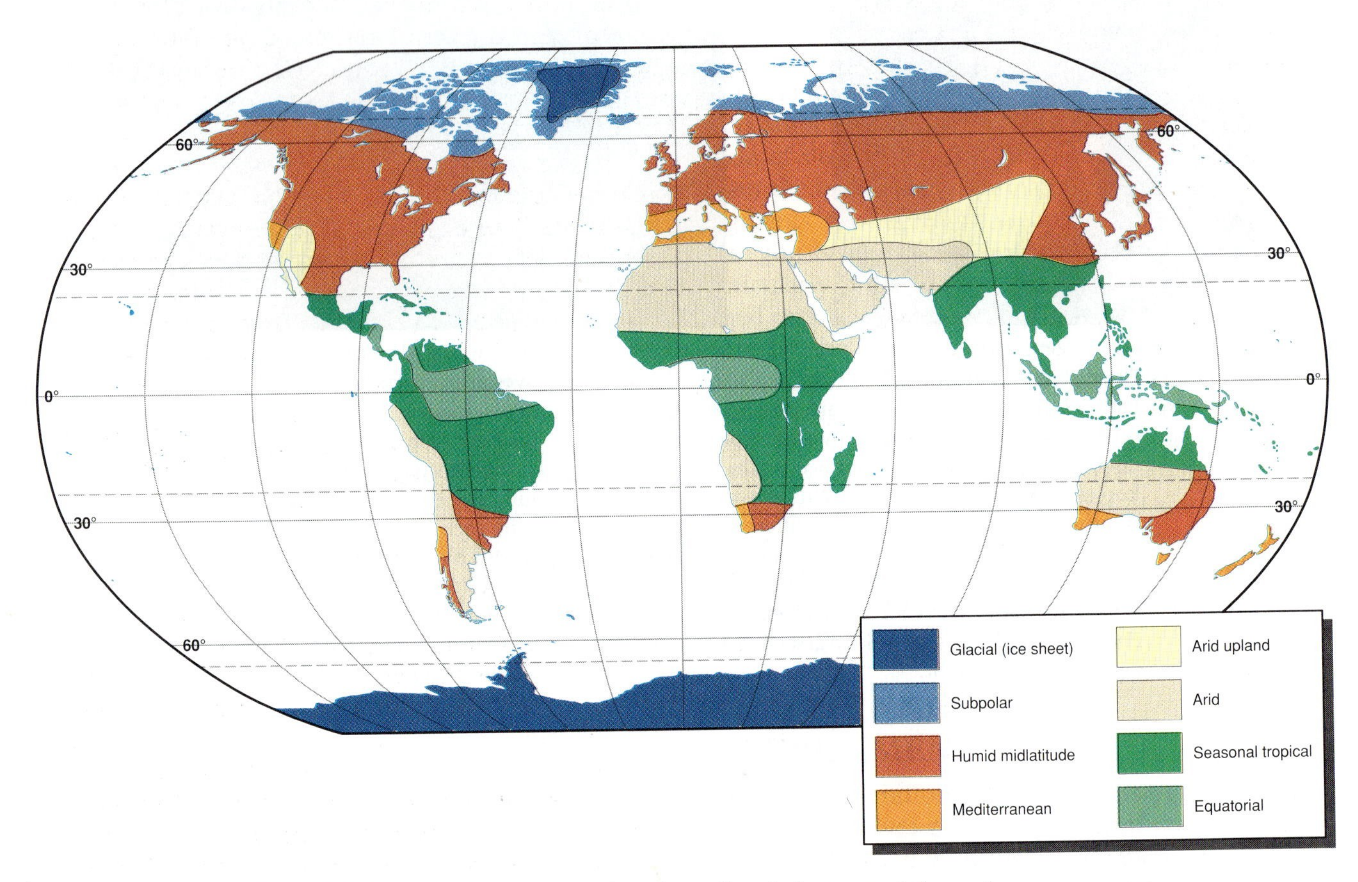

Figure 18-6 World landform regions based on climatic factors and the surface processes at work.

The Most Active Surface-Relief Environments

In the active periglacial zone and the seasonal tropical zone, geomorphic processes act rapidly. They lower the land surface in different ways.

The *active periglacial* zone occurs in polar regions where there is considerable precipitation but no permanent ice cover. It is affected by frost weathering and mass movements, which provide a continuous supply of rock fragments that can be transported away. Enhanced streamflow results from snowmelt, and in regions where there are heavy falls of snow, valleys are often cut rapidly. Studies on some of the hilly islands to the north of Canada, and especially on parts of Spitzbergen (see Chapter 14), have demonstrated the rapid change in the landscape of periglacial areas caused by stream activity fed by meltwater. Some high-latitude land areas not covered by ice are not subject to rapid change, because they do not receive heavy precipitation. Wind action can also be effective in moving quantities of finer weathered particles from outwash plains and braided river beds and depositing them as loess.

In active periglacial areas, rock is broken into fragments by weathering through exposure at the surface and through changes in ice volume in the upper part of the permafrost zone. Sections through hillsides and beneath valley floors show rock interleaved by ice well below the active layer. When rivers erode their channel floors, there is plenty of fragmented rock available for transport. Hillslope mass movements, such as solifluction, combine with stream action on the steeper hill slopes in spring and summer to move a constant supply of debris to the slope foot. Streams are frozen all winter, but heavy flows following snowmelt during the spring and fall cause downstream transport of gravel-sized load, corrasion of the stream bed, and excavation of the ice-broken rock beneath. High rates of downcutting up to several meters per 1000 years occur.

The *seasonal tropical climatic zone* is typified by widespread lowering of the surface, as opposed to the concentration of erosion along valleys. Seasons in the tropics are determined by rainfall rather than temperature fluctuations. The equatorial climatic environment receives rain throughout the year, but zones to north and south are characterized by high-sun rains and low-sun drought. Tropical South America, Africa, Australia, and southern India have extensive zones that follow this seasonal pattern.

Where the high-sun rains last for 6 to 8 months, water remains in the weathered layer. The high tropical temperatures combine with the continuous availability of water to favor chemical activity, which is enhanced by high levels of organic activity. In some of the soils, as much as 25 percent of the volume consists of insects that secrete acids. Most areas have a weathered layer of regolith several meters deep consisting of highly weathered clays. Red iron oxides are common, a sign of the effectiveness of chemical activity. Solid rock is rapidly weathered to fine regolith, as shown by the sharp transition from unweathered rock to regolith.

The dry season has an important influence, although it is too short (less than 3 to 4 months) to stop weathering processes. During the dry period, the vegetation dies back and the soil surface dries out and cracks, reducing the cohesion of the particles. At the beginning of the wet season, sheetwash and streamflow rapidly remove the dried particles from the ground. It appears that the rate of surface removal roughly equals the rate of weathering of solid rock at the base of the regolith, since the regolith layer retains its thickness as the whole landscape is lowered.

The landforms of seasonal tropical areas are extensive plains or plateaus covered by deep regolith (Figure 18-7). Bare rock is exposed on individual hills, known as *inselbergs,* and on steep slopes between plateau levels. These slopes are too steep to develop or retain soil. Their slopes are maintained by undercutting at the base by chemical weathering below soil level and rockfalls from above.

The active periglacial and seasonal tropical environments give rise to contrasting assemblages of landforms (see Figure 14-26). However, certain common features result in rapid surface changes in both types of area. First, both environments have intensive weathering that produces as much rock debris as mass movements and stream processes can remove. Second, stream erosion is enhanced by high flows from snowmelt in active periglacial regions, and by the early wet-season runoff over dry ground in the seasonal tropics.

The Least Active Surface-Relief Environments

At the other end of the spectrum, in some surface-relief environments there is less rapid change at the present time because weathering and erosion are slower. Three environments—ice-covered, tropical arid, and humid midlatitude—are in this category.

The central areas of Antarctica and Greenland are covered by *ice sheets* that are mainly frozen to the underlying rock or have slow internal flow. In places where it is so thin that it does not flow, the

Figure 18-7 A seasonal tropical landscape in Masvingo Province, Zimbabwe. The bare hills rise sharply from the regolith-covered plain.

ice acts as a protective cover and preserves delicate preglacial landforms and regolith (see Chapter 15). Weathering is inactive beneath thick ice. Where rocks are exposed, weathering is slow because there are few alternations of freezing and thawing in the subzero temperature conditions. Toward the margins of ice sheets, and of the ice caps in Iceland, warmer atmospheric conditions and the production of meltwater give rise to more rapid ice flow and occasional catastrophic floods.

In *tropical arid regions* such as much of the Sahara, surface relief is currently changing very slowly. Weathering rates are slow, wind action is ineffective against resistant rock, and the rainfall sparse, so many of these landscapes are being preserved rather than altered. Rainfall and the consequent landform change by stream processes are concentrated in the mountainous areas of such deserts. The movement of sand is concentrated by the wind in large ergs, but although their surface form changes they do not lower the landscape. The main process that lowers tropical arid region surfaces is the deflation of clay and silt particles and their removal outside the arid region. Desert conditions appear to be expanding into semiarid areas outside true deserts at present, helped by human activities that remove vegetation cover or reactivate old dunes on margins.

Arid zone landforms experience more rapid change when they are adjacent to mountains in middle latitudes. These mountains attract winter precipitation that produces high flows of water during the early summer. This is common in the southwestern United States, in Algeria (Figure 18-8), and in Iran. Such features as alluvial fans and temporary lakes often occur in areas of internal drainage.

It may be surprising that *humid midlatitude regions* are another zone of slow landform change. Humid midlatitude regions currently experience little valley cutting. They typically have surface streams that flow continuously, but these streams produce little erosion. Rainfall seldom reaches the high levels of intensity in individual storms so common in the tropics. Furthermore, in the natural state the land is covered by vegetation, which intercepts precipitation and reduces the peaks of waterflow through the watershed. High streamflows that result in major geomorphic changes are unusual, so the work carried out by streams is modest and slow. Most

Figure 18-8 An arid landscape in southwestern Algeria.

streams transport only solutes, clay, or sand, which do not allow them to accomplish much erosion.

Weathering and mass movements are also slow in these regions. The solid rocks are covered by regolith almost everywhere, and although it may be only 1 meter deep and kept moist, the temperatures remain moderate to low, and so regolith protects against very active weathering except for the action of acidic water on limestone. In higher areas, temperatures are low enough to encourage frost action, but on lower elevations they are not high enough to produce rapid chemical change. Slow weathering produces small amounts of weathered debris, which is soon removed by flood flows.

As with other regions where landform change is naturally slow, parts of humid midlatitude regions experience higher rates of erosion. Higher rates of erosion occur in mountainous areas, where seasonal wet-dry alternations are marked, and where surface materials are of low resistance. For example, the Mediterranean lands of southern Europe are mainly hilly areas where the long, dry summers alternate with wet, stormy winters. When humans removed vegetation to extend farming, these regions eroded rapidly, creating rocky slopes, deep valleys, and extending deltas at the river mouths. The soft loess deposits of central China have been among the most rapidly eroding parts of the world since farming methods were intensified, and the Yellow River is filled with suspended sediment eroded from their surface.

Of increasing significance in humid midlatitudes are intensive human activities, which may reduce or enhance weathering, mass movements, and stream processes. Engineering works are built to provide stable channels for navigation and to protect against flooding. The extraction of water for farming and domestic and industrial uses also reduces flows in streams. Human activities can also enhance natural processes. Examples include soil erosion and the reactivation of mass movements, often caused by the removal of surface vegetation cover; dams built across the upper Mississippi-Missouri tributaries and the main rivers reduce the amount of sediment reaching the mouth so that the sinking rate of the delta now exceeds the deposition rate; and sinkholes in Florida that are often reactivated by removing groundwater for human uses.

Climatic Change and Relict Features

The present major landform environments each contain a number of landforms that cannot be explained by the surface processes acting there today. Many of the unexplained features can be accounted for by the advances of ice sheets and shifts of climatic belts during the Pleistocene glacial phases. Figure 18-9 shows how the major landform environments fluctuated in area during this period. For 200 million years the globe was dominated by tropical environments, but it began to cool following the Eocene epoch. During the Pleistocene, ice sheets advanced and retreated, causing all the climatic zones to shift and narrow toward the equator.

Relict landforms are those produced in one climatic environment that remain behind after the environment changes and are not removed by the new set of processes. Figure 18-10 depicts two such changes. The first *(A)* depicts the shifts at an ice sheet margin as the climate warms and leads to melting and retreating. It leaves behind deposits and erosional landforms. The area once covered by ice is first subject to periglacial conditions that cause solifluction and deepen stream valleys, and then to less erosionally active humid midlatitude conditions. The second *(B)* shows how the desertification of a seasonal tropical area leaves behind inselbergs, but separates the regolith into fine clay (blown right away), sand (forming dunes), and boulders (a lag deposit).

Virtually all parts of the continents outside of the ice sheets on Antarctica and Greenland contain relict landforms. Other areas that have not been affected by climatic change only occur in narrow zones close to the equator where tropical rainforest has been maintained. The landforms produced under one regime remain for some time as the next is established, and influence the processes in the new climatic regime.

The speed at which the new environment removes traces of relict features varies. Active periglacial and seasonal tropical zones contain few relict features, since the processes there remove them quickly. Arid and humid midlatitude zones, on the other hand, are often dominated by relict landforms, since processes in these environments work too slowly to have produced major changes since the last ice recession some 10,000 years ago.

TIME SCALES AND THE MANNER OF CHANGE

Landforms change over time because of the action of internal Earth processes and of the atmosphere-ocean forces at work on them. Different processes cause changes at different rates.

In the time scale of tens to hundreds of millions of years, the major events are the formation of mountain systems by the crushing and uplift of sections of Earth's crust. Superimposed on these events involving the solid-earth environment are the effects of surface processes and the influences of changing climates. Climatic change acts over longer and shorter time scales. There have been periods of a hundred million years or so when climates changed little and conditions around the globe were relatively uniform. The uniform, low-activity conditions have been interrupted by periods of 20 to 50 million years, during which an ice age occurs, causing major fluctuations of climate and rapid changes in the surface-relief environment. The long-term sequence of climatic changes may be influenced by the movements of continents and major tectonic processes (see Chapter 12).

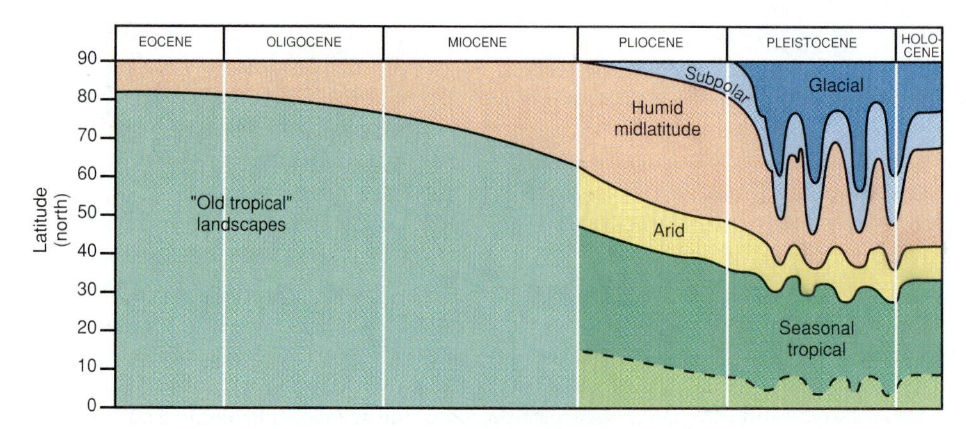

Figure 18-9 The changing positions of the world climatic zones. The cooler climates expanded after the Eocene, and there were very rapid fluctuations during the Pleistocene.

Shorter climatic fluctuations, such as the advance and retreat of ice sheets within an ice age, occur over time scales of 100,000 years or so. Those in the Pleistocene had major impacts on the present landscapes. They not only caused the extension of glacial conditions in phases lasting 100,000 years, but the fluctuations produced many of the relict landforms found in present landscapes.

Over periods of a few decades, landforms change little. For example, a stream channel form is adjusted to the average flow of water through it and any change takes place within that context. Erosion at one point on a stream channel is compensated for by deposition elsewhere. If the processes at work do not change, the relationship between form (stream channel) and process (streamflow) remains in balance, or **steady-state equilibrium.** When new conditions begin to affect the processes at work, the landforms will respond. The overall equilibrium shifts to comply with the new conditions. For example, gradual channel-widening will occur if flow is increased, or if the stream is supplied by more load. A stream flowing past a mine erodes rock fragments from spoil heaps, and this additional load causes its channel below the mine to become wider and more braided than before. Where streams have been dammed and water extracted, their channels often respond by getting narrower.

It might be assumed that the changes described above occur gradually and smoothly, but it appears that stresses resulting from changes more commonly build up until a particular **threshold,** or level of activity, is passed, after which there is sudden change. Thus, the Cimarron River (see Chapter 14) maintained its narrow channel with vegetated banks until a major flood in 1914; after this, the channel changed rapidly.

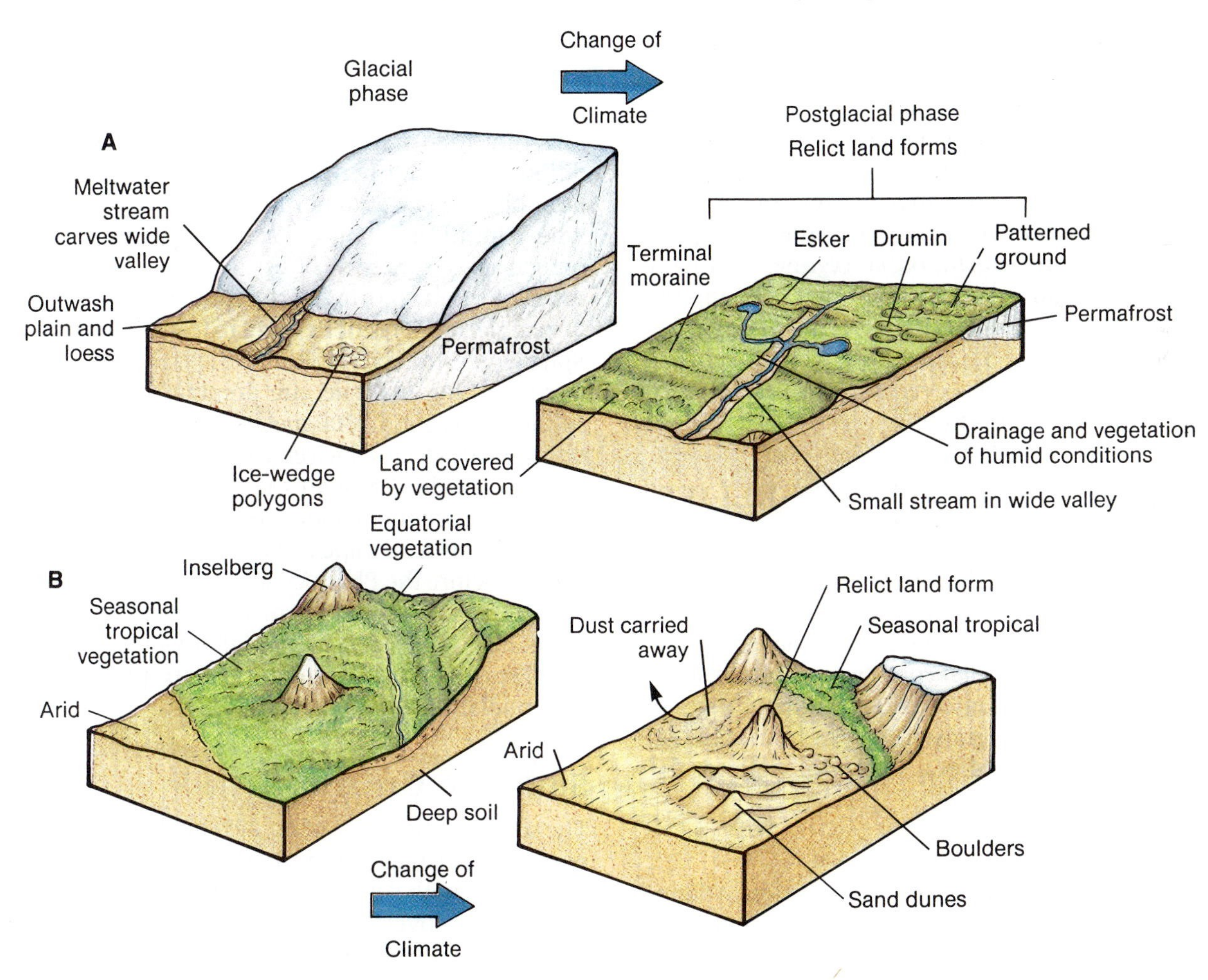

Figure 18-10 Relict landforms are produced by changing climates. Landforms produced under one set of climatic conditions may be out of harmony with another. In **(A)** the changes resulted from the melting of an ice sheet and the retreat of the permafrost. In **(B)** aridity increased and inselbergs formed in a seasonal tropical environment became relicts in an arid area.

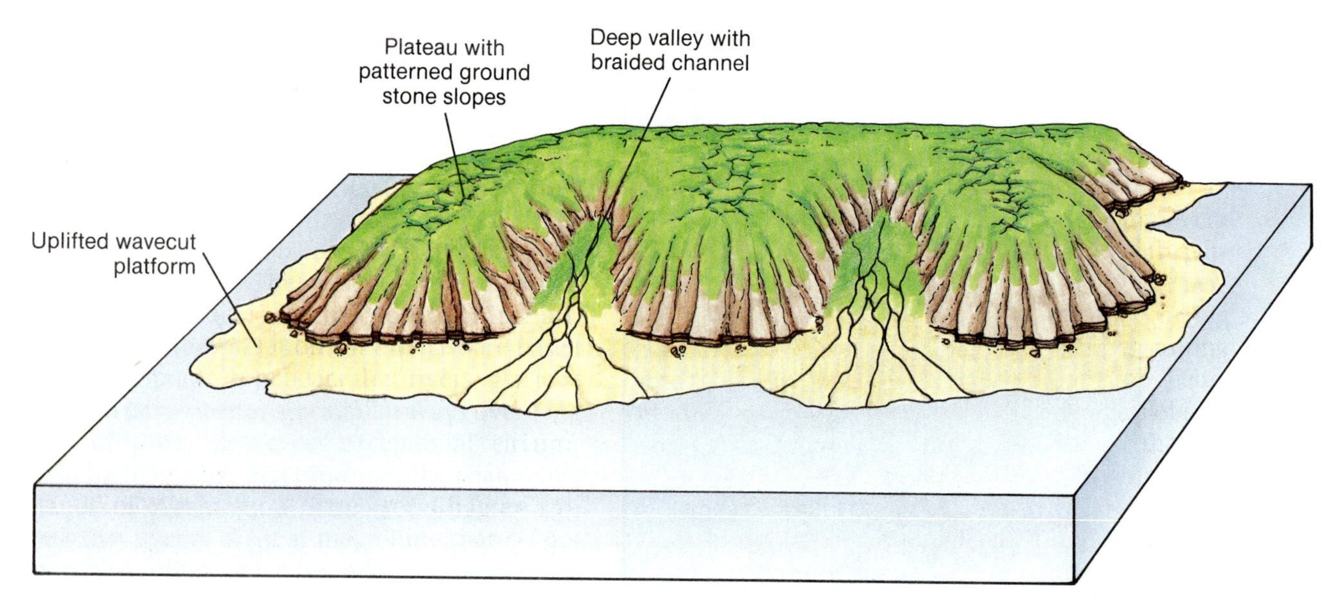

Figure 18-12 Some of the typical features of Spitzbergen that are forming at the present time.

vealed by the glacial removal of overlying soil in the northern section of this escarpment, and on the limestones bordering the southern margin of the Shield.

The entire region was covered by ice sheets during the Pleistocene glaciations. As they advanced, they scraped soils off upstanding ridges and deposited till on the ground beneath the ice. Drumlin hills and moraines were common depositional landforms from this phase.

As the ice melted and retreated across the area 15,000 to 12,000 years ago, the rock fragments it carried were left behind. Meltwater ran off the surface, redistributing some of the debris. Huge lakes of meltwater, connected by wide channels, formed in the depressions left by the melted ice. In many of these channels, the direction of flow changed as isostatic uplift of the land followed the melting of the ice sheets. The meltwater lake levels fluctuated, depositing clay and silt on top of the glacial deposits when the lakes were higher than at present (Figure 18-15).

Postglacial streams have narrow meandering channels and low discharges for much of the year, but spring meltwater flow in southern Ontario is sufficient to carry out channel erosion. Much land remains poorly drained, especially between drumlins and on the margins of meltwater channels. The resultant landscape is a mixture of ice-eroded bare rock, glacial deposits of both active and disintegrating ice, and meltwater deposits. The present stream channels are narrow, but flow in very wide valleys that were clearly not created by present flows.

Human activities have had local impacts. Large stone quarries have been excavated in the Niagara Escarpment, and gravel pits in the outwash deposits. The retreat of Niagara Falls has slowed by an order of magnitude as the flow of water has been reduced by the construction of diversion canals to hydroelectric facilities.

This environment resulted from the advance of ice sheets and the postglacial reestablishment of humid processes on thick glacial deposits. Similar situations are found in other parts of North America, such as the Upper Midwest, and northern Europe, such as northern Poland.

Southeastern United States

The southeastern United States has experienced fewer climatic fluctuations than the landform environments already described. During the Pleistocene, some higher parts of the southern Appalachians had a colder climate in which frost action was significant, but the region was not covered by ice sheets. The region is an example of one that has continued to be covered with natural forest vegetation and has been subject to almost uninterrupted chemical weathering and stream processes over long periods.

The region has a varied relief. The younger rocks

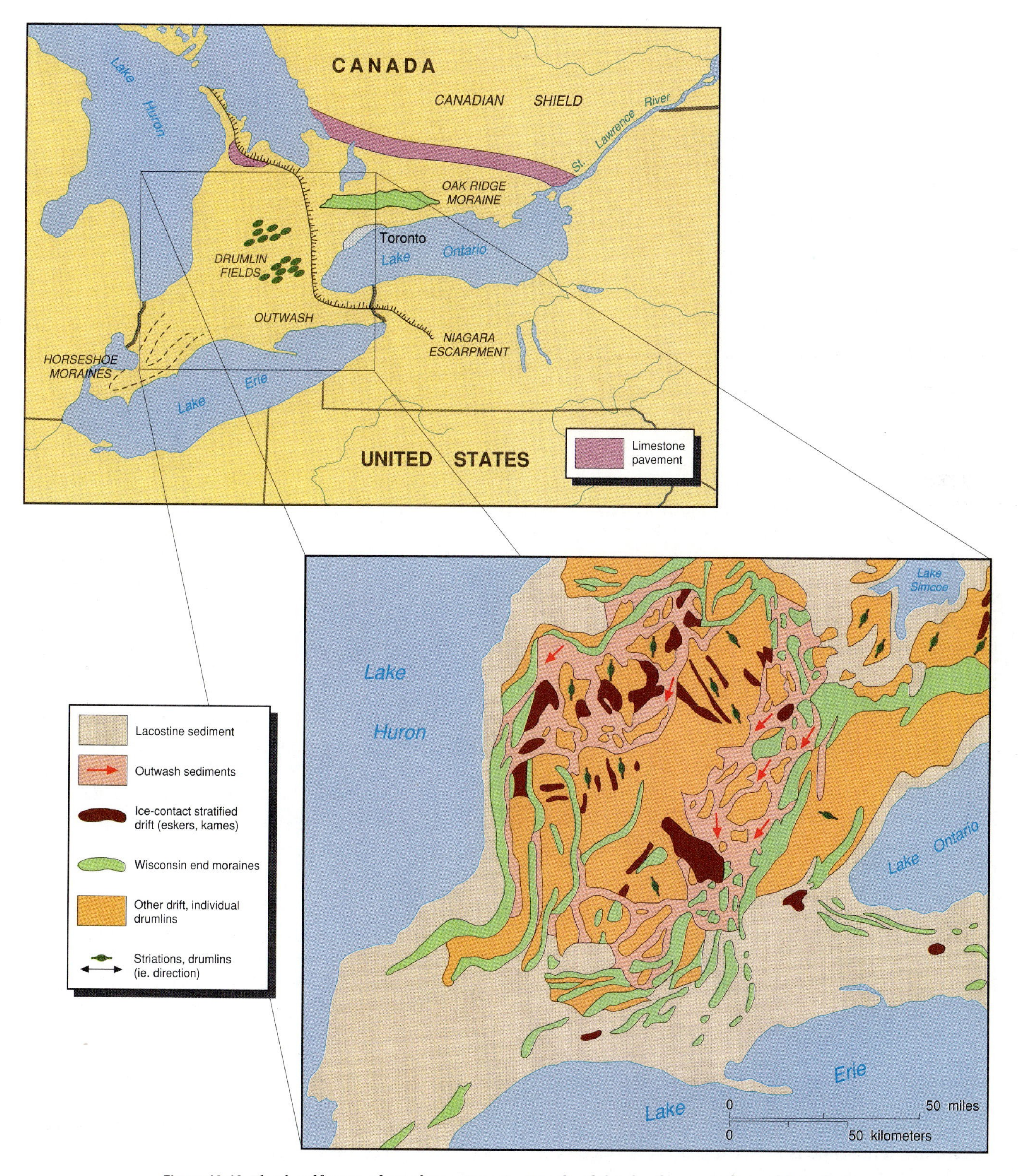

Figure 18-13 The landforms of southern Ontario. Much of this landscape is formed by relict glacial deposits.

Figure 18-14 Glacial and stream landforms in southern Ontario.

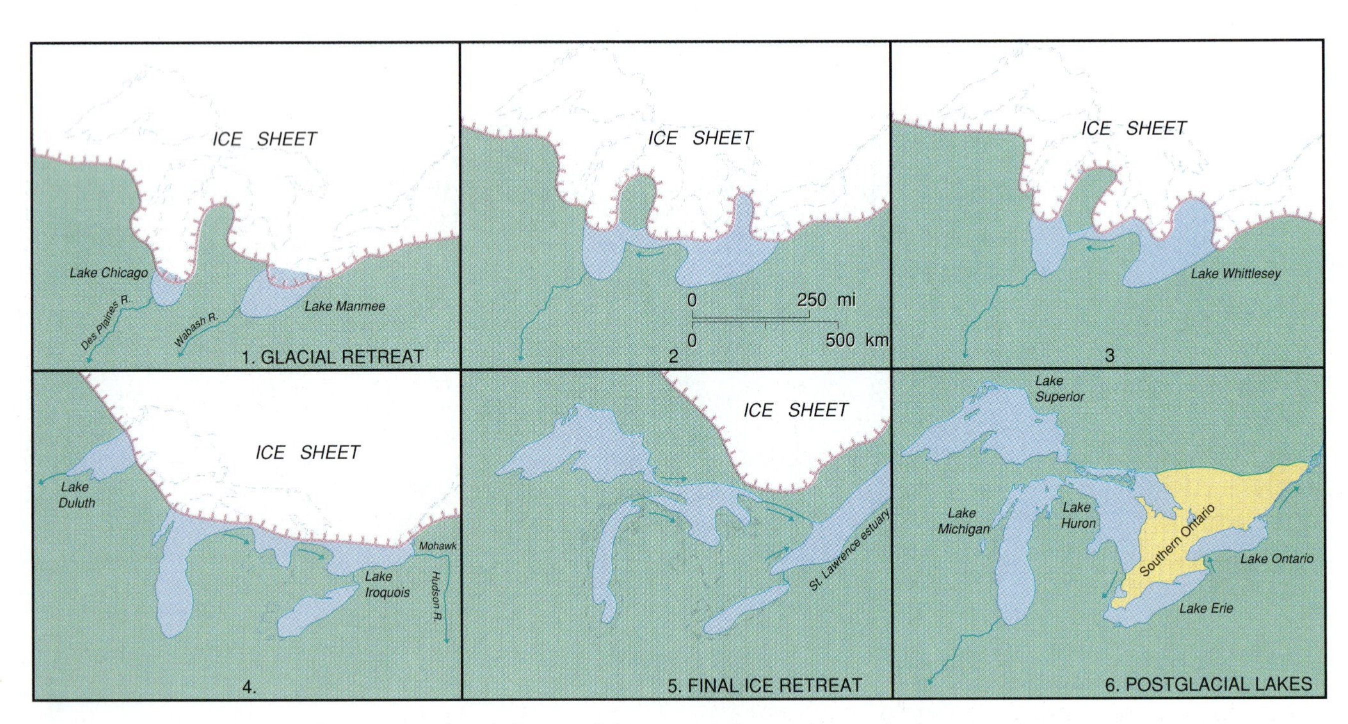

Figure 18-15 The evolution of the Great Lakes as the ice sheet retreated.

of the coastal plain flank the older rocks of the southern Appalachians. The younger rocks are tilted toward the Atlantic Ocean and form a series of parallel ridges and eroded valleys with trellis stream patterns (Figure 18-16). Southern Appalachia is composed of the piedmont—a plateau eroded across resistant ancient rocks—and the Blue Ridge, also formed of ancient rocks, which rises to over 2000 meters (6000 feet). Watersheds in the higher regions are separated by sharp divides, since the streams cause erosion to the upper limits of the slopes.

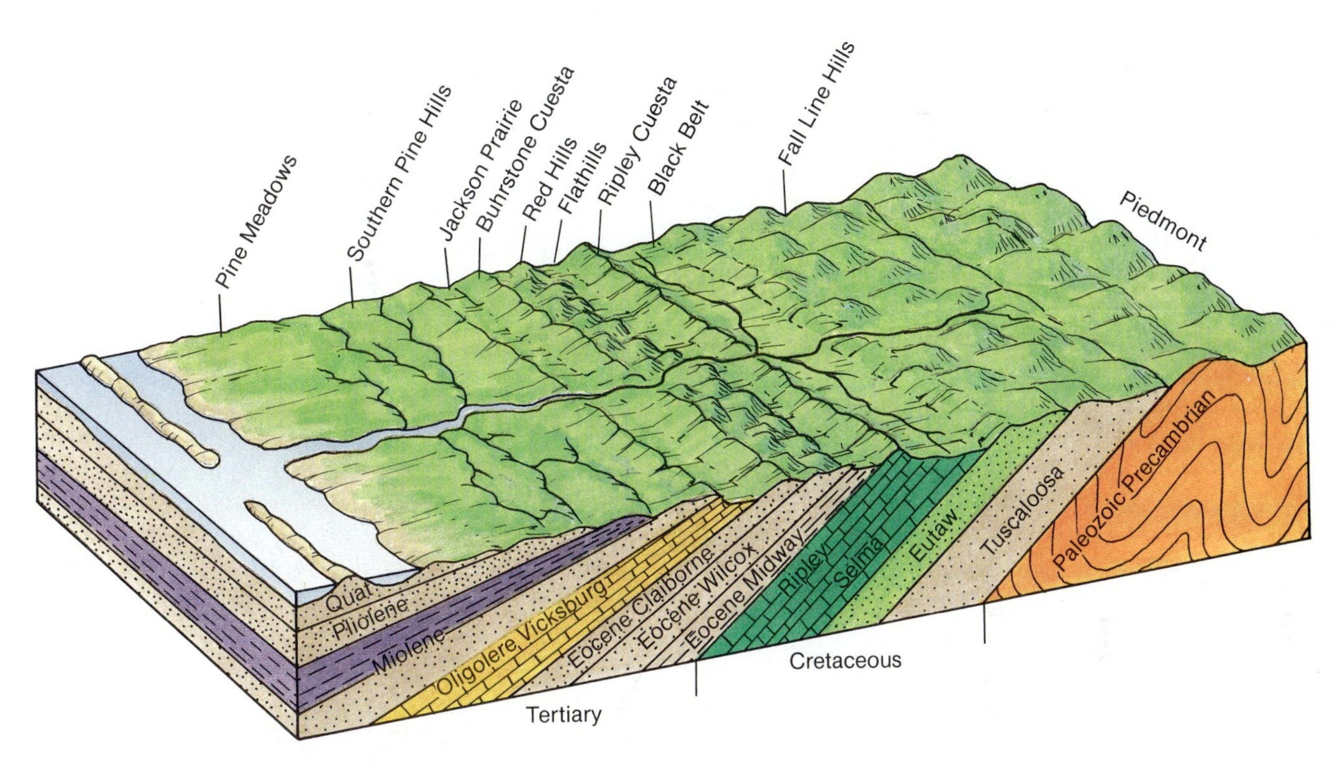

Figure 18-16 Southeastern United States: rock structure and relief. The tilted rocks of the coastal plain form ridges on the more resistant sandstones and limestones and low-lying areas on the clay rocks.

The flatter slopes of moderate height, such as the piedmont and uppermost parts of the coastal plain escarpments, are the site of some of the most deeply weathered rocks in North America. The soils are red and yellow, typical of those produced under abundant rainfall and the influence of intense chemical weathering. Water flowing through them has removed the soluble matter, and they are part of a weathered zone that extends to depths of several tens of meters. Most of this deep weathered layer is a relict feature that may have been formed during warmer climatic conditions before the Pleistocene. The surface of the piedmont may thus be a former seasonal tropical landform environment. Such deep soils farther north in North America became the source for much glacial till.

An important effect of the glacial period was the lowering of the Atlantic Ocean surface as global sea-level fell. The streams cut deeper valleys across the coastal plain to reach the lower glacial sea level, and their tributaries eroded the softer clay belts between the ridge-forming sandstones and limestones. The postglacial rise in sea level drowned the lower parts of the coasts, producing extensive deep estuaries; subsequent coastal processes have created offshore barrier beaches and islands.

From around 1750, human activity initially increased rates of erosion. Natural vegetation was removed, and row crops such as cotton were planted, exposing the surface. The amount of sediment moved along the streams to the Atlantic declined beginning in the early part of the twentieth century, when much former cotton-growing land was allowed to return to woodland, and many reservoirs that store water and trap sediment were constructed (Figure 18-17).

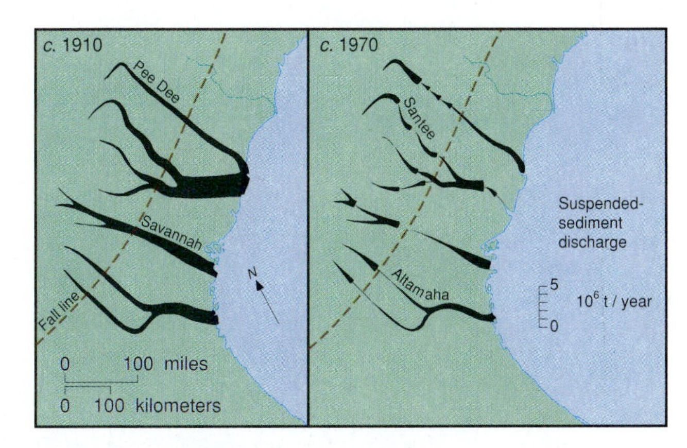

Figure 18-17 The amount of sediment moving in the streams of southeastern United States reflects soil erosion (1910) and reduced cotton farming and increased dam construction (1970).

The Sahara, Northern Africa

The Sahara is a desert region larger than the United States. Much of it has been arid for a long time, although the degree of aridity increased considerably some 5000 years ago. The Sahara contains a great variety of landforms (Figure 18-18), with extensive rocky areas (hammada), sand seas (ergs), and gravel plains (reg).

The central part of the Sahara is the most arid. Rainfall is highest on the upland areas such as the Tibesti and Ahaggar massifs, but it is extremely irregular. Surface runoff lasts only for an hour or so before the water flow evaporates or sinks into surface deposits. The landforms in solid rock are currently changing very little. There are relatively consistent winter rains on the northern margins and rather irregular summer rains on the southern margins.

The marginal areas show evidence of climatic change and past surface-relief processes in shrunken lakes, vegetation-covered dunes, and ancient drainage lines. The Sahara is probably a desertified version of a formerly more humid landscape. The plains and mountain massifs resemble the wide plains, plateaus, and inselbergs of tropical seasonal environments. Desiccation ended the chemical weathering that formed these landforms and they thus became relict landforms. The regolith dried out and became subject to wind action. Most of the regolith had been chemically weathered to clay and dried out to fine dust, which continues to be blown from the desert area. Convergent windflows concentrated the coarser sand into sand seas. This major desertification occurred during the shift in climatic zones that led to the Pleistocene. It has been resumed since the last glacial retreat.

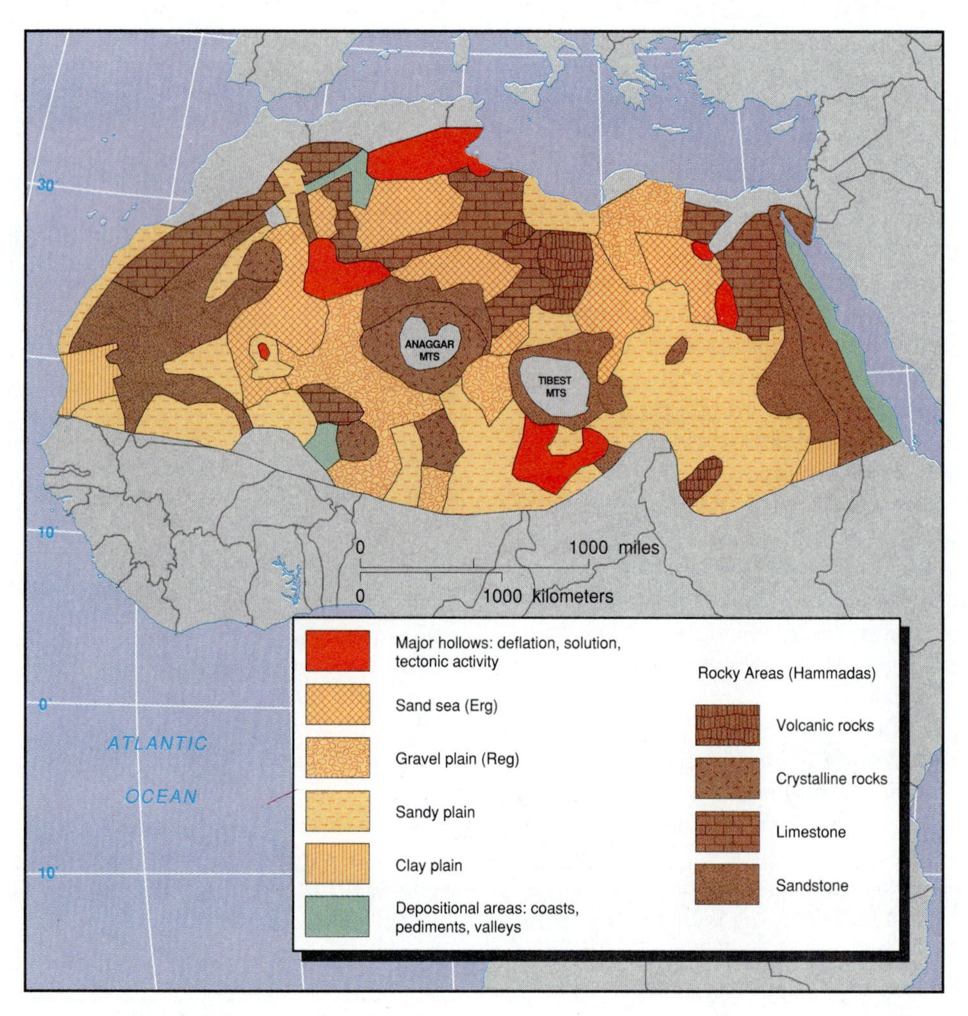

Figure 18-18 The main types of landform in the Sahara, northern Africa.

Some of the recent desertification has resulted from human activities, such as growing grain on the Mediterranean margins during Roman times and modern attempts to extend commercial peanut cultivation in the semiarid southern margins of the Sahara. These activities have caused overpopulation, overgrazing, and overuse of groundwater.

The Amazon Basin, Brazil

The Amazon River is the world's greatest river in terms of the amount of water it delivers to the ocean. The water comes from heavy rainfall at all seasons over most of its huge drainage basin (Figure 18-19). Additionally, the headwaters that rise in the Andes mountains bring early summer snowmelt, which swells these tributaries with both water and sediment. The tributaries that carry a lot of sediment in suspension are called "whitewater" streams. Other tributaries of the Amazon rise on the ancient crystalline rocks of the Guiana and Brazilian highlands. These bring smaller flow inputs spread across a longer period and a load dominated by solutes in clear streams. A third group of tributaries rises in the wide flood plains, where the soils are highly organic. Such streams are known as "blackwater" streams. The total flow is thus composed of water from different sources.

The landforms of the Amazon Basin can be divided most simply into those of the surrounding uplands and those of the broad expanses of flood plain. The uplands are covered by deeply weathered soils, protected by dense forest. There are extensive relict plains, plateaus, and inselbergs. The flood plains are dominated by the wide, deep stream channels; the main channel is nearly 2 km wide for its last 1500 km (1100 miles). Alluvial deposition takes place on the flood plain surface from annual floods. Stream erosion is retarded along the clearwater and blackwater streams but more active along the whitewater streams, where channel shifts are common following snowmelt. Chemical weathering is active throughout the basin. However, the protective forest cover slows the wearing down of the landscape.

Relict features are indicated by some areas of angular gravel, which are foreign to an environment with such intense chemical weathering and the plains and inselbergs typical of seasonal tropical environments. They may be the result of drier conditions during recent glacial phases. Further evidence of disruptions during the Pleistocene has been identified in the anastomosing stream pattern on the Brazil-Venezuela border. It has been suggested that arid conditions led to the disruption of a former stream network, and the new one has not fully established itself on the former arid landscape.

The Amazon was little modified by human activities until the last 50 years. Since then timber cutting has caused large-scale soil erosion and stream incision and added to the sediment load and downstream deposition. The building of hydroelectricity stations is altering the flow of some of the Amazon tributaries. At the same time, gold mining and other forms of mineral extraction are adding unusual solute and suspension loads to the streams.

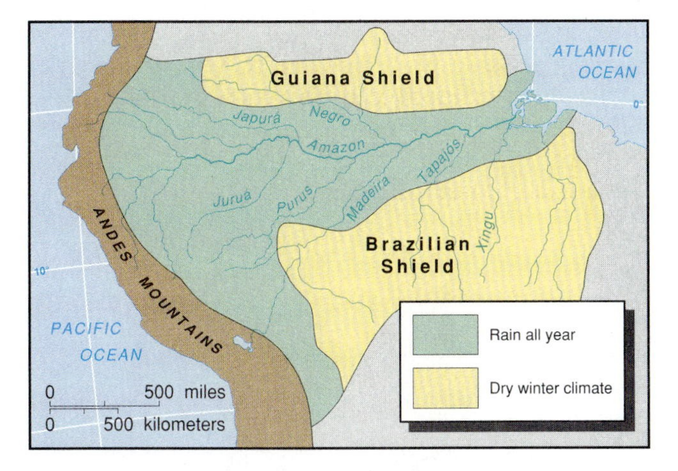

Figure 18-19 The Amazon basin, South America. The tributaries draining from the Andes bring large amounts of meltwater and sediment. Those from the Guiana and Brazilian shield areas bring solute load. Those rising in the flood plain carry fine particles and organic matter.

New Zealand

New Zealand is an example of a landscape where most of the features have been caused by internal earth forces acting rapidly. It has been estimated that if the present mountains were reduced to sea level, new mountains would rise to the same height within half a million years—the same length of time as that over which the most intense Pleistocene glacial processes have operated.

New Zealand is built on a fragment of Gondwanaland that broke away from Australia some 80 million years ago. Most of the rocks that make up the two islands were formed following erosion of this continental fragment and were folded into mountain ranges. Over the last 50 million years, New Zealand has been located on the convergent margin between the Pacific and Indian plates. Subduction zones extend from Tonga to central New

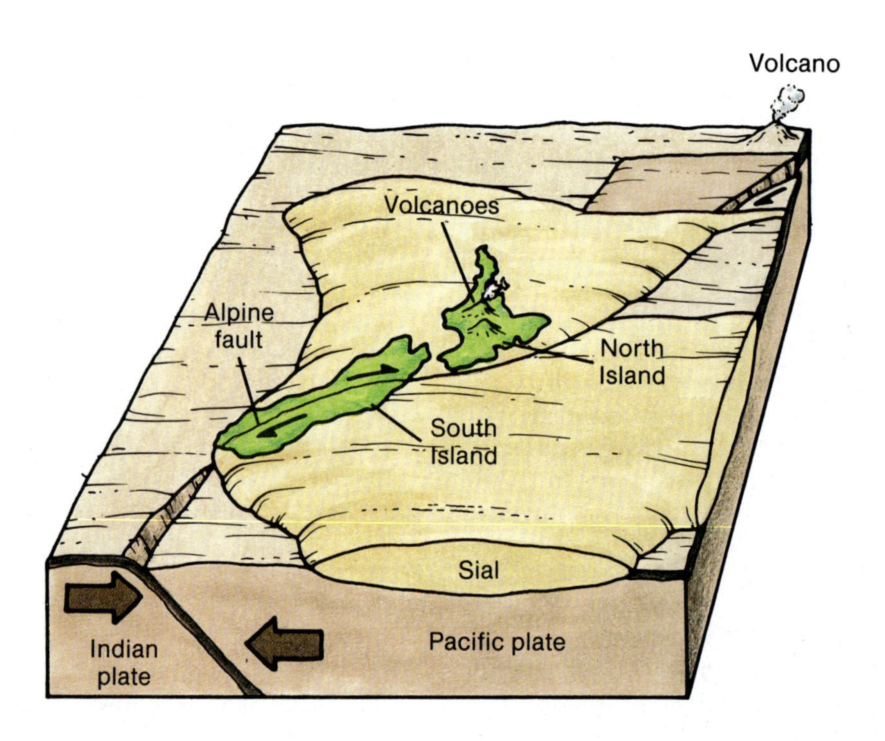

Figure 18-20 The major landform features of New Zealand, emphasizing the importance of solid-earth processes in causing rapid uplift. New Zealand is formed on a continental (sial) fragment that split from Australia and Antarctica. The two islands lie on a line where the Indian and Pacific plates converge. The Alpine fault cuts through South Island and its frequent movements cause earthquakes and uplift.

Zealand (where the Pacific plate is subducted), and from southwest New Zealand to Antarctica (where the Indian plate is subducted). In between there is a transform fault (Figure 18-20). The plate activity produces volcanic activity on North Island and continuous uplift of the western part of South Island. The Southern Alps are still actively rising: annual uplift averages 20 mm near Mount Cook.

While the tectonic processes have been causing uplift, surface-relief processes have also been active. The Southern Alps have been heavily glaciated (Figure 18-21), producing U-shaped valleys and fjords draining to the west coast. West of the mountains, raised beaches testify to the uplift, and east of them there are old river plains and lake floors that have been raised to form tilted plateaus, now deeply incised by rivers.

New Zealand was inhabited by very few people until the twentieth century, and the landforms show few signs of human influence. Local examples of soil erosion, mining excavations, and valley-floor sedimentation occur.

Figure 18-21 A fjord and mountains on South Island, New Zealand.

FRONTIERS IN KNOWLEDGE
The Evolution of Landforms

This chapter discusses approaches to the evolution of landscapes at different geographic scales: it seeks to identify general patterns. This aspect of landform studies has been somewhat neglected in the last 20 to 30 years, with researchers having focused on the operation of specific processes. One of the major challenges facing geomorphologists is to apply their increasing understanding of processes and of changes at small geographic scales to interpretations of regional and global landforms.

SUMMARY

1. The study of landform origins must take into account interactions between internal and external processes, variations in surface processes within climatic environments, and changes over time.
2. Slopes are basic units of landform surfaces and reflect the interactions between external processes and the slope materials.
3. At present, landform change outside mountains is most rapid in two zones—the active periglacial regions, and the seasonal tropics.
4. The least active surface-relief environments are the glacial, arid, and humid midlatitude.
5. Some regions that were formerly less active, such as the humid midlatitudes and the margins of arid areas, are changing more rapidly owing to human activities.
6. Climatic changes alter surface-relief processes. Relict landforms remain when one climatic environment replaces another.
7. Climatic changes produce both gradual and sudden changes in the landforms of a region.

KEY TERMS

slope, p. 482
relict landform, p. 490
steady-state equilibrium, p. 491
threshold, p. 491
uniformitarianism, p. 492
catastrophism, p. 492

QUESTIONS FOR REVIEW AND EXPLORATION

1. Construct a summary chart showing the interactions of the forces that produce landforms.
2. Select a local area you are familiar with, or examine the photos in Figure 18-5, and make a study of the slopes on a hillside. Suggest how you would design a program of field measurements to test a hypothesis about the form of the slope (e.g., the relationship between the slope angle and the slope materials, or between the slope angle and the position on the hillside).
3. List some of the evidence that might indicate that an area had been subject to several landform environments over time.
4. How do the events of the Mount St. Helens eruption (Box 1-2) relate to the discussion of uniformitarianism and catastrophism?

FURTHER READING

Bird JB: *Natural landscapes of Canada*, Wiley (1980). Detailed discussion of the evolution of regional landforms in Canada.

Büdel J: *Climatic geomorphology*, Princeton University Press (1982). One analysis and explanation of landform origins on a global scale.

Parsons AJ: *Hillslope form*, Routledge (1988). A study of the factors controlling slope forms.

Selby MJ: *Earth's changing surface*, Oxford University Press (1985). A geomorphology text containing several chapters that cover the matters discussed in this chapter.

Summerfield MA: *Global geomorphology*, Longman/Wiley (1991). Chapter 14, Climate, Climatic Change and Landform Development, and Chapter 18, Long-term Landscape Development, expand on the ideas covered in this chapter and cover additional aspects of this realm of geomorphology.

THE LIVING-ORGANISM ENVIRONMENT

Tallgrass Prairie
Plants in midsummer
White Co. IND

Mark Wright/Photo
Researchers, Inc.

Ecosystem Structure and Process

19

CHAPTER

X had marked time in the limestone ledge since the Palaeozoic seas covered the land. Time, to an atom locked in a rock, does not pass.

The break came when a bur-oak root nosed down a crack and began prying and sucking. In the flush of a century the rock had decayed, and X was pulled out and up into the world of living things. He helped build a flower, which became an acorn, which fattened a deer, which fed an Indian, all in a single year.

From his berth in the Indian's bone, X joined again in chase and flight, feast and famine, hope and fear. He felt these things as the little chemical pushes and pulls that tug timelessly at every atom. When the Indian took his leave of the prairie, X moldered briefly underground, only to embark on a second trip through the bloodstreams of the land.

This time it was a rootlet of bluestem that sucked him up and lodged him in a leaf that rode the green billows of the prairie June, sharing the common task of hoarding sunlight. To this leaf also fell an uncommon task: flicking shadows across a plover's eggs. The ec-

static plover, hovering overhead, poured praises on something perfect: perhaps the eggs, perhaps the shadow, or perhaps the haze of pink phlox that lay on the prairie.
When the departing plovers set wing for the Argentine, all the bluestems waved farewell with tall new tassels. When the first geese came out of the north and all the bluestems glowed wine-red, a forehanded deermouse cut the leaf in which X lay, and buried it in an underground nest, as if to hide a bit of Indian summer from the thieving frosts. But a fox detained the mouse, molds and fungi took the nest apart, and X lay in the soil again, foot-loose and fancy free.

Aldo Leopold, "A Sand County Almanac," 1949

The living-organism environment, also called the **biosphere,** supports all living things on Earth. This environment is created by the interacting margins of the atmosphere-ocean, solid-earth, and surface-relief environments. The combination of these host environments is unique in the solar system; no other planet is thought to support a living-organism environment. Physically, the biosphere is a thin layer that extends underground (as soil) and into the atmosphere and oceans. It is a precariously small part of the total Earth environment, but is home to an astounding diversity of plants, animals, and microscopic organisms. Humankind is a small but important part of this living environment and depends closely on its other elements for survival. Understanding the processes that operate in the living environment and their linkages with the atmosphere-ocean, solid-earth, and surface-relief environments can help humans monitor and control their impact on the whole Earth system.

This chapter first explains the basic structure and processes of the living-organism environment. The environment is organized as a series of *ecosystems,* through which energy and materials move. The flows of energy and materials are vital to the survival of humans in the ecosystem and also link the living-organism environment to the other Earth environments. The next sections examine in more detail how organisms interact with their environment and with each other to control the distribution of plants and animals in an ecosystem. The nature of these interactions changes over time so that ecosystems grow and develop. The final sections show how the current global distribution of plants and animals depends on the process of evolution and on the movements of continents and oceans over millions of years.

STRUCTURE AND PROCESS IN THE BIOSPHERE

Living things in the biosphere intercept the flows of energy and materials from one host environment to another. For example, to build body tissue green plants use solar energy, atmospheric gases, and chemical elements weathered from rock. The soil intercepts and holds water, acting as a storage reservoir in the hydrologic cycle and providing water to living things. The opening excerpt shows how materials weathered from rock can become part of a living organism. Organisms in turn release energy and materials back to the host environment through processes such as transpiration (Chapter 6). The movement of energy and materials through the biosphere controls how plants and animals are arranged in the ecosystem.

The Ecosystem, Ecology, and Biogeography

The study of spatial pattern and process in the biosphere is an important subdivision of physical geography called biogeography. Biogeographers examine the structure of the living-organism environment, its processes, its changes over time and space, and its dependence on and linkages with the physical conditions of the other Earth environments. Biogeography is closely linked to the scientific discipline of ecology, which is the study of organisms and their relationships with their living and non-living environment. The main difference between these two disciplines is the geographer's emphasis on spatial pattern and geographic distribution. The **ecosystem** is the central concept of both ecology and biogeography and has been defined precisely by Eugene P. Odum, an eminent American ecologist, as "any unit that includes all of the *organisms* in a given area interacting with the *physical environment* so that a *flow of energy* leads to *exchange of materials* between living and non-living parts within the ecosystem." The components of an ecosystem can be grouped into two parts (Figure 19-1). First, the organic part of the system is made up of living plants, animals, and microorganisms, their waste products, and the decaying remains of organisms that were once alive. Second, the organisms live in an inorganic or physical environment characterized by particular conditions of rock, soil, relief, and climate. Ecosystem structure refers to the type, amount, and distribution of each component in the ecosystem. The tropical rainforest ecosystem, for example, contains different types of plants and

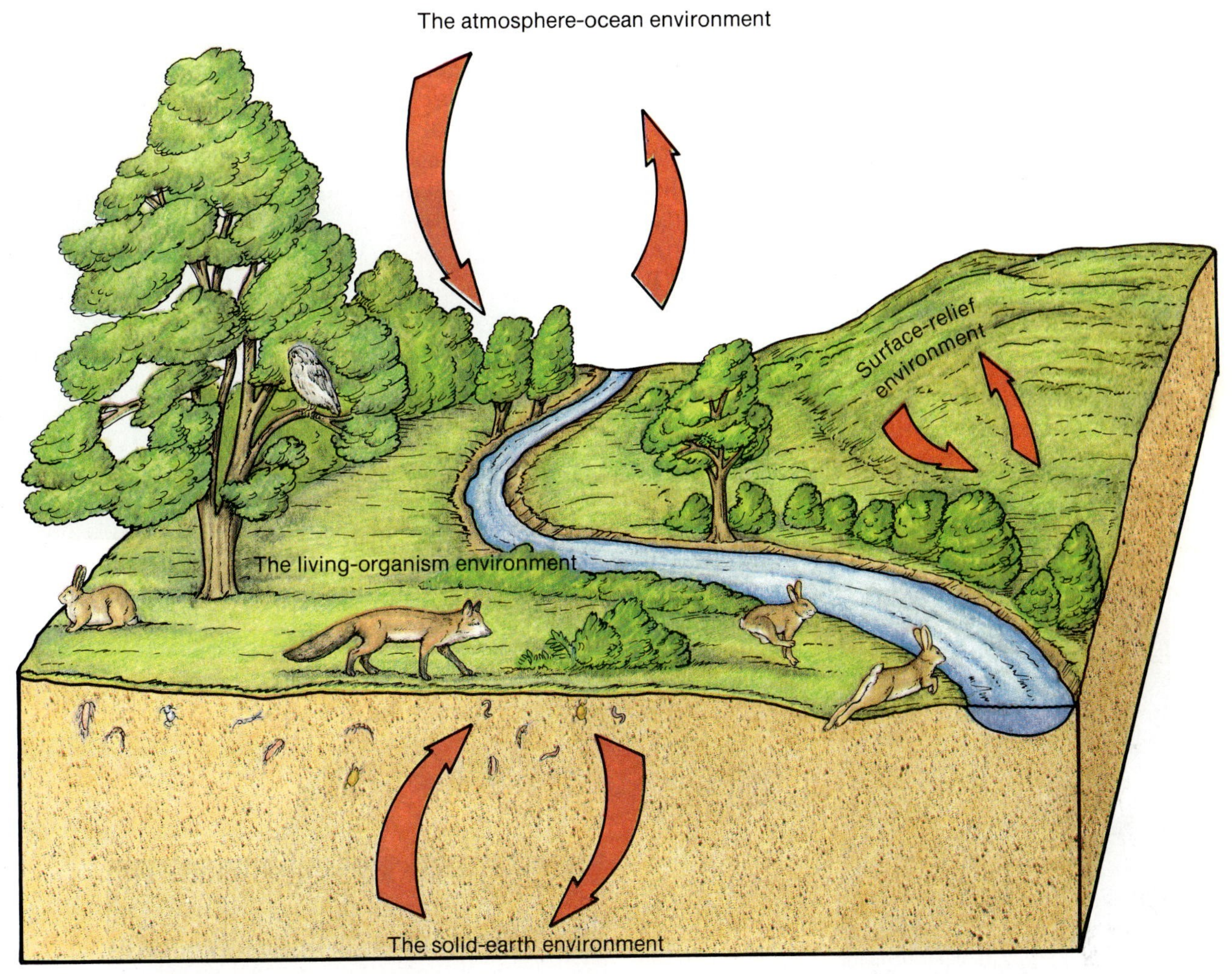

Figure 19-1 Part of a wooded ecosystem in the humid midlatitudes. The plants and animals above the ground, and the burrowing animals and microorganisms in the soil, make up the organic part of the ecosystem. The physical environment of solar energy, atmospheric gases, climate, water, soil, and topography in which they live is the inorganic environment.

animals and has a different physical environment compared to prairie grassland. The ecosystem processes (flows of energy and materials) hold the organic and inorganic components together as a system and link them to other living and nonliving systems. To understand the spatial distribution of ecosystems, the biogeographer must have a firm understanding of both ecosystem process and ecosystem structure. This is equivalent to studying atmospheric structure and heat transfer before tackling global circulation, or precipitation patterns and river flow before analyzing fluvial landforms.

Scales of Study

The ecosystem is a useful framework in which to study structure and process in the living-organism environment. It can be applied at any scale. Earth's living-organism environment is made up of a series of nested ecosystems, as shown in Figure 19-2. The top level is the whole of Earth's biosphere, the global ecosystem. The global perspective now available from satellite imagery has revolutionized study of the structure and processes of the biosphere at this scale.

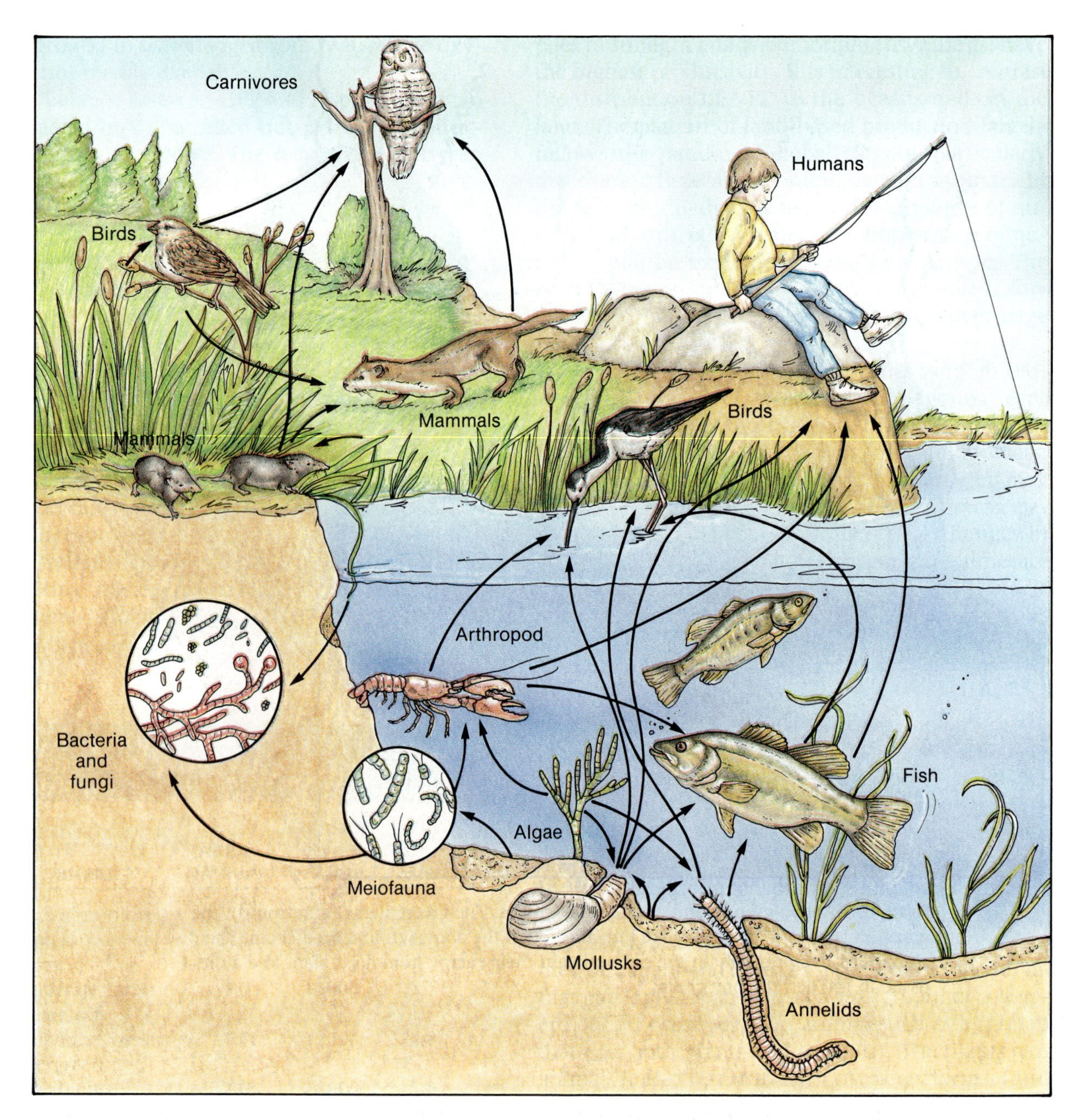

Figure 19-6 The food web in a pond ecosystem. The human and some of the birds are omnivores, taking energy from a variety of trophic levels. The different arrows distinguish energy flows to different trophic levels.

trophic level. Figure 19-6 shows how some organisms, particularly omnivores, can take energy from a variety of trophic levels. This flexible approach to feeding is a strategy for minimum risk. Such variation in food pathways protects individual species and preserves the stability of the ecosystem as a whole.

The transport of useful energy from the sun through the ecosystem is not highly efficient. The sun radiates an intense amount of energy to Earth, the equivalent of one million Hiroshima-sized bombs every day. About half of this is filtered out by the atmosphere or is reflected by the clouds or Earth's surface (Chapter 4), and the autotrophs can use only a fraction of the energy that remains. The overall efficiency of energy transfer between the sun and the autotrophs is about 0.2 percent. Further, only a small part of the energy available at each

Figure 19-7 The giant panda, symbol of the Worldwide Fund for Nature. There are now thought to be fewer than 1200 pandas living in the wild. The panda's natural habitat is the bamboo forests of southwestern China and the bamboo is virtually its only source of energy. Human use of the forest has restricted the panda's range, and when the bamboo flowers (as described in the text) the pandas face starvation.

trophic level is passed directly to the next level. Much of the energy assimilated into plant or animal tissue through feeding is released through respiration as heat energy, and is not available to the next trophic level. In addition, much of the energy originally captured above ground ends up in the detrital food chain, as plants and animals excrete waste products or die before they are eaten. As a rule of thumb, only about 10 percent of the energy fixed at each trophic level is passed on to the next level. This low efficiency of transfer explains why food webs usually have only three or four trophic levels, and why the greatest biomass is usually found at the lowest trophic levels (Figure 19-8).

Humans are at the top of many three- and four-level food webs. Clearly this is not an "efficient" way for people to tap the solar energy fixed by the autotrophs. It would be much better, in terms of energy transfer, to eat grasses and grain plants, or more efficient animal converters such as earthworms, than to feed plants to large animals, which we then eat. Meat is an important part of the diet of the developed world, but in many less developed countries it is less readily available. Food chains in these countries generally have fewer trophic levels and lower energy subsidies such as tractors and artificial fertilizers. They are therefore much more efficient in the use and transfer of energy. However, a vegetarian diet is only healthy if certain groups of foods are eaten together, to ensure a reasonable intake of vitamins and protein. In developing countries, poverty may exclude this choice, and the monotonous diet may be unbalanced and debilitating.

The quantitative study of energy flow through an ecosystem is enormously complex, not least because of the problems in defining the detailed structure of the food chain. Many of the statistics available are estimates. The detailed studies that have been carried out tend to suggest that most of the energy flow in a mature ecosystem is through the detrital food web. The biologic transfers of energy visible at the surface are therefore only a small fraction of the activity going on underground. Accurate data on energy flow are important in understanding, for example, the amount of oxygen and carbon dioxide exchanged between living things and the atmosphere-ocean environment, or the amount of heat released during respiration. The need to provide these data on a regional or global scale is a considerable challenge to the biogeographer and the ecologist.

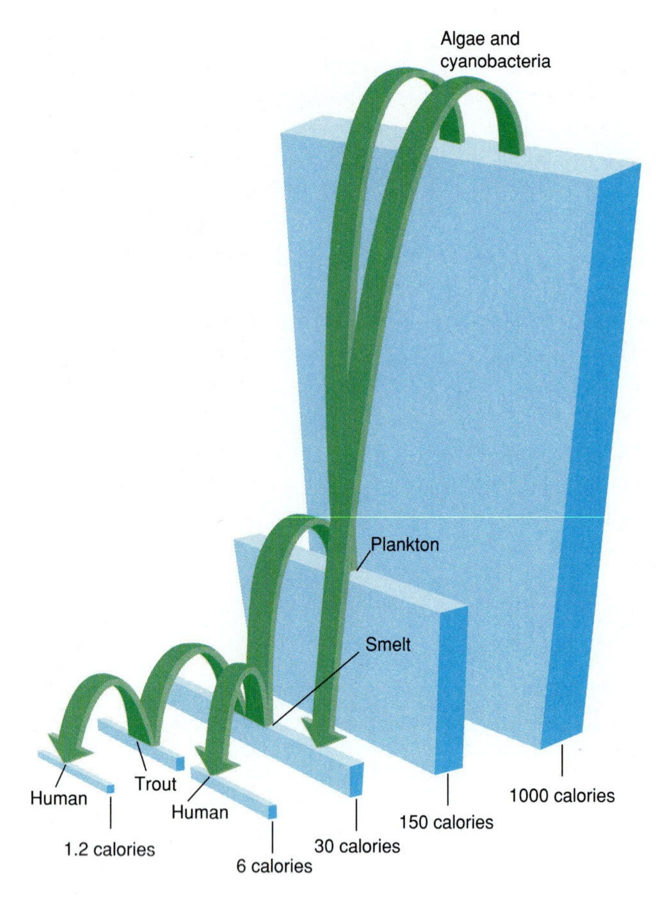

Figure 19-8 The transfer of food energy through the food web of Lake Cayuga, one of the Finger Lakes in New York state. The photosynthesizing algae are food for plankton, and smelt feed on both algae and plankton. The smelt are harvested directly by humans and are also eaten by trout. Most humans would rather eat trout than smelt, but, as the diagram shows, this is a less efficient way of using energy.

Table 19-2 Common Nutrients and Their Amounts in the Human Body

ELEMENT AND SYMBOL	APPROXIMATE % OF EARTH'S CRUST BY WEIGHT	% OF HUMAN BODY BY WEIGHT
Oxygen (O)	46.6	65.0
Silicon (Si)	27.7	trace
Aluminum (Al)	6.5	trace
Iron (Fe)	5.0	trace
Calcium (Ca)	3.6	1.5
Sodium (Na)	2.8	0.2
Potassium (P)	2.6	0.4
Magnesium (Mg)	2.1	0.1
Hydrogen (H)	0.14	9.5
Manganese (Mn)	0.1	trace
Fluorine (Fl)	0.07	trace
Phosphorus (P)	0.07	1.0
Carbon (C)	0.03	18.5
Sulfur (S)	0.03	0.3
Chlorine (Cl)	0.01	0.2
Vanadium (V)	0.01	trace
Chromium (Cr)	0.01	trace
Copper (Cu)	0.01	trace
Nitrogen (N)	trace	3.3

NUTRIENT CYCLING

The Nature of Nutrients

The movement of energy through the food web drives the second important process in the ecosystem. This is the circulation of chemical materials or nutrients between the organic and the inorganic environments and among the trophic levels. Ninety-two chemical elements occur in nature. Of these, between 30 and 40 seem to be necessary for life, although it is not always clear what role they play. The chemical elements needed for life are called **nutrients.** Carbon, oxygen, hydrogen, and nitrogen make up a high proportion of living tissue and are therefore needed in large quantities by all living things (Table 19-2). The combination of oxygen (O) and hydrogen (H) as water (H_2O) is particularly important. In the human body atoms found as water outnumber those in all other molecules, and life stops very quickly if the water supply is cut off. Nutrients needed in large quantities are called **macronutrients.** At the other extreme some essential nutrients are needed only in very small quantities. These are called **micronutrients** or trace elements, and are often scarce in nature. Micronutrients important to humans are, for example, cobalt (a part of vitamin B_{12}), zinc, and molybdenum (both parts of enzymes).

There is a fundamental difference between the pathways of energy and of nutrients in the ecosystem. The movement of energy in the ecosystem is one-way. The supply is constantly replenished by the sun. Energy is fixed into the system, used by the organisms in it, and ultimately released to the atmosphere as heat. This heat energy cannot be reused by the organisms. In contrast, the supply of chemical elements in Earth environments is finite and must be recycled.

Biogeochemical Cycles

Nutrients in the ecosystem move in a continuous cycle from the nonliving to the living environment and back again (Figure 19-9). The pathways along which they move are called **biogeochemical cycles;** the "bio" refers to the living phase of the cycle in organisms; the "geo" refers to the nonliving phase in soils, rocks, and water.

In the processes of respiration and photosynthesis most organisms exchange some nutrients directly with the atmosphere-ocean environment. Other nutrients are weathered from the exposed rocks of the solid-earth environment. They may enter the soil directly or take a longer route through the action of wind, water, or ice in the surface-relief environment. The nutrients available in the ecosystem are taken up by organisms and distributed around the system, initially in the above-ground food web. This shared pathway of nutrients and energy is followed most clearly by carbon, hydrogen, and oxygen, which are the raw materials for the carbohydrate manufactured in photosynthesis. During this passage through the food web some nutrients are exchanged with the atmosphere-ocean environment and with water in the soil. The remaining nutrients enter the detrital food chain either as waste materials excreted from living things or as dead plant or animal tissue. The decomposers break down the organic materials into simpler inorganic building blocks, which lie in the soil "foot loose and fancy-free," as described in the chapter-opening excerpt.

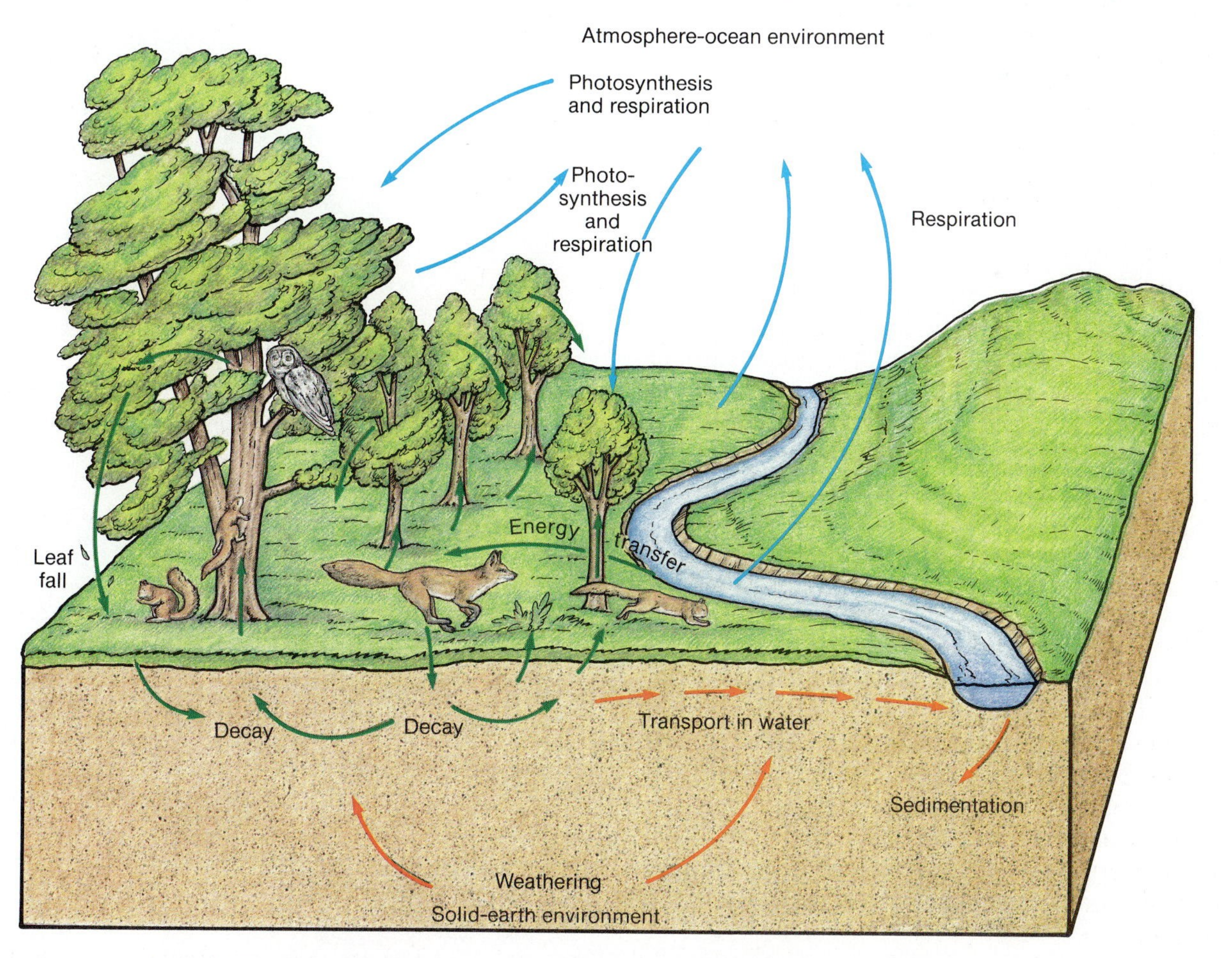

Figure 19-9 Biogeochemical cycles. Nutrients are exchanged among the atmosphere-ocean environment, the biosphere and the solid-earth environment. In this diagram blue arrows represent transfers to and from the atmosphere-ocean environment and red arrows transfers to and from the solid-earth environment. Green arrows represent transfers within the living-organism environment.

At this point the nutrients are available for recycling and follow one of two routes. First, the nutrient may reenter the organic phase of the cycle, when, for example, it is taken up from the soil by a plant root. In the opening excerpt this occurred when the atom left the Indian's body and joined a bluegrass leaf. Second, the nutrient may start the long journey back to solid rock in the solid-earth environment, through the processes of transport, deposition, and sedimentation. This pathway shows how the processes of rock formation (Chapter 10), mountain building (Chapter 12), and weathering (Chapter 13) are linked to each other, and to living organisms. Sediments are laid down and consolidated to form solid rock. The rocks are uplifted by tectonic activity, exposed to the atmosphere, and weathered. The molecules and atoms released by weathering find their way into the soil and sometimes into the bodies of autotrophs. Then they may be distributed to other living things. For example, the atoms that make up your bones may once have been part of a mountain range.

Nutrient Stores: Reservoirs and Pools

The geographer's main concern in studying nutrient cycles is to find out how the nutrients get from one part of the ecosystem to another, so that supplies reach the living organisms in sufficient quantity and on time. The success of the delivery is determined by two factors: first, the amount of each nutrient available in the environment; second, the rate at which each can be moved from its current location to where it is needed. Both characteristics depend largely on the location of the nutrients in the biosphere and in the wider Earth environments.

The atmosphere-ocean and particularly the solid-earth environments can be considered *reservoirs* of materials, because they offer long-term storage to large quantities of nutrients and are not easily depleted. The more dynamic parts of the ecosystem, the plants and animals themselves, and the soil and litter (dead plant and animal material in the process of decomposition) are *pools* of nutrients. They hold smaller and more variable amounts of nutrients for a shorter time. Their contents can be made available quite rapidly, as when one organism eats another or a plant draws water from the soil.

Gaseous and Sedimentary Cycles

The details of the cycles of different nutrients vary, depending on their main source in the biosphere. In general, nutrient pathways can be divided into *gaseous cycles* and *sedimentary cycles*. As the names suggest, gaseous cycles involve exchange primarily between living things and the atmosphere-ocean environment, while sedimentary cycles involve exchange between living things and the solid-earth environment.

Gaseous cycles act on a global scale; a temporary imbalance in local conditions is balanced rapidly by interchange with the broader atmosphere-ocean system. Nutrients that move in a gaseous cycle are therefore approximately equally abundant across the globe. The gaseous nutrient cycle of nitrogen is shown in Figure 19-10, *A*. Figure 3-8 illustrates the carbon cycle.

Oxygen is usually exchanged directly between living organisms and the atmosphere-ocean envi-

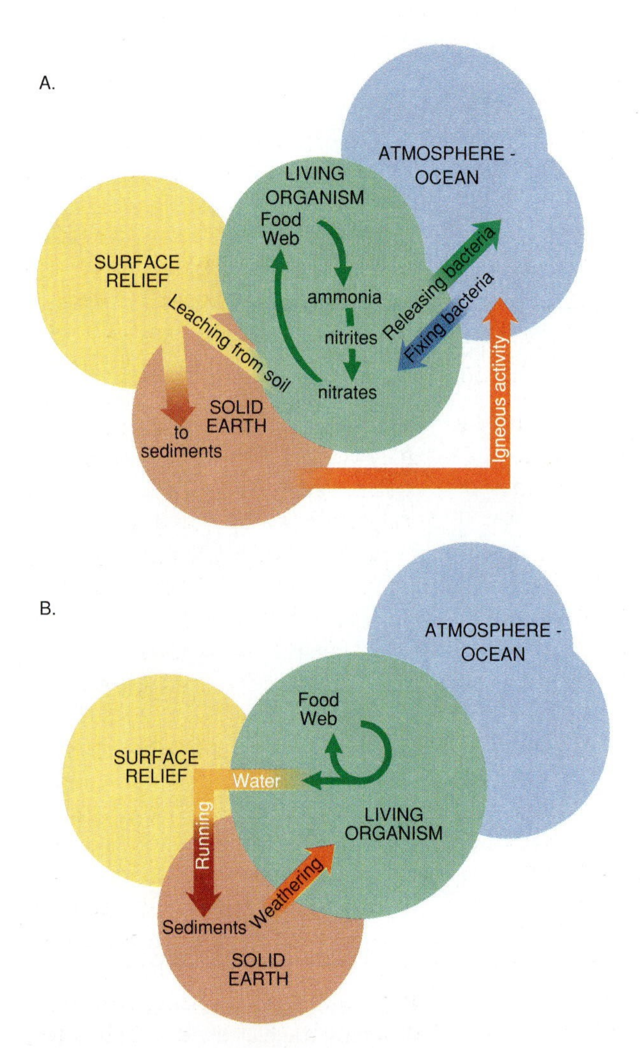

Figure 19-10 **A,** The nitrogen cycle, an example of a gaseous nutrient cycle. **B,** The phosphorus cycle, an example of a sedimentary nutrient cycle.

ronment, either as the gas oxygen or in water. In contrast, atmospheric nitrogen cannot be used directly by most living organisms. Its movement between the organic and the inorganic parts of the ecosystem depends on several groups of highly specialized bacteria. Atmospheric nitrogen is converted to a usable compound by nitrogen-fixing bacteria in the soil. Some of these bacteria live on the roots of *legumes* such as peas, alfalfa, and clover. The bacteria fix more nitrogen into the soil than these plants will need. Legumes are therefore often planted in rotation with other crops to avoid or decrease the need for nitrogen fertilizer. The nitrogen taken up by plants is eventually returned to the soil as dead plant or animal tissue. It is broken down to a chemically usable form again by a second group of specialized bacteria. The nitrogen can then be reused by plants, released to the atmosphere by a third group of bacteria, or, in terrestrial ecosystems, washed out of the soil to enter an aquatic ecosystem. When too much nitrogen is washed into streams, perhaps as a result of overapplication of nitrogen fertilizers to cropped land, the natural balance of aquatic ecosystems can be upset. This situation is discussed further in the section of this chapter about human impact on ecosystems.

Sedimentary cycles involve exchange primarily between living things and the solid-earth environment. Such cycles are characteristic of micronutrients such as phosphorus, copper, and zinc. By definition, sedimentary cycles act over a longer time scale than gaseous cycles, because they are controlled by the processes of sedimentation and weathering (Chapters 12 and 13). They also act on a more localized spatial scale, because the mineral content of rock and the rate of weathering vary from place to place. This means that there may be clear spatial patterns of excess and deficit in these nutrients. Figure 19-10, *B*, shows the phosphorus cycle as an example of a sedimentary nutrient cycle. Phosphorus is often a limiting factor on the growth of crops and may be added artificially in fertilizer. The ultimate source of phosphorus is crystalline rock, from which it is weathered at a very slow rate into the soil. It is easily washed from the soil by percolating rain water, and so returns quite rapidly to the oceans and to the sedimentary rocks of the solid-earth environment. Oceanic phosphorus can be brought back to the land surface through fish and seabirds (Figure 19-11). Current estimates suggest, however, that phosphorus is being returned to the solid earth more quickly than it is being released. This has led to attempts to circumvent the natural weathering process by mining phosphate-rich deposits directly to produce fertilizer. The use by farmers of larger quantities of phosphate fertilizers than are needed hastens the return of phosphorus to the rock environment.

Figure 19-11 Phosphate-rich sediments are formed by the concentration of *guano* (bird droppings) over hundreds of years. These pelicans are on a small island in the Gulf of California, Mexico. The phosphorus in their droppings comes from the fish they eat; the fish in turn gain phosphorus from the ocean waters.

The process of nutrient transfer can be thought of as a continuous series of interchanges among the nutrient pools and reservoirs, driven by the movement of energy at a global and at a local scale. On the local scale, materials move through the biomass, litter, and soil pools as the energy converted to chemical form by photosynthesis passes from one organism to another. On the global scale, nutrients are weathered from the solid-earth environment by physical and chemical reactions partially controlled by the atmosphere.

LINKAGES

Nutrient cycles show how transfers of matter as well as energy link the living and the nonliving parts of the ecosystem into a unified, functioning whole. Nutrients are drawn from the ***atmosphere-ocean*** *and* ***solid-earth*** *environments and may travel to living organisms through the rivers and streams of the* ***surface-relief*** *environment.*

THE INTERACTION OF PLANTS AND THEIR ENVIRONMENT

Autotrophs, the green plants, blue-green algae, and phytoplankton that build organic matter through photosynthesis, are the foundations of the ecosystem. The amount and diversity of life in the system are determined by the autotrophs' variety and productivity. The autotrophs in turn rely on the physical environment to provide the conditions for life such as sunlight and water. The ecosystem is controlled at this most basic level by the interaction of each living autotroph with its nonliving environment.

The principal requirements for plant life are sunlight for photosynthesis, water, nutrients, an appropriate temperature, and physical support. These resources are supplied through the atmosphere-ocean system and the soil. They may be modified by other organisms. Tall trees, for example, shade the plants growing beneath them, and organisms compete with each other for resources.

For each environmental variable, such as temperature or available nutrients, there is a clearly defined range in which a plant can survive. This range is called its **tolerance** (Figure 19-12). The plant grows most strongly in a narrower **optimum** range, typically toward the center of its tolerance. Here conditions for growth and reproduction are ideal. A plant can survive brief periods outside its tolerance range, during an unexpected drought or a late frost, for example, but prolonged exposure or very severe conditions will kill it. Hence, plants that thrive in the tropics will not live in arctic temperatures. A plant's optimum and tolerance ranges can change with growth and maturity. The late frost that kills new shoots would, for example, be tolerated by the dormant buds.

In practice the plant's environment is made up of a diverse set of interacting gradients, each of which can be measured and plotted to produce a graph such as Figure 19-12, *B*. The width of the tolerance range can be different for each environmental gradient. Plants that can withstand variable conditions of moisture and temperature, for example, may have very specific requirements for one nutrient. In this case, adding the missing nutrient will cause a spectacular growth increase. In natural systems it is usually difficult to isolate a single factor that restricts growth. This is because most successful plants have a wide tolerance range, and because conditions in the environment do not act independently, but interact with each other. For example, where low rainfall threatens the health of a plant,

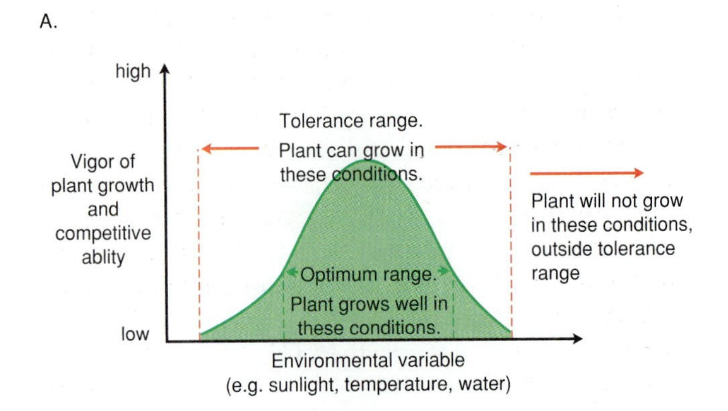

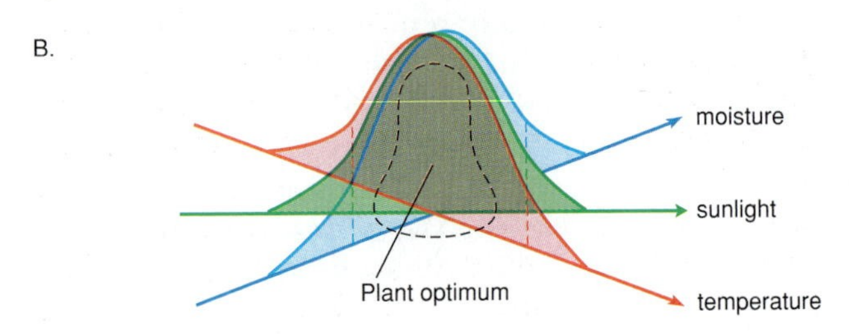

Figure 19-12 Optimum and tolerance ranges. **A,** A plant will grow best where the environmental conditions fall within its optimum range. **B,** The plant's environment is in reality a set of interacting variables; the plant grows best where each variable is independently within the plant's optimum range, or where the variables interact to create optimum conditions.

the lack of rainwater can be offset by moderate temperatures that slow evapotranspiration or by soil that retains moisture and makes it available to the plant. Conversely the effects of low rainfall can be aggravated by high temperatures or by rapid drainage. The distribution of prairie grasses in North America is controlled primarily by available water. Blue grama grass *(Bouteloua gracilis)* grows successfully in areas with less than 40 cm (16 inches) annual precipitation. The little bluestem grass *(Andropogon scoparius)* is normally excluded from this zone but may be present where the soil retains water. These two species are part of a continuous gradation of grass species from the east to the west of the United States, a distribution controlled by the interaction of temperature, precipitation, and local soil conditions. Temperature and precipitation largely determine the regional ecosystems or biomes discussed in Chapter 21.

A plant's optimum and tolerance ranges define the environmental conditions under which it can live. All plants, however, face **competition** from other organisms that can make as good or better use of the environment's resources. Competition usually confines the plant to a restricted part of its range. A plant for which the current environment is optimal, for example, will drive out a plant for which conditions are near the edge of its tolerance range. This process tends to a situation where all plants grow in their optimum ranges. Where the optimum ranges of competing plants are similar, one plant will dominate and push the other toward its tolerance limits, where it is vulnerable to further competition and may be excluded from the ecosystem completely. Each plant therefore grows where the environmental conditions are tolerable and where it can compete successfully with other plants. This set of conditions makes up the plant's **ecologic niche.** The importance of competition varies from one plant and one environment to another. In the northeastern United States, for example, maples, beech, and white pines all grow best on the good nutrient-rich soils. The white pines are, however, confined to nutrient-poor soils because they are unable to compete on the better soils with the other species. The nutrient conditions in the poorer soils are within the tolerance range of the white pines but not of the maple or beech.

The principle of ecologic niches extends throughout the ecosystem. So far, only the first step in building the ecosystem, the interaction of each autotroph with its living and nonliving environment, has been considered. Further up the food web, the autotrophs play an important role in shaping conditions for other organisms. The ecological niches for the heterotrophs are determined by the energy and nutrients supplied by the autotrophs and by the physical environments they create.

The patterns of ecologic niches formed by the Earth environments lead to corresponding patterns in the distribution of plant species and the other organisms they support. A plant association, or plant community, is a grouping of plants that is repeated in the landscape where a set of environmental conditions recurs. Box 19-1: Living on Water, identifies the main plant associations found in the Everglades National Park, Florida, and shows how they are related to the living and the nonliving environment. The plant association is an important and well-established tool in describing and analyzing vegetation patterns.

A relatively new field of investigation in biogeography is the ecotone, the dynamic zone of transition between one established plant association (and therefore one ecosystem) and another. In the ecotone, organisms from the surrounding associations are living at the limit of their tolerance or their competitive ability. Some species can exist only in the unique conditions of such boundary zones. Any stress or change should therefore be apparent in the fragile and diverse ecotone before it is seen in the main communities. Some biogeographers advocate that management of vegetation communities should concentrate on monitoring and maintaining the health of the ecotone, rather than that of the neighboring ecosystems.

Ecologic Succession

Ecologic succession is the growth and development of an ecosystem over time. It is driven by changes in the plant's physical environment. This change can occur in two ways. First, succession can be prompted by a change in the environment that is unrelated to the plant's own activity. Typically this is a sudden or catastrophic event, such as a forest fire, that results in new conditions for plant growth. Other examples are the creation of a new land surface by a landslide or glacier retreat and ecosystem changes caused by human activity such as forest clearance or drainage of wetlands.

Second, succession can be prompted by the plant's own action in modifying its surroundings. A plant adds organic matter to the soil while growing, thus altering the ability of the soil to hold water and nutrients. Local conditions of wind flow, sunlight, temperature, and humidity change as the plant canopy develops. The plant's action on its environment often makes conditions less suitable for its own offspring, and more suitable for those of a competing species. A plant that colonized an open area of well-drained soil, for example, would have a lower competitive advantage when its site becomes shaded and damp as the plant grows and spreads. Birch trees spring up rapidly, but are outdone by taller, shadier, and longer-lived trees. A colonizing species, therefore, has somewhat perversely, a built-in capacity for self-destruction. More succinctly, "colonization initiates succession." As the environment changes, the original species are replaced by those with tolerance and optimum ranges better suited to the new conditions.

Ecological succession is a combination of these two methods of altering the environment. Once set in motion, plant-guided succession may be interrupted and restarted by further catastrophic changes. Regular fires, whether accidental or deliberate, are a particularly important control on succession in many forest and grassland ecosystems.

The sequence in which succession progresses depends on the mix of species nearby when a space on the ground appears. The colonizing or replacing

this section. A plant association is made up of many individual plants, each interacting individually with the environment and with other plants. There is no direct dependence between "organs" as Clements suggested. However, Clements' view of the climax community as a stable, self-perpetuating system in a stable environment is consistent with modern thinking, and the terminology is still in use.

Young and developing ecosystems are markedly different from climax communities. A young system is characterized by low species diversity and short, simple food chains. The species found in immature ecosystems are adapted to survival and rapid reproduction in an exposed and unstable environment. Much of their energy is used to produce copious amounts of seeds. In a young ecosystem the energy fixed by photosynthesis exceeds that used up in respiration by a wide margin. There is therefore a high net primary production (NPP), and the system rapidly accumulates biomass.

In the climax ecosystem the living and nonliving environments contain a greater variety of ecologic niches. There is therefore a greater species diversity and a more complex food web. The energy fixed in organic matter is more closely matched by the energy needed to keep it functioning, and the addition of biomass slows down. Individual species tend to be larger, have a longer life cycle, and produce fewer seeds. A higher proportion of energy is channelled through the detrital food chain, and there will be a considerable store of nutrients in the biomass, litter, and soil.

These characteristics of the mature ecosystem are effective self-regulatory or **homeostatic** mechanisms. They maintain the ecosystem in a state of equilibrium by providing some flexibility in the transfer of energy and nutrients. For example, each heterotroph typically has several sources of energy in the diverse food web. The store of nutrients acts as a buffer against disruption of supplies from the other Earth environments. If the disruption of energy or nutrient transfers is severe or prolonged, the ecosystem gradually moves toward a new equilibrium. This has happened in the oceans around Antarctica (Figure 19-14). Most species in the grazing food chain share the whales' food of tiny shrimplike organisms called krill. Overfishing of whales has led to an abundance of krill in recent years. This has in turn increased the population of seals and penguins, leading to a new equilibrium among the trophic levels in the sea. On land, however, the growing number of Antarctic fur seals is starting to alarm

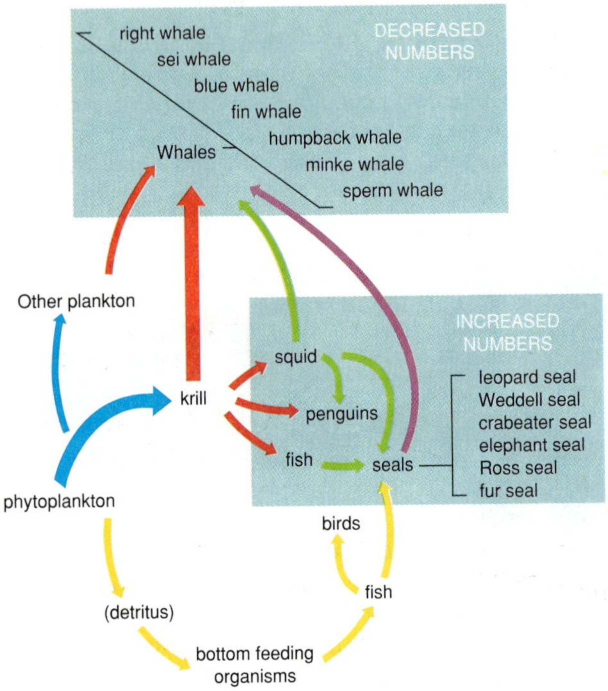

Figure 19-14 A simplified representation of the food web in the Southern Ocean. Overfishing of the whale stock has led to a readjustment of biomass in the ecosystem. The decrease in whales has been compensated for by increased numbers of penguins and seals although, as described in the text, this has led to problems elsewhere.

Figure 19-15 Southern fur seals on the South Orkney islands in the South Atlantic. The large numbers of seals coming ashore on the island are damaging the delicate moss and lichen carpet that covers parts of the ground surface.

ecologists who work on the islands of the southern Atlantic. On the South Orkney islands, between the Antarctic Peninsula and South America, large numbers of seals now come ashore in the summer as their more usual resting places fill up. The seals trample the delicate mosses and lichens that cover the coastal areas and destroy the ecosystem associated with them (Figure 19-15). In this example the linked terrestrial part of the ecosystem has yet to reach the new equilibrium.

Ecosystem Structure: The Distribution of Species

Most plants and animals seem ideally suited to their environments. Chimpanzees are at home in the trees of the tropical forest. The shaggy white coat of the polar bear makes it difficult to spot among the ice and snow and protects it from the severe cold of the arctic climate. But there are surprising gaps and anomalies in this distribution; for example, the apes found in Africa are absent from the forests in South America, and polar bears live in the Arctic but not the Antarctic. Conversely, penguins live around the South Pole but not the North. The pouched marsupials are found mainly in Australia and South America, although they could fit in niches elsewhere. The biogeographer is left with two contradictory observations. First, species are clearly suited to the environments in which they live. Second, they are not present in very similar environments from which they are geographically separated.

This pattern in the development and distribution of species is the result of two interrelated processes. The first is the evolution of different forms of life over time. The second is the isolation of groups of plants and animals by barriers such as land, sea, deserts, and high mountains as Earth's surface changes.

To understand these processes it is necessary to know a little basic biology. There are several million different types of living things in the world. To make sense of this rather chaotic assemblage, humankind has given names to the types of life that have been discovered and differentiated. This process of classification is called *taxonomy*. About a million and a half types of living things have been named, and scientists know that there are many more, especially in diverse wilderness areas such as the tropical rainforests.

The system of naming living things is very much like other classifications. It is made up of a hierarchy of groupings; all the living things under a heading at each level have some kind of unifying characteristic. Humankind's proper name is *Homo sapiens*. *Homo* (from the Latin for "man") is the name of the genus, *sapiens* (from the Latin for "wise") is the name of the species. Used together, the genus and the species names uniquely describe a species—a type of living form that is able to interbreed with others of the same species and produce fertile offspring. The importance of taxonomy in biogeography is that species grouped together as one genus are often variations on the same basic theme. Similar species on opposite sides of the globe may be descended from a common ancestor.

Diversity and Evolution

The great diversity of life forms we know today is thought to have a common ancestor, formed under the hostile conditions at the birth of the planet. Clues about Earth's history are available from the study of the solid-earth environment, paleoclimatology, and observations of the rest of the solar system, but scientists are unlikely ever to know under exactly what conditions the first forms of life evolved. Earth's early atmosphere probably had no oxygen and therefore no stratospheric sunscreen of ozone (Chapter 3). Ironically, scientists think that short-wave solar radiation, now screened out by ozone and so harmful to living things, may have provided the energy that prompted the association of chemicals into an organic or living form. For the first billion or so years of its existence, Earth was a violent and hostile environment for living things. They were simple and small and their survival must have been tenuous. Because of the lack of protective ozone, the primitive forms of life existed only where they were sheltered by a layer of water.

The first big step toward the living-organism environment as we know it today was the development of the photosynthesizing blue-green algae, over a billion years ago. Photosynthesis releases oxygen into the atmosphere as a by-product. The gradual buildup of available oxygen and, consequently, of the protective ozone layer, is thought to have made conditions in the biosphere more hospitable and allowed rapid development of other forms of life. The effect of the early cyanobacteria on the formation of Earth's atmosphere demonstrates how the Earth environments were interdependent from an early date.

It is still a big step from the early primitive forms of life to the diversity of forms today. The development of variety is the result of evolution. This differs from ecologic succession, discussed earlier in this chapter. Succession gives existing species the chance to become part of an ecosystem. **Evolution** is the development of new species and new forms of life. Each generation of a species passes on its distinguishing characteristics to its offspring in its genetic material. A mistake in copying the patterns or codes carried by the genes can lead to the birth of a slightly different form of life. If this new form of life is successful in its environment, it will survive and flourish, possibly to the detriment of other preexisting species over which it has a competitive advantage. The new species will pass on its advantages to its own offspring. This process, which favors forms of life that can best make use of their environment, is called **natural selection.** Charles Darwin, a scientist and naturalist who lived in the 1800s, was the first important theorist of the principles of evolution. He gave the graphic name *survival of the fittest* to the process of natural selection. As the environment changes, so will the most desirable characteristics; survival of the fittest leads to changes in existing species and the development of new ones that best match the changing environment.

Darwin made a series of classic studies on the animal species of the Galapagos Islands, demonstrating in particular how different species of finch in the islands had evolved from a common ancestral stock on the South American mainland. The finches on different islands had evolved different beak forms to match the variety in their food sources. A similar process has been documented for the birds of the Hawaiian Islands (Figure 19-16).

Darwin's work implied that the genetic process by which natural selection works is gradual and continuous. More recently, scientists have developed the theory of *punctuated equilibrium*. This suggests that species remain virtually unchanged for

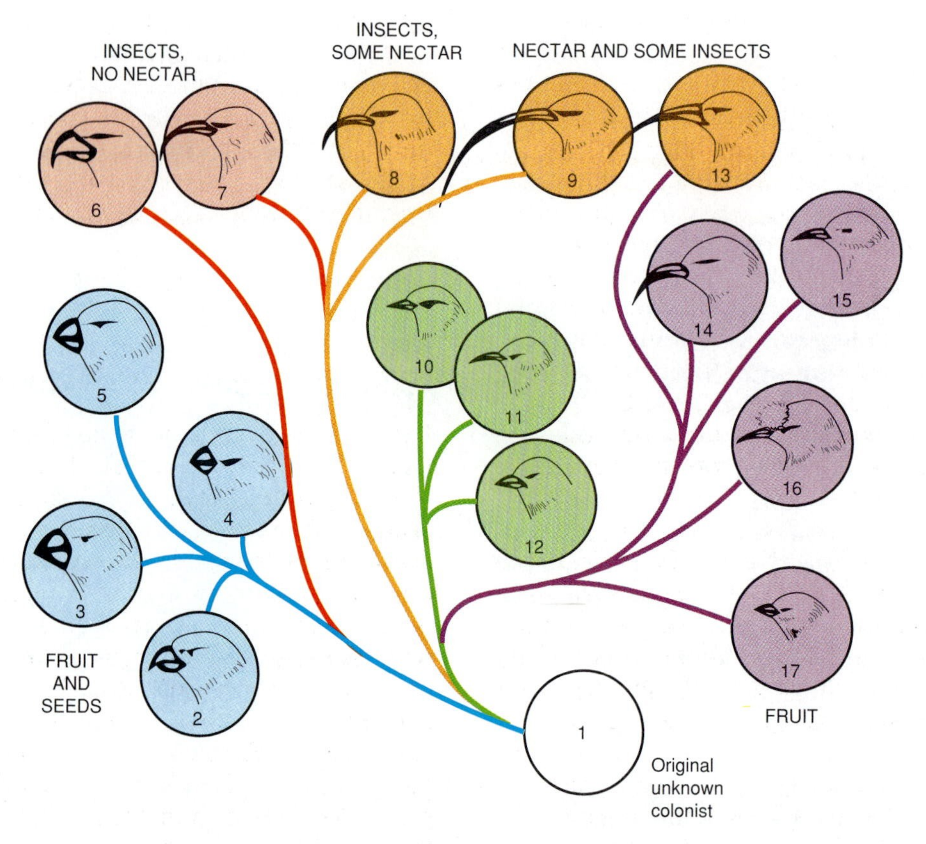

Figure 19-16 All the species of Hawaiian honey-creepers are thought to have evolved from the same ancestor. This diagram shows how the modern species can be grouped together on the basis of their main energy source. Different groups have evolved different beak shapes to match the physical characteristics of their food.

long periods of time. When changes do occur they are sudden and radical, causing dramatic differences over a few generations. Scientists supporting and opposing this argument both claim evidence from the fossil record, which shows both gradual evolution of species and sudden changes with "missing links" between species and their descendants. The fossil record also shows evidence of *mass extinctions,* points in time at which large numbers of species were wiped out. The dinosaurs, for example, were exterminated in a mass extinction about 65 million years ago. The cause of mass extinctions is a controversial subject in modern science, and current theories range from climatic change to the impact of meteorites. The debates about the causes of mass extinctions and the mechanisms of punctuated equilibrium are unlikely to be solved in the near future, in part because only a tiny fraction of living things end up as fossils.

Continental Movement and Species Migration

The pattern of species seen in the world today is the result of the evolutionary process, coupled with the changing pattern of continents and oceans through Earth history as a result of plate tectonics and climatic change. Research in this area has produced a fascinating account of change and interdependence in all Earth environments over time.

Both animals and plants are able to migrate from one area to another if the intervening barriers are not too harsh. Terrestrial animals cannot, however, cross large expanses of water, and fish cannot cross land barriers. The polar bear, an animal adapted to cold climates, could not cross the temperate and tropical conditions on the way to the South Pole. The seeds and pollen of plants are more easily distributed through wind and ocean currents and via the more mobile animals such as birds. Present day distributions of plants and animals are governed by the conditions of isolation and the opportunities for migration that arose from the changing distribution of the continents discussed in Chapter 11 and the emergence and submergence of land bridges in the ice ages, discussed in Chapter 9. The pouched marsupials (such as the kangaroo) are thought, for example, to have evolved in South America and travelled to Australia when these continents were still linked via a relatively warm Antarctica. The marsupials left behind in South America and in other parts of the world were largely overcome by the greater competitive ability of the placental mammals (whose young develop within the mother's body) that evolved later. By this time, Australia was a separate continent and the placental mammals could not reach it. The Australian marsupials had no competition and have survived and flourished. Siberia and Alaska were connected across the Bering Straits in the last glaciation, providing a migration route for Arctic species between the American and Eurasian continents. This is thought to be the route by which *Homo sapiens* entered the North American continent.

The main centers of evolution and dispersion of plants and animals can be recognized by mapping the biogeographic "realms" of plant and animal species (Figure 19-17). Each realm contains a different combination of species, showing that the areas evolved separately at some time in the past. The biogeographer Alfred Russell Wallace, observing patterns of bird distribution in Asia and Australasia in the nineteenth century, demonstrated the importance of changing sea level in separating the Oriental and Australian biogeographic realms (Figure 19-17, *C*). During the Pleistocene glaciation, the islands to the north and west of Wallace's line were connected to each other and to the mainland, allowing free dispersion of species in this area. Similarly, New Guinea and the surrounding islands were joined to Australia. The deeper water between the two continental shelves remained an effective barrier to migration. The animal species of the two realms have therefore evolved separately.

HUMAN IMPACT ON ECOSYSTEMS

Human Impact on Natural Systems

Human beings typically disrupt the natural ecosystems they come into contact with, altering the flow of energy or nutrients. As shown in Figure 19-14, humans have altered the transfers of energy in the ecosystem of the southern oceans. In this case the original equilibrium could be regained if whale stocks were allowed to recover. A much more serious threat comes from recent attempts to harvest the tiny shrimplike krill directly. Krill are a fundamental link in the food web, and an adjustment to compensate for their removal is probably beyond the capacity of the system's homeostatic mechanisms. Species higher up the trophic structure could therefore be wiped out.

Eutrophication is the natural enrichment of a water body by nutrients received through runoff in the watershed. Cultural or **accelerated eutrophi-**

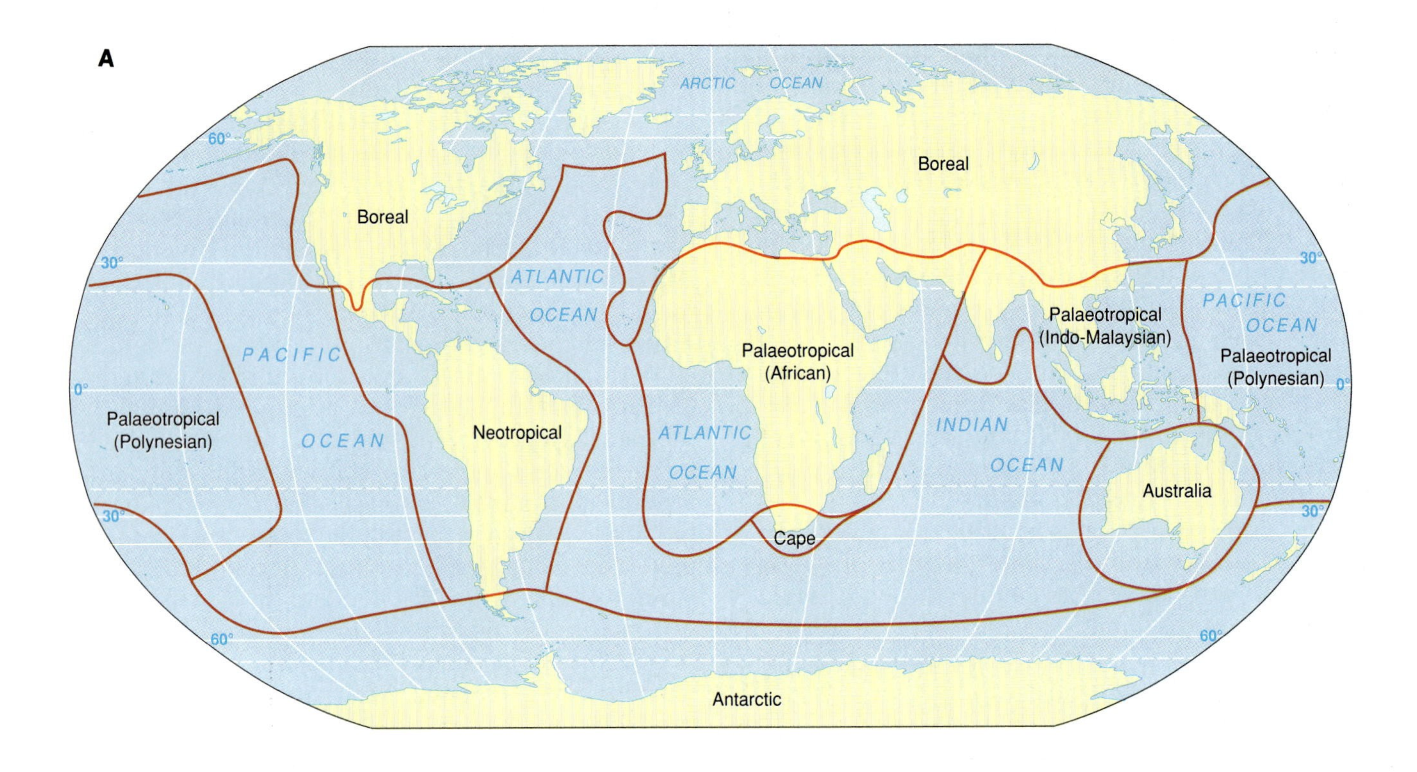

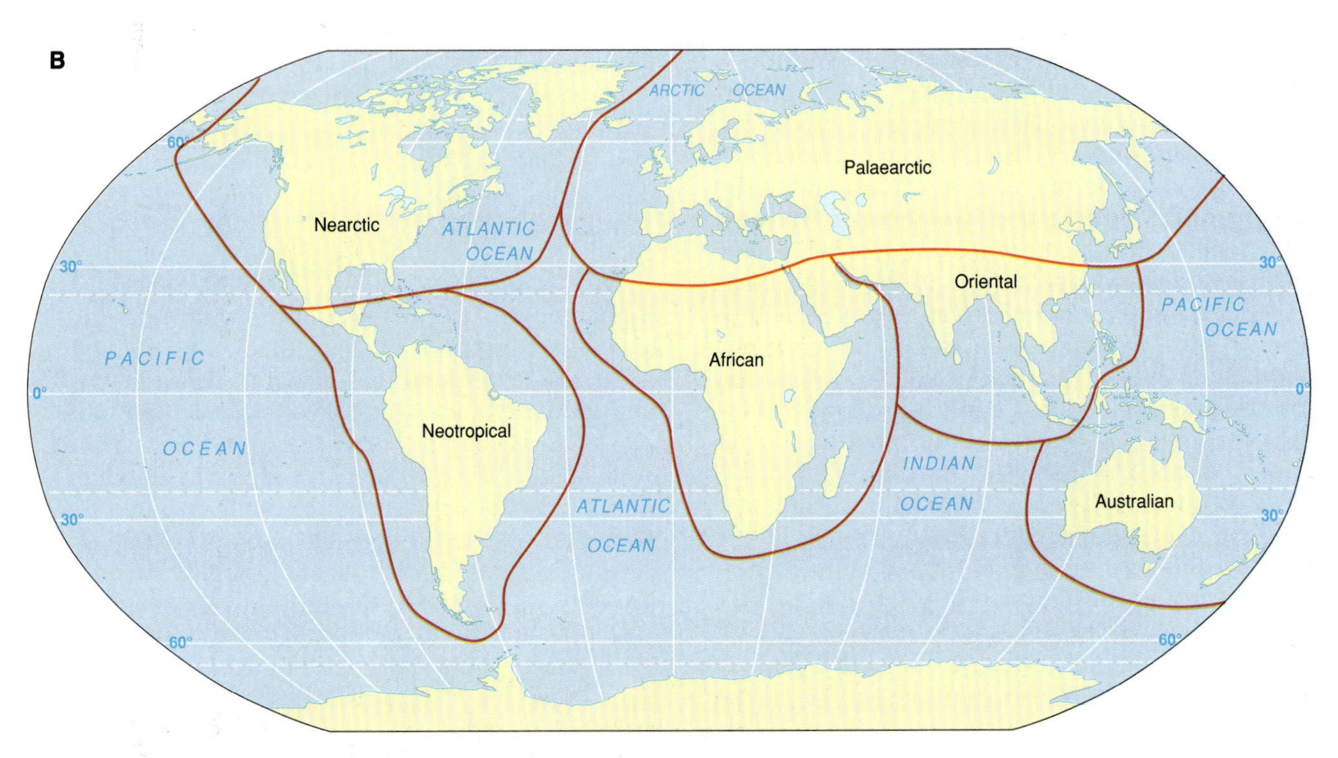

Figure 19-17 **A,** The biogeographical realms of plants. **B,** The biogeographical realms of animals.

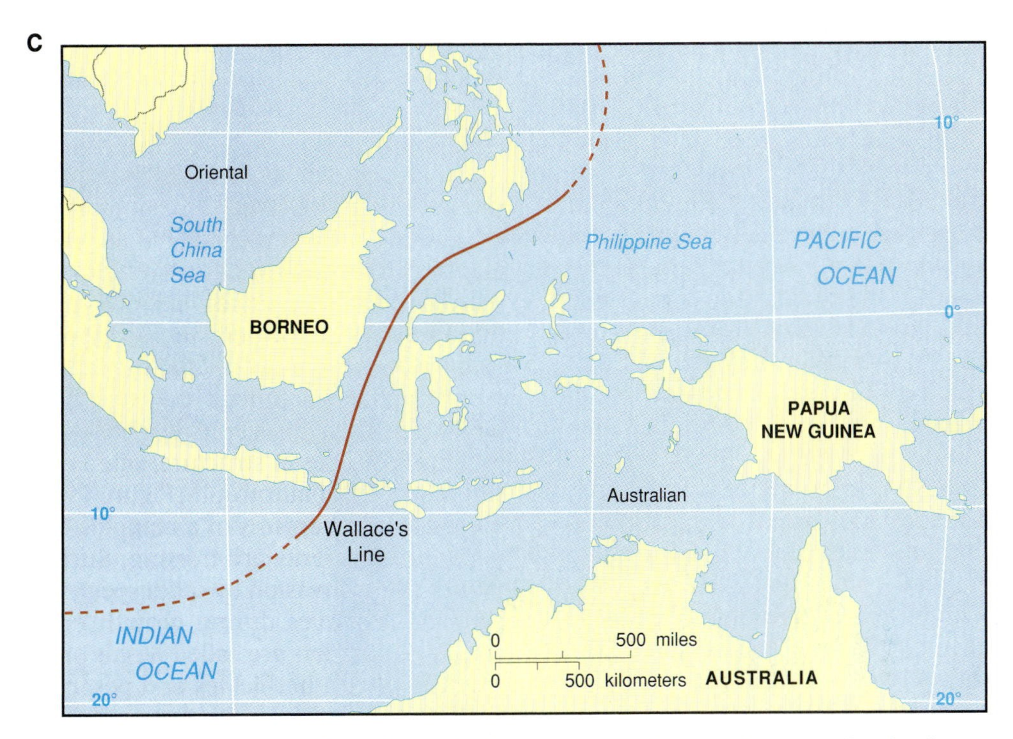

Figure 19-17, cont. C, Wallace's line, which separates the fauna of the Oriental realm from that of the Australian realm. Wallace's line divides two island groups, each of which was once a more or less continuous land surface but was not connected to the other.

cation occurs where large doses of nutrients are added over a short period as a result of human activity. Phosphorus, a naturally scarce nutrient, is added in sewage and domestic waste. Some of the nitrogen applied as fertilizer to agricultural crops is washed out of the soil and into rivers and lakes. Adding these nutrients removes the limits to growth of the blue-green algae that are the primary producers in aquatic ecosystems. We see the subsequent population explosion as an algal bloom (Figure 19-18). The living algae alter the physical conditions of the ecosystem by excluding light from the lower water layers. When they die, the decomposers that feed on them use up most of the oxygen in the water, so other forms of life are suffocated. These changes in the environment affect how energy and nutrients can be moved in the ecosystem, and a change in the food web may follow.

A further example of human impact on nutrient transfer is the addition of synthetic compounds to the biogeochemical cycle. This addition can be the result of undesired by-products, as in the case of pollutants, or it can be deliberate, as in the application of pesticides and herbicides to agricultural crops. The *organochlorine* or *chlorinated hydrocarbon* compounds used as herbicides or pesticides (such as DDT) are particularly harmful. They are chemi-

Figure 19-18 An algal bloom resulting from the enrichment of fresh water by nutrients.

cally stable and may persist as toxins in the ecosystem for long periods. A study of the fate of pesticides in soil found that 39 percent of the DDT applied was still in the soil 17 years later. More importantly, these chemicals become increasingly concentrated in the body tissue at each trophic level. DDT and its derivatives, now widely banned, are thought to be responsible for reducing the populations of fish-eating birds such as the osprey and the bald eagle. The concentration of the chemical in the bird's body is thought to interfere with the intake of calcium, resulting in the production of fragile eggshells easily crushed by the incubating parent.

As was stated at the beginning of this chapter, humankind is just another species in the biosphere, dependent on its resources for survival. The human abilities to analyze and plan should tell us when human impacts on a natural ecosystem exceed its ability to adjust, but the altered food chain of the southern oceans, the effects of cultural eutrophication, and the impacts of DDT were largely unforeseen. At a global scale, humankind is only just starting to comprehend the enormity of its impact on the atmosphere-ocean system through acid rain, destruction of the ozone layer, and the human input to climatic change.

The concept of **conservation** centers on sustainable use of the ecosystem's resources so that the system remains in equilibrium, and its long-term value is maintained. Some people feel that humans have a moral obligation to sustain other species on Earth. There are also more practical reasons based on resources. Each species is part of a food web. Its extinction will therefore alter the trophic structure of an ecosystem from which humans might draw energy. It is also part of the **gene pool,** the natural variation in genetic material from which human medicines and domestic plants and animals have been drawn. Losing a species means losing its genetic code, potentially useful in the future. The tropical rainforests, for example, known to contain many undiscovered species, are being cleared much faster than they can be replaced naturally. Potentially valuable species are becoming extinct even before they are discovered. An understanding of sustainable cropping rates, based on the productivity of the natural system, would allow a more rational approach to the use of tropical rainforests.

Human-Dominated Ecosystems

As well as altering natural ecosystems, for thousands of years humans have replaced natural ecosystems with their own. Many scientists suggest that no completely natural systems remain on Earth because human effects on the atmosphere-ocean environment have carried their impact across the globe.

Human impact on the biosphere is most complete in agricultural systems. The annual harvest of grain crops constantly restarts ecologic succession. The autotrophs reach their maximum production in one season and are harvested. The soil is never modified by dead plant tissue because the crop and much of its residue are removed from the land. In terms of growth and development, most agricultural systems never get further than their childhood. They are extremely simp
single species c
meostatic mec
a store of nutr
vulnerable to i
pete for resou
crop. The invac
eradicated wit

Chemical fe
iting factors to
The apparentl
crops is therefc

artificial nutrients. The energy subsidies include the use of fossil fuels (stored photosynthesis) to drive the tractors that prepare the seed bed and to pump water into irrigation canals. Energy is also used to manufacture the fertilizers, pesticides, and herbicides, and these elements then enter the system's biogeochemical cycles.

The difference between the natural and the agricultural system is highlighted by the description of the journey of a nutrient, a continuation of Leopold's writings that opened this chapter.

"It is the nature of roots to nose into cracks. When Y was thus released from the parent ledge, a new animal had arrived and began redding up the prairie, to fit his own notions of law and order. An oxteam turned the prairie sod, and Y began a succession of dizzy annual trips through a new grass called wheat.

The old prairie lived by the diversity of its plants and animals, all of which were useful because the sum total of their cooperations and competitions achieved continuity. But the wheat farmer was a builder of categories; to him only wheat and oxen were useful. He saw the useless pigeons settle in clouds upon his wheat, and shortly cleared the skies of them. He saw the cinch bugs take over the stealing job, and fumed because here was a useless thing too small to kill. He failed to see the downward wash of over-wheated loam, laid bare in spring against the pelting rains. When soil-wash and cinch bugs finally put an end to wheat farming, Y and his like had already traveled far down the water shed."

Figure 19-19 Agricultural ecosytems. **A,** The agricultural ecosystem shown here consists of one cereal crop. Its growth is subsidized by inputs of energy and nutrients and further energy is used to harvest and transport it to where it can be sold. **B,** In less developed countries single crops are still grown but, in most cases, over smaller areas and with fewer subsidies of energy and artificial nutrients. The crops are also typically used close to where they are grown.

The urban ecosystem is powered by a variety of sources, such as fossil fuels and the kinetic energy of falling water in hydroelectric schemes. The direct use of the sun's radiant energy in this ecosystem is very small. Typically food is transported from other areas to the city; photosynthesis is reserved for the small ecosystems in parks and gardens.

The ecosystems in the biosphere are therefore the result of two fundamental sets of processes. First, the ability of green plants to trap the sun's energy and pass it on to heterotrophic organisms permits life, and fuels the interaction of plants with their environment, the transfer of nutrients, and the growth of the ecosystem. Second, the historical interaction of the Earth environments allowed life to evolve, fueled the process of evolution, and stimulated the continuing activity of the solid-earth environment that has distributed the different forms of life around Earth. In our short history on Earth, humankind has made considerable inroads and alterations to natural ecosystems by destabilizing, simplifying, exploiting, and removing many of them.

FRONTIERS IN KNOWLEDGE

Ecosystem Structure and Process

One of the greatest challenges of studying an ecosystem is to express accurately and quantitatively the movement of energy and nutrients through it. If this can be done on a local scale, the figures can be extrapolated to a regional or biome scale, to reveal processes operating in the global biosphere. This sounds straightforward, but even the most simple ecosystem, such as a handful of soil or a single tree, has a complex structure and linkages that may be difficult to describe exhaustively. The situation is exacerbated in systems such as the tropical rainforest, where scientists have not yet even cataloged all the species.

In recent years, conservationists have attempted to recreate natural habitats, a process known as restoration ecology, and this subject continues to cause considerable controversy. Scientists who believe that the "new" ecosystems are equivalent in structure and process to the original ones claim that this is a mechanism for redressing the balance of human impact. Others are more skeptical, taking the view that the complexity of natural systems cannot be adequately mimicked by humans. This is an area of current research and more work is needed.

SUMMARY

1. The biosphere is the inhabitable part of Earth, formed by the interaction of major Earth environments.
2. The study of spatial patterns and processes in the biosphere is called biogeography.
3. The basic unit of biogeography is the ecosystem, a system of living things interacting with their nonliving environment.
4. The parts of the ecosystem are linked together by the flow of energy and nutrients.
5. Energy flow is one-way. Short-wave radiant energy from the sun is captured by the autotrophs in the process of photosynthesis and used to turn simple chemical elements into carbohydrates.
6. The amount of energy fixed by the autotrophs is the gross primary production. Some is used in respiration, leaving a balance of net primary production, which accumulates in the ecosystem as biomass.
7. Energy is passed to herbivores, carnivores, and omnivores through the food web. The transfer of energy through the different levels is not efficient. Only about 10 percent is passed on at each stage.
8. The supply of nutrients on Earth is finite, and nutrients are recycled. The pathways along which they travel are called biogeochemical cycles. Nutrients held mainly in the solid-earth environment are exchanged in sedimentary cycles. Gaseous cycles involve nutrients that are held mainly in the atmosphere-ocean environment.
9. The distribution of plants is controlled by the conditions of the physical environment and by competition with other plants.
10. Ecosystems change over time, through sudden catastrophic change and by slower, self-governed change. The end-point of change is a steady-state community that has well-developed homeostatic mechanisms.
11. New species form as a result of natural selection.
12. The distribution of species around the world is governed by evolution and by opportunities for migration at different times in Earth's history.
13. The human impact on ecosystems is to disrupt their stability, by altering their relationships with major Earth environments, and by altering the movement of energy and nutrients within them.

KEY TERMS

biosphere, p. 508
ecosystem, p. 508
biome, p. 511
photosynthesis, p. 512
autotrophs, p. 512
chlorophyll, p. 512
carbohydrates, p. 512
primary producers, p. 513
gross primary production, p. 513
respiration, p. 513
net primary production, p. 514
biomass, p. 514
heterotroph, p. 514
herbivore, p. 514
carnivore, p. 514
omnivore, p. 514
food chain, p. 514
detrital food chain, p. 514
decomposers, p. 515
food web, p. 515
trophic level, p. 516
nutrients, p. 518
macronutrients, p. 518
micronutrients, p. 518
biogeochemical cycles, p. 519
tolerance, p. 522
optimum, p. 522
competition, p. 522
ecologic niche, p. 523
ecologic succession, p. 523
climax, p. 527
homeostasis, p. 528
evolution, p. 530
natural selection, p. 530
eutrophication, p. 531
accelerated eutrophication, p. 531
conservation, p. 534
gene pool, p. 534

QUESTIONS FOR REVIEW AND EXPLORATION

1. Draw a sketch showing how the major Earth environments interact in forming the biosphere.
2. What are the differences between ecology and biogeography?
3. Show how the ecosystems in your area fit into the nested sequence described in Figure 19-2.
4. What are the similarities and differences between the flow of energy and the flow of nutrients in an ecosystem?
5. Construct a diagram like Figure 19-12 for the oxygen cycle. Show how it differs from the nitrogen and the phosphorus cycles.
6. How and why do ecosystems change over time?
7. At what stage of succession are ecosystems in your locality? How do you think they will change over time?
8. Show how humankind *(Homo sapiens)* fits into a classification of all living things.
9. What is natural selection?
10. Describe how the processes of evolution and isolation have interacted to lead to the pattern of species distribution in the world today.

FURTHER READING

Alvarez F, Asaro F, Courtillot V: What caused the mass extinction? *Scientific American*, October 1990, 42-61. A debate between scientists proposing two theories for the mass extinction of the dinosaurs. Alvarez and Asaro propose the "meteorite theory," Cortillot the "volcanic eruption theory."

Coleman G, and Coleman WJ: How plants make oxygen, *Scientific American*, February 1990, 45-51. This paper describes in some detail how oxygen is generated as a result of photosynthesis.

Dale VH: Mount St. Helens, *National Geographic Research and Exploration*, 1991, 7(3) 328-341. This paper describe the revegetation of the Mount St. Helens debris slide, and the processes of ecological succession it initiated.

Duplaix N: South Florida water: paying the price, *National Geographic*, July 1990, 81-113. An account of the agricultural and urban use of water in this region, and its effects on the Everglades National Park.

Odum EP: *Ecology and our endangered life-support system*, 1989, Sinauer. A good overview of the structure and processes of the ecosystem.

THE LIVING-ORGANISM ENVIRONMENT

Spain, Castille and Leon plowed fields west of Salamanca

Robert Frerck/Odyssey/Chicago

Soil Structure, Processes, and Environment

TER

…orms a patchy covering
…ls our own existence and
… of the land. Without soil,
…n could not grow and
… could survive.
…-based life depends on the
…at soil depends on life, its
…nance of its true nature
…g plants and animals. For
… of a marvelous interaction
…t materials were gathered
… fiery streams, as waters
…ents wore away even the
… and ice split and shat-
… to work their creative
…rials became soil.

Life not only formed the soil, but other living things of incredible abundance and diversity now exist within it; if this were not so the soil would be a dead and sterile thing. By their presence and by their activities the myriad organisms of the soil make it capable of supporting the earth's green mantle.

Rachel Carson, Silent Spring, 1962

terial. The soil-forming processes, particularly translocation, tend to differentiate the soil profile into a series of horizontal layers, or **soil horizons** (Figure 20-3, *A*). These are given letters according to their genesis and their relative position in the profile. Not all the horizons described here are present in every soil. The O horizon consists of organic material at or near the soil surface. This material is in the process of being broken down and is not yet incorporated into the soil. The A horizon, beneath the O horizon, is what a gardener or farmer would call the topsoil. It is the rooting zone for most plants. Below this comes the B horizon. The depth at which the A horizon grades into the B horizon varies from one soil to another; typically the two horizons are identified by the action of translocation. In many soils formed in humid climates, for example, nutrients and fine particles are washed out of the A horizon and deposited in the B horizon. In this case the A horizon is described as an E (for eluviated) horizon, and the B horizon is assigned additional descriptive letters to show the source of its added material. The B horizon grades into the C horizon, which is largely undifferentiated parent material.

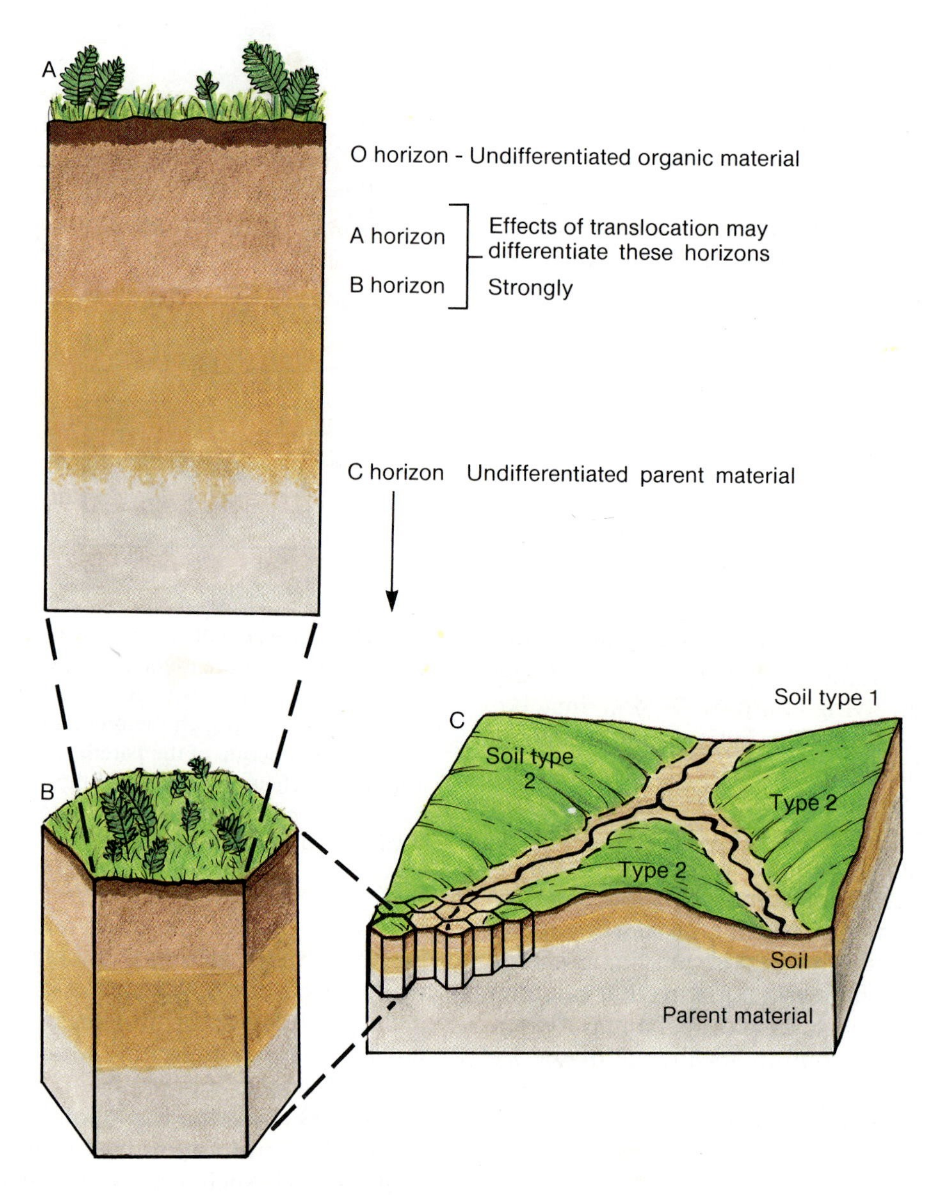

Figure 20-3 **A,** A soil profile, differentiated into soil horizons by the translocation of materials. **B,** The soil pedon, a three-dimensional representation of the soil profile. **C,** Pedons fit together to form the soil landscape and soil types.

Soil is a three-dimensional body that varies continuously over the landscape. The concept of a soil profile is extended to three dimensions using the concept of a **soil pedon,** an identifiable unit of more or less uniform soil. Soil pedons are grouped together to form polypedons and soil types for the purpose of soil survey and mapping (Figure 20-3, *B, C*). The variability of soil types at Earth's surface is caused by variations in the soil-forming processes and the materials that they have to work on.

SOIL COMPONENTS

A soil is made up of four elements: the inorganic or mineral fraction (derived from the parent material), organic material, air, and water. The abundance of each component and its importance in the functioning of the soil system vary from horizon to horizon and from one soil to another.

Mineral Fraction

The mineral fraction is the remains of the weathered parent material, broken down to its individual grains. It is the relatively static physical framework of the soil and in most cases is by far the largest constituent. Its physical and chemical characteristics are largely governed by those of the parent material. The terms *sand, silt,* and *clay* are given to mineral grains that fall into specific size ranges (Table 20-1). Hard minerals within the parent material end up as large, durable sand grains; softer minerals weather down to smaller silt and clay fragments and to individual chemical elements. The relative proportions of sand, silt, and clay in the soil determine the important property of soil texture, discussed in the section on soil characteristics below. The very finest particles, called **colloids,** are particularly important in keeping nutrients in the soil and making them available to plants.

Table 20-1 The Relative Sizes of Sand, Silt, and Clay Particles

PARTICLE	GRANULES OF COMPARABLE SIZE	SIZE OF GRANULES IN MM
Gravel	Hazelnuts	>2
Coarse sand	Builder's sand	0.2-2
Fine sand	Castor sugar	0.02-0.2
Silt	Confectioner's sugar	0.002-0.02
Clay	Putty	<0.002

Organic Matter

The organic matter in the soil has two components. The first is the waste products from living plants and animals and the body tissue of dead ones. It is in various stages of being broken down to an inorganic form by the action of the organisms in the detrital food chain. The nutrients held in the biomass, litter, and soil pools are constantly being cycled and exchanged with other Earth environments in this process (Chapter 19). The second component is the millions of living plants, animals, and bacteria that move materials around the soil. Organic matter in the soil usually represents about 2 to 6 percent by volume in the surface horizons but is much more important than this figure suggests.

The end product of the breakdown of dead organic material is **humus,** a structureless, dark brown or black jelly found beneath the soil surface. It varies in type and distribution from one soil to another, depending on the sort of organic material and the conditions within the soil. The tiny organic particles bind to the mineral colloids to form the **clay-humus complex.** The clay-humus complex is important in controlling soil fertility, as is discussed in the section on soil chemistry below.

Soil Water

Soil water, its dissolved elements, and suspended particles make up the **soil solution,** the liquid part of the soil. The soil solution, together with soil air, fills up the spaces or pores among the mineral and organic components of the soil. It is the main agent of translocation, carrying dissolved chemical elements and small particles through the soil. Soil water is required for plant growth, for the functioning of soil organisms and microorganisms, and for the continued weathering of the parent material.

Water is held in the soil by the attraction of water molecules to each other and to soil particles. The same powers of cohesion hold a droplet of water together as it clings to the side of a glass. Soil water exists in one of three states in the soil. The amount held in each state changes over time, which affects the availability of water to plants and the potential movement of materials and nutrients. **Hygroscopic water** is held tightly, close to the surface of individual mineral grains (Figure 20-4). It is effectively unavailable to plants because the attraction between the water and the mineral grain is greater than the "sucking power" of plants' roots. A soil in which the water is nearly all hygroscopic is said to be at its **wilting point** because under this condition, plants can obtain no moisture from the soil.

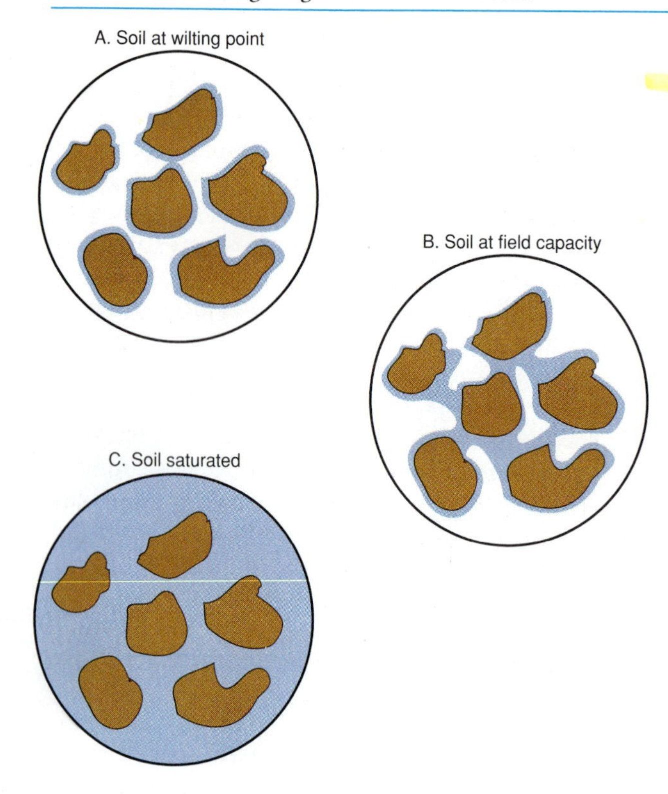

Figure 20-4 Water in the soil. **A,** At wilting point the remaining water, mainly hygroscopic, is held tightly to the soil particles and cannot be drawn away by plant roots. **B,** At field capacity capillary water fills the small pore spaces in the soil and air fills the larger pore spaces. Capillary water moves slowly through the soil in response to pressure gradients between wet and dry areas. **C,** In a saturated soil all pore spaces are filled with water and air is exluded from the soil. A proportion of this water will normally drain from the soil as gravitational water, leaving the soil at field capacity.

Other water molecules can link up with the hygroscopic water attracted by the mineral grain (Figure 20-4, *B*). These molecules are held less tightly; they are called **capillary water**, and they can move in any direction in the soil and are available to plants. Capillary water remains in the soil because the combined attraction of the water molecules to each other and to the soil particles is greater than the force exerted by gravity. It moves from a wet to a dry area in the soil in response to variations in pressure similar to those found in the atmosphere. The most common movement is upward, toward roots and the soil surface, where water is lost by evaporation and transpiration.

The third type of soil water is **gravitational water,** which is not held in the soil at all but moves freely under the influence of gravity (20-4, *C*). It is available to plants, but only temporarily. It is present immediately after rain but drains downward quickly. A soil with its maximum quota of capillary water but no excess gravitational water is said to be at **field capacity**—an ideal state for plant growth. The amount of water held in each state of availability depends on conditions in the soil, not on the properties of the water itself. The most important variable is the size of the soil particles and therefore the size of the pores in between them. Water drains rapidly from the large pores in a coarse-grained, sandy soil, but is held close to the soil particles in the smaller pores of a fine textured clay. For the same reason, fine-grained soils will hold more water at wilting point.

Soil Air

Soil air occupies the pore spaces not currently filled by water. A soil at field capacity contains a good balance of air and water. Air in the soil performs much the same function as it does in the free atmosphere. It brings oxygen to the soil organisms, including plant roots, and removes carbon dioxide. The **aeration** of soil is the ability of the soil air to perform this exchange of gases and can be critical in plant growth. Aeration is affected primarily by the balance of air and water in the soil, but also by the speed with which soil air can be exchanged for new atmospheric air. The composition of soil air differs from that of free atmospheric air because soil air is "compartmentalized" by intervening fragments of soil and water. It may vary considerably in composition over very short distances, depending on the biochemical processes it is involved in. Plants that grow in constantly waterlogged soils may evolve elaborate mechanisms to collect air. Mangroves, for example, are trees that grow in fine sediments at the coast. They have developed roots that grow upward through the sediment to reach tidal water and air. The roots have special pores that are permeable to air but not to water.

SOIL CHARACTERISTICS

The nature and distribution of the four soil components and some of the processes that act on them produce a number of soil characteristics including texture, structure, color, and chemistry (including acidity). The soil scientist uses these characteristics to diagnose the soil's ability to support plant growth, and therefore to supply energy and nutrients to the living-organism environment. For example, a light-colored and acid sandy layer near the top of the profile, underlain by a rich brown clay horizon, shows that material is being washed from the upper (depleted) horizon to the lower (enriched) horizon. A soil with these characteristics is unlikely to support grain crops, but could be used for coniferous trees with their deeper roots.

Texture

Soil texture is a description of the relative proportions of sand, silt, and clay in the soil. Figure 20-5 shows the terms used internationally for soils with different textures. Soil texture affects the soil's ability to hold moisture and nutrients and make them available to plants. The presence of the colloids, the smallest clay particles, is particularly important; so much so that the term "clay" is included in the name of any soil composed of more than about 25 percent clay. Soils with extreme textures are poor media for plant growth, as demonstrated in many suburban yards. The large grains and pore spaces of sandy soils mean that they drain rapidly, and can hold little capillary water or nutrients for plants. In contrast, clay soils are quite easily waterlogged or have all their water in hygroscopic form, preventing the movement of oxygen to plant roots and effective respiration in the detrital food chain (Chapter 19). Clay soils are also compact, presenting a physical barrier to plant roots. The ideal soil for plant growth is a loam, which has a high proportion of silt and little clay. Texture is a fundamental characteristic of the soil; it is not easily altered by management. It is usually assessed by specialized equipment in the laboratory, but, with practice, it can be estimated quite easily by touch in the field.

Soil Structure

Soil structure describes how individual grains of soil are bound together into clumps or **peds.** Soil aggregated in this way is a better medium for plant growth than a structureless mass of single grains, because it is more stable and less easily moved by wind or water. Of these the ideal for plant growth is a crumb structure, because it gives plant roots

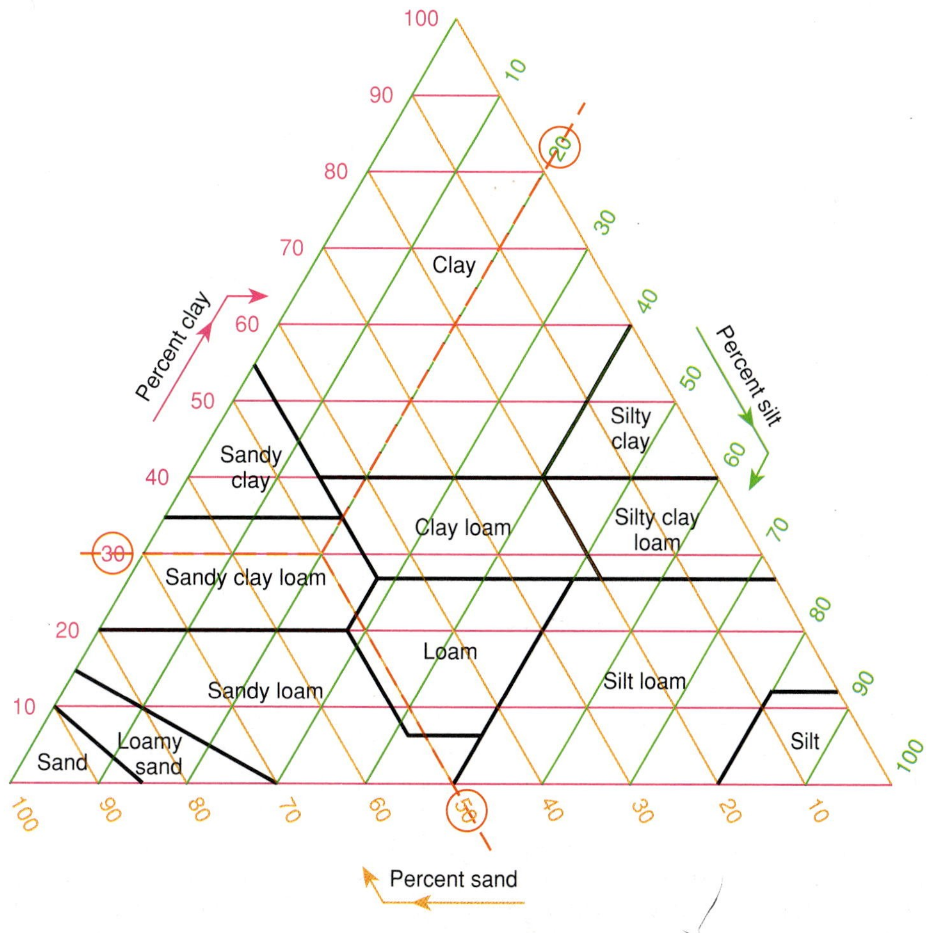

Figure 20-5 The percentage of sand, silt, and clay in the major soil texture classes. A soil with 20% silt, 30% clay, and 50% sand is a sandy clay loam, as highlighted on the diagram. In which texture class does a soil with 10% sand, 10% clay, and 80% silt fall?

ready access to both water and nutrients and allows ample aeration of the soil. The mechanisms that control soil structure are complex and incompletely understood. Unlike soil texture, though, soil structure *can* be influenced by human management. Poor structure that hampers plant growth is often the result of mismanagement. Together, structure and texture control both the size and distribution of pore spaces in the soil (soil porosity) and how quickly water can move through the profile (soil permeability).

The most important factor in promoting good structure is to maintain a reasonable level of organic matter in the soil. Humus helps to bind together the individual grains into aggregates. Other factors influencing soil structure are the processes that break apart rocks in weathering (Chapter 13). They include cyclical wetting and drying (and therefore expansion and contraction) of the fine clay grains, alternate freezing and thawing of the soil, and the stirring action of roots and soil animals.

Soil Chemistry

Soil chemistry describes the type and abundance of chemical elements in the soil. Two interdependent parts of soil chemistry are particularly important to soil's role in supporting the ecosystem. These are soil fertility, which is the ability of the soil to supply nutrients to plants, and soil acidity, which affects the action and composition of soil microorganisms, the rate of breakdown of organic material, and the availability of nutrients in the soil.

Soil nutrients are held at three sites in the soil. First, most nutrients are unavailable to plants, locked up in the remaining unaltered regolith and in intact organic material. Second, the soil nutrients available to plants are mostly held on the clay-humus complex. Third, a smaller amount is available in the soil solution. Soil fertility depends on the distribution of useful nutrients between the available and unavailable forms. The main inputs of available nutrients come from the breakdown of parent material and decaying organic matter. The main outputs are the uptake of nutrients by plants and their loss in water that drains out of the soil. The clay-humus complex largely controls the flow from input to output.

The tiny particles, or colloids, of mineral and organic matter that make up the clay-humus complex carry electrical charges on their surfaces, which are mainly negative. Most nutrient ions released from weathering of parent materials, such as calcium (Ca^{++}), potassium (K^{+}), sodium (Na^{+}), and magnesium (Mg^{++}), are positively charged ions called *cations*. The cations are attracted to the clay and humus particles. Negatively charged nutrient ions, called *anions*, such as sulfur, phosphorus, and nitrogen (which can also exist as a cation), come from the breakdown of organic material. These are also held by the clay-humus complex, at the smaller number of sites with positive surface charges.

Both anions and cations can be exchanged among the clay-humus complex, the surrounding soil water, and the plant roots. Some cations are less strongly attracted to the colloids than others. For example, potassium and sodium are easily dislodged, and are replaced by ions of aluminum or hydrogen, which bind more strongly to the particle of clay or humus. The replacement of other nutrient ions by hydrogen is particularly common under conditions of heavy rainfall, because rainwater (especially acidic rainwater) provides abundant hydrogen ions. The ability of the soil to hold nutrients is measured by the **cation exchange capacity,** commonly abbreviated to CEC. The CEC measures how many cations can be adsorbed by a unit of soil. A soil with a high CEC has a high fertility. An anion exchange capacity can be calculated in the same way. Soils that contain colloids of clay and humus tend to have higher CECs than soils that contain only large grains.

Soil acidity, measured on the pH scale, reflects the concentration of dissolved hydrogen ions in the soil. It ranges from 0 to 14, where 0 is the most acid (having the highest concentration of hydrogen ions) and 14 the most alkaline (lowest concentration). Distilled water has a neutral pH reading of 7. Most plants grow best in a pH of between 6 and 7, although some species flourish under more extreme conditions. Figure 20-6 shows how the pH readings of agricultural soils compare to those of common household products. Soil acidity is an important soil characteristic because it affects the chemical state and mobility of other nutrients and materials.

Acid soils occur where there is a poor supply of nutrient ions from the mineral fraction and organic material, or where rainfall is high and the nutrient ions are replaced by hydrogen and physically washed out of the soil in solution. Where both conditions occur, the soils are very acidic. Under very acid conditions the action of the detrital food chain is slowed down and a deep and incompletely broken down organic layer can accumulate, resulting in the formation of *peat*.

Alkaline soils are typical of arid and semiarid regions. In these areas water is in short supply and so therefore are hydrogen ions; the limited amount

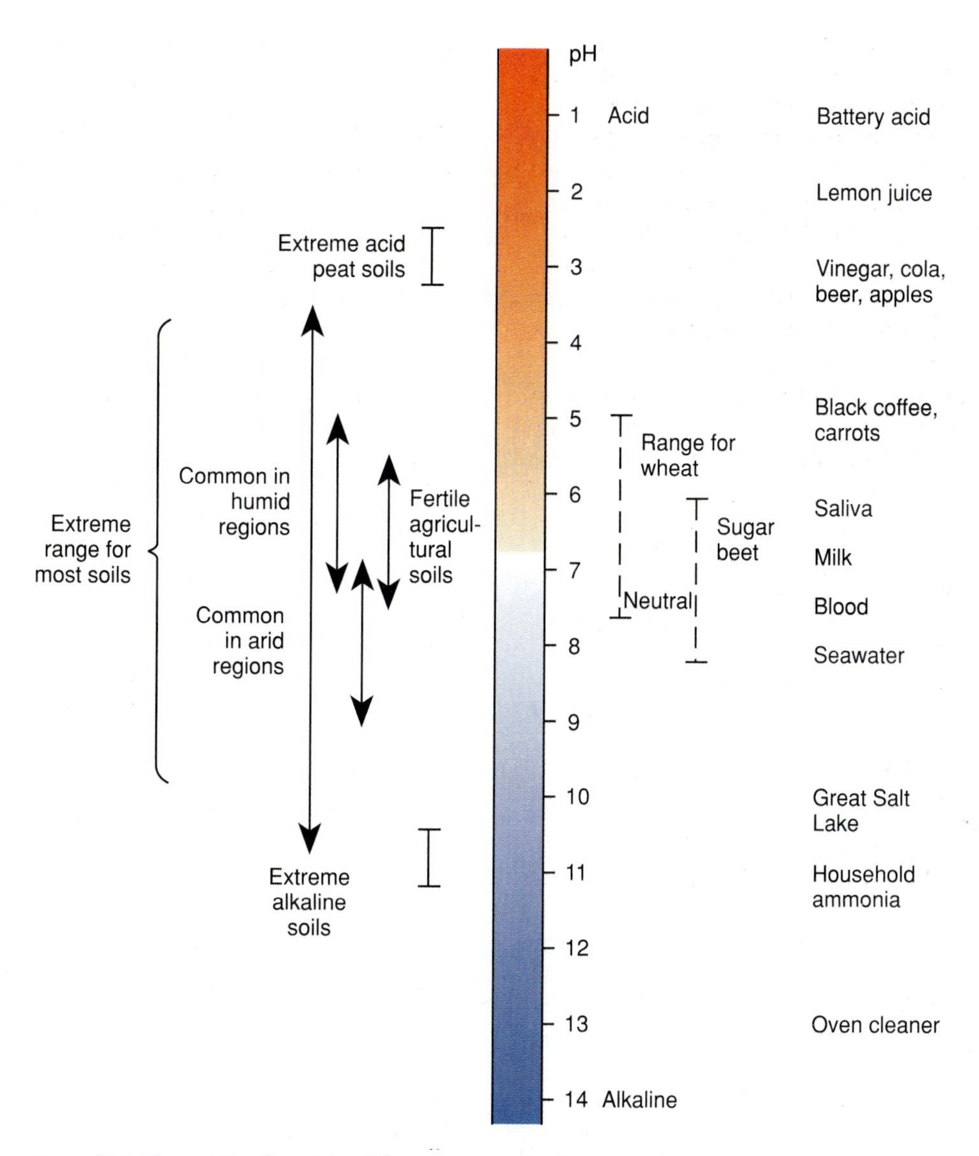

Figure 20-6 The acidity found in different types of soils and the tolerance range of some common crops. The pH values of some household goods are included for comparison.

of water present may in fact move upward through the soil in response to evaporation from the surface. Nutrients are not washed out of alkaline soils but can accumulate in concentrations that are dangerously high for plant growth.

Soil Color

Color is one of the more obvious soil characteristics and is a good indicator of the processes operating in the profile. Most soils can be described as black, brown, orange, red, yellow, gray (including gray-blue or gray-green), and white. To make sure that everyone describes soil color in the same way, an international descriptive code has been set up, based on a series of colored chips similar to those used to choose paint (Figure 20-7). Soil color is controlled by the distribution of organic material in the profile and by the state of any iron present. Soils with little organic matter are white or gray; those with organic matter are brown or black. Where iron has been oxidized (Chapter 13), the soils have a red tint. This can vary from a slight warming of a basically brown color to a full-blown red soil. Where iron is not oxidized, the dominant colors are gray, gray-blue, and gray-green. These colors are particularly common in waterlogged soils.

The description of soil color itself is not partic-

Figure 20-7 **A,** A Munsell soil color chart. Each colored chip on the page has three index figures and a short written description that identify it uniquely. Separate pages take the scientist through shades of yellow, red, gray, and green. The chart is named for Albert Munsell, who developed the naming system in 1905 as a way of systematically describing the colors used in an atlas. **B,** This handful of soil shows the considerable variations that are possible in the "brown" color of soil.

ularly instructive. Its importance lies in the clues it can give about the processes operating in the soil. For example, as noted at the start of this section, a bleached white topsoil or E horizon underlain by a rich brown or red-brown B horizon shows that organic material and iron have been moved from the upper horizon to the lower. In the sections below on pedogenic regimes and soil classifications, note the differences in color among different soils and among horizons of the same profile.

FACTORS AFFECTING SOIL FORMATION

The differences in soil characteristics found across the world, or even within the United States, demonstrate that the processes of soil formation vary in intensity and interact in different ways on different raw materials in different places. This section examines the factors that control the soil-forming processes and how they interact.

V.V. Dokuchaiev, working in the late 1800s, was among the first to recognize that soils form as the result of the interaction of the Earth environments. Building on this work in the 1930s and 1940s, Hans Jenny stated that soils were the product of the direct and indirect interaction of climate, parent material, topography, organisms, and time (Figure 20-8).

Climate

Climate is the dominant factor in soil formation and affects all three soil-forming processes. First, it controls how fast weathering of parent material proceeds and the degree to which fragments of rock are broken down to their constituent minerals. In general, high temperatures and high precipitation increase the rate of chemical weathering. Second, climate is a major influence on the type of ecologic niches available in the ecosystem, and therefore on the number and type of organisms in the aboveground and the detrital food webs. This in turn

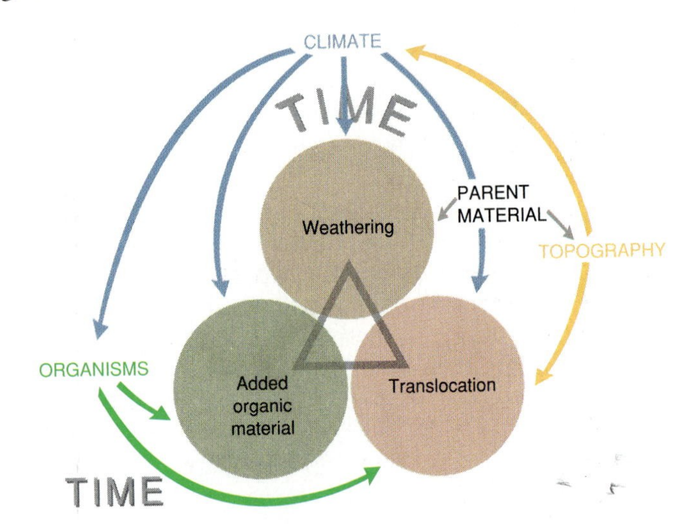

Figure 20-8 The effect of the soil-forming factors (climate, parent material, topography, organisms, and time) on the soil-forming processes (weathering, translocation, and the addition of organic matter). The soil-forming factors interact among themselves, as do the soil-forming processes. Virtually all elements of this diagram influence, or are influenced by, the others. Only the main linkages are shown here.

determines the type and amount of dead organic material added to the soil and how well it is broken down and moved around in the soil. Third, climate determines the balance between precipitation and temperature, influencing the vertical movement of water and soluble nutrients in the soil.

When rainfall is high and potential evapotranspiration (Chapter 6) is low, rainwater moves down through the profile. Hydrogen ions in the soil solution exchange with nutrient cations on the clay-humus complex and the nutrients are washed out of the soil. This process is so common in soil development that it has been given the special name of **leaching.** Where the downward movement of water is strong, the finest clay particles may be suspended in the soil solution and washed down the profile. The physical transport of particles in suspension is termed **eluviation** (from the Latin *ex* or *e*, meaning out, and *lavere*, meaning to wash). **Illuviation** (from the Latin *il*, meaning in, and *lavere*) refers to the redeposition of materials elsewhere in the profile. In the humid midlatitudes, leaching typically results in a nutrient-deficient or *eluviated* upper horizon and an enriched or *illuviated* B horizon. The eluviated horizon is often called simply the E horizon. The extent to which the horizons are differentiated depends on the intensity of leaching, which is controlled by the amount and intensity of rainfall, the type of vegetation, and the drainage of the soil and parent material. In the heavy rainfalls of the humid tropics, leaching can be so severe that nutrients are washed straight out of the soil, with no redeposition.

Where rainfall is low and potential evapotranspiration (PET) is high, any water that infiltrates the soil will later be drawn upward toward the surface and toward plant roots. Nutrients are also drawn upward and deposited near the surface. In both cases the result is a clear differentiation between the upper and lower horizons in the soil profile.

It is difficult to think of any characteristic of the soil or its processes that is not in some way controlled by climate. The global distribution of soils therefore closely matches the distribution of major climatic regimes and world vegetation types (discussed in Chapter 8 and Chapter 21, respectively). On a local scale soils vary with the microclimates formed by altitude and aspect.

Parent Material

The parent material for soil is regolith weathered from an exposed rock surface. Soils may develop directly over the weathered surface, or after the fragments have been transported and deposited by water, ice, or wind (Chapters 14 through 16). The nature of the parent material affects the soil chemistry and the soil texture. Parent materials that contain a high proportion of nutrient cations (such as calcium, sodium, potassium, magnesium, and iron) maintain a reasonable soil fertility even if the nutrients are regularly washed out of the soil in rainwater. Under the same climatic conditions parent materials without these cations, such as dune sand and sandstone, produce an acidic soil, high in hydrogen ions but low in other nutrients. Sandstone is coarse-grained and therefore produces a sandy soil with a low colloidal content that drains rapidly. Some igneous and metamorphic rocks weather to complex clay minerals that contribute to the clay-humus complex and help to hold nutrient ions and water in the soil. In a given climate, parent materials that weather easily produce large amounts of regolith and eventually a deep soil. Resistant rocks produce shallow and often stony soils.

Landforms

The detail of the surface-relief environment at a specific site acts as a stage for the major soil-forming factors. The principal variable is slope, which affects the movement of both materials and water. Soils that develop on steep slopes are subject to soil erosion through surface wash and creep. This results in thin soils and, on a vegetated surface undergoing creep, a characteristic stepped surface. Water tends to follow the angle of slope and move downhill at the surface or through the upper horizons. Vertical percolation of water through the profile is reduced. Soils at the foot of a slope collect both materials and water from above. The waterlogged soils of wetland areas are typically the result of such local variation in topography. They have low aeration and therefore support specialized and sometimes unique biologic communities. Once drained, wetlands may produce fertile soils. The details of surface relief at a site can lead to a considerable change in soil conditions over a short distance. This is well illustrated by a soil *catena*, a sequence of soils that develops on a slope under uniform conditions of climate and parent material (Figure 20-9). The elevation and aspect of a site also affect soil processes by influencing the effects of climate.

Soil Organisms

Soil organisms are the living part of the soil ecosystem. Their presence and activity are controlled by their physical environment and by the flows of energy and nutrients they receive. Some soil organisms, particularly the bacteria and algae, live almost completely within the soil. They are part of

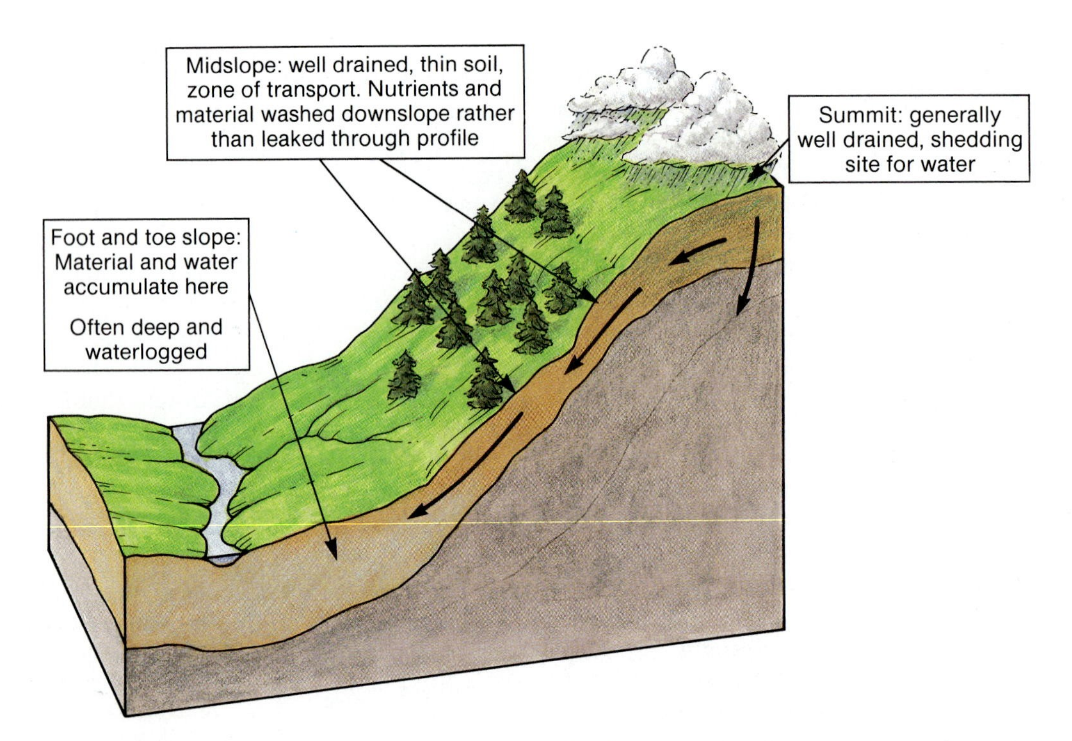

Figure 20-9 An example of a soil catena, in which soils at different positions on a gradient develop markedly different profiles. Differences in soil development due to topography are important controls on the local-scale distribution of vegetation in an ecosystem.

the detrital food chain, gaining energy from the carbohydrate molecules in the dead plant and animal tissues that fall to the ground as waste material. Their action on the soil is mainly chemical, in that they break down the dead organic matter in place. Some of the larger soil fauna have a physical as well as a chemical effect on the soil. They dig and burrow and move both organic and inorganic material about. Earthworms, for example, literally eat their way through the soil. Up to 6 tons of soil per hectare (15 tons per acre) may pass through their bodies in a year. In most soils, earthworms, termites, ants, and the soil microflora are vital to the soil-forming processes of translocation, and the incorporation of dead organic material. The higher plants growing at the soil surface also influence the soil through the addition of organic matter and the penetration of roots. Grasses probe the soil with thousands of small roots; trees dig deeper with fewer, larger roots.

Time

The soil-forming processes of weathering, addition of organic material, and translocation operate over time. As shown in Figure 20-2, beginning with unaltered parent material, a soil typically develops through the continued action of the soil-forming processes over hundreds and thousands of years, to a "mature" profile. If undisturbed, the developed soil exists in a dynamic equilibrium with its environment. It will change only if the soil-forming processes alter substantially in response to change in the controlling factors described here. In this sense the developed soil is comparable to any other steady-state system such as a stream channel or an ice mass.

Over time, the influence of each of the factors listed above changes. Initially the interaction between climate and parent material (or between the atmosphere-ocean environment and the solid-earth environment) is dominant. Through the process of weathering, this interaction controls the nature of the regolith and how quickly it accumulates. Soon the rock surfaces are colonized by living things and the living-organism environment starts to contribute, providing the detrital food chain and the beginnings of a nutrient cycle, thus ensuring that organic material will be added through the process of ecological succession. The surface-relief environment influences proceedings throughout. It provides a location for the regolith to accumulate or be deposited, and for soil to form, but its main influence is in controlling the movement or translocation of materials through the soil in water. As the soil develops, the process of weathering becomes less dominant because the parent material is protected beneath the soil. The dominant processes are then the addition, incorporation, and cycling of organic material and the translocation of materials up and down the profile. Climate, soil organisms, and landforms therefore have a greater day-to-day influence on changes in the developed soil than the nature of the parent material. The parent material continues to influence soil development because it determines the texture and availability of nutrient cations.

Young soils are widespread in Earth's living-organism environment, and in any classification are usually grouped together in their own category. They include soils forming on recently exposed surfaces such as glacial deposits, recent flood plains, or fresh lava. In areas unaltered by tectonic activity, or where processes in the surface-relief environment act slowly, developed soils may have existed for hundreds of thousands of years. The most dramatic changes experienced by these soils have been some of the climatic changes described in Chapter 9, and more recently and more dramatically, the effects of human activity.

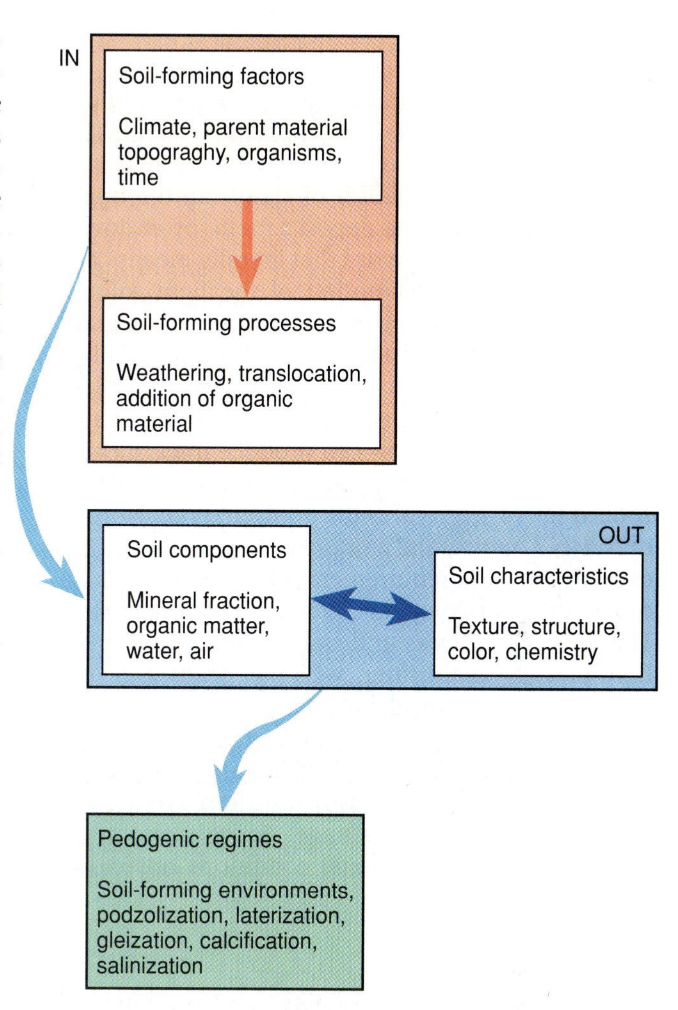

Figure 20-10 The linkages between the elements of soil science studied in this chapter. The soil-forming processes and soil-forming factors work on the soil components, producing a range of conditions recognizable through the soil characteristics. The different types of conditions can be summarized by a limited number of pedogenic regimes.

PEDOGENIC REGIMES

Figure 20-10 shows how the subjects examined so far in this chapter fit together, and how this examination of soils will be completed. The three soil-forming processes are controlled by five factors, one each from the atmosphere-ocean, solid-earth, surface-relief, and living-organism environments, all acting over time, the fifth factor. Together, the soil-forming processes and the factors that influence them determine the basic nature of the soil. The initial result of their interaction is a soil made up of four components—mineral fraction, organic matter, water, and air. In a well-developed soil the soil-forming processes have worked on these components to produce soil characteristics—texture, structure, color, and chemistry. The components and characteristics of soil develop together and are interdependent. For example, soil chemistry and the changing proportions of water and air in the soil depend on texture. Soil structure is largely dependent on texture and on the amount of organic matter in the soil.

Given the possible variations in the factors that control soil formation, there is a tremendous potential for variety in soil characteristics at Earth's surface. In practice, though, the soil-forming factors and processes fall naturally into a limited set of soil-forming environments, called **pedogenic regimes**, which produce soils with similar characteristics around the world. There are five major pedogenic regimes: podzolization, laterization, gleization, calcification, and salinization, each of which is described below. Soil scientists and physical geographers can examine the characteristics of a soil in the field, and, through their knowledge of soil-forming processes, work backward through the sequence shown in Figure 20-10 to understand its origin.

Figure 20-13 A gleyed soil from Brazil. In this example air has entered the soil to allow the oxidation of iron, resulting in red and yellow mottling of the otherwise gray profile.

matches potential evapotranspiration. However, there is still some downward movement of water, particularly after rainfall or snowmelt. This gentle leaching displaces calcium carbonate from the upper horizons and redeposits it at depth as a *calcic* horizon. The other nutrient ions remain in the upper horizon. The calcic horizon is recognizable in the profile by its light color. The degree of differentiation between the A horizon and the calcic horizon, and their depth, depend on the strength of leaching.

Calcification is typical of the soils found under midlatitude grasslands. Compared to podzolization and laterization it is a moderate process and creates fertile soils. Most of the nutrients are retained in the upper horizons, the climatic regimes are favorable to soil organisms, and organic material is readily incorporated into the profile. The best-known examples of calcified soils are found in the eastern part of the Great Plains of the United States, the Ukraine, and in similar steppe or prairie landscapes of the middle latitudes. Figure 20-14 shows an example of calcified soils and the ecosystem that it supports.

Salinization

The final major pedogenic regime is **salinization,** which, as the name suggests, is characterized by the accumulation of salts, mainly compounds of sodium, calcium, and magnesium. Unlike the other regimes, it is dominated by *upward* translocation of materials in water. Salinization is typical of arid regions where evaporation is high; it is particularly common in low-lying or flat areas where subsurface drainage is poor. In this environment, moisture is quickly evaporated from the surface and the upper horizons, creating a moisture gradient between the surface and the lower horizons. Capillary water is drawn up from supplies lower in the profile by capillary action, in much the same way as kerosene moves up the wick of a lamp. This water carries dissolved salts upward, and when it evaporates, it leaves salts concentrated near the surface. This process ends in the development of a salty or *salic* horizon, which is usually white and clearly recognizable (Figure 20-15). If evaporation is particularly strong, the salts are deposited at the soil surface.

A

B

Figure 20-14 **A,** A calcified prairie soil from Texas. The soil has a loamy texture and the deep brown color shows its high organic content. This example has formed over calcareous deposits. **B,** The typical grassland community supported by calcified soils.

In most cases a concentration of salts in the soil is toxic to plants and to soil organisms. Salic (salty) soils in their natural state are therefore virtually useless for agriculture. However, the soil chemistry can be improved by irrigation and drainage, which flush the salts out of the soil. Ironically, irrigation without adequate drainage only exacerbates the problem and is a major cause of salinization in arid areas. The added water has three effects. First, the water itself contains traces of salts, and these are added to the soil. Second, irrigation ensures a continuous supply of water for evaporation, so that salts continue to move up the profile and accumulate near the surface. Third, irrigation may raise the water table to the extent that the plant roots are in a saline solution. Under these conditions normal biologic processes are reversed. Instead of drawing water from the soil, plant roots may lose water to the soil solution, causing the desiccation and death of the crop.

The first records of salinization caused by irrigation are about 5000 years old and come from present day Iraq (ancient Mesopotamia). In 3500 B.C. the Sumerian people grew two main crops, wheat and barley, in almost equal proportions. By about 2500 B.C. only about 15 percent of the land

Figure 20-15 A salt-affected soil from Botswana. The upper horizons and soil surface are dominated by salt.

grew wheat, and by 1700 B.C. no wheat was grown at all. Wheat is not as tolerant of salinization as barley, and its decline marks the onset of this condition in the soil. In the United States the problem dates from the early 1900s and is particularly serious in the Central Valley of California.

SOIL CLASSIFICATION

Each of the pedogenic regimes, or soil-forming environments, summarizes a particular range and combination of conditions and processes in the soil (Figure 20-10). The regimes represent the five extreme soil types that make up Earth's surface. In practice, soils vary gradually and continuously within and among the regimes, creating a detailed mosaic of soil types at Earth's surface, each with slightly different characteristics. Given the importance of soils in the ecosystem and their impact on human activity, it is important that physical geographers make some sort of order out of this variation and communicate information about the nature and distribution of soils in the landscape. The geographer does this through soil classification.

The classification of soils is similar to that of any other naturally occurring feature. The variability along one or more factors is gauged and broken down into manageable blocks, giving each block a name and a set of identifying characteristics. Classifications used in the United States in the 1930s and 1940s were genetic, based on considerations of each soil's origins. The names used in this classification linger on in the terminology of the pedogenic regimes. Podzolization led to "podzols," laterization to "latosols" and "laterites," and calcification to "calcisols." A major problem with this classification, however, was that it depended on agreement about the soil's origins, which were not always certain. The genetic classification of the thirties and forties has been replaced by the *United States Comprehensive Soil Classification System,* also called **Soil Taxonomy,** developed by the Soil Survey staff of the United States Department of Agriculture since the 1950s.

Soil Taxonomy is based on the present-day characteristics of the soil. No assumption is made about the soil's formation. However, as outlined in Figure 20-10, process and origin can usually be inferred from a description of the soil. Virtually all the physical, chemical, and biologic characteristics described earlier in this chapter are used to classify the soil under this system.

Soil Taxonomy divides the soils of the world into eleven **soil orders,** primarily on the basis of closely defined diagnostic horizons. Each horizon is recognized by its position in the profile and by characteristics such as thickness, color, chemistry (particularly the type and amount of nutrients held in the soil), texture (particularly the amount of illuviated clay), and structure. Figure 20-16 shows the distribution of the soil orders across the world and in the United States. The soil orders are further divided into 47 suborders, and then into about 230 great groups, 1200 subgroups, 6000 families, and 13,000 series. Each series can be further subdivided into a few minor variants or phases. Soil maps used at a local scale show the distribution of soil series and phases; the descriptions at this level have stayed more or less the same through different classifications.

The nomenclature of Soil Taxonomy is an important and occasionally bizarre characteristic of the system. Each syllable of the soil's name, at any level, provides information about its characteristics. As far as possible, the root of most syllables is drawn from Latin or Greek so that they will be recognized in many languages. Table 20-2 shows how this system works for the first level division into soil orders. Each soil order contains one descriptive syllable, a linking "i" or "o," and the syllable "sol" from the Latin *solum,* meaning "soil." Names for the lower levels of the hierarchy are constructed by adding more descriptive syllables at the beginning of the name, with additional descriptive words at the family level. The series is given a local geographic name. Table 20-3 gives some examples of full soil names. The language of Soil Taxonomy is logical, exhaustive, and unambiguous. Furthermore, the soil names convey graphic and detailed information relating to their subject. An outline of each of the soil orders and some of their characteristics is given below.

Entisols

Entisols show little or no evidence of pedogenic processes. They are found on steeply eroding slopes, on flood plains that are regularly swamped with new material, or in any place where the normal soil-forming processes do not have, or have not yet had, a chance to act. This includes newly exposed surfaces, where it is simply a matter of time before a soil develops, and soils on old but resistant parent materials such as sand dunes, where soil formation is slow or nearly nonexistent (Figure 20-17). Most of the Entisols in the United States are found in the west and in areas with steep slopes, low rainfall, and recently exposed or deposited parent material. The saline soils of the Central Valley and Imperial Valley in California are in this order. The saturated soils of the lower coastal plain in Southern Florida

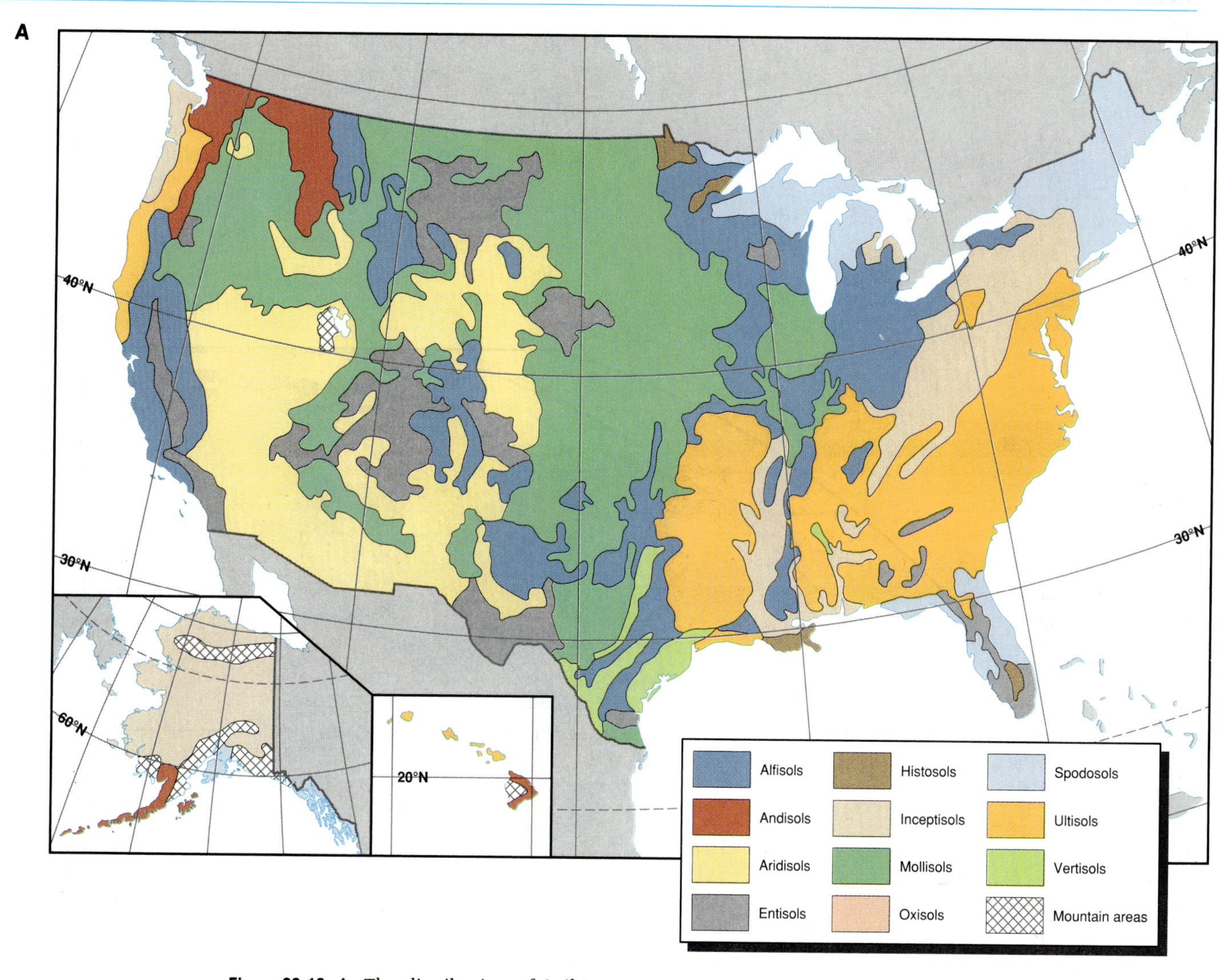

Figure 20-16 **A,** The distribution of Soil Taxonomy's 11 soil orders in the United States.

Table 20-2 Derivation of the Names of the Soil Orders in Soil Taxonomy

ORDER	SYLLABLE	DERIVATION AND MEANING	ORDER	SYLLABLE	DERIVATION AND MEANING
Entisol	Ent	(No derivation); recent soil	Spodosol	Od	*Spodos* (Greek) wood ash; ashy (podzol) soil
Andisol	And	*Ando* (Japanese) volcanic	Aridisol	Id	*Aridus* (Latin) dry; dry soil
Inceptisol	Ept	*Inceptum* (Latin) beginning; young soil	Mollisol	Oll	*Mollis* (Latin) soft; soft soil
Vertisol	Vert	*Verto* (Latin) to turn; inverted soil	Alfisol	Alf	(No derivation); aluminum (Al) and iron (F) soil
Histosol	Hist	*Histos* (Greek) tissue; organic or bog soils	Ultisol	Ult	*Ultimus* (Latin) last; ultimate (of leaching)
			Oxisol	Ox	*Oxide* (French) oxide; oxide soils

Table 20-3 Some Examples of Full Soil Names in the Soil Taxonomy System

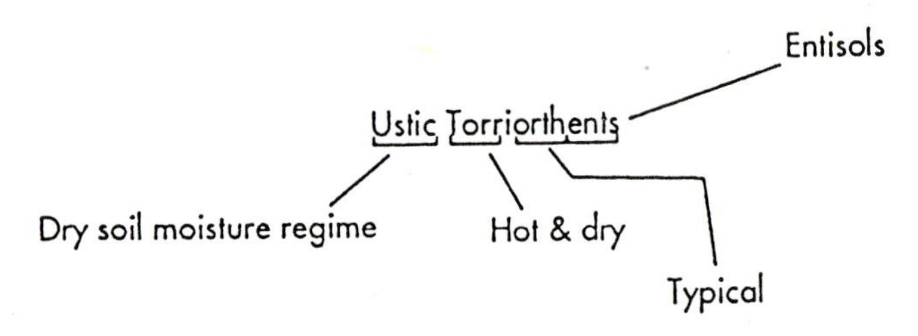

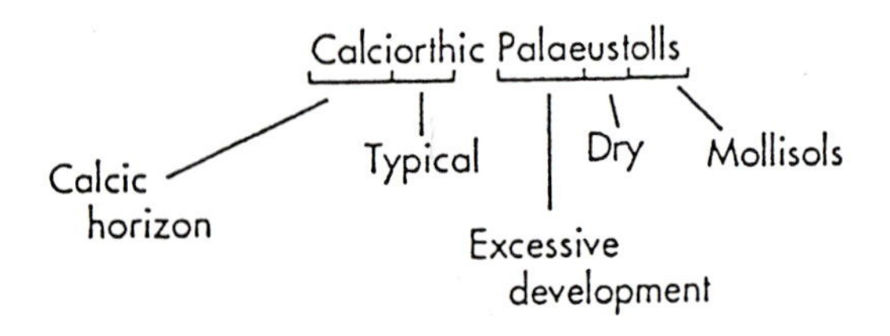

are Entisols, as are the sandy soils found in the Sand Hills of Nebraska. The latter are members of the great group Ustipsamments (on a sandy parent material with a dry climate). The Sand Hills retain some features of a stabilized dune system, are mainly grass-covered, and are used extensively for beef grazing.

Andisols

The soil order **Andisols** was introduced to Soil Taxonomy in 1990 to describe soils that develop on parent materials of volcanic origin, including ash, tephra, and solid rocks such as basalt. These soils were previously described as Entisols. They are found on or near active volcanoes, and their global distribution therefore mirrors the distribution of volcanic activity. The main distinguishing feature is the nature of the mineral fraction. They tend to also have high organic content and a high cation exchange capacity. Over time Andisols may develop to Ultisols or Oxisols.

Figure 20-17 The Great Sand Dunes National Monument, USA. The dry conditions and coarse parent material create a barren landscape. Few organisms can survive these conditions and a soil does not develop.

Inceptisols

Inceptisols are young soils that have reached a more advanced state of development than Entisols. They can be considered an early stage in the development to a "mature" soil. Although they have some degree of horizonation, weathering is incomplete and they still contain primary minerals released from the parent material but not yet broken down. Leaching occurs in most of these soils but has not been operating long enough to differentiate eluviated and illuviated horizons. Inceptisols are as varied an order as the Entisols, and they have a wide geographic distribution. Inceptisols are found on recent flood plains and over extensive areas of the cold tundra near the poles (Chapter 21). In these areas the pedogenic regime is dominated by the cold, wet conditions, and the soils have many characteristics of gleization.

Vertisols

Vertisols have a high clay content; this means that they contract and crack open when dry. Material from the surface falls into the cracks and is eventually wetted by rain (Figure 20-18). The soil is dampened because of the cracks and therefore expands from the bottom upward. This creates pressures and tension that result in the upward movement of the soil and the gradual inversion of the profile. For this reason Vertisols are also known as "self-swallowing" soils. The pressures of expansion and contraction tend to produce a typical microrelief, a few centimeters high and a meter or so wide, of alternating depressions and ridges. In common with the Entisols and Inceptisols, the Vertisols have a wide geographic distribution that reaches all climatic environments. The division into suborders, however, is defined by climate, which determines

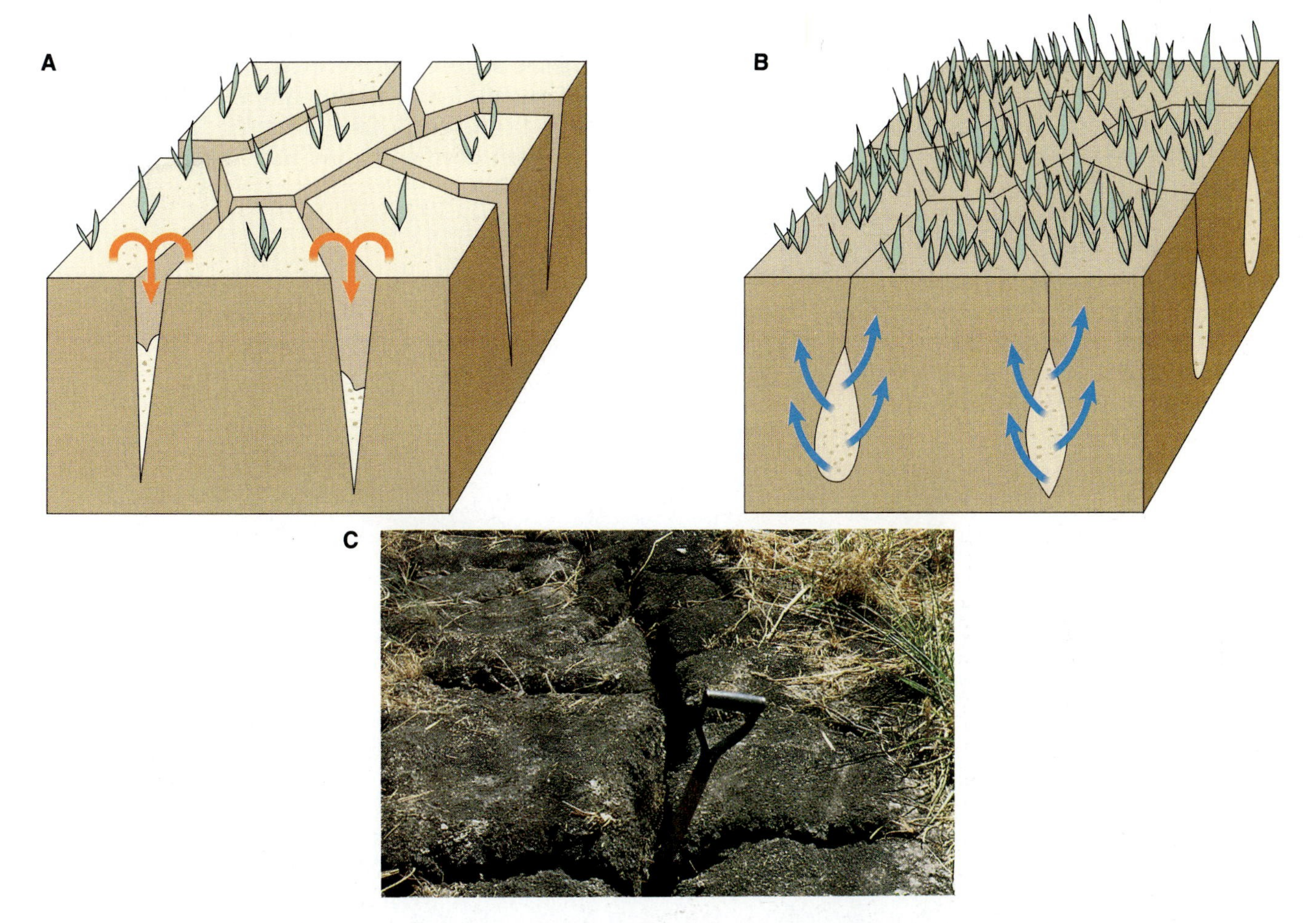

Figure 20-18 The development of Vertisols. **A,** In the dry season the soil dries out and cracks apart. Material from the surface falls down the cracks. **B,** In the wet season the soil swells and the cracks disappear. The surface material is now trapped at depth, and the bottom of the soil is wet where rainwater has entered the cracks. The "bottom-up" wetting sets up pressures that result in the upward movement of soil. Thus over time the soil is gradually inverted. **C,** A Texas Vertisol in the dry season with open cracks.

the number of days in a year the cracks in the soil will be open. The main area of Vertisols in the United States is in Texas, where they cover 10.4 percent of the state and are used for agricultural crops, including cotton and grain sorghum. The soils are commonly plowed or otherwise turned over several times a year, to break down the massive clay aggregates into a more hospitable seed bed.

Histosols

The **Histosols** are the organic or peat soils, composed mainly of organic materials in various stages of decay. They form where organic matter accumulates more quickly than it can be broken down. This is typical of waterlogged conditions, which promote acidity and inhibit activity in the detrital food chain. Histosols will develop at any site, under any climatic regime, where these conditions are met, although they are clearly most common in areas of moderate to high precipitation. The most typical Histosols are the blanket bogs and peats of northwestern Europe (Figure 20-19). The plant remains preserved in the peat profiles can be used to chart changes in vegetation at and around the site through time. These data can give valuable insights into past climatic change and the impact of early human agriculture on the natural vegetation.

In North America the largest areas of Histosols are in Alaska and Minnesota. Their distribution is widely scattered, however, and they are also found in Michigan, on the deltaic plain of the Mississippi River, and in Florida and Hawaii. Much of the Everglades in Florida is underlain by Histosols in the suborder Saprists. In these soils the organic material is almost completely decayed. Where drained, particularly near Lake Okeechobee, they are fertile agricultural soils and are used intensively.

Spodosols

Spodosols form most readily on freely draining parent materials under conditions of high rainfall and heavy leaching. In general they are the product of the pedogenic regime of podzolization and were called podzols under earlier classification systems. A diagnostic feature is the presence of a B horizon in which organic material, aluminum, and (usually) iron have been redeposited from the E horizon. Spodosols can develop in any humid climate, but

Figure 20-19 A Histosol landscape from northern Wales, UK. Here rainfall is high and acid conditions dominate. The plant species are small herb and shrub species. Deep peats may form in valleys and between the rocky outcrops.

are most often found under coniferous forest vegetation.

The Pilgrims came across Spodosols at their early settlements in Plymouth County, Massachusetts. Their attempts at plowing, the typical agriculture of Western Europe, mixed the organic surface horizon with the infertile leached horizon to create a poor soil that could not support extractive agriculture. Even without plowing, the soil would be a poor medium for plant growth because of its sandy texture and high acidity. Many of these farms were abandoned with the promise of a new start in the West, and the vegetation has thus reverted to pitch pine and scrub oak.

Aridisols

Aridisols are the shallow soils of arid areas. The most important factor affecting their formation is the lack of moisture. The natural vegetation is typically sparse and shrubby (Figure 20-20), so there is little organic input to the soil and a high risk of wind erosion. Although the low precipitation precludes much downward translocation, in some profiles the high pH encourages the solution and movement of silica, which may be redeposited at deeper levels. In other profiles a calcic horizon may develop as water moves up through the profile from a parent material rich in calcium. Under normal climatic conditions, there is little if any translocation of clay particles. Aridisols are common in the irrigated west of the United States and are used for grazing and for small patches of irrigated agriculture. Where the water table is close to the surface they are prone to salinization. The sparse and shrubby vegetation may seem unsuitable for grazing, but with careful introduction of plant species, it provides adequate forage. Aridisols cover more of Earth's surface than any other soil orders (18.8% globally, 11.5% of the United States).

Mollisols

The **Mollisols** are the soils of the world's grassy plains, prairies, and steppes and are the most extensive soil order in the United States. They are productive soils, used nowadays to grow grain and to raise livestock. The dominant soil-forming process is calcification. The diagnostic characteristic is the presence of a thick dark brown or black A ho-

Figure 20-20 The sparsely vegetated landscape of the Joshua Tree National Monument, USA. The soil has few colloids and therefore has a low water- and nutrient-holding capacity.

rizon rich in organic material. The organic content decreases with depth.

The suborders of the Mollisols are defined by the climatic regimes of moisture and temperature. Waterlogged Aquolls develop on the flood plains of the Mississippi and Missouri rivers. The Udolls are generally well-drained and form in more humid climates. In the United States they are dominant in Oklahoma, Kansas, Iowa, Illinois, and the adjacent parts of Nebraska and Missouri. The original vegetation of these Udolls was the long prairie grass that the early settlers found so difficult to plow. Today, this area is part of the corn belt and virtually all of the land is farmed, particularly for soybeans and corn as fodder (Figure 20-21). Soils of the suborder Ustolls are typical of drier climates, where drought is common. Ustolls predominate west of the Udolls on the Great Plains, before giving way to the Aridisols in Utah and Arizona. The Ustolls and Udolls were the main casualties of the Dust Bowl of the 1930s.

Alfisols

The final three soil orders, Alfisols, Ultisols, and Oxisols, all develop under damp to wet conditions. They are differentiated primarily by the temperature regime under which they form. **Alfisols** are usually found in the coolest conditions. They are an intermediary stage in soil genesis between the Mollisols and the Spodosols. The Alfisol suborder Udalfs is the most widespread in the United States. They are found over glacial deposits and over calcium-rich parent materials, and typically have, or have had, a vegetation of deciduous forest. Because they combine high nutrient content with moderate temperature and moisture regimes, these soils are fertile and have been used extensively for agriculture. Udalfs are common neighbors to the Udolls (suborder of the Mollisols) in the American Midwest. Their yield is slightly less than that of the Udolls because of the lower organic content. In the cooler conditions of Wisconsin and New York, Udalfs are the basis of the dairy industry.

Figure 20-21 A deep, rich Mollisol from Iowa. The dark brown color shows the concentration of organic matter near the surface.

Ultisols

Ultisols are found in the warmer temperatures of the mid to low latitudes. They are strongly weathered and leached and are therefore nutrient-poor and acidic. A subsurface horizon with high amounts of clay particles is usually present. Ultisols are an intermediary step from Alfisols to Oxisols and are often found with one or another of those soil orders in the landscape. The difference between Alfisols and Ultisols is in the degree of leaching and in their nutrient content; their profiles are otherwise similar. Without management, Ultisols are poor agricultural soils, mainly because the release of nutrients by weathering is more or less equal to their removal by leaching, and the nutrients in the organic material are rapidly recycled. Early settlers in Virginia and Maryland ran into problems in cropping these soils, and like the Spodosol areas of New England, many areas were abandoned in the exodus to the West. In 1840, however, a Virginia landowner named Edmund Ruffin introduced the practice of adding lime to the soil to counteract acidity. Lime is a general term that describes various compounds of calcium and magnesium. Ruffin's original lime was marl, a soft calcareous rock dug straight from the ground. The calcium and magnesium can replace some of the hydrogen and aluminum ions on the clay-humus complex and so reduce the soil's acidity and make more cations available to plants. Liming had an immediate and dramatic effect on corn yields on Ruffin's land, and began a reversal in the area's fortunes that has continued into the twentieth century.

Oxisols

The **Oxisols** are the result of intense laterization. They are the extremely weathered and leached soils of the tropics and do not occur in the continental United States. Oxisols are the end process of a pedogenic regime starting at Entisol and moving through Inceptisol, Alfisol, and Ultisol. In general they have very little horizonation but may contain hardened laterite, as discussed in the section on laterization. The suborders are defined mainly by moisture and temperature regimes. The most typical are those in the Orthox suborder, which are usually associated with equatorial rainforests. The lush vegetation belies the inadequacy of the soil as a medium for plant growth. Most of the nutrients are rapidly cycled between the biomass and the litter, and the paucity of available nutrients in the soil means that replacement crops, particularly grasslands used to rear beef cattle, require a continuous artificial subsidy of nutrients.

SOILS AND HUMAN ACTIVITY

As noted at the beginning of this chapter, soils are a fundamental part of the living-organism environment. They control two of the most basic characteristics of a plant's ecologic niche, the supply of nutrients and the supply of water. They also provide a habitat for the organisms of the detrital food chain. By providing an anchorage for plants, they influence the flow of energy in the biosphere, and, through the process of transpiration, they have an impact on the atmosphere-ocean environment. Most importantly, from the point of view of human activity, they help to dictate what plants can be grown where and in what quantity, thus deciding the nature, size, and geographic distribution of the energy supply available to humans. Efficient use and conservation of the soil resource has become a key issue in physical geography as the world's population grows and demands for food increase.

Only a limited range of soils can support extractive agriculture in which little or no organic material returns to the soil. Soils suitable for cropping are typically formed by the pedogenic regime of calcification, or the less extreme forms of podzolization, and even these soils need occasional periods of rest and recuperation. These fertile soils cover only 11 percent of Earth's ice-free land surface, and their distribution is a major constraint to human activity and the distribution of population. Optimistic agronomists predict that up to 24 percent of Earth's ice-free surface could be cultivated, but this would mean bringing more marginal land into production.

The distribution of fertile soils is uneven across the world. In North America, 22 percent of the land is potentially cultivatable, although only 13 percent is actually used; much of the balance is used for buildings and transportation networks. In Europe 36 percent of the land could be cropped and almost all of it (31 percent) is in production. The situation is different in some less developed countries, where population growth places heavy demands on the agricultural system. In southern Asia, 20 percent of the land is suited to agriculture but 24 percent is cropped. A similar situation occurs in southeast Asia (14 percent suitable and 17 percent in production). These figures show that marginal lands are being cultivated, often through irrigation. There is no doubt that irrigation has boosted the productivity of agriculture in dry lands, to the benefit of millions of people in the developed and the less developed worlds. Improperly managed, however, irrigation can lead to salinization, as discussed in the section above, and the sterilization of large areas of land.

The most widespread impact of cultivation on soil is soil erosion. Many people associate soil erosion with agriculture in less developed countries, or with historic events such as the American Dust Bowl of the 1930s. It is seen as a problem that does not affect developed countries or that has occurred in the past but can now be mitigated by modern technology. This is not the case. Soil erosion occurs in all continents, although it is true that some of the highest rates occur on the marginal lands of developing countries, where areas with high rainfall and steep slopes and areas prone to drought are intensively cultivated (see also Box 21-1: Desertification). In the developed world, modern farming practices tend to exacerbate rather than solve the problem. Soil is renewed naturally at a rate of about one ton per hectare per year. Erosion rates are currently estimated at between 18 and 100 times this rate. In the United States the average annual soil loss is 18 tons per hectare. In states with intensive agriculture, such as Iowa and Missouri, the rate is over 35 tons per hectare.

Soil erosion, acting through the surface-relief agents of wind and water, is a natural process. Cultivation hastens the process by loosening the topsoil and exposing it directly to the agents of erosion. The effect on the soil is to remove the finest and the lightest mineral particles, and much of the organic matter that is usually concentrated in the upper horizons. This in turn reduces the nutrient content and the nutrient and water-holding capacity of the soil. The soil particles are eventually deposited elsewhere, which can lead to silting in irrigation and water conservation works.

The Dust Bowl on the Great Plains of the United States in the 1930s is a prime example of soil erosion on a massive scale, and it provides a useful lesson to contemporary soil scientists. It was caused by three factors. First was the agricultural economy of the time. During the First World War (1914-1918), international wheat prices soared, which provided an incentive to extend the acreage of wheat on the Great Plains. The extension was achieved by investment in increased mechanization, particularly the use of tractors and combine harvesters. In time this created a vicious circle, in which continued high wheat production was needed to pay off the loans for the machinery, especially when wheat prices fell after the war. Second, as a result of relentless cropping to maintain production, the soil was deeply plowed and became depleted of organic matter. The result was a loose and friable topsoil susceptible to erosion. The third factor was the climate of the area; the Great Plains are naturally dry, with strong winds and a tendency for severe drought. A group of drought years between 1926 and 1934 led to massive wind erosion of the topsoil, as described in the opening to Chapter 16. In May 1934, dust from the valleys of the Mississippi and Missouri rivers reached New York, blocking the sunlight. In Washington, D.C., the dust filtered into a congressional hearing in which officials from the Department of Agriculture were pleading for funds to halt the erosion. Since this period a policy of soil conservation has been introduced to the United States, with some limited success. Soil erosion can be physically slowed or halted by minimally disturbing the soil surface in plowing, by using crops that have a high vegetation cover, by covering the soil

A

B

C

D

Figure 20-22 Methods of preventing or controlling soil erosion. **A,** Beans planted in wheat straw in Kansas. **B,** Strip cropping, in which different types of crop are planted in narrow parallel belts. This example is in Montana. **C,** Building terraces helps to prevent erosion on slopes. **D,** Ploughing along the contours instead of up and down the slope reduces erosion by water.

surface with mulch, and by plowing along the contours or using terraces on steep slopes (Figure 20-22). The real problem, however, lies in persuading agricultural communities in all countries that this long-term investment in soil fertility is preferable to higher profits or a better short-term standard of living.

Human activity has other impacts on the soil, including the use of chemical fertilizers instead of organic material to replace the nutrients extracted in crops. This has a number of effects. First, it reduces the activity of the detrital food chain because the supply of energy from dead organic material is cut off. Second, heavy leaching of the fertilizer can lead to the enrichment of surrounding water bodies and nitrogen contamination of groundwater. Third, natural organic matter does more than add nutrients to the soil; it helps to maintain a good soil structure, and, through the clay-humus complex, improves the cation-exchange mechanism and water-holding capacity of the soil. In less developed countries, much of the dung that would normally be left to fertilize the soil is collected and burnt for fuel, because of shortages in the supply of firewood. This type of degradation, the progressive loss of organic matter, is therefore widespread among very different agricultural communities.

A final human impact on soil is the addition of toxic substances, through deposition from air (in acid rain, for example—see Box 6-2, p. 150), seepage from hazardous wastes in landfills or deep wells, and the insecticides, fungicides, and pesticides applied to agricultural crops, golf courses, and lawns. The full impact of these substances on the soil ecosystem is still largely unknown, and is a topic of widespread and urgent research in physical geography.

In summary, a healthy soil is vital to the functioning of the ecosystem and to the capture of energy for human beings. The limited distribution of fertile soils at Earth's surface restricts the distribution and activity of humans. Given the current and future demands for food production, it would seem appropriate to conserve and nurture this limited resource. The trend, however, seems to be in the opposite direction, with 6 million hectares (roughly the size of West Virginia) going out of production in the world each year because of soil erosion, salinization, and other degradation. Cropping becomes uneconomical on a further 20 million hectares each year for the same reason.

FRONTIERS OF RESEARCH
Soil Science

Most research in pedology has been geared to improving agricultural productivity. The efficiency in applying fertilizers, for example, would be improved if scientists better understood how nutrients are processed in the soil. However, in an increasingly industrialized society, studies of the pathways and ultimate fate of pollutants and contaminants in the soil, including those used in agriculture, are taking over as a dominant issue. Particular topics include the residency time of nuclear fallout, such as that from the Chernobyl accident, in the soil, and the movement of toxic substances from landfill sites.

SUMMARY

1. Soil forms as a result of the interaction of the major Earth environments in the biosphere. It exists in a dynamic equilibrium with these environments.
2. Soil controls the movement of energy and nutrients in the ecosystem and is an ecosystem in its own right.
3. Soil is made up of mineral matter, organic matter, water, and air.
4. Soil is formed by the action of the three interrelated processes of weathering, addition of organic matter, and translocation.
5. The soil-forming processes are governed by climate, parent material, topography, organisms, and time.
6. The action of the soil-forming processes creates a variety of distinctive soil profiles, which may be differentiated into horizons. Each horizon can be described in terms of its texture, structure, color, and chemistry.
7. The different soil characteristics found across the world are the product of different conditions of the soil-forming processes and the factors that control them. These can be summarized as a limited number of pedogenic regimes.
8. The pedogenic regimes are podzolization, laterization, gleization, calcification, and salinization.
9. Soil scientists have developed methods of soil classification to describe and order the variability of soils. The system in use in the United States is called Soil Taxonomy or the United States Comprehensive Soil Classification. It is based on a wide range of current characteristics of the soil profile, some of which are grouped together as diagnostic horizons.
10. Soil Taxonomy groups the soils of the world into 11 soil orders, and breaks these down into suborder, great group, subgroup, family, series, and phase. The name given to the soil at each level reflects its characteristics and its position in the hierarchy.
11. The 11 soil orders of Soil Taxonomy are Entisols, Andisols, Inceptisols, Vertisols, Histosols, Spodosols, Mollisols, Aridisols, Alfisols, Ultisols, and Oxisols.

KEY TERMS

soil, p. 540
translocation, p. 541
soil profile, p. 541
soil horizon, p. 542
soil pedon, p. 543
colloids, p. 543
humus, p. 543
clay-humus complex, p. 543
soil solution, p. 543
hygroscopic water, p. 543
wilting point, p. 543
capillary water, p. 544
gravitational water, p. 544
field capacity, p. 544
aeration, p. 544
soil texture, p. 545
soil structure, p. 545
ped, p. 545
cation exchange capacity, p. 546
leaching, p. 549
eluviation, p. 549
illuviation, p. 549
pedogenic regime, p. 551
podzolization, p. 552
laterization, p. 552
laterite, p. 553
gleization, p. 553
calcification, p. 553
salinization, p. 554
Soil Taxonomy, p. 556
soil order, p. 556
Entisols, p. 556
Andisols, p. 560
Inceptisols, p. 561
Vertisols, p. 561
Histosols, p. 562
Spodosols, p. 562
Aridisols, p. 563
Mollisols, p. 563
Alfisols, p. 564
Ultisols, p. 564
Oxisols, p. 564

QUESTIONS FOR REVIEW AND EXPLORATION

1. Draw an annotated diagram to show how Earth environments interact to form soil. Use Figures 20-1 and 20-15 as a starting point.
2. What does physical geography gain from and contribute to pedology?
3. Outline the stages in the development of a soil profile.
4. Explain the difference between soil structure and soil texture and describe the importance of each as a soil characteristic.
5. Using the photographs of soil profiles in this chapter, show how useful soil color is in showing what processes are going on in the profile.
6. What are the similarities and differences between the pedogenic regimes of podzolization and laterization?
7. Outline the relative importance of each of the soil-forming processes and factors (climate, parent material, etc.) in each pedogenic regime. Draw up your results as a table.
8. Decide which, if any, pedogenic regime is dominant in each of the soil orders in Soil Taxonomy. Comment on how well the classification reflects soil genesis.
9. Which is more useful, a classification based on observable characteristics or one based on soil genesis? Why?
10. Using a soil map, find out about the soil series and phases in your local area. Write out their full classification according to Soil Taxonomy.

FURTHER READINGS

O'Hare G: *Soils, vegetation, ecosystems* (Oliver and Boyd), 1990. A good introduction to the structure and processes of soils.

Hodgson B: North Dakota: tough times on the prairie, *National Geographic*, 321-347, July 1987. This paper describes the economics of prairie agriculture, with particular reference to the issues of soil erosion and taking land out of production for conservation purposes.

New York Times: Back issues for May 1934 (especially May 12) describe conditions when a dust storm from the dry western states hit the city.

Simpson K: *Soil* (Longman), 1983. Another good introductory text, this time dealing with soils as a medium for crop growth.

THE LIVING-ORGANISM ENVIRONMENT

Giant Sequoia—Sequoia National Park, California
David Muench

Biomes

21 CHAPTER

I stayed two days close to the bodies of the giants and there were no chattering troupes with cameras.

There's a cathedral hush here. Perhaps the thick soft bark absorbs sound and creates a silence. The trees rise straight up to zenith; there is no horizon. The dawn comes early and remains dawn until the sun is high. Then the green fern-like foliage so far up strains the sunlight to a green gold and distributes it in shafts or rather in stripes of light and shade. After the sun passes zenith it is afternoon and quickly evening with a whispering dusk as long as was the morning.
Thus time and the ordinary divisions of the day are changed . . . nearly the whole of daylight is a quiet time. . . . Underfoot is a mattress of needles deposited for over two thousand years. No sound of footsteps can be heard on this thick blanket. To me there's a remote and cloistered feeling here.

John Steinbeck, Travels with Charley, 1962

The biosphere is made up of a series of nested ecosystems, each linked to others at the same scale and neighboring scales by the movement of energy and nutrients (Figure 19-2). This chapter examines the ecosystems at one level in this hierarchy, that of the regional ecosystem or biome. Some examples of biomes are the prairie grasslands of the Great Plains and the Ukraine steppes, the rainforest of South America, Africa, and Indonesia, and the hot deserts of the Sahara, the Peruvian coast, and the southwestern United States.

A *biome* is a large ecosystem in which relatively uniform climatic conditions lead to uniformity in the living-organism environment, particularly in the type, or growth form, of the primary producers. **Growth form** describes the physical size and shape of the dominant vegetation and is the main characteristic by which biomes are differentiated. The opening excerpt shows how the giant redwood trees dominate the Pacific coast forest of the United States. The most common division of life forms is into trees, shrubs, and herbs. **Trees** are plants with a single woody stem (trunk) that branches at some height above the ground, raising most of the green vegetation above the ground surface. **Shrubs** are also woody plants, but are smaller than trees and have stems that branch nearer the ground. **Herbs** are small plants, such as grasses, that have no woody tissue—they form the layer of vegetation closest to the ground. Plants such as vines and lianas can grow to considerable heights, but have little woody tissue and must use other plants for support.

Individual biomes are recognized by the dominant growth form of the vegetation and by its more detailed characteristics. Forest biomes, for example, share a dominance by trees but contain differing proportions of **deciduous** species, which lose their leaves completely at some time of the year, and **coniferous** species, which are in leaf all year.

Different growth forms of vegetation support different sorts of animals in food webs of different sizes and complexity. Biomes are not therefore just assemblages of primary producers, but fully developed biologic communities interacting with their inorganic environment.

CLASSIFICATION OF BIOMES

The study of biomes has a well-developed history in physical geography, but scientists disagree over such topics as the names for different biomes and even whether some of them exist at all. The emphases of individual researchers are influenced by the regions in which they have worked, where different biomes are dominant in the landscape. An important point to remember is that, like all classifications, a division of the biosphere into biomes is a human simplification of a complex set of interactions.

The most important control on the growth form of vegetation is climate, particularly the annual and seasonal temperature regime and the availability of moisture. Figure 21-1 shows how some biomes, such as the tropical forests, extend over a wide range of annual rainfall amounts, but have quite specific temperature requirements. In contrast, grasslands are dominant over a wide range of temperatures, but are confined by their need for specific amounts of water. The distribution of each biome reflects the tolerance and optimum ranges of the primary producers in its plant community. As noted in Chapter 20 climate also exerts a major influence on soil formation. The biomes studied in this chapter are shown in Table 21-1. Figure 21-2 shows the distribution of the biomes across the globe, and in more detail within the United States. The reasons for this distribution are explained in this chapter.

Mountain environments are something of a special case. This grouping contains an enormous variety of ecosystems and is considered separately. Some elements of the mountain climate, such as day length, are the same as in the surrounding lowlands, while temperature, wind, and precipitation may vary dramatically with elevation. The latitudinal variation in these factors and the concentrated vertical changes mean that mountain environments vary dramatically over short distances.

Land masses are the main areas of interest for humans and therefore for physical geographers. However, 71 percent of Earth is covered by ocean ecosystems, which contribute about 34 percent to the global primary productivity. The ocean is also an important control on atmospheric processes and is valuable in terms of food and waste disposal for humankind. The structure and function of marine ecosystems are therefore briefly discussed in this chapter. Marine ecosystems are very different from those in freshwater or on land, mainly because of the constant movement of large volumes of salt water and the generally reduced influence of human activities. They are a fascinating object of study in their own right but lie outside the main focus of this book.

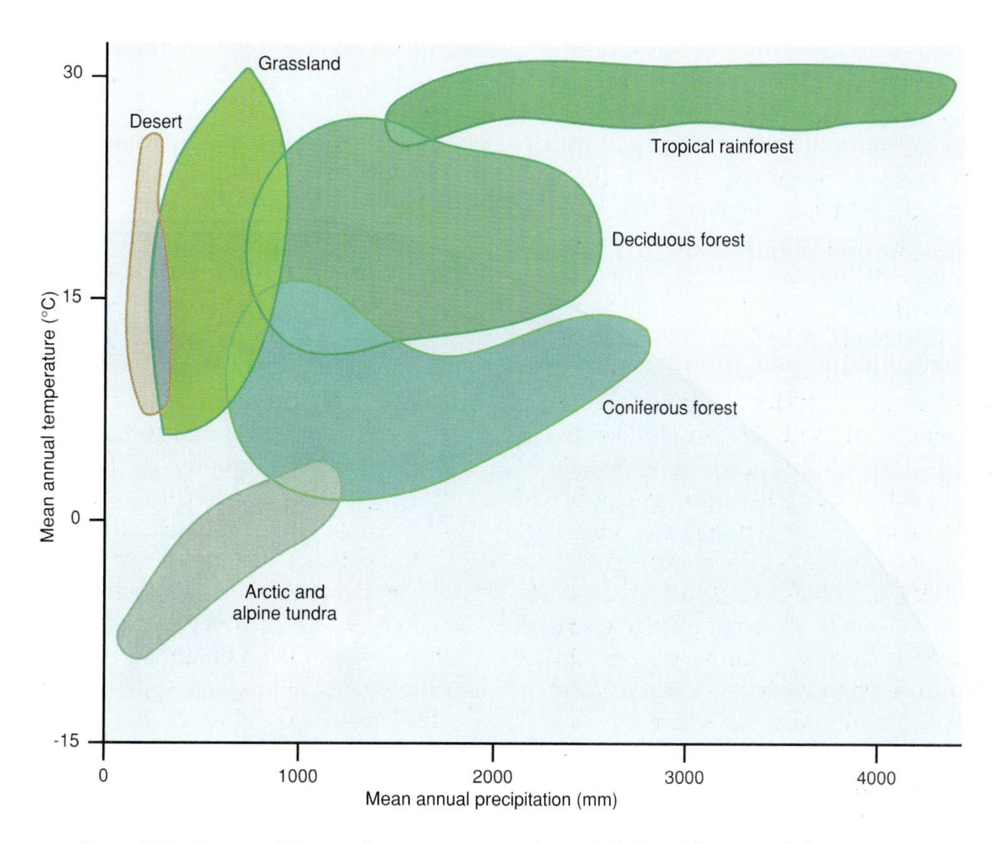

Figure 21-1 The conditions of temperature and rainfall that delimit each biome.

GLOBAL DISTRIBUTION OF BIOMES

The biomes are distributed around the Earth as broad bands roughly parallel to the equator; the patterns in the north and the south hemispheres are approximate mirror images of each other. Within the broad latitudinal bands, the detailed pattern is determined by major relief features and the size and shape of continents.

Table 21-1 Biomes Described in this Chapter

Forest Biomes Tropical rainforest Tropical seasonal forest Midlatitude forest Boreal forest
Grassland Biomes Tropical grasslands (savanna) Midlatitude grasslands
Tundra **Mountain Biomes** **Ocean Biomes**

Biomes, Climate, and Soil

World maps of biomes, soils, and climatic environments generally agree, as might be expected because of the interrelationships among climate, vegetation, and soils. Discrepancies among the maps arise because different criteria and levels of detail are used in each classification of the biosphere, and because these maps are taken from a "freeze frame" in Earth history. A full-length feature film would show the birth of the planet, the building of continents, the opening and closing of the oceans, the evolution and dispersal of living things, and the belts of climate moving north and south as the ice ages came and went. Because changes in Earth environments work at different rates, they may well be out of sync with each other when the "picture" is taken. For example, many present-day vegetation communities are still adjusting to the climatic changes since the Pleistocene Ice Age. The processes of pedogenesis, or soil formation, work on an even longer time scale than vegetation succession. If Earth environments were to remain stable for a few million years, the processes would most likely catch up with each other and the maps would become more alike.

B

Figure 21-2,B The distribution of biomes in the conterminous United States.

Ecotones and Variation Within Biomes

The classification and mapping of biomes give the impression that they are uniform over their full extent and that they are clearly delimited on the ground. According to Figure 21-2, a person could, in the southwest United States, stand at a boundary line with one foot next to a cactus in a blazing desert and the other next to a grazing bison in a temperate grassland. To the north, one could straddle a line between grazing bison and the cool shade of deciduous woodland. Of course, this is not possible. The change from one biome to another is marked by an ecotone, a gradual transition between vegetation communities. The biomes are best thought of as a series of intergrading plant communities; the most rapid change occurs in the ecotones between them.

Change in vegetation is not confined to the ecotones, however. The conditions of climate within a biome are not completely constant or uniform. Temperature, for example, usually varies gradually between the southern and the northern limits of each biome. East-west variations in precipitation across a continent also occur. The "type" climate, vegetation, and soils of a biome can be thought of as the midpoint in its range of environmental conditions; the biome becomes progressively less typical farther from the midpoint. In the context of these gradual changes, the nature of the ecosystem that develops at any one place is controlled by the detailed local interaction of Earth environments. Where these local conditions are extreme, the local ecosystem is a widespread and semi-permanent departure from the biome type. The changes in elevation on mountain ranges cause similar local variation.

The map of world biomes also ignores the effect of humans on the natural vegetation, particularly their removal of large parts of the natural vegetation and its replacement by agricultural or industrial systems. Figure 21-2 and the descriptions that follow outline the nature and distribution of world biomes as they would exist in a purely natural state, without past or present human intervention.

In summary, Figure 21-2 gives a general impression of the world biomes. It is, however, a generalization at the present time and at a high level in the hierarchy of ecosystems described in Chapter 19; the conditions at a particular place in the biosphere still depend on the dynamic balance of Earth environments at that place.

FOREST BIOMES

Forests are the most complex and productive of the world biomes. They make heavy demands on the nonliving parts of the biosphere, particularly soils. Trees have a large amount of woody material, in the form of twigs, branches, and trunks, that requires energy but is not able to capture it directly by photosynthesis. Trees therefore only grow where the supply of the raw materials for photosynthesis, such as moisture, nutrients, and sunlight, enables their leaves to meet these heavy demands for gross primary production.

Tropical Forest Biomes

The tropical forest biome is usually divided into two more-or-less distinct biomes. These are tropical rainforests, where there is little seasonal change in climate and the trees are in full foliage all year, and tropical seasonal forests, where there is a pronounced dry season and the trees are deciduous. The "evergreen" nature of the rainforest is a result of the unvarying climatic conditions; individual leaves are short-lived, but the trees grow and shed their leaves continuously all year round. Forests at the boundary between these two types contain both deciduous and "evergreen" plants; the first sign of a change to drier conditions is the appearance of deciduous trees in the uppermost layer of the canopy.

Tropical Rainforests

Tropical rainforests are one of the oldest ecosystems on Earth; they are thought to have evolved more or less continuously since the Tertiary period that started more than 60 million years ago. Tropical rainforests occur at and around the equator, the "best" environmental niche on land for photosynthesis, where the raw materials of sunshine, water, and carbon dioxide are present in abundance all year. In this area the trade winds converge, producing strong uplift and heavy and frequent rain. The winds bring in a constant supply of moisture, the sun is always high in the sky, and the day is about 12 hours long throughout the year. The productivity of tropical rainforests is therefore high. They contain a luxuriant and apparently chaotic wealth of primary producers and support a tremendous variety of life at higher levels in the food web (Figure 21-3). It is unusual to find stands of trees with a high proportion of a single species, and the seedlings are often different species from the mature trees in the area. Conservative estimates of the number of tree species range from 40 to 100 species per hectare and some scientists give figures of over 200 tree species per hectare. This mixing of species is one factor that until recently made it difficult to extract timber from the forest. The constancy of climatic conditions is particularly important in supporting the rainforest ecosystem. Energy flow is constant throughout the year, allowing a high gross primary production. There are also no seasonal highs and lows to govern reproduction in either plants or animals.

The primary producers in rainforests are arranged into three or four reasonably distinct vertical layers as they compete for light (Figure 21-4). The top layer of the canopy is discontinuous and consists of the highest trees, or *emergents*, which stick up like umbrellas from the surrounding canopy. The lower, continuous, tree canopy can be roughly divided into two layers. The upper layer includes *epiphytes*, plants that use the trees as a platform to reach the sun's energy but gain nutrients and moisture from roots that hang freely in the air, or tap water and dead organic material falling from the upper layers. Recent research suggests that epiphytes may also supply nutrients to the trees they perch on. Climbing plants, such as *lianas*, are rooted in the ground and use the trees as platforms by which to reach the sunlight. In the lower layer the trees are densely packed and may have large leaves to trap the light. The shrub and herbaceous layers near the ground tend to be patchy, limited by the small amount of light that filters down to the forest floor, which may be as little as 2 percent of that received at the top of the canopy. Common house plants, such as the African violet, grow wild in these gloomy conditions. The popular view of an impenetrable jungle undergrowth in the rainforest is accurate only where light can penetrate to the forest floor, in clearings and along riverbanks.

The animal life of the forest is also stratified with the vegetation. The upper layers are inhabited by birds and bats and the middle layers by different

A

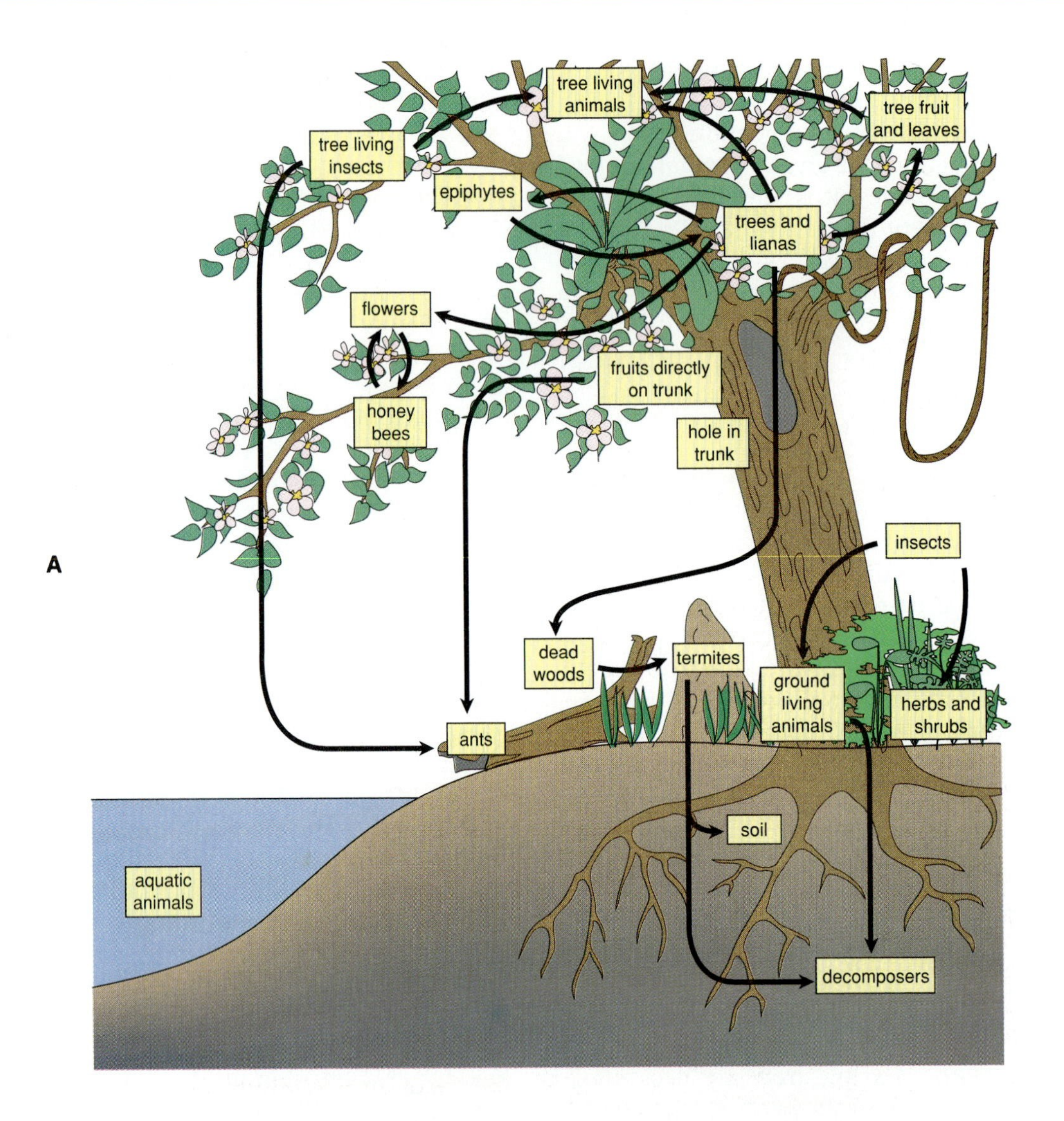

B

Figure 21-3 **A,** Some of the interrelationships between plant and animal species in a small part of the rainforest. In reality the linkages extend vertically and horizontally in a complex food web. **B,** The varied and luxuriant nature of the rainforest vegetation is apparent in this example from Madagascar.

B

Figure 21-4 Stratification of vegetation in the rainforest. **A,** Species at different levels require different amounts of solar energy; those in the lower canopies may also develop large leaves to trap the maximum amount of available light. **B,** Densely packed trees forming the rainforest canopy.

birds and by climbing animals such as squirrels, sloths, and monkeys (Figure 21-5). Deer and pigs live on the ground layer, eating fruits and materials that drop from the upper layers. Hundreds and even thousands of species of insects and invertebrates are found at each level.

Many species have evolved ways to make the best use of the environment's resources. There is often excessive moisture at lower levels in the forest, and some plant species have evolved *drip-tips* to move water quickly off their leaves (Figure 21-6, *A*). Another characteristic of many rainforest trees is *buttress roots* (Figure 21-6, *B*). Their origins and purpose are rather obscure. Scientists originally thought they were purely for mechanical support, but recent experiments show that they may have a function in extracting nutrients directly from the litter layer.

Figure 21-5 Animal life in the tropics. **A,** The three-toed sloth pictured here uses its claws to hang from a tree, sound asleep, for about 18 hours out of 24. Its ungroomed coat is a home to green algae, which are food in turn to moths. **B,** In contrast to the sloth, monkeys are quick-witted and agile, able to move rapidly from tree to tree in search of food. **C,** Birds of paradise are among the colorful bird species found in the upper canopy of the rainforest.

The three-dimensional diversity of the rainforest makes it immensely difficult to study scientifically. Most knowledge of the structure and function of this ecosystem is fragmentary; in attempting to describe it scientists are trying to put together a jigsaw puzzle with most of its pieces missing. Most scientists agree on two points, however. First, they know very little about this biome, so human use of its resources must proceed carefully until more pieces of the jigsaw are in place. Second, the diversity of species, many still undiscovered and unnamed, means that tropical rainforests are the last great reservoir of genetic diversity, a gene pool of species that may someday be useful to humans. Humans extract an amazing variety of products from the forest, including woods and pharmaceutical drugs—an estimated 40 percent of prescribed drugs in the United States have their origin in rainforest species. This includes extracts from a flower called the rosy periwinkle that are used to combat some forms of leukemia in children. Taken together, these are strong scientific arguments for reducing or ceasing the wholesale clearance of this biome that is occurring in some countries (see Investigation: Stop the Burning).

The key to human use of this productive system lies in understanding the relationship between its structure and the movement of both energy and nutrients. The high productivity of the rainforests suggests that the wood reserves could withstand frequent logging. However, research has shown that a large proportion of the primary production is used to renew leaves, stems, and other nonwoody material. The wood production of tropical forests is similar to and in some cases less than that of temperate forests; the current high rate of felling is therefore far above the rate of regeneration.

The lush appearance of the rainforest suggests that the land on which it grows could support an equally productive agricultural system, and large amounts of forest have been cleared for this purpose. A current trend is to replace forest with grassland that is used to raise beef cattle. The soils underlying tropical rainforests are typically of the order Oxisols, which contain few nutrients. The luxuriant natural vegetation is maintained by very rapid recycling of nutrients between the biomass and the litter pool. Very little organic material finds its way into the soil profile, and when it does, it is rapidly broken down and reabsorbed by the shallow plant roots, or leached from the soil. Some roots are even thought to take nutrients directly from the litter layer. This type of adaptation for rapid recycling is the result of the long evolutionary period of this biome; plants imported into the biome can only be supported by massive doses of artificial nutrients. Soil erosion, initiated by the removal of the protective vegetation cover, can also lead to the exposure of the resistant laterite layer at the surface.

A

B

Figure 21-6 **A,** Plants in the lower level of the canopy have "drip-tips" on their leaves that allow rapid run-off of water. **B,** Buttress roots are a common feature of rainforest trees, although scientists are unsure of their function.

The indigenous agriculture of forest-dwelling people is a form of shifting cultivation. In this system a small plot is cut or burnt and planted for a few years. When the natural fertility of the soil runs out, the farmer moves on and clears a new patch, leaving the original clearing to recover. Even this type of low-impact agriculture can have a considerable effect on the vegetation, because the secondary regrowth of the forest may be less rich than the original.

Tropical Seasonal Forest

Tropical seasonal forest is a zone of transition between tropical rainforest and tropical grassland or savanna and typically borders one or both of these biomes. It is characterized by a gradual change in dominance from the "evergreen" trees typical of the rainforest to almost completely deciduous stands in drier areas. The change in vegetation is related to two changes in climate. First, total rainfall decreases away from the equator. Second, seasonality in both temperature and rainfall is more pronounced, which tends to limit the growing season and may produce a distinct dry season. This seasonality is caused by movement of the intertropical convergence zone (ITCZ) rain belt with the overhead sun. Various subdivisions of this biome have been recognized. Examples include *semi-evergreen rainforest,* where the upper layers are deciduous and the lower canopy is evergreen, and *deciduous seasonal forest,* where virtually all the trees are deciduous. In the driest areas the cover of trees is discontinuous, so the light can penetrate to the shrub and herb layer. The light promotes growth in the lower plant layer and *tropical scrub forest* is formed.

The structure of the forest ecosystem changes gradually through the transitional stages, as the trees adapt to survival through the increasing length or intensity of the dry season. In semi-evergreen forest, the species diversity and vertical structure of the rainforests are present in a reduced form. There are very few emergents, and typical features such as buttressing and epiphytes are less common. The forest canopy is generally more open, resulting in more growth in the lower strata, particularly in the herb layer. At the drier end of the spectrum, lack of moisture, or some unpredictability in its availability, produces stunted trees grouped together in dense thickets. Here, rather than losing leaves completely, the trees have developed small hard leaves that can withstand water deficit and are retained all year round. The accumulation of dry plant material in the drier parts of this biome creates a high fire risk; resistance to fire is thought to be an important determinant of the species structure of these plant communities.

The Midlatitude Forests

In comparison to the rainforests, the **midlatitude forest** biome has been greatly affected by human activity in historical time. Midlatitude forests in their natural state remain only in pockets of the biome's potential distribution throughout Europe, in eastern and western North America, and eastern Asia. Smaller areas occur in the southern hemisphere in South America, southern Africa, Australia, and New Zealand. In the lowlands the forest has been cleared to make way for croplands; in upland areas it has been cut for lumber or firewood. Where remnants remain they form a refuge for rare plants and animals and may be protected by law.

In general this biome occupies the climatic niche defined by mean annual temperatures of 0°C to 20°C (32°F to 68°F) and annual precipitation of 500mm to 2300 mm (20 inches to 92 inches). It therefore overlaps in moisture requirements with the tropical forests, but has markedly lower annual temperatures, which change dramatically with the seasons. Most of the trees in this biome are deciduous, dropping their leaves at the onset of cold weather, often with spectacular displays of color (Figure 21-7). In the tropical seasonal forest, the deciduous state is controlled by drought; in temperate forests it is controlled by the low levels of incoming solar energy in winter. In winter frosts, the moisture in leaf tissues would freeze and expand, breaking up the leaf cells and killing the plant. The tree must also reduce transpiration and conserve moisture because it is less able to pull water from the cold soil in winter. Trees therefore drop their leaves and remain dormant until warmer temperatures promote new growth. Buds that must survive the winter are protected by scales. This marked seasonal change extends to all trophic levels; all organisms must find a way to survive the winter months when the available energy is low. Some animals, particularly birds, migrate to a warmer biome; others stay put and hibernate or live on food stored from the summer months.

In contrast to the tropical forests, the midlatitude forests contain fewer tree species. Deciduous forest ecosystems on different continents are similar in growth form, structure, and function, but they contain different species. The hickory and maple of North America, for example, are largely absent from Europe, where oak and beech are dominant. Elm was also common in Europe until the onset of Dutch elm disease. Local variation in climate, soils, topography, and historical use gives rise to distinct associations of species, such as the beech-maple forest in parts of the north of the United States, the oak-hickory association on the Piedmont plateau east of the Appalachians, and what were the oak-chestnut forests of the Appalachians before chestnut blight removed the chestnuts. Stands of different species in this biome, however, all have more or less the same structure. Compared to the tropical forests, there is little vertical stratification within the tree layer and the canopy is more open, allowing sunlight through to the forest floor (Figure 21-8).

Figure 21-7 Leaf-fall in midlatitude forests creates a spectacular display of red and yellow shades. The leaf colors change with changes in the type and concentration of pigments in the leaf.

Figure 21-8 The open canopy of these deciduous trees allows sunlight through to the forest floor.

The ground layer is at its richest in spring; it dies back as the trees increase their foliage into summer. Figure 21-9 compares the vertical distribution of biomass among leaves, roots, and branches in a temperate and a tropical forest.

The concentrated leaf fall in autumn from the deciduous trees adds a large amount of organic matter to the soil, providing a source of energy for soil organisms. The temperate climate means that leaching, although it occurs, is not as heavy in these soils as in the tropics; the organic material is therefore broken down and the products are kept in the soil in readiness for plant growth. The result is a fertile soil, typically an Alfisol, with a good distribution of colloidal material and a good structure for plant growth. Midlatitude forests will also form on the poorer Ultisols. The combination of good soils, hardwoods, and opportunities for hunting made these woodlands an attractive option to early settlers and farmers once they possessed the technology to cut trees on a large scale.

Conifers occur in this biome where local conditions of soil or climate allow them to compete with deciduous broadleaf trees. Compared to the flat leaves of deciduous trees, pine needles transpire slowly. Conifers also need fewer nutrients than deciduous trees to achieve the same growth. The conifers are therefore widespread in dry and cold conditions outside the tolerance range of the broadleaf trees, or on sandy soils that are acid, have a low nutrient content, or are rapidly drained. Examples in the midlatitude forest biome include the pitch and long-leaf pine on the sandy soils along North America's Atlantic coast plain, some evergreen eucalyptus in eastern Australia, and the cypress and pine of the southeastern United States. The pine woodlands of Florida and other parts of the East coast would develop through ecological succession to deciduous forest, if the fires that regularly sweep through these regions were stopped. The pines have thick cork-like bark that protects them from fire, whereas the young deciduous trees and other competing plants have no such protection and are burnt back in each fire. The pine community therefore depends on a regular, natural interruption of succession for its survival.

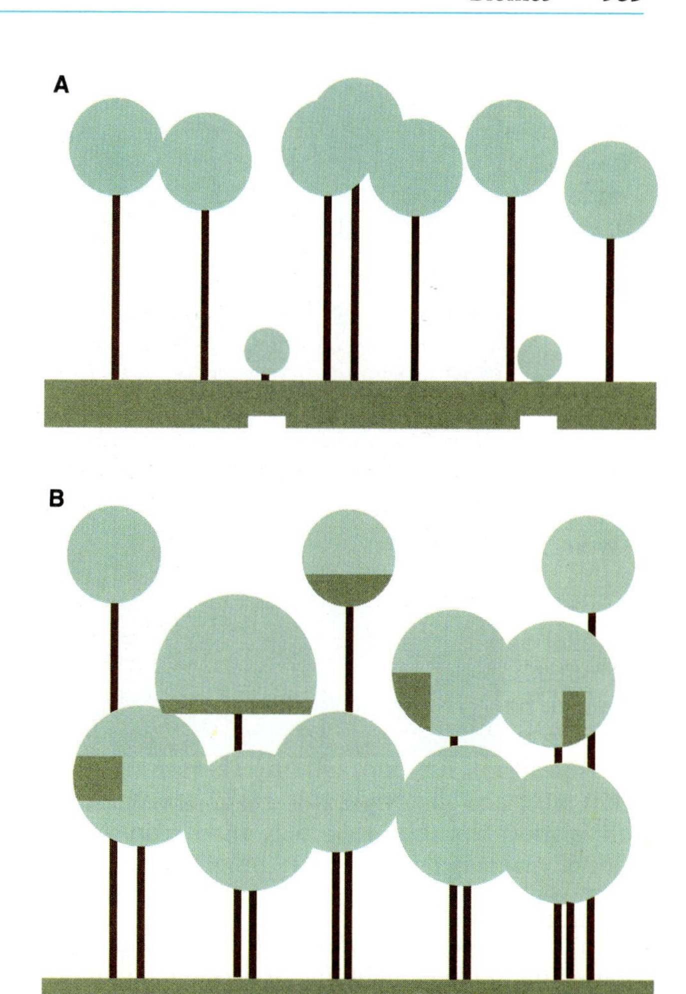

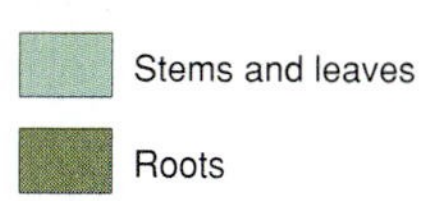

Figure 21-9 The distribution of biomass in **A,** a midlatitude woodland, and **B,** a tropical rainforest. In the rainforest profile the biomass is more evenly distributed and the physical demarcation of root (in the ground) and leaf (above the ground) layers is less rigid.

Mediterranean Forest and Scrub

As the name suggests, the "type" area for the **Mediterranean forest and scrub** biome community is the area fringing the Mediterranean, but it also occurs in parts of California, South America, southern Africa, and southwest Australia. It develops in climatic conditions where winters are mild and rainy compared to the midlatitude forest, and summers are hot and dry. These summer conditions are caused by the poleward extension of subtropical high pressure zones. Where trees are present, species such as the Aleppo pine and evergreen oaks are dominant. These have small, toughened leaves that help to reduce water loss. Where the climate precludes tree growth, the vegetation is made up of shrub species that can form dense thickets (Figure 21-10). These scrub communities are called **chaparral** in America, *maquis* (or *garrigue* in a degraded

better climates. Compared to the boreal forests to the north, these forests have a longer growing season, with higher temperatures and more available water.

GRASSLAND BIOMES

Grasslands occur where the combination of temperature and moisture regime precludes tree growth, but the limitations are not severe enough to produce desert. A seasonal moisture deficit is the main climatic control on the distribution of both temperate and tropical grasslands. However, climate alone does not explain the distribution of the grassland biome. Many researchers consider that both past and present human activity are important determinants of its nature and extent. For example, the limits of the prairie grasslands in North America are thought to have been extended and maintained by deliberate burning and, to a lesser extent, by the trampling effects of grazing domesticated animals. Recent experiments have shown that, when protected from burning and grazing, prairie grassland at the eastern limits of its tolerance range develops rapidly into woodland. This is almost certainly the case in the tropics, where the savanna grasslands are extensive and in some areas there is a long history of human activity.

Tropical Grasslands (Savanna)

The term **savanna** covers a great variety of types of vegetation. All savannas are dominated by grasses, but the proportion of tree and shrub species varies. The different combinations give rise to a range of more or less distinct subtypes, from pure grass communities, through open grassland with isolated trees, to areas that have a nearly continuous tree cover above the grass layer. In some areas, particularly southwest Africa, there is a continuous succession of subtypes from the wetter savannas bordering the tropical seasonal forest to drier areas bordering the desert (Figure 21-13). In general, however, the relationship between vegetation type and climate is less clear in the tropical grasslands than in other biomes. In broad terms, the typical climate for savanna vegetation has constantly high temperatures (above 20°C) and moderate total rainfall (500 to 2000 mm, or 20 to 80 inches per year), which is usually concentrated into a few months. The long dry season, up to 8 months, is the most easily recognizable characteristic. In southwest Africa, however, one type of savanna receives rainfalls as low as 185 mm (7.5 inches) per year, and in contrast, savanna is found in some areas where rainfall figures would suggest a cover of tropical forest. To complicate things further, savanna grassland can also be found where there is no appreciable dry season at all, in places such as the Amazon basin and parts of the African rainforests.

Clearly, factors other than climate have a strong effect on the nature and distribution of the vegetation, and finding them has become a research issue in biogeography. These studies give a clearer insight into how different Earth environments interact at a particular place in the biosphere. They are particularly important in the case of the tropical grasslands because this biome has a high primary production, in areas of the world where population growth is outstripping food supply. Some researchers suggest that the surface-relief and the soil environments control the availability of water and thus control the vegetation. For example, in the case of southwest Africa described above, the low rainfall is offset by a readily accessible supply of groundwater, which maintains a grass cover in a very arid climate. A more complex theory attempts to explain the common distribution in which woodland grows on slopes and grasses dominate the plateaus in between, in terms of geomorphologic evolution.

Other researchers point to the effects of fire. The accumulation of dead grass in the dry season means that savannas are vulnerable to fire, started naturally or accidentally or deliberately by humans. Burning removes the old dead plant tissue and promotes the growth of young and nutritious grass shoots. It is therefore advantageous to the grazing animals and to the carnivores, including humans, that hunt them. Recent research suggests that the savanna ecosystem was in place before the arrival of *Homo sapiens,* but regular burning, initiated by human settlement of these areas, has clearly influenced the pattern of vegetation that exists today. In summary, the variations in vegetation within the savanna biome are the result of human activity added to the complex interactions of Earth environments in the biosphere.

The grass species in savanna vegetation are typically coarse and adapted to dry conditions. The leaves die off rapidly at the start of the dry season. However, the sheaths from the dead leaves help to protect the tissue that produces the new shoots, which grow rapidly once the rain comes. The grasses usually grow to about 80 cm (32 inches) in height, but there are exceptions, such as the elephant grass *(Pennisetum purpureum),* which forms thickets and reaches heights of up to 5 meters (16 feet). The trees and shrubs are usually different species from those in the neighboring tropical forests. They have also adapted to the dry conditions by

Figure 21-13 **A,** The grassland and trees of the savanna biome in the Maasai Mara Game Reserve, Kenya. **B,** The distribution of forest and savanna in west Africa is closely related to climate.

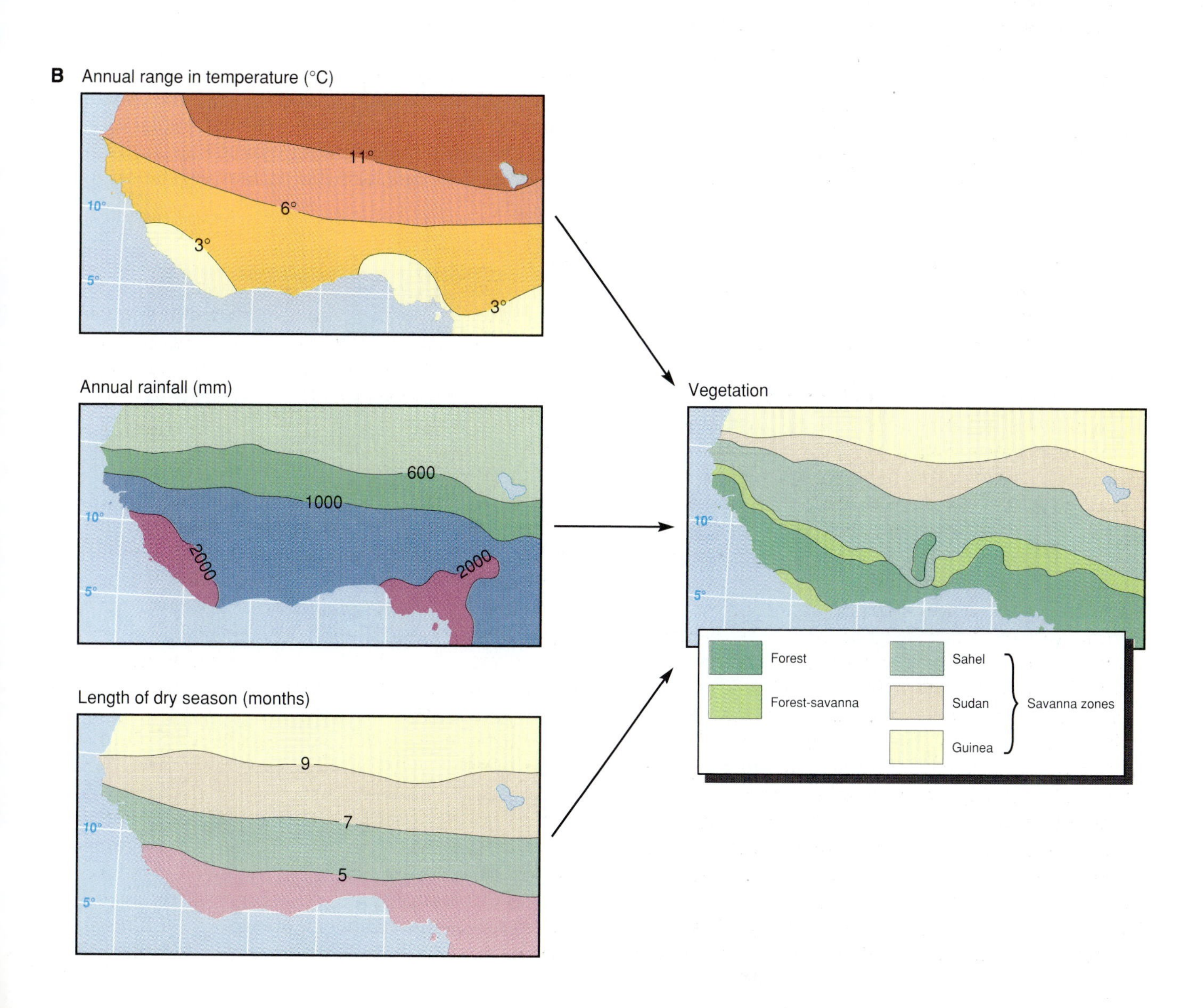

being deciduous or by having small, tough leaves. Trees and shrubs need a greater water supply than the grasses and have a deep, dense root network; their presence in very dry areas usually indicates a source of water at depth. The typical soils of the tropical grasslands are Ultisols and occasionally Alfisols.

The primary producers in the savanna are less numerous and less diverse than those in the tropical forests. There are therefore fewer ecologic niches for species at the higher trophic levels, and the structure of the ecosystem overall is much simpler. However, the total energy flow into the systems is high enough to support very large numbers of certain species, such as the great herds of grazing animals (zebra, antelope, giraffe, and wildebeest) and attendant carnivores (hyenas, and particularly lions) in Africa. In other continents the species have evolved differently; the marsupials, particularly kangaroos and wallabies, are dominant in Australia, and in South America there are fewer hoofed grazing animals but many more birds (Figure 21-14). The seasonal migration of grazing animals in the Serengeti National Park shown in Figure 21-15 demonstrates how each species uses a slightly different element of the primary production as a source of energy. Competition is also reduced in tree savanna as different species graze at different levels (Figure 21-16).

Current human activities in the savanna involve cultivation and more intensive grazing centered around large and relatively permanent settlements, in contrast to the hunting, gathering, and extensive herding of stone age people. Overgrazing and overcultivation in dry conditions, particularly in the grassland-desert ecotone, lead to loss of vegetation and soil erosion, both of which are characteristic of **desertification** (see Box 21-1: Desertification), the process by which semiarid lands become deserts. The naturally erratic rainfall in these areas combined with current human activities has contributed to widespread famine in recent years.

Midlatitude Grasslands

The **midlatitude grassland biome** includes the natural vegetation communities of the prairies of North America, the pampas of South America, the steppes of Eurasia, and the grassveld of South Africa (Figure 21-17). In common with the midlatitude deciduous woodlands, and in contrast to the tropical grasslands, much of the natural vegetation in

A

B

Figure 21-14 Wildlife in the savanna. **A,** Zebras, wildebeests, and gazelles are typical herbivores of the African grasslands. They are food for the carnivorous lions. **B,** In Australia the dominant herbivores are marsupials, of which the kangaroo shown here is a common example.

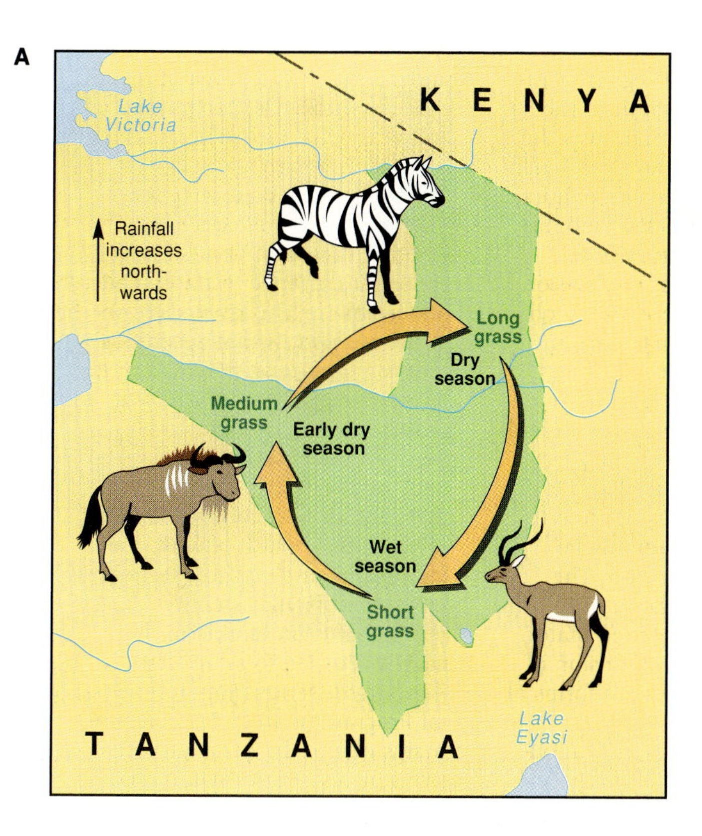

Figure 21-15 The seasonal migration of animals in the Serengeti National Park. **A,** Grazing animals spend the wet season in the short grass area to the south, moving progressively west and north with drier weather. **B,** Different groups of animals move at slightly different times and eat different parts of the grass canopy. The zebras move first and graze the upper grasses, wildebeests just behind them graze the middle grasses that are now exposed, and finally the gazelles eat the shortest grasses. The graph shows the timing of peak animal numbers of each group in the western Serengeti.

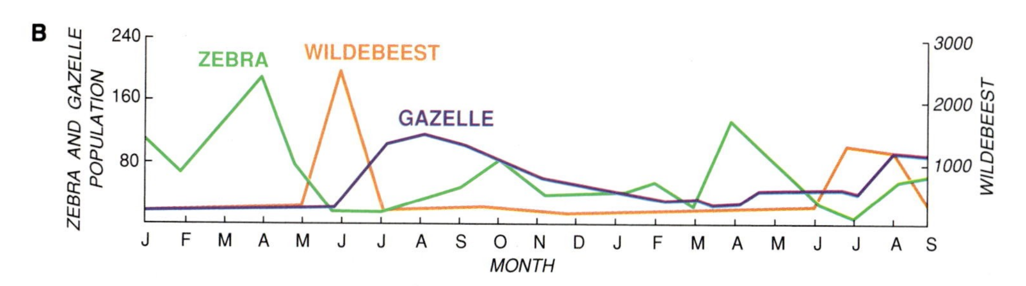

this biome has been converted to farmland, particularly for grain crops. Both temperature and rainfall vary seasonally, and, except at the eastern limit of the North American prairie, there is not enough water for tree growth. The winter can be cold but it is comparatively short and the growing season is long. Precipitation more or less balances potential evapotranspiration over the year as a whole, although there is often a water deficit in the warm summer months. The soil moisture is gradually recharged in the cooler winter months when evapotranspiration is reduced.

The temperate grasslands of Eastern Europe have developed over a more or less uniform topography and bedrock. The soils show a classic gradation from northwest to southeast, a pattern that reflects the decreasing availability of water and increasing insolation. The gradual change in climate is also mirrored in what remains of the natural vegetation.

Figure 21-16 Sharing food sources in the tree savanna. Giraffes browse among shoots that other species cannot reach.

A

B

Figure 21-17 **A,** Midlatitude grassland in its natural condition. Most of this biome has been converted into cropland, as shown by the extensive wheatfield in **B**.

This area is the "cradle" of modern soil science. V.V. Dokuchaev, a famous Russian soil scientist, used it as a study site for the first investigations into soil formation in the late 1800s. He demonstrated that soil is a dynamic body, responding to changes in factors such as climate, living organisms, and surface relief, a viewpoint that revolutionized the study of soils and their role in the ecosystem.

The prairies of North America are similar to the steppes of Eastern Europe, but are controlled by a more complex set of gradients. In North America, the grasslands extend further south, their elevations rise from about 150 meters (450 feet) in the east to about 1600 meters (5250 feet) in the west, and precipitation decreases from east to west. There is therefore a clear east-west sequence of tall-grass, mixed, and short-grass prairie that is related to moisture, and a north-south species gradient within each moisture zone that is related to temperature. The east-west zones correspond roughly to the U.S. corn belt (tall-grass), grain belt (mixed), and open range (short-grass) areas. The tall-grass prairie is a zone of transition between the temperate forest to the east and the mixed-grass prairie; the grasses are about 1.5 to 2.4 meters (60 to 90 inches) in height and are interspersed with patches of oak-hickory woodland. The mixed-grass prairie is the most "typical" prairie vegetation; it contains a mixture of short, medium, and tall species, averaging about 0.6 to 1.2 meters (24 to 46 inches) in height. Short species dominate where grazing pressure is high. The short-grass prairie at the western edge of this zone is thought to be at least partially the result of heavy grazing of the taller grasses, coupled with the effects of reduced rainfall. Drought-resistant or **xerophytic** plants, such as sagebrush and prickly pear (Figure 21-18), are also found in the short-grass prairie. These plants have adaptations to maximize their use of available water and to minimize its loss. The adaptations include deep roots to reach underground supplies of water or shallow roots that spread over a large horizontal area. To reduce transpiration, plants have small or needle-like leaves, which may also have a waxy covering. Experiments have shown that trees will grow on some of these soils when the grasses are removed. Their absence from the prairie is attributed to the effects of the periodic but fairly regular drought periods, the competitive advantage of grasses after burning, and grazing and trampling by the bison, antelopes, and domesticated animals that used to roam the area.

Before widespread farming, the dominant animals of the American prairies were the bison and the pronghorn antelope, both of which ranged far across the grasslands in large herds. Both animals were hunted close to extinction and are now protected. The burrowing gopher (or "prairie dog") is a food source for carnivores such as the coyote (a sort of wild dog), foxes, and birds of prey. The gophers' concentrated grazing around the entrance to their burrows keeps out invading trees and shrubs.

The most extensive area of temperate grassland in the southern hemisphere is the Argentinian pampas. Here the rainfall range is 500 mm to 1000 mm (20 to 40 inches), rather higher than that in the United States or Eurasia and more evenly spread across the year. Potential evapotranspiration is also high, however. As in the American prairies, the gradation in rainfall from east to west is mirrored in the height of the vegetation. Other temperate

Figure 21-18 Xerophytic plants such as the prickly pear, shown here, are adapted to the dry conditions of the short-grass prairie.

grasslands are found in southeastern Australia and on the high veld of South Africa.

The soils of the temperate grasslands are typically fertile Mollisols, which are now cultivated for grain crops. In North America the east-west moisture gradient is matched by a change in suborder from Udolls in the east, through Ustolls, to the drier order of Aridisols in the west.

Tundra

Tundra is a Finnish word that means "barren land." It is used to describe the stunted and restricted vegetation communities that form poleward of the boreal forest, where the constant cold precludes normal tree growth. In the northern hemisphere, tundra extends across the northernmost parts of Eurasia and North America, mostly above 60° N. It is also found in isolated fragments in the southern hemisphere, on the southernmost Antarctic islands, and on the northern tip of the Antarctic continent. Some authors recognize subdivisions of high arctic, middle arctic, and low arctic tundra, based on decreasing severity of environmental conditions, increasing diversity of species, and increasing productivity. Alpine tundra is a similar community found in the cold temperatures of high mountains. The climate statistics for the tundra areas describe a grim environment for life: the precipitation is low, at around 400 mm (16 inches) per year, the subsoil is permanently frozen, and there are only about 50 days in the year when the surface temperature is above freezing. In addition, the grasslands are frequently swept by gale-force winds, which lowers the temperature even further.

Not surprisingly, the tundra has a low diversity of primary producers. Their productivity is, however, surprisingly high in the short summer. The vegetation cover is predominantly mosses and lichens, with some very hardy grasses, herbs, sedges (plants similar to grasses), and shrubs, with occasional dwarf trees. These species grow close to the ground to escape the wind and have small leaves to reduce water loss (Figure 21-19). All the plants must make good use of the short growing season. Some plants are evergreen, so that they can start photosynthesis as soon as temperatures rise; others make at least some parts of the plant, such as shoots and flower buds, ready at the end of the summer for immediate growth the following spring. The buds are protected by the snow during the winter and can start to grow early in the spring.

The structure of the higher trophic levels in the tundra ecosystem is controlled by the pronounced seasonality in production. Some grazing animals, such as the caribou, feed here only in the summer

Figure 21-19 The low-growing vegetation of the tundra biome.

months, returning to the boreal forest during the winter. Smaller, burrowing animals, such as lemmings, can survive all year round.

The soils of the tundra ecosystem are dominated by the cold conditions, and particularly by the frozen ground or permafrost. The surface layer above this, which thaws in the brief summer, is termed the *active layer.* During the thaw the water table remains at or near the surface because the permafrost prevents drainage. This has a number of effects. First, the lack of oxygen in the waterlogged soil means that there is only a very small soil flora and fauna. The breakdown of organic material is correspondingly slow, leading to acid conditions and the formation of peat. Second, the whole active layer is mobile and may move under gravity; the action of freezing and thawing leads to further physical disturbance of the soil. "Normal" pedogenic processes of gleization and podzolization do occur in the tundra, but they are often masked by the physical effects of disturbance to the active layer.

Until recently human activity in the tundra was restricted to the hunting, herding, and fishing economy of the indigenous peoples. This system was adapted to, and had little impact on, the structure and function of the natural system. However, recent advances in technology have opened the tundra to mineral exploitation and the establishment of military bases. A high-energy society has been moved into this fragile, species-poor ecosystem. Overcoming the problems of carrying out human activities on permafrost has taxed human ingenuity (see Box 15-1: Problems of Living on Permafrost) and has led to disruption of the tundra ecosystem. As discussed in Chapter 19, a common effect of human activity is to simplify the food web, by removing one or two links, either deliberately or accidentally. In an ecosystem as naturally simple as the tundra, disturbance at one level may have catastrophic implications for the rest of the system.

DESERT BIOMES

Desert biomes exist in areas where potential evapotranspiration far exceeds inputs from precipitation. Such areas are most extensive around the tropics due to the high temperatures and calm descending air. Deserts can also form in continental interiors, in any area of rain shadow, and along coasts that have an upwelling cold current offshore. In most cases the lack of moisture precludes a full vegetation cover, and all species must be adapted in some way to the low levels of available moisture. The effects of low and irregular precipitation can be offset to some extent by groundwater storage or the concentration of runoff by local topography. Within an otherwise barren landscape, there may therefore be small patches of complete and even lush vegetation cover. In general, however, and in comparison to the other world biomes, the desert supports few species and has a low productivity.

The study of adaptations to drought is a fascinating topic in biology and biogeography and can be of use to agriculture in arid and semi-arid lands. The typical conception of desert vegetation is of drought-resistant succulent plants such as cacti, which take up water when it is available and store it for use in the dry season. Other drought-enduring or xerophytic plants, described in the section above on midlatitude grasslands, find the very small amounts of water they need almost continuously through wet and dry seasons. Another set of plants either complete their whole life cycle within the space of one wet season or simply lie dormant in the most severe conditions and grow only when the moisture regime allows it. These species are responsible for the dramatic "blooming" of the desert just after rain (Figure 21-20). The natural animal life of desert areas is restricted, by the low productivity of the primary producers, to insects, reptiles, birds, and burrowing animals such as gerbils and gophers. The temperature drops sharply with depth below the soil surface; burrows are therefore a means of escaping the daily highs and lows of sur-

Figure 21-20 Succulent plants, such as this teddy bear cactus, survive dry periods by storing water in their tissues. They also have a tough outer skin that reduces transpiration.

face temperature that are typical of desert areas. A burrow as little as 50 cm (20 inches) below the surface has a more or less constant temperature. The species found in deserts around the world have evolved separately but show remarkably similar types of adaptation to drought.

The soils of the desert biome are typically poorly developed and of the order Entisols or Aridisols. The dominant soil-forming factors are parent material and climate. Where the water table is near the surface, salic horizons may develop. Desert soils are not necessarily barren. The lack of natural eluviation means that weathered nutrients remain in the soil despite a low colloidal content. With careful irrigation the soils can support agriculture.

MOUNTAIN BIOMES

The world biomes described so far are based on the study of ecosystems in a narrow layer close to sea level. High mountains force the biogeographer to rethink this classification and to consider the three-dimensional nature of the land surface and of climate. The foothills of mountain ranges are located in one of the forest, grassland, tundra, or desert biomes described so far. On the upper slopes, however, the environment may be radically different; temperature usually falls with elevation, and wind and precipitation patterns may be altered. The drop in temperature associated with a climb of 100 meters (310 feet) is roughly equivalent to moving 100 km (65 miles) north at sea level in North America; the atmosphere changes much more rapidly vertically than it does horizontally.

Given that temperature decreases with altitude, a scientist might expect the changes in vegetation, climate, and soils from pole to equator to be mirrored on the climb up a mountain. In broad terms this is true, but there are many complicating factors such as the local exposure to wind and rain and whether the mountain is an isolated peak or part of a range. Figure 21-21 illustrates some typical sequences of change in vegetation with altitude in different parts of the world. The changes are most marked in tropical areas, where a climber can go from tropical rainforest to alpine tundra in the space of a few (horizontal) kilometers. In the tropics, highland areas may provide a more equable climate for human habitation and agriculture than the surrounding lowlands. They are therefore relatively densely populated and intensively farmed. This is the case in the Andes of Peru and Colombia, where the mountains offer better resources than the rainforests to the east and the dry coastal plain to the west. The traces of the ancient Inca civilization that remain in this area show the importance of these "islands" of suitable climate for human settlement.

Some of the vegetation belts on mountains are the result of particular mountain microclimates and have no clear parallel near sea level. Examples of this are the mountain forests of the western cordillera of North America, and, more spectacularly, the cloud forest that forms at high altitudes in tropical areas. The cloud forests form at the height where water condenses into cloud. The environment is therefore wet, creating a rich forest with many mosses and epiphytes, a stark contrast to the dry forest and grassland below it. In the desert biome the air contains so little moisture that the harsh drought conditions may persist even at higher altitudes; typically, however, the desert grades eventually into Mediterranean-type scrub or into tropical grassland.

LINKAGES

*Mountains, the product of processes within the **solid-earth environment,** are acted on by the **atmosphere-ocean environment** to form the **surface-relief environment.** The topography that results affects the microclimate at a particular place, thus controlling the nature of the **living-organism environment** that develops there.*

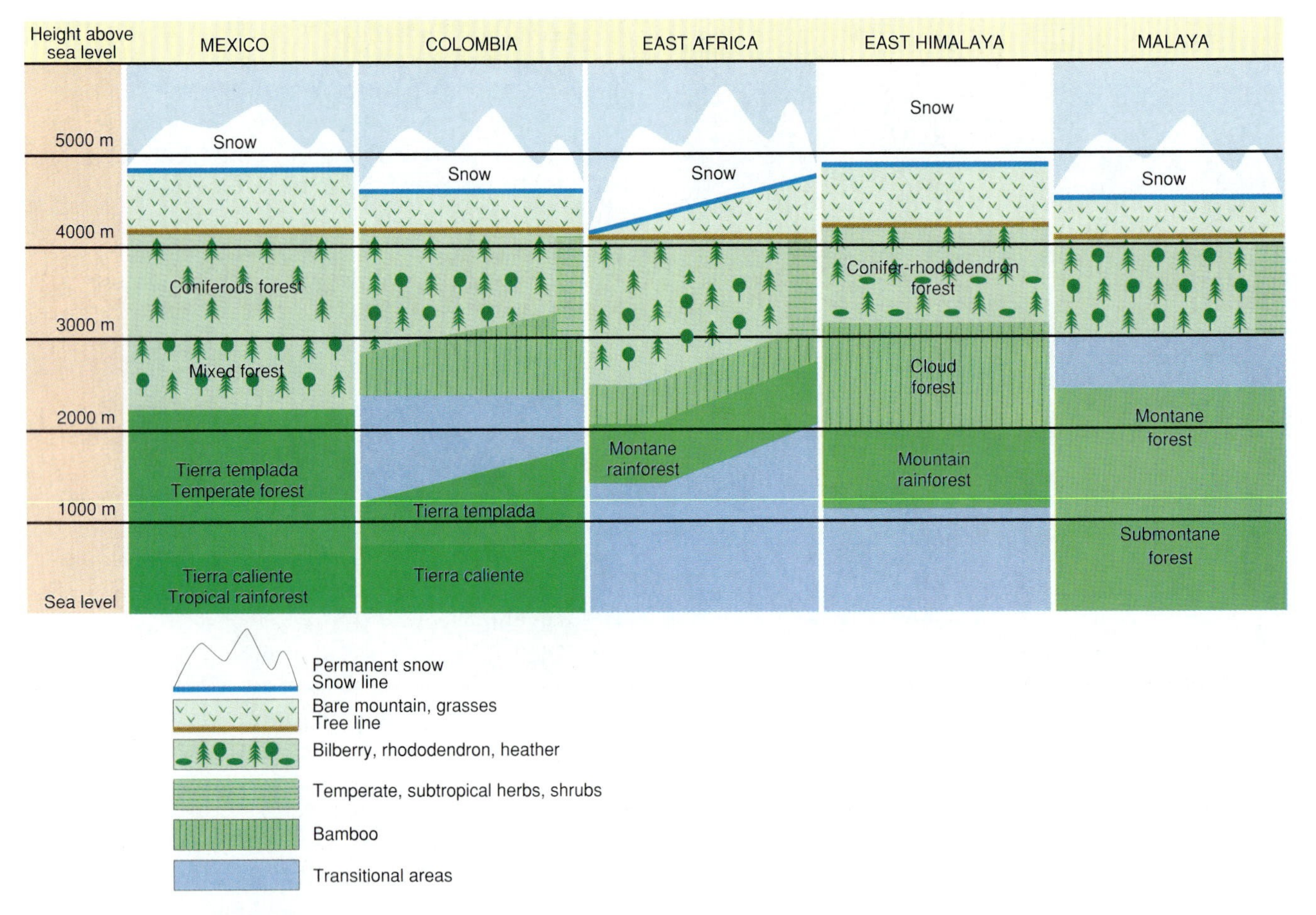

Figure 21-21 The sequence of change in vegetation with height on some mountain ranges.

OCEAN ENVIRONMENTS

On the land, most organisms are confined to a very narrow layer—the bottom few tens of meters of the atmosphere and the top meter or so of the soil. In the ocean, organisms are vertically distributed much more extensively. Although most living things are concentrated near the surface to catch the sun's rays, they can be found at the bottom of even the deepest ocean trenches. On land the living organism interacts with the other Earth environments to control its own development, through the process of ecologic succession. In the ocean the physical environment constantly moves and changes and the living organisms have very little control over it. This constant movement and mixing make it difficult to define any distinctive biomes within the oceans; in fact the whole concept of biomes tends to break down when applied to this environment.

However, some patterns are apparent. The temperature gradient from equator to pole in the atmosphere also applies to the oceans (Figure 21-22); the total biomass is greatest in the midlatitudes because of the high seasonal plant production in this area. Superimposed on this gradient are patterns relating to shallow continental shelves and deep or open ocean, and the presence of warm or cold masses of water. Salinity variations are also important. A number of more or less identifiable ecosystems can be pulled out from these intermingled and dynamic gradients.

Warm water continental shelves are the richest zones in terms of diversity of ecologic niches and plant and animal species. Distinctive and productive communities are found on mud, sand, rock, and reef beds. Spectacular and beautiful coral reefs form in this environment (Figure 21-23), in shallow waters where the temperature is regularly above 20°C (68°F). The reefs are formed by tiny living animals, which feed on even smaller plankton living in the water. When the corals die, their skeletons accumulate and are gradually cemented together to form the mass of the reef; the living corals are only a thin veneer at the surface. The corals are sensitive to both sunlight and salinity. Reefs form only in clear,

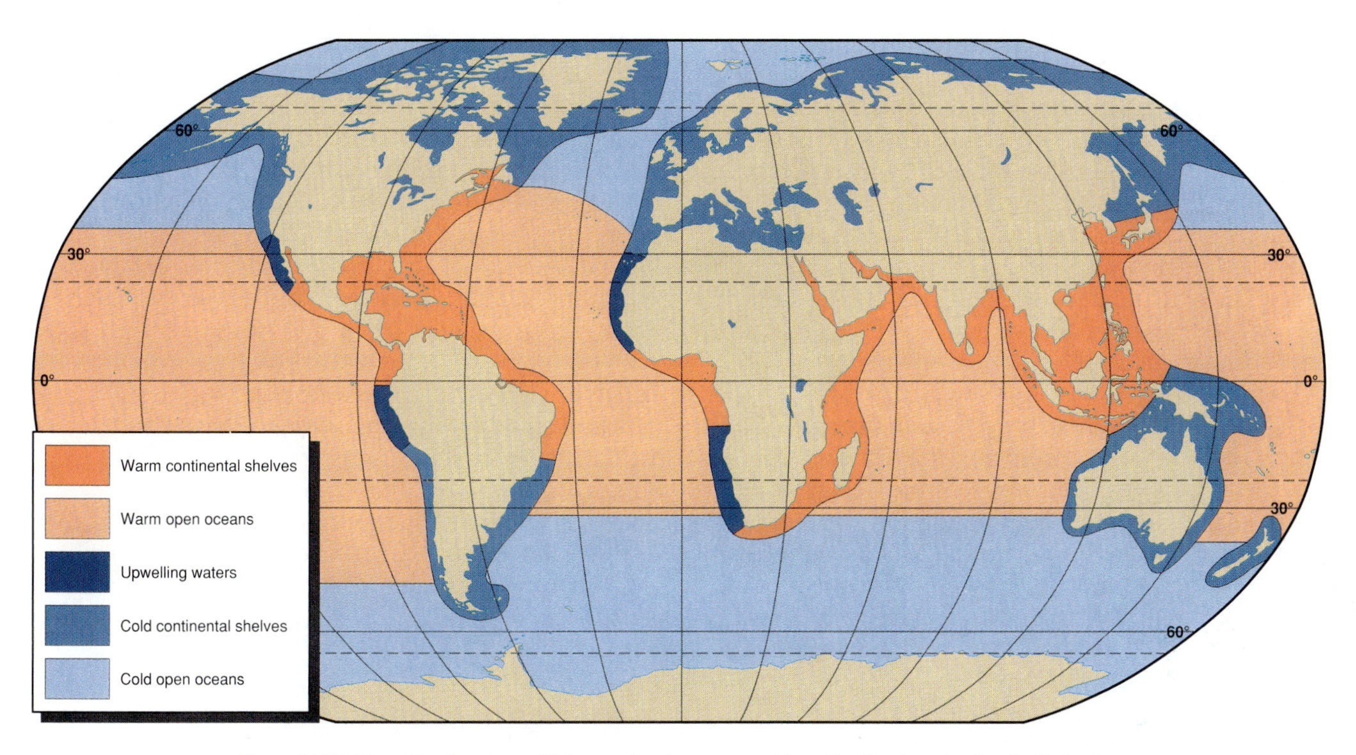

Figure 21-22 The distribution of biomes in the ocean. The distribution is closely linked to latitude, to the presence of deep and shallow water, and to upwelling currents.

Figure 21-23 An aerial view of a coastal coral reef. The coral itself and the diverse ecosystem it supports are attractive to divers. In some areas so many souvenirs have been removed that specimen collecting is now banned.

shallow water, with high salinity. They are largely absent from the eastern coast of South America, for example, because salinities are low due to freshwater discharge from the Amazon and the Orinoco rivers. The corals die if smothered by sediment; the reefs are therefore usually found in areas of vigorous water movement. The coral reefs are "islands" of high productivity in otherwise nutrient-poor waters. They create many ecologic niches and support a complex, diverse, and colorful food web that has attracted marine biologists and amateur divers for generations.

Where *upwelling cold waters* rise to the surface off the west coasts of continents they bring mineral nutrients to the upper layers of the ocean, as, for example, along the coast of Peru. The combination of nutrients and the strong insolation produces high levels of productivity, with complex food webs that involve phytoplankton, zooplankton, fish, and birds. The high productivity may also support a prosperous fishing industry.

Warm, open ocean waters experience few seasonal changes, so there is little vertical mixing of the waters and nutrients are in short supply at the surface. Although the surface waters support a reasonably diverse food web, the population of each species is small and the total biomass is smaller than in the midlatitudes. At greater depth the biomass decreases and the food web is even less diverse, with very small numbers of the carnivores and detritus feeders due to the scarcity of food.

Midlatitude coastal waters are cooler and have fewer species than the warm coasts in the tropics, but the population of each species is larger. Most of the world's major fisheries are situated in these areas. *Cool, open ocean waters* have seasonal phytoplankton growth, as do cool shallow waters, but nutrient supply is more limited. Primary productivity is highest in springtime as the temperature rises, especially where vertical mixing brings nutrients up from the bottom waters.

In general, humans should have less impact on the marine ecosystems than on terrestrial ecosystems, because the oceans are vast and most of their area is remote from human activity. However, as outlined in Box 3-2, human rubbish can turn up on isolated beaches thousands of miles from the nearest landfall. Since the oceans are not easily accessible, the scale of human impact remains largely unknown. The greatest effects are apparent in the altered food webs of fishing grounds, as outlined for the southern oceans in Chapter 19, and the accumulation of waste material in the narrow coastal zone. The coast, especially the area regularly covered and uncovered by the tide, is an ecotone between terrestrial and marine ecosystems; it is often diverse and productive and its species occupy very specialized ecologic niches. Coastal waters, however, are also a convenient dumping ground for human waste; all domestic, industrial, and agricultural materials discharged into a river will eventually reach the coast, and much is discharged directly into coastal waters. Coastal organisms face a range of pollution hazards, from eutrophication to physical smothering to poisoning. Coastal and particularly estuarine locations are also prime sites for industry and cities, and many productive coastal ecosystems are simply removed so that the land can be "reclaimed" for human use. Much of the Netherlands coast has been reclaimed in this way for agriculture.

LINKAGES

The distribution of living things in the ocean is closely related to climate, particularly temperature, and to oceanic circulation. Climate is controlled by the ***atmosphere-ocean*** *environment. Ocean circulation is controlled by the* ***solid-earth*** *environment (distribution of continents and oceans) and by atmospheric circulation. Cold offshore currents can also create dry conditions in the neighboring terrestrial environment, thus influencing processes in both the* ***surface-relief*** *and the* ***living-organism*** *environments.*

HUMAN IMPACT ON BIOMES

The natural ecosystems of some biomes are of little value to humans, and conditions of climate or soil preclude their being used for agriculture. The natural systems of deserts and the tundra are therefore relatively undisturbed. In contrast, biomes such as the midlatitude forests and grasslands have been almost completely cleared for agriculture. Other biomes are partially controlled by human activity. For example, the extent of both tropical and temperate grasslands is thought to be related to past and present burning by humans. The tropical forests have been undisturbed for a long time but are now being cleared at a rapid rate.

Some of the characteristics of human impact on ecosystems were discussed in Chapter 19. Typically humanity's use of natural resources is selective; humans have domesticated only a very small number of animals and plants, and we use the words "weed" and "pest" to describe other, natural components of the ecosystem that compete with domesticated species for energy and nutrients. Humans attempt to maximize the transfer of energy through their simple food webs by shortening the natural process of succession and by subsidizing the ecosystem with energy and materials. The natural vegetation communities of the midlatitudes have generally been replaced by these "fast-food" ecosystems. Where forests have been replanted, the favored species tend to be conifers, which can support a higher rate of harvesting than the native broadleaf trees. The current alarm over the rate of disappearance of the tropical forests (see Investigation: Stop the Burning) is based on an increased understanding of the structure and function of this biome and the knowledge that it cannot sustain a high rate of harvesting.

In areas where human use is less intensive, the original climax vegetation has been cleared and replaced by a semi-natural community. Here the species are those that occur naturally, but their dominance as a community is governed by human intervention. These communities are typical of the open grass and heather moors in northwestern Europe, which replaced midlatitude forest on higher ground. In some areas the natural succession to woodland is prevented only by vigorous management, either by maintaining a high grazing pressure, or by cyclical cutting or burning. In others the time scale over which the semi-natural community has existed means that the soils have adapted to a new equilibrium and a reversion to woodland is unlikely.

INTERACTIONS IN THE BIOSPHERE

The Earth environments interact to create a living environment, the biosphere. The nature of the biosphere at any particular place, whether it is the whole Earth, the savanna biome, one valley, or a small patch of woodland, is controlled by how, in detail, the Earth environments interact at that place.

Biomes are the visible expression of interaction between Earth environments at the regional scale. At this scale climate controls the nature of the biosphere. Temperature and precipitation control the global distribution of the primary producers and therefore of all living things. The solid-earth environment contributes by distributing the continents, oceans, mountain ranges, plateaus, and plains for climate to work on, and by supplying the raw materials for soil.

The surface-relief environment controls the detailed form of the biosphere within each biome. The different physical conditions created, for example, by steep and gentle slopes, support slightly different ecosystems so that a "uniform" biome is also a mosaic of smaller ecosystems. At yet a finer scale, each plant competes with its neighbors for light, air, and nutrients supplied by Earth environments in the few meters or so around it. Similarly, animals compete for food, water, and territory.

The interaction of Earth environments therefore controls the distribution of all living things at all scales in the biosphere. The biomes are one level of description in this hierarchy. They are a natural framework for study in physical geography—the description and explanation of the natural environment at a particular place and time.

FRONTIERS IN KNOWLEDGE

Biomes

Perhaps the most pressing problem in the study of biomes is understanding the extent to which human actions are responsible for delimiting the patterns of vegetation on a global scale. The case of the ecotone between tropical grassland and desert in the Sahel is a grim reminder of the importance of this type of study to human life, and concern about the disappearance of the tropical rainforest is increasing and vocal. The solution to the problem lies in a thorough understanding of the current interaction of Earth environments and how they continue to be influenced by past climatic change. Only when scientists have achieved such an understanding will they be able to assess the effects of the human input and start to plan to mitigate its effects. Only when politicians realize the severity of the human impacts will the plans be put into effect.

INVESTIGATIONS in Physical Geography

Stop the Burning

> *"Delight... is a weak term to express the feelings of a naturalist, who, for the first time, has wandered by himself in a Brazilian forest. The elegance of the grasses, the novelty of the parasitical plants, the beauty of the flowers, the glossy green of the foliage, but above all the general luxuriance of the vegetation, filled me with admiration. To a person fond of natural history, such a day as this brings with it a deeper pleasure than he can ever hope to experience again,"*

Charles Darwin recorded these thoughts on February 29, 1832, when H.M.S. *Beagle* reached the port of Salvador (latitude 12° 58′S) on the east coast of Brazil. Darwin's successors continued to marvel at the luxuriance of the tropical rainforests but now also realize the importance of this biome in controlling the nature of the atmosphere-ocean environment and in providing resources for human use.

Tropical rainforests exchange water, carbon dioxide, and oxygen with the atmosphere-ocean environment. The air around the forests is constantly wet because of transpiration from the dense vegetation. This water returns to the forests as rain. Scientists suggest that up to 50 percent of the rainfall in the Amazon basin, Brazil, the world's largest expanse of rainforest, is recycled directly from evapotranspiration in this way. Tropical rainforests cover less than 10 percent of the land surface, but account for about 40 percent of the total terrestrial plant biomass. The forests absorb carbon dioxide from the atmosphere in the process of photosynthesis and release oxygen. They are therefore a major pool of terrestrial carbon; the forest biomass and soils contain 20 to 100 times as much carbon as is held in croplands and pasture. Tropical rainforests therefore help to regulate the concentration of carbon dioxide in the atmosphere.

The rainforests are also astoundingly rich in species; scientists estimate that about 2.5 million species still

Burning in the aftermath of forest clearance.

await discovery. Any one of these species may hold the genetic key to advances in medicine or agriculture, or the potential to evolve into an intelligent form of life. The first humans are thought to have developed from rainforest species. The rainforests are species-rich because they have evolved, undisturbed, over literally millions of years. The heart of the existing forests is thought to have survived the Ice Ages. In these ancient refuges, some species occupy a very specialized niche, existing in only one valley, on one island, or one mountainside. Scientists' estimates of species richness are based on field study and this knowledge of the biome's evolution. Each new area that is explored reveals new species, so the vast unsearched areas probably contain many more.

Tropical rainforsts are therefore a valuable part of the biosphere. They help to regulate the composition of the atmosphere-ocean environment and provide a genetic stock useful to humans and the basis for future generations of living things. The forests, and the lands on which they grow, also have more obvious commercial attractions. They contain exotic and beautiful hardwood trees that can be cropped and sold for timber. They occupy large areas of land where both water and sunshine, the two main ingredients for photosynthesis, are in abundant supply. They therefore appear to be ideal for colonization and agricultural expansion. The value of the rainforests for logging and cropping now means they are disappearing at an alarming rate, to the detriment of their current and future value in the global biosphere. Deforestation in the tropics is a major environmental issue that has united conservationists around the world. The driving forces of deforestation, however, the severity of its impact, and precisely who and what are affected, are still matters of debate.

Deforestation is the conversion of a forest into a barren area or into another type of vegetation such as grassland or crops. Rates of tropical deforestation vary in different parts of the world and reliable data are hard to find. The general consensus, however, is that overall rates of deforestation are increasing. In 1979, about 75,000 square km (29,000 square miles) were cleared. In 1989 about 142,000 square km (55,000 square miles), about the size of Illinois, were cleared, an increase of 90%. The current rate of clearance is the equivalent of five Central Parks (New York City) of forest disappearing every hour of every day. About 8 million square km (3.1 million square miles) of tropical rainforest now remain out of the original 14 million square km, and 51.5% of the residue is concentrated in just 3 countries; Brazil (28%), Zaire (12.5%), and Indonesia (11%). The 10 countries that have lost the most forest area are listed in the accompanying table. Elsewhere annual rates of deforestation range from a high of 15.6% on the Ivory Coast to a low of 0.12% in the Guianas, with an average of 1.8%.

The undergrowth, cut branches, and trees not removed for logging are burned to clear the ground for agriculture, roads, or buildings, releasing carbon directly to the atmosphere. Using mathematical models, scientists have estimated that carbon emissions from the rainforests now stand at 1.4 billion metric tons per year, an increase of 75 percent since 1979. Other scientists predict that the reduced transpiration caused by deforestation means that, even if clearance stopped tomorrow, the region's rainfall has decreased to such an extent that the forests may not regrow. Ecologic teams are working literally in front of the bulldozers to catalog the flora and fauna of the forests before they are cut down. Scientists, environmental pressure groups, and politicians from the developed world have joined to say: stop the burning.

The reasons for deforestation are complex. There are three main agents of deforestation; the commercial logger, the cattle rancher, and the small-scale farmer.

Countries with high rates of deforestation

COUNTRY OR REGION	ORIGINAL EXTENT OF FOREST (THOUSANDS KM²)	EXTENT OF FOREST IN 1989 (THOUSANDS KM²)	DECREASE TO 1989 (THOUSANDS KM²)	% DECREASE TO 1989 (%)	CURRENT ANNUAL DEFORESTATION (THOUSANDS KM²)
India	1600	165	1435	89.5	4
Brazil	2860	2200	660	23	50
Colombia	700	278.5	421.5	60	6.5
C. America	500	90	410	82	3.3
Thailand	435	74	361	83	6
Indonesia	1220	860	360	29.5	12
Burma	500	245	255	51	8
Zaire	1245	1000	245	19.5	4
Mexico	400	166	234	58.5	7
Philippines	250	50	200	80	2.7

INVESTIGATIONS in Physical Geography—cont'd

The logging and ranching activities are typically financed by overseas investment and have attracted considerable media attention. The small-scale farmers are usually native or from neighboring countries, and collectively they account for more deforestation than the loggers and the ranchers combined.

Until the early 1960s, the small-scale farmer was typically a member of a native tribe, who practiced rotating or "shifting" cultivation of small clearings made in the forest. This had a minimal impact on the forest because it involved a small number of people. From the 1960s onward, the indigenous peoples have been joined by people who migrated or have been moved to the marginal forest lands in the hope of a better life. The migrants move because they are poor, part of a rapidly growing population, and because they own no land, among other reasons. Most of these conditions are beyond the control of individual farmers. The newcomers practice a more settled form of agriculture, causing a widespread and permanent loss of forest cover. Deforestation by new settlers is concentrated in the frontier regions opened up by new roads. It is analogous to the widespread deforestation of Western Europe and the settlement of the American West, both still seen by many as a necessary and, at times, glorious expansion of the nation. The main differences between the past and the present are that scientists now better understand the long-term value of forests in the global ecosystem, and that the destruction occurs more quickly.

The starting point in tackling deforestation is reliable information on the rates and locations of clearance. As recently as 1987, a scientist researching deforestation concluded that "available estimates on forest resource and conversion rates are utterly confusing, obsolete, unreliable, and often contradictory." The situation is improving rapidly as the value of remotely sensed images, an objective and repeatable form of data gathering, is realized. Some scientists go so far as to say that remote sensing is the only way to resolve discrepancies and provide up-to-date information. Satellite images were used successfully in the late 1980s to adjudicate between Amercian researchers, the Brazilian government, and the World Bank, all of whom proposed different rates of deforestation in Brazil. Several sorts of imagery are useful. An exhaustive study of 2325 photographs from the Space Shuttle and earlier lunar missions has detected an increase in the extent of smoke on the southern edge of the Amazon basin. Much of the activity is centered around the road BR 364, which links the three Brazilian states of Mato Grosso, Rondonia, and Acre, and BR 429 to the south (shown on the accompanying map). Landsat images confirm these findings, and images taken a few years apart show the effect of the surge of migration that followed the paving of BR 364 in Rondonia. When available, the Landsat data pinpoint the new clearings. They are not, however, the best data to give regular updates, because it takes over 200 scenes to cover a large area like the Amazon basin, and the frequent cloud means that good data are not always available. Scientists from the Laboratory of Terrestrial Physics at Goddard Space Flight Center found that data from weather satellites gave the best results, because they cover a large area and many more images are available. They used a waveband in the thermal part of the spectrum to distinguish between cleared and forested areas. A comparison among weather satellite data, Landsat, and ground survey showed that the estimates of deforestation from the weather satellite data were accurate to about 1000 square km (about 10 percent).

It is important to keep a reasonable perspective among the most pessimistic scenarios of forest loss. The role of the tropical forests in the carbon cycle is undeniably important, but it is minuscule in comparison to the fluctuations in oceanic carbon. Disturbing the forest will almost certainly lead to extinctions of large numbers of species, assuming that current estimates of species richness are correct. But mass extinctions, such as those at the beginning of the Ice Ages, have happened naturally and will doubtless happen again. Some species, particularly those limited to specific niches, are on the edge of natural extinction and could be quite casually blotted out by natural processes. Deforestation will give opportunities to other species and may allow the evolution of new ones. According to some biologists, the future potential of rainforest species is exaggerated. Humans have prospered using only a handful of the 1 million species that we already know about; we are unlikely to need

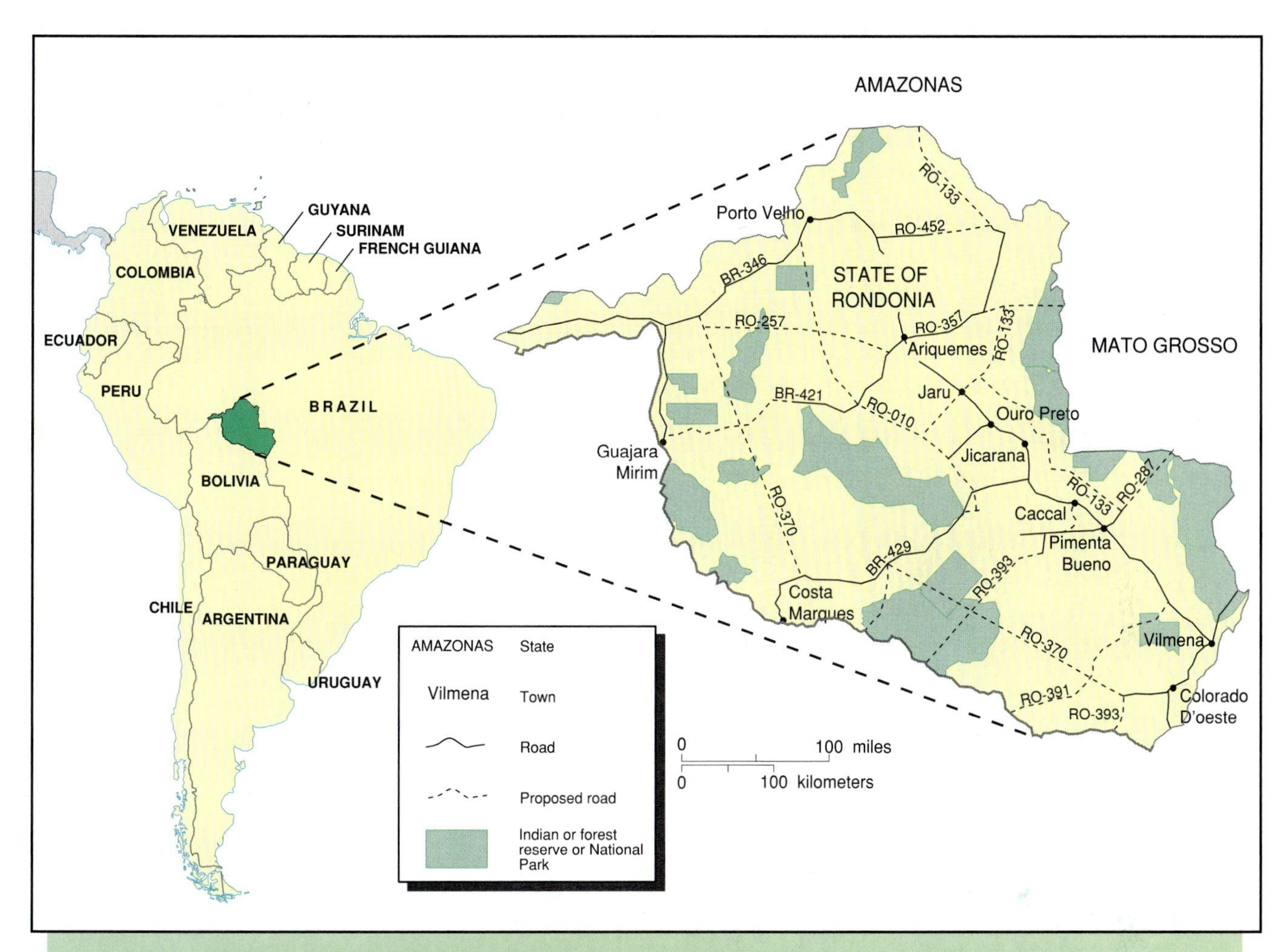

New and planned roads in Rondonia and Acre, southern Brazil. Road-building opens up previously inaccessible forest areas, which are rapidly settled.

another 2.5 million. The environmental concerns of well-housed and well-nourished populations in the developed world must be set against the grinding poverty that is widespread elsewhere and that exploitation of the rainforests is designed to offset.

The future of the tropical rainforest is an international issue. The financial and ecologic crises in the developing countries in which the forests are concentrated are partly due to the willingness of industrialized, developed nations to invest in their development plans. It is ironic that these countries, having cleared most of their own forests and thinned out the remainder through acid rain, are now insisting on conservation in the tropics, while still expecting payment of debts incurred when development was encouraged. Contemporary deforestation is not confined to the tropics; in the Pacific Northwest of the United States, for example, coniferous forests are undergoing widespread deforestation. Both deforestation and conservation in the tropics hit hardest at the people least able to control their future, the indigenous forest tribes and the small-scale migrant cultivators. Tropical deforestation is a complex issue that has global causes, global implications, and requires global action to solve.

SUMMARY

1. A biome is a large area over which the conditions of vegetation, climate, and soil are more or less uniform. Biomes are one level in the nested hierarchy of ecosystems that makes up the global biosphere.
2. The main characterizing feature of the biome is its natural vegetation, defined by growth form rather than by species.
3. Studies at the biome level highlight the importance of the atmosphere-ocean environment in shaping the biosphere at a regional scale, and the local pattern formed by the detailed interaction of Earth environments.
4. The world biomes studied in this text are tropical rainforest, tropical seasonal forest, midlatitude forest, boreal forest, midlatitude grasslands, tropical grasslands, tundra, deserts, mountains, and oceans.
5. The map of world biomes is similar to that of world soils and world climate. The discrepancies are mostly due to different rates of adjustment after climatic change and the effects of human impact.
6. Conditions in each biome vary gradually over its extent; biomes grade into each other over ecotones.

KEY TERMS

QUESTIONS FOR REVIEW AND EXPLORATION

1. List the different effects of Earth environments and their interaction on two contrasting biomes.
2. For one or two specific biomes, determine the nature of the local departures from the ecosystem "type." What causes them? How important are these local variations in giving a distinct appearance to the biome?
3. Find out what the natural vegetation of your area is. What type of local variation would you expect to find in a circle of 5 miles radius from your home or college?
4. Explain the global distribution of tree growth.
5. Prepare arguments for and against the idea that the equatorial areas are the best ecologic niche for plant growth.
6. Compare the energy flow and nutrient cycling of tropical, temperate, and boreal forest.
7. What are the similarities between deserts and the tundra? How have organisms in the two biomes adapted to the harsh conditions?
8. What is the effect of a pronounced dry or cold season on vegetation growth?
9. Using Figure 21-2 as a base, describe the types of ecotones that may be found in North America.
10. How do the changes in vegetation with elevation in a mountain range differ from changes with latitude?
11. Explain why the concept of biomes is not particularly helpful in a study of the oceans.

FURTHER READINGS

Calder N: *Spaceship Earth* Viking/Channel 4, 1991. Originally designed to accompany a TV series, this book examines current human impacts on the biosphere with particular emphasis on the benefits of a vantage point in space.

Ellis W: Africa's stricken Sahel, *National Geographic,* August 1987. Outlines the problems of desertification in this marginal zone.

Findley R: Endangered old-growth forests, *National Geographic,* 106-136, September 1990. Findley describes the felling of Pacific coast and montane conifers in North America, deforestation that is continuing "at home" while attention is focused on the tropics.

Lanting A: Botswana, a gathering of waters and wildlife, *National Geographic,* 5-33, December 1990. A graphic photo-essay including material on the savanna environment.

Repetto R: Deforestation in the tropics, *Scientific American* 262(4), 1990. This paper describes the causes, rates, and consequences of tropical deforestation, paying particular attention to the role of government policy.

Glossary

The terms included in the glossary are those listed as Key Terms at the end of Chapters 2-21. The chapter where they are first defined is indicated in brackets.

ablation zone The zone near the margin of a glacier or ice sheet where melting and evaporation exceeds accumulation. (15)

absolute humidity The mass of water vapor per volume of air (grams per cubic meter). (6)

absorption The retention of radiated energy, such as insolation, instead of scattering or transmitting it. The process of absorption transforms the radiated energy to another form, often heat. (4)

abyssal plain Wide, relatively flat areas on ocean floors between 3500 and 4000 meters deep that are covered by layers of fine sediment. (12)

accelerated eutrophication The enrichment of a water body through the addition of large doses of nutrients over a short period, typically as a result of human activity. (19)

accumulation zone The zone on a glacier or ice sheet where snow accumulates and is converted to ice faster than melting removes it. (15)

acidity The number of hydrogen ions present in a solution; a pH value of less than 7. (6)

acid rain Rain that has a pH value of less than 5.5 where human agency has contributed to the acidity. (6)

adiabatic expansion The cooling of a rising parcel of air without loss of heat to its surroundings. As the air rises it expands in air of lower density, and the heat energy it contains is spread through a greater volume, lowering the temperature. (6)

adiabatic lapse rate The rate at which temperature falls in a rising body of air. (6)

advection The horizontal movement of air or water, as in the movement of air and surface ocean currents from low to high latitudes. (4)

advection fog Fog that is formed by warm, moist air flowing over a cold surface. (6)

aeolian Wind-based processes. (16)

aeration The exchange of gases between atmospheric air and air in the soil. In a well aerated soil, oxygen and carbon dioxide are transferred rapidly from one site to another so that the soil air has approximately the same composition as the atmosphere. In a poorly aerated soil, local deficits of oxygen occur and excesses of carbon dioxide may build up. (20)

aerial photograph Remotely sensed image acquired by a camera mounted in an airplane. (2)

aerosols Suspended minute particles of dust, sea salt, liquid droplets, carbon, and metals in the atmosphere. (3)

aggradation The building up of the land surface by deposition from running water, flowing ice, wind, and the sea. (13)

air mass A large body of air in contact with the ground and having relatively uniform internal conditions of temperature and humidity. (7)

albedo The proportion of insolation that is reflected; it varies according to the reflecting surface. (4)

Alfisols One of the 11 soil orders defined under the Soil Taxonomy classification system. Alfisols are widespread in the United States and are typically associated with a natural vegetation of deciduous forest. (20)

alluvial fan A fan- or cone-shaped deposit of a stream. (14)

alluvium Sedimentary deposits by rivers, including flood plains and alluvial fans. (13)

altitude Height above sea level. (4)

anastomosing channel Stream channels that divide and rejoin in a netlike pattern. (14)

Andisols The most recently introduced of the 11 soil orders defined under the Soil Taxonomy classification. Andisols develop on parent soils of volcanic origin, including ash and tephra. (19)

angle of repose The maximum slope at which a pile of unconsolidated material (talus) will remain stable. Coarser material has a higher angle of repose than finer material. (13)

anticline An upfold in rocks, like an upturned boat or saucer. (10)

anticyclone A weather system in which there is high atmospheric pressure at the center. The winds diverge in clockwise patterns in the northern hemisphere (counterclockwise in the southern hemisphere). (7)

aquiclude A rock layer of low permeability. (14)

aquifer A permeable rock layer that has sufficient porosity to hold water and allow it to be withdrawn for use. (14)

argon (Ar) One of the three major atmospheric gases; the least chemically active. (3)

Aridisols One of the 11 soil orders defined under the Soil Taxonomy classification. Aridisols are the shallow soils of arid areas. (20)

aridity Dry conditions at a place as the result of low precipitation, or of an evaporation rate that exceeds the precipitation. (6)

aspect The direction in which a slope faces. (4)

asthenosphere A layer between 100 and 200 km below Earth's surface in which rocks may melt partially and flow. (10)

atmosphere A thin envelope of gases, dust, and water droplets surrounding Earth, formed largely of nitrogen, oxygen and argon, and divided into layers by temperature. (3)

atmospheric pressure The force exerted by the weight of the atmosphere. (4)

atom The smallest particle of an element that can exist alone or combine with atoms of other elements. (3)

autotroph A plant that is able to manufacture carbohydrates in the process of photosynthesis. (19)

backswamp A marshy area outside the channel on a river flood plain. (14)

backwash The movement of sea water down a beach under gravity after a wave breaks. (17)

barchan A crescent-shaped sand dune. (16)

barrier beach An elongated beach form lying parallel to the coastline. (17)

basalt An igneous rock that is formed of small crystals of mafic minerals and occurs in lava flows at Earth's surface. (10)

baseflow The groundwater contribution to streamflow. (14)

base level The lowest level to which denudation works. (13)

batholith A large igneous intrusion up to 1000 km across that occurs in the cores of mountain ranges and is often composed of granite. (10)

beach Depositional landforms in the coastal zone. (17)

bed load The coarsest fraction of a stream's load, usually moved in contact with the stream bed. (14)

Bergeron-Findeison theory of precipitation The theory that involves the formation of ice crystals in clouds in order to produce particles that are large enough to be precipitated. (6)

biogeochemical cycle A pathway along which nutrients move between the living and nonliving parts of the ecosystem. (19)

biomass The amount of plant and animal tissues in the ecosystem, usually measured as dry weight per unit area. Total biomass can be divided into plant and animal components or by its distribution among trophic levels. (19)

biome A regional ecosystem in which the climatic conditions, resulting soil development, and dominant growth-form of vegetation are more or less uniform. (19)

biosphere The living-organism environment, created by the interacting margins of the atmosphere-ocean, solid-earth, and surface-relief environments. (19)

boreal forest A forest biome dominated by coniferous trees and found in cold regions. The boreal forest is also called the *taiga* and, in the United States, northern coniferous forest. Winters are long and harsh and summers are cool. (21)

boundary current An ocean current that flows close to the coast on the eastern or western margins of an ocean. (5)

boundary layer The layer of the atmosphere close to the ground, in which conditions are closely affected by the contact. (8)

braided channel A stream channel in which the waterflow divides around channel bars. (14)

breaker zone The inshore zone along which waves break. (17)

calcification A soil-forming or pedogenic regime in which gentle downward percolation of rainwater washes calcium carbonate into the lower horizons but leaves other nutrients near the surface. (20)

caldera A crater left by a very large volcanic explosion. (11)

calorie The amount of heat energy needed to raise the temperature of one gram of water by 1°C at sea-level pressure. (4)

capillary water Molecules of water that are held loosely in the soil against the force of gravity. Capillary water travels through the soil in response to pressure gradients and is the main source of water for plants. (20)

carbohydrate An energy-rich organic compound of carbon, hydrogen, and oxygen created in photosynthesis. (19)

carbonation The process of chemical weathering in which rainwater (a weak carbonic acid) reacts with calcium carbonate (limestone). (13)

carbon dioxide (CO_2) A minor atmospheric gas capable of absorbing long-wave radiation; important in photosynthesis. (3)

carbon monoxide (CO) A minor atmospheric gas capable of absorbing long-wave radiation. (3)

carnivore An organism, usually an animal, that eats only meat. (19)

cartography The techniques of map-making. (2)

catastrophism A view of landform development that regards the processes as acting mainly in sudden, high-magnitude events. (18)

cation-exchange capacity A measure of the number of nutrient cations that can be absorbed by a unit of soil, commonly used as an indicator of soil fertility. (20)

cavern A cave (underground cavity) system in limestone rock. (14)

chapparal The local name for dry scrublike vegetation in California and Mexico that is part of the Mediterranean forest and scrub biome. (21)

chelation A process of chemical weathering that is dependent on organic matter and acids. (13)

chinook wind The warm, dry wind that blows east from the Rockies. The air is subject to warming by adiabatic compression. (6)

chlorofluorocarbons (CFCs) Minor atmospheric gases produced by human activities; capable of absorbing long-wave radiation and of being involved in ozone depletion in the stratosphere. (3)

chlorophyll The green pigment in plant leaves that receives sunlight in the first step of photosynthesis. (19)

cirque An amphitheatre-shaped rock basin eroded by a glacier, usually just below the peaks in high mountains and often containing a lake if not occupied by a glacier. (15)

cirque glacier A glacier that occupies a cirque but does not flow farther down valley. (15)

cirriform cloud High-level cloud composed of ice particles. (6)

clay-humus complex A mixture of tiny particles of clay and humus in the soil. The particles have surface electrical charges that attract and hold nutrient ions. (20)

cliff A steep rock face, common in coastal zones where wave attack is strong and the rocks are cohesive. (17)

climate The long-term condition of the atmosphere as measured by average conditions and the range of weather experienced. (8)

climate classification A division of the world into climate regions in each of which some climatic elements (e.g., temperature, precipitation) and conditions (e.g., weather systems) are similar. (8)

climatic change A shift in the average climatic conditions. (9)

climatic environment The surrounding conditions that produce the measured characteristics of the climate of a region, including heat balance, water budget, surface type, and weather systems. (8)

climatic variability Fluctuations of climatic elements around seasonal and yearly averages of conditions in the atmosphere. (9)

climax A plant community that is in stable equilibrium with prevailing conditions in the physical environment, particularly climate. (19)

climograph A diagram that includes graphs and illustrates the general climatic conditions of a place. (8)

cloud Condensed masses of water droplets, ice particles, or both in the atmosphere above the ground. (6)

cloudburst More than 1 millimeter of rain falls per minute for at least 5 minutes. (8)

cloud cluster A group of cumulonimbus clouds near the Equator. Cloud clusters are responsible for a high proportion of tropical rainfall. (7)

coastal zone The zone from just below low tide to just above high tide where marine processes have an important influence on landforms. (17)

cold-core low A nonfrontal midlatitude cyclone that forms following the occlusion of a frontal cyclone. (7)

cold front A front at the leading edge of an advancing cold air mass. (7)

cold ice sheet, glacier An ice sheet or glacier where the ice remains at very low temperatures, tens of degrees below freezing. (15)

collision-coalescence theory of atmospheric precipitation The theory of precipitation that is based on the fusing of cloud droplets to form larger drops that are able to fall. (6)

colloid The smallest fragment of mineral material in soil. Colloids form a major part of the clay-humus complex. (20)

competition The struggle among individual organisms to obtain the limited resources of energy, nutrients, and water supplied by the physical environment. (19)

compound A substance formed by the bonding of different elements, e.g., water (H_2O). (3)

condensation The process by which water vapor changes to liquid water. (6)

condensation nuclei Particles in the atmosphere around which condensation occurs. (6)

conditional instability Instability in the atmosphere that is conditional on condensation taking place, causing a change from the dry to the wet adiabatic lapse rate. (6)

conduction The transfer of heat energy from a hot to a cool substance by molecular contact, and without movement within or between the substances. It is most effective in solids. (4)

coniferous Plant species that are in leaf all year and generally bear cones and have needlelike leaves. (21)

conservation Sustainable use of an ecosystem's resources, particularly energy and materials, so that the system remains in equilibrium and its long-term value is maintained. (19)

continental rise An area on the margin of an ocean basin with a surface slope up toward the continent. It is a wedge-shaped deposit composed of debris washed from the continent that thins on the ocean side. (12)

continental shelf The top surface of a wedge of sediment formed on a continental margin, usually covered by less than 130 meters of water. (12)

continental slope The outer slope of the continent, leading from the edge of the continental shelf to the ocean floor. (12)

continuity method of weather forecasting The approach to weather forecasting that is based on the progress of a midlatitude cyclone through the stages of the theoretical model. (7)

contour line A line on a map connecting points of equal height above or below a fixed base level, usually sea level. (2)

convection The transfer of heat energy within a substance by circulatory movement resulting from differences in temperature (and density). It is most effective in liquids and gases. (4)

convergent margin A plate margin where plates are colliding and plate material is destroyed. (11)

converging air flow Winds that blow toward one another, causing surface air to pile up in the middle and rise. (5)

coral reef A rock structure composed of limestone that is formed just below sea level by colonies of coral animals. (17)

cordillera-type folded mountains Folded mountains that are formed by the collision of a continental and an oceanic plate. (12)

Coriolis effect A deflective force caused by the rotation of Earth: it is equal and opposite to the pressure gradient force. (5)

corrasion The mechanical erosion of rock as rock fragments are moved across it by running water, flowing ice, wind, or the sea; the rock is abraded. (14)

corrosion Denudation of rock by chemical weathering. (14)

counterradiation Long-wave radiation from the troposphere gases and clouds back to Earth's surface. (4)

crust The outer layer of the solid-earth environment, composed of sial (continental crust) and sima (denser ocean-floor crust). (10)

cumuliform cloud Separate, heaplike clouds formed by the convection of bubbles of air. (6)

cyclone A weather system in which there is low atmospheric pressure at the center. The winds converge in counterclockwise patterns in the northern hemisphere (clockwise in the southern hemisphere). (7)

deciduous Plant species that lose their leaves completely at some time of the year, usually associated with a cold or dry season. (21)

decomposers Organisms that gain their energy by digesting dead plant and animal tissues in the detrital food chain. Most decomposers are small or microscopic organisms living in the soil. (19)

deflation Wind action that removes unconsolidated clay, silt, or sand from a land surface. (16)

deflation hollow A basin or depression formed by deflation. (16)

delta An alluvial deposit at a river mouth, where fluvial and marine processes interact. (17)

density The concentration of matter; mass per unit volume (e.g., kg/m^3). (3)

denudation Erosion processes that lower Earth's continental surface relief. (13)

deposition The dropping of particles transported by running water, flowing ice, wind, or the sea in low-energy environments. (13)

descriptive statistics Statistics such as a mean or average that convey the most important features of a data set in a compact and manageable form. (2)

desert biome An ecosystem that develops in areas where potential evapotranspiration far exceeds input from precipitation. (21)

desertification The degradation of land to the extent that it is unable to support vegetation and becomes desertlike. (20)

desert pavement An area of bare stones forming a crust in an arid environment. (16)

detrital food chain The food chain in which waste or dead plant and animal tissues are decomposed and recycled. (19)

dew Condensation of water vapor in contact with the ground when air is cooled below saturation point. (6)

dew point The temperature at which condensation begins in a cooling parcel of air; the maximum temperature at which a relative humidity of 100 percent is reached. (6)

diffused light Indirect solar radiation received at Earth's surface after scattering, especially through clouds. (4)

digital satellite image Remotely sensed image acquired by a digital sensor mounted on a satellite. (2)

dike A sheetlike igneous intrusion that cuts across layers of other rocks. (10)

dip The steepest slope of a rock layer measured at right angles to a horizontal line across the layer. (10)

dip-slip fault A rock displacement that moves rocks up or down across the fault plane. (10)

discharge The flow of water in a stream, measured in volume per unit of time. (4)

dissolved load The portion of a stream's load carried in solution. (14)

divergent margin A plate margin where two plates move apart and new plate material forms in the gap between. (11)

diverging air flow Winds that blow outwards, or away from a common central area. (5)

doldrums Light winds or calms, mainly over the oceans near the equator. (5)

doline A surface depression on limestone rock in a karst landscape. (14)

downburst A strong downdraft of air in a thunderstorm. (7)

drainage divide The perimeter of a watershed, or drainage basin. (14)

drizzle A form of atmospheric precipitation characterized by small raindrops. (6)

drumlin A streamlined, elongated hill formed usually of glacial till. (15)

dry adiabatic lapse rate The adiabatic lapse rate in unsaturated air (i.e., with a relative humidity of less than 100 percent). (6)

dune A deposit of wind-blown sand forming a mound or ridge. (16)

earthquake A shock, or series of shocks, generated by sudden movement in Earth's crust or upper mantle. (11)

earthquake magnitude The size of an earthquake, measured on the Richter Scale. (11)

easterly wave A tropical weather system on the equatorial margin of the subtropical anticyclone that is characterized by a wavelike feature in the isobars. (7)

easterly winds, currents Winds or currents that blow or flow from the east. (5)

ecologic niche The set of environmental conditions in which an organism can survive and compete successfully with other organisms. (19)

ecologic succession The growth and development of an ecosystem over time. (19)

ecosystem An organized system made up of plants, animals, and the nonliving environment to which they are linked by flows of energy and materials. (19)

Ekman spiral The impact of winds on the surface ocean flow is demonstrated by a spiral showing the greatest effect at the surface and a declining effect with depth. (5)

electromagnetic spectrum The range of wavelengths in which electromagnetic waves are propagated, from the shortest wavelengths (gamma rays) to the longest (radio waves). (4)

electromagnetic waves Energy that is radiated in waves of different lengths, in which the electric and magnetic fields vary. (4)

El Niño current A warm ocean current that flows south off Colombia and Ecuador between Christmas and Easter, and occasionally extends farther south for longer periods (the enhanced El Niño current). (8)

eluviation The physical removal of particles in suspension from one part of the soil to another, usually from upper to lower horizons. (20)

emergent coast A coastline that shows evidence of the sea level falling relative to the land or the land rising relative to sea level. (17)

empiric classification A classification (e.g., of climate regions) based on measurements of climatic elements (e.g., temperature, precipitation). (8)

Entisols One of the 11 soil orders defined under the Soil Taxonomy classification. Entisols are found in areas where the normal soil-forming processes do not have, or have not yet had, a chance to act. (20)

environmental lapse rate The rate at which temperature decreases with increasing height in the troposphere. (4)

epeirogenesis Movements of Earth's crust that produce broad upward or downward warping of the surface. (11)

epicenter The point at Earth's surface directly above an earthquake focus, and usually the place where maximum damage occurs. (11)

epoch A division of geologic time; several epochs make a period. (10)

equinox The days (March 21, September 21) when the sun is overhead at the equator and all parts of the world have 12 hours of daylight and 12 hours of night. (4)

era A major division of geologic time. (10)

erg An extensive region of sand dunes (or sand sea), particularly in the Sahara. (16)

erosion The wearing away of the land surface and removal of rock debris by agents such as running water, flowing ice, wind, and the sea. (13)

esker A narrow winding ridge of sand

and gravel formed in a stream flowing beneath ice. (15)

estuary A river mouth that widens toward the sea and is influenced by tidal changes. (17)

eustatic sea-level change A worldwide change in sea level caused by the formation or melting of ice sheets, or by earth movements that change the shapes of ocean basins. (17)

eutrophication The natural enrichment of a water body by the addition of nutrients received through runoff from the watershed. (19)

evaporation The process by which a liquid, such as water, is changed to a gas (e.g., water vapor). (6)

evapotranspiration The total transfer of liquid water to water vapor at Earth's surface: evaporation plus transpiration. (6)

evolution The gradual development of new species through genetic shift and natural selection. (19)

exfoliation The process by which layers of rock break into sheets on exposure at the surface following unloading. (13)

extrusive igneous rock An igneous rock that forms at Earth's surface following the eruption of magma as lava or ash. (10)

fall A type of rapid mass movement in which blocks of rock fall freely without requiring lubrication. (13)

fault A form of rock deformation in which the layers are broken and displaced. (10)

fault-block mountain A mountainous area where the rocks have been uplifted between faulted boundaries. (12)

fetch The distance over open water across which a wind blows. (17)

field capacity The condition in which a soil contains the maximum amount of capillary water but no excess gravitational water. (20)

fjord A narrow inlet of the sea in a mountain region formed by the drowning of a glaciated valley. (17)

flood plain The floor of a river valley, adjacent to the channel, that is covered by flood water. (14)

flow A type of mass movement in which water or ice lubricates the regolith being moved. Varieties include earthflows, mudflows, and avalanches. (13)

focus The point at which an earthquake occurs, usually below the surface. (11)

foehn wind The dry, warm wind that blows south from the Alps in Europe. (6)

fog Condensation in the air in contact with the ground so that visibility is reduced to under one kilometer (0.62 miles). (6)

fold A form of rock deformation in which the rock layers are contorted, but not broken. (10)

food chain A linked sequence of plants and animals in the ecosystem through which energy is transferred. (19)

food web A number of interlinked food chains. (19)

freezing rain Rain that freezes on a cold surface. (6)

friction layer The layer of the atmosphere that is in contact with the ground and in which friction with the surface reduces the windspeed. (5)

front A narrow zone in the atmosphere between two contrasting air masses. (7)

frost Condensation on surfaces that are cooled below freezing point. (6)

frost wedging A form of physical weathering in which the freezing and thawing of water in rocks exerts pressure and splits the rock apart. (13)

gene pool The pool of natural variation in genetic material represented by all living things. Humans have drawn on a small part of the gene pool to derive medicines and domestic animals. (19)

general circulation model A mathematical model based on a two-dimensional grid of surface climate observations, and sometimes including vertical layers as well. Such models demand very powerful computers, and are used to predict future climate. (8)

genetic classification A classification (e.g., of climate regions) based on explanatory factors that have produced the measured differences. (8)

Geographic Information System (GIS) A computer database used to collect, store, retrieve, analyze, and display spatial data. (2)

geomorphology The science that studies the nature and origin of landforms. (10)

geostationary orbit A high orbit (36,000 km, 22,500 miles, above Earth's surface) in which a satellite moves at the same speed as Earth and therefore appears to be stationary to an observer at the surface. (2)

geostrophic wind A wind that flows parallel to the isobars as a result of a balance between the pressure gradient force and Coriolis force. (5)

glacial phase A time in the course of an ice age during which it gets colder and ice sheets advance. (9)

glacier A body of flowing ice largely confined within a valley. Examples include cirque glaciers, valley glaciers, and piedmont glaciers. (15)

gleization A soil-forming or pedogenic process characteristic of waterlogged and poorly aerated soils. Little oxygen is available and this restricts the presence and activity of soil organisms. (20)

global warming A trend toward rising temperatures in the atmosphere-ocean environment. (9)

graben (or rift valley) A down-faulted section of crustal rock between two horsts. (12)

granite An igneous rock composed of large crystals of felsic minerals that forms beneath Earth's surface in batholiths. (10)

gravitational water Soil water that is not held in the soil but moves freely under the influence of gravity. (20)

greenhouse effect The relationship between Earth's atmosphere and insolation, in which the atmosphere is transparent to incoming short-wave radiation, but traps outgoing long-wave heat radiation. (4)

gross primary production The amount of energy fixed into organic compounds by an ecosystem's primary producers. (19)

ground moraine A flat to irregular surface on glacial till following deposition at the base of a glacier or ice sheet. (15)

groundwater Water that accumulates in aquifer rocks below the water table. It is the main source of stream baseflow. (14)

growth-form The physical size and shape of vegetation. The growth-form of the dominant vegetation is the main characteristic by which biomes are differentiated. (21)

Gulf Stream ring An eddy that

breaks away from the Gulf Stream current, and may have a relatively warm or cold core. (5)

Hadley cell The vertical convectional cell in the tropical atmosphere, incorporating the surface trade winds, ascending air above the intertropical convergence zone, northerly flow aloft and descending air in the subtropical high-pressure zones. Named after G. Hadley (1735). (5)

hail The form of atmospheric precipitation that is composed of lumps of ice that may have an internal structure of concentric layers. (6)

hammada A flat, bare-rock surface in a desert, that may be partly covered by a lag deposit. (16)

hanging valley A tributary valley that joins a main valley at a height well above the floor of the main valley. Hanging valleys are common in formerly glaciated uplands. (15)

heat Heat is the total molecular movement in a substance: the greater the molecular movement, the greater the heat energy and the higher the temperature. (4)

heat balance The balance between insolation and terrestrial radiation in the atmosphere. (4)

heat island An area in which there are increased temperatures compared to the surrounding area: normally used in connection with urban-rural contrasts. (8)

heave A form of mass movement in which wetting and drying, or freezing and thawing, of the ground surface causes it to rise and fall. It gives rise to patterned ground on flat surfaces and creep on slopes. (13)

helium (He) One of the lightest of the minor gases in the atmosphere; with hydrogen, it dominates the outermost layer. (3)

herb A small plant that has no woody tissue and forms the layer of vegetation closest to the ground. (21)

herbivore An animal that eats only plants. (19)

heterotroph Describes an organism that cannot photosynthesize and must rely on autotrophs and other heterotrophs for an energy supply. (19)

Himalayan-type folded mountains Folded mountains that are formed by the collision of two continental plates. (12)

Histosols One of the 11 soil orders defined under the Soil Taxonomy classification. Histosols are made up mainly of organic materials in various stages of decay. (20)

homeostasis The concept of self regulation, in which an ecosystem maintains a stable equilibrium with its environment by adjusting to any disruption in flows of energy or materials. (19)

horst A fault-block mountain bounded by several faults. (12)

hot spot Zones away from plate margins where magma rises toward the surface and often erupts to create volcanoes. (11)

humidity The amount of water vapor present in the atmosphere. (6)

humus A structureless brown or black jelly that is the final product of the breakdown of organic material in the soil. The tiny organic particles join with clay colloids to form the clay-humus complex. (20)

hurricane A tropical cyclone that occurs in the northern Atlantic Ocean. (7)

hydrogen (H) The lightest of the minor gases in the atmosphere; with helium, it dominates the outermost layer. (3)

hydrology The study of the distribution and properties of water, including its circulation from ocean to atmosphere and from atmosphere back to the surface. (14)

hydrolysis A form of chemical weathering in which water combines with rock minerals to form clay minerals and sand particles. (13)

hygroscopic nuclei Nuclei that dissolve when water condenses on them. (6)

hygroscopic water Water held tightly to the surface of individual mineral grains in the soil. It is unavailable to plants because the attraction between the water and the mineral grain is greater than the "sucking power" of plants' roots. (20)

hypothesis A short and testable statement that forms the basis of an experiment in physical geography. (2)

ice abrasion The erosion of rock by ice in which rock fragments contained in the ice scrape rock beneath and along the sides of the glacier or ice sheet. (15)

ice age A time when ice sheets form over the polar areas and spread out to cover high-latitude regions. (9)

iceberg A mass of floating ice in a sea or lake that has broken away from a glacier or ice sheet. (15)

ice cap A small ice sheet. (15)

ice plucking The erosion of rock by ice in which the basal ice freezes around blocks in well-jointed rock and pulls them free. (15)

ice sheet A large continuous layer of ice that may cover all or most of a continent. (15)

igneous rock Rocks that are formed by the solidification of molten rock after it has migrated from the point of melting to the point of cooling. (10)

illuviation The redeposition of materials in a soil following transport in water. (20)

Inceptisols One of the 11 soil orders defined under the Soil Taxonomy classification. Inceptisols are young soils in a more advanced stage of development than Entisols. (20)

inferential statistics Statistics that use the laws of probability to extrapolate from a smalll sample of measurements to the full population of all such measurements. (2)

infiltration The process by which water enters the soil. (14)

insolation The flow of radiant energy from the sun, mainly in shorter wavelengths. (4)

instability in the atmosphere The state of a rising body of air when it remains warmer than the surrounding air and so continues to rise. (6)

interception The process by which precipitation is prevented by vegetation from direct impact on the ground. (14)

interglacial phase A time during an ice age when temperatures rise and ice sheets retreat. (9)

intertropical convergence zone The line, close to the Equator, where northeast and southeast trade winds converge. (5)

intrusive igneous rock An igneous rock that cools below Earth's surface and occurs intruded between other rocks. (10)

ion Atoms or molecules that have a positive or negative charge; they occur in solutions and in the outer atmosphere. (3)

island arc An arc-shaped line of volcanic islands on one side of an ocean trench, formed where two oceanic plates collide. (12)

isobar A line drawn on a map joining places of the same atmospheric pressure. (5)

isostatic change Changes in the height of Earth's continental surface as a result of adding or removing loads such as sediment or ice. (11)

isotherm A line drawn on a map joining places of the same atmospheric temperature. Isotherms are usually surface lines, and may be corrected to a sea-level value. (4)

isotope A variety of an element that has the same chemical properties as the element but a different atomic weight (e.g., oxygen-16 and oxygen-18). (9)

jet Strong wind near the top of the troposphere. (5)

joint A crack in a rock where the two sides are not displaced. Joints occur in igneous rocks following cooling and in sedimentary rocks following drying and compaction. (10)

kame A landform resulting from deposition by glacial meltwater along the sides or margins of a glacier or ice sheet. (15)

karst landscape A landscape on limestone rock that is characterized by surface depressions and underground caverns. (14)

kettle An enclosed depression in glacial till formed by the burial and melting of a detached block of ice. (15)

knickpoint A change of slope on a stream course, in which a gentler upper gradient gives way to a steeper lower one. (14)

Köppen classification of climate An empiric climate classification based on the average monthly and annual temperature and moisture characteristics that delineate natural vegetation regions. (8)

lag deposit A deposit commonly found in arid regions, in which coarser rock fragments are left when clay, silt, and sand have been removed by the wind. (16)

lapse rate The rate of temperature change with height—a vertical temperature gradient. (4)

latent heat The amount of heat absorbed or released when a body changes its state or phase without any change of temperature in the body. In water heat is absorbed as latent heat during the melting of ice and evaporation of liquid water; it is released during condensation and freezing.

laterization A soil-forming or pedogenic regime in which strong leaching and aggressive weathering create a deep, red soil that is acid and nutrient deficient. (20)

latitude The angular position of a place north or south of the equator. Used with longitude to give a precise location to all places on Earth's surface. (2)

lava flow A layer of molten igneous rock moving across Earth's surface. (10)

leaching The transport of materials in solution through the soil profile, typically a downward washing of nutrients in rainwater. (20)

lee cyclone A nonfrontal midlatitude cyclone that forms in the lee of a mountain range, where descending air produces a low-pressure center. (7)

lightning A visible flash that results from electrification within a thunderstorm. (7)

linear dune A sand dune that is oriented parallel to the main wind direction. (16)

lithosphere The solid layer above the asthenosphere that incorporates the crust and uppermost mantle rocks. (10)

loess A deposit of silt and clay that has been carried by wind. (16)

longitude The angular position of a place east or west of the prime meridian, an arbitrary reference line running from North Pole to South Pole. Longitude is used with latitude to give a precise location to all places on Earth's surface. (2)

longshore current A movement of water in the breaker and surf zones, approximately parallel to the shore. It is induced by strong wave currents. (17)

longshore drift The movement of sand and pebbles along the shore when waves approach at an angle to the shore. (17)

long-wave radiation Electromagnetic waves with a length of 4 micrometers or greater, including heat (infrared) waves, microwaves, and radio waves. (4)

macronutrients Nutrients such as oxygen, hydrogen, and carbon that are needed by organisms in large quantities. (19)

magma A mobile mixture of liquid and gaseous minerals that cools to form an igneous rock. (10)

mangrove colony An area in the subtropics and tropics that is covered by mangrove plants. Mangroves are able to send roots down through shallow water and have a growth of leaves above the water. (17)

mantle The layer of the solid-earth environment between the crust and core. It is mostly solid, but includes the weak asthenosphere near its upper boundary. (10)

map projection A mathematical construction used to depict the spherical globe on a flat piece of paper. (2)

marine abrasion Erosion by the sea that is carried out by waves dashing pebbles and boulders against cliffs. (17)

mass movement The movement of rock and unconsolidated regolith downhill largely by the influence of gravity. (13)

meandering channel A winding channel in which erosion of the outside of bends causes steep banks, while deposition occurs on the inside bank. (14)

Mediterranean forest and scrub A dry and scrublike vegetation that develops where winters are mild and humid and summers are hot and dry. This vegetation type is called chaparral in the southwestern United States.

meridian Imaginary reference lines around the globe, from the north to the south pole, that join places of the same longitude. Each meridian is half a great circle. (2)

mesoclimate A climate region that covers part of a major climate region and includes several microclimates (e.g., the climate of an upland region). (8)

mesopause The upper boundary of the mesosphere. (3)

mesosphere The layer of the atmosphere between the stratosphere and the thermosphere, in which there is a decrease of temperature with increasing height. (3)

metallic ore A mineral from which a metal can be produced, is in demand, and occurs in quantities that are economic to mine. (12)

metamorphic rock Rock that is formed by the alteration of minerals under extreme heat or pressure, but without complete melting or migration. (10)

methane (CH_4) A minor atmospheric gas that absorbs long-wave radiation. (3)

microclimate The smallest geographic scale of microclimates, ranging in size from a few square centimeters to a few square kilometers. (8)

micronutrients Nutrients such as cobalt, zinc, and molybdenum that are needed by organisms in only small quantities. Excess quantities of micronutrients can be toxic. (19)

midlatitude anticyclone A high-pressure weather system that occurs in middle latitudes in association with cyclones, or as a larger "blocking" form. (7)

midlatitude cyclone A low-pressure weather system that occurs in middle latitudes. Most have fronts as the result of converging contrasting air masses. (7)

midlatitude forest A forest type dominated by deciduous trees. Midlatitude forest is the natural vegetation of most of western Europe and parts of the eastern United States. It has mostly been cleared for agriculture. (21)

midlatitude grassland Grassland vegetation found in the drier parts of the midlatitudes, the natural vegetation of the American prairies and the Eurasian steppes. Most of this biome has been replaced by arable agriculture. (21)

midlatitude westerlies Winds that blow from the west between 40° and 60° of latitude. (5)

millibar (mb) A unit of atmospheric pressure, based on the idea that average atmospheric pressure equals one bar. A millibar equals 1000 dynes per square centimeter, or 100 Newtons per square meter. (3)

mineral Naturally occurring combinations of chemical elements bonded together in orderly crystalline structures. (10)

molecule A combination of atoms held together by electric bonds. (3)

Mollisols One of the 11 soil orders defined under the Soil Taxonomy classification. Mollisols are the fertile calcified soils of the midlatitude grasslands and now support grain crops. (20)

monsoon wind Winds that reverse direction between summer and winter, from ocean-continent to continent-ocean respectively. (5)

moraine A landform produced by the deposition of glacial till. Examples include terminal moraines and ground moraine. (15)

mountain biomes The ecosystems associated with high mountains. The nature of the ecosystem changes dramatically with height in response to changes in the physical environment, primarily climate. (21)

mud flat An intertidal area on which mud is deposited in an estuary or lagoon. (17)

natural levee An elevated bank along the margins of a river channel that is formed during overbank flooding and the deposition of the coarsest stream load. (14)

natural selection The process in which individual plants and animals die because they cannot compete with other, stronger, individuals that are better adapted to the environment. The successful individuals pass on their characteristics to the next generation. Also known as "survival of the fittest." (19)

net primary production The balance between gross primary production and respiration. (19)

nitrogen (N) The most common gas in Earth's atmosphere. (3)

nitrogen oxides (N_x0) Minor constituents of the atmosphere that are being added from human activities, especially automobile exhausts. (3)

nonfrontal low A midlatitude low-pressure weather system without fronts. (7)

normal fault A dip-slip fault in which the rocks are pulled apart and one side slides down over the other. (10)

nutrients The chemical elements that are needed for life. (19)

occluded front A front in a midlatitude cyclone where the cold front has caught up with the warm front and lifted the entire warm sector off the ground. It forms at the end of the life cycle of a midlatitude cyclone. (7)

ocean basin A structural depression in Earth's crust that is filled by water and floored by mafic igneous rock. (12)

ocean biome The regional ecosystems found in the oceans. They are less easily differentiated than biomes on the land, but can be delimited roughly by water depth and temperature. (21)

ocean current Faster moving flows of surface water in the oceans. (5)

ocean drift A slow-moving flow of surface ocean water. (5)

ocean gyre A circular pattern of ocean surface currents. (5)

ocean ridge Raised areas along divergent plate margins where ocean floor is created. (12)

ocean trench An elongated deep in the abyssal plain formed along a convergent plate margin; the deepest part of an ocean basin. (12)

omnivore An organism that eats both plants and animals. (19)

ooze Fine deposits of calcium carbonate or red clay on the ocean floor. (12)

optimum The range of an environmental variable within which conditions for growth and reproduction are ideal and in which an organism grows most strongly. (19)

orogenesis The process of mountain formation. (11)

outwash plain A depositional landform created by the meltwater deposition of sand and gravel in front of an ice sheet or ice cap. (15)

overland flow The surface movement of water derived from precipitation, usually in a shallow sheet. (14)

oxbow lake A small crescent-shaped lake formed by a river cutting through the neck of a meander. (14)

oxidation The process of chemical weathering in which oxygen dissolved in water reacts with iron and other rock minerals to form oxides. (13)

Oxisols One of the 11 soil orders defined under the Soil Taxonomy classification. Oxisols are the extremely weathered and leached soils found in the tropics. (20)

oxygen (O_2) The second gas of the atmosphere; an important product of photosynthesis. (3)

ozone (O_3) A minor atmospheric gas present near the surface, but concentrated in the upper stratosphere, where it absorbs ultraviolet rays. (3)

paleoclimatology The study of ancient climates. (9)

paleomagnetism The study of the magnetism of ancient rocks. (11)

parabolic dune A crescentic dune in which the arms point upwind, usually formed by the blowout of vegetated dunes. (16)

parallax The apparent displacement of a stationary object when viewed from two different angles. Parallax allows measurement of ground height from stereoscopic aerial photographs. (2)

pedogenic regime A particular combination of soil-forming factors and processes that gives rise to a distinctive soil profile. (20)

peds The aggregates formed by the clustering together of individual mineral grains in the soil. (20)

percolation The process by which water moves downward through soil and into rock via pores and joints. (14)

periglacial environment The area that lies in front of an ice sheet in polar latitudes, where the climate and landform processes are controlled by very low temperatures. (15)

period A division of geologic time; several periods make an era. (10)

permafrost Permanently frozen ground, occurring in very cold parts of the world. (9)

permeability The rate at which liquids or gases will pass through a substance; often used in relation to water passing through soil or rock. (14)

persistence method of weather forecasting The approach to weather forecasting that is based on information about the speed of travel of a weather system and assumes it will continue at the same rate. (7)

photogrammetry The precise measurement of vertical and horizontal distance from aerial photographs. (2)

photointerpretation The techniques of interpreting aerial photographs based on shading, texture, size and shape. (2)

photosynthesis The manufacture of carbohydrates from water and carbon dioxide in the presence of sunlight. (19)

piedmont glacier A glacier formed where several valley glaciers coalesce on emerging from a mountain range onto a coastal plain or lowland. (15)

pixel The "footprint" of ground over which a satellite sensor records reflected or emitted energy. The sensor records no detail within the pixel. (2)

plate A large, virtually rigid section of lithosphere. (11)

plate tectonics The idea that subsurface convection currents cause plates to move and so cause changes in the positions and relief of the continents and ocean basins. (11)

pluvial lake A lake in an arid region that formerly covered a greater extent when the climate was more rainy. (9)

podzolization A soil-forming or pedogenic regime characterized by acid conditions and excessive downward movement of water, causing a leached and nutrient-deficient upper horizon and an illuviated lower horizon. (20)

point bar Deposits of sand or gravel on the inside of a meander bend in a stream channel. (14)

polar easterlies Winds that blow from the east polewards of 60° North and South. (5)

polar front The zone where cold polar air converges with tropical air. (5)

polar-front jet The jet that forms above the polar front. (5)

polar orbit A low orbit (approximately 900 km, 500 miles, above Earth's surface) used by remote sensing satellites for Earth observation. (2)

polar wandering The idea that Earth's magnetic poles moved their position over time. (11)

pollution Materials that are added to the ocean or atmosphere by human activities and that degrade the quality of the medium. (3)

porosity The proportion of void space in a rock or soil. The voids may be filled with air or water. (13)

porphyry An igneous rock that is composed of mixtures of large and small crystals. (10)

potential evapotranspiration The amount of evapotranspiration that would take place if a supply of water was available at all times. (6)

precipitation in the atmosphere The deposition of atmospheric water on Earth's surface as rain, snow, hail, or sleet. (6)

pressure gradient The difference in pressure per kilometer between high and low centers of atmospheric pressure. (5)

pressure gradient force A force operating in the atmosphere that causes air to move from places of high pressure toward places of low pressure. (5)

primary data Data recorded at first hand by the experimenter, usually in the field or the laboratory. (2)

primary producers Organisms able to trap their own energy through photosynthesis. (19)

radiation The transfer of energy by means of electromagnetic waves. (4)

radiation fog Condensation in the air near the ground as a result of radiative cooling of Earth's surface. (6)

rain The form of atmospheric precipitation that is composed of large drops of liquid water. (6)

rainshadow An area of low atmospheric precipitation in the lee of a mountain range. (6)

reduction The process of chemical weathering in which oxygen is removed from minerals, usually in conditions where groundwater contains little oxygen. (13)

reflection Radiation that bounces off a substance without changing its wavelength or affecting the substance. (4)

regolith A layer of loose rock fragments, sand, and clay particles formed as the result of rock weathering or deposition following mass movement, streamflow, ice movement, or wind action. (13)

relative humidity The amount of water vapor in the atmosphere, expressed as a percentage of the total that could be held at a specific temperature and pressure. (6)

relict landform A landform that was formed by different processes than those operating on it at present. (18)

relief The physical form of Earth's surface. It is made up of a range of landforms of varying size, height, and shape. (10)

remote sensing The acquisition of information about an object without touching it. In physical geography, remote sensing refers mainly to the use of aerial photographs or satellite images to gather information about Earth environments. (2)

representative fraction A fraction

showing the number of units on the ground represented by one unit on a map. (2)

resistant rock A rock that withstands erosional processes. (13)

reverse fault A dip-slip fault in which compression forces one side to be pushed upward over the other. (10)

ripple A small elongated deposit of sand formed at right angles to the wind direction, or a beach landform produced by wave action in the breaker zone. (16)

rock A compact, consolidated mass of mineral matter; the minerals may be interlocked following crystalization, or rock fragments cemented together. (10)

Rossby wave Wave form in the polar-front jet. (5)

salinity The proportion of dissolved salts in water, measured as parts per thousand. (3)

salinization A soil-forming or pedogenic regime in which water is drawn to the soil surface, depositing salts in the upper horizon and at the surface. (20)

saltation Sand transport by the wind or particles being moved along the bed of a stream in a bouncing pattern. (16)

salt-crystal growth The weathering process in arid regions in which the growth of salt crystals exerts pressure on the rock or sediment in which they are located. (13)

salt marsh A vegetation community that develops on mud flats in intertidal conditions; the plants tolerate saline and fresh water at high tide and exposure to air at low tide. (17)

saturated adiabatic lapse rate The adiabatic lapse rate in saturated air. (6)

saturated air Air that contains the maximum water vapor for a specific temperature and pressure. Any further addition will be balanced by condensation. (6)

savanna Grassland vegetation found in the tropics. (21)

scale The mathematical relationship between a distance on the ground and its representation on a map. (2)

scattering The dispersal of radiation in all directions after hitting a substance without changing the wavelength of radiation or affecting the substance. (4)

sea-floor spreading The idea that the ocean floor is generated along mid-ocean ridges and then moves toward ocean trenches. (11)

sea ice The formation of ice by the freezing of the sea surface. (15)

secondary data Data used in a study but not collected specifically for it. Examples of secondary data sources are published reports, maps, and the results of previous experiments. (2)

sediment A disconnected mass of solid particles, grains, and fragments of rock that have been worn from older rocks, transported, and deposited. (10)

sedimentary rock Rock that is formed from products of the breakup of other rocks, or from organic debris, or from minerals precipitated out of solution; hardened sediment. (10)

seismic sea wave A large wave form in the Pacific or Indian ocean that is caused by an earthquake shock or volcanic eruption. (17)

seismic wave Earthquake shock wave that is transmitted by rocks. (10)

sensible heat Heat energy in the atmosphere that changes the temperature of the atmosphere when it is transferred by convection or conduction. (4)

shield A section of continental crust composed of mainly metamorphic rocks older than 570 million years. (12)

shore platform A coastal landform that is horizontal or slopes gently seaward and is formed by wave action, weathering, or deposition of calcium carbonate. (17)

short-wave radiation Electromagnetic waves shorter than 4 micrometers, including reflective infrared, visible light, ultraviolet, x, and gamma rays. (4)

shrub A woody plant that, in comparison to a tree, is small and has a stem that branches close to the ground. (21)

sill A sheetlike igneous intrusion that forms between layers of other rocks so that the boundaries are approximately parallel. (10)

sleet Atmospheric precipitation that forms as raindrops freeze to ice pellets on falling through cold air. (6)

slide A form of mass movement that involves movement of a section of rock along a slide plane. (13)

slope An inclined surface. Landform slopes are individual units having an area, slope angle, and orientation. (18)

snow The form of atmospheric precipitation that is composed of ice crystals. (6)

snow line The altitude of the lower limit of permanent snow cover. (8)

soil horizon A horizontal layer in the soil differentiated from other horizons by its contents, characteristics, and processes. (20)

soil pedon A conceptual three-dimensional unit of more-or-less uniform soil used as a mapping unit. (20)

soil profile A vertical slice through a soil from the surface down to the parent material. (20)

soil solution The liquid part of the soil, consisting of soil water, dissolved elements, and suspended particles. (20)

soil structure A description of how individual grains of mineral material are bound together into clumps or peds. (20)

Soil Taxonomy The method of soil classification developed by the Soil Survey Staff of the United States Department of Agriculture, also known as the United States Comprehensive Soil Classification System. (20)

soil texture The relative proportions of sand, silt, and clay in a soil. (20)

solar constant The intensity of solar radiation at the outer limit of Earth's atmosphere. (4)

solstice The days in the year when the sun is overhead at the Tropic of Cancer (June 21) and Tropic of Capricorn (December 21). (4)

solution The process of chemical weathering by which matter is changed from a solid or gaseous into a liquid state by combining with a solvent such as water. (13)

specific heat The amount of heat required to change the temperature of one gram of a substance by one degree Celsius. (4)

Spodosols One of the 11 soil orders defined under the Soil Taxonomy classification. Spodosols are the result of the pedogenic regime of podzolization. (20)

spring A point at which groundwater flows out on the surface. (14)

stability in the atmosphere The state of a rising body of air when it becomes cooler than the surrounding air and falls back toward the ground. (6)

steady state equilibrium The condition in which processes operate within a narrow band that does not change over a period of time. (18)

storm wave A high-energy sea wave that is generated in the midlatitude storm belts. (17)

strata Layers of sedimentary rock. (10)

stratiform cloud Layer clouds forming a structureless gray sheet that covers the sky. (6)

stratopause The upper boundary of the stratosphere. (3)

stratosphere The layer of the atmosphere above the troposphere and below the mesosphere. The upper part contains a concentration of ozone that absorbs incoming ultraviolet rays. (3)

stream capacity The total load that a stream can carry at a particular velocity. (14)

stream channel The landform in which a stream flows, defined by its bed and banks. (14)

stream competence The largest particle that a stream can move at a particular velocity. (14)

stream hydrograph A graph showing the relationship between precipitation and streamflow. (14)

stream load The material that a stream carries, including dissolved, suspended, and bedload fractions. (14)

strike A horizontal line that is perpendicular to the dip of a rock surface. (10)

strike-slip fault A fault in which the displacement is horizontally on either side of the crack. (10)

subduction The process by which one plate is forced downward into the mantle beneath a lower density plate. (11)

submergent coast A coastline that results from a rise of sea level or sinking of the land: valley mouths are drowned. (17)

subsidence The mass movement process in which the land surface is lowered by removal of underlying rock or soil. (13)

subtropical anticyclone (See subtropical high-pressure zone.) (7)

subtropical high-pressure zone Semipermanent areas of high pressure centered over the Atlantic, Pacific, and Indian oceans between 20° and 30° North and South. (5)

sulfur dioxide (SO_2) A minor gas of the atmosphere, added by volcanic eruptions and the burning of fossil fuels. (3)

supercontinent cycle The idea that there has been a regular, repeated pattern of events that first lead to the formation of a single world continent and then split it apart into a series of smaller continents. (12)

supercooled water Cloud droplets that do not freeze at temperatures below freezing point, often because of a lack of freezing nuclei. (6)

supersaturation A relative humidity of over 100 percent. This occurs in air cooled below the dew point without condensation occurring, often because of a lack of condensation nuclei being available. (6)

surf zone The zone between the breaker zone and the shore in which water moves up the beach in swash and down the beach in backwash. (17)

surface creep The slow mass movement process in which soil moves downslope by heaving. (16)

swash The movement of a layer of water up a beach following the breaking of a wave. (17)

swell wave A wave in the sea characterized by regular occurrence and a long wavelength relative to its height. (17)

syncline A downfold in rocks that resembles a boat or saucer. (10)

temperature The measure of heat contained in a substance. (4)

temperature gradient The rate of temperature change with distance or height (see lapse rate). (4)

temperature inversion A lapse rate that increases with increasing altitude for a section of the troposphere (instead of decreasing). (4)

tephra Particles of solidified igneous material erupted from an explosive volcano and ranging in size from fine ash to large bombs. (11)

terrestrial radiation Long-wave radiation from Earth's surface (heat). (4)

thermal (warm-core) low A nonfrontal midlatitude cyclone that is produced by surface heating. (7)

thermocline The zone in the oceans separating surface warm water and deep cold water. (3)

thermosphere The outer layer of the atmosphere above the mesosphere, where atmospheric density is least and temperature increases with altitude. (3)

threshold A condition marking the transition from one state to another that may be irreversible. (18)

throughflow The movement of water downslope through regolith. (14)

thrust fault A fault that displaces one set of rocks almost horizontally over another. (10)

thunder The sound generated by the discharge of electricity when lightning occurs in a thunderstorm. (7)

thunderstorm Single or multiple cumulonimbus clouds which produce heavy rain, thunder, lightning, and strong downdrafts of air. (7)

tide The rise and fall of the ocean water level resulting from the gravitational attraction of the sun and moon on Earth. (17)

tidal current A flow of water in response to tidal changes, either in-and-out of a river mouth or around an island. (17)

tidal range The difference in height between high tide and low tide levels. (17)

till Sediment deposited by glaciers and ice sheets. (15)

tolerance The range of an environmental variable within which a plant can survive. (19)

tornado A small, violent rotating storm that forms in extremely violent thunderstorms and is characterized by a funnel-shaped vortex descending from the cloudbase to the ground. (7)

trade winds Surface winds in the tropics, blowing from the northeast in the northern hemisphere and the southeast in the southern hemisphere. (5)

transform margin A plate margin where two plates move sideways and little plate material is either formed or destroyed. (11)

translocation The movement of materials in the soil, primarily by water but also through the action of soil organisms. (20)

transpiration The process by which plants pass water vapor into the atmosphere. (6)

transport The movement of rock fragments, particles, and dissolved matter by mass movements, rivers, glaciers, the wind, and the sea. (13)

transverse dune A sand dune with an asymmetric cross-section that stands at right angles to the wind direction. (16)

tree A plant with a single woody stem (trunk) that branches at some height above the ground, raising most of the green vegetation above the surface. (21)

trophic level The position of an organism in a food web. Trophic level also describes the organism's "distance" from the autotrophs. (19)

tropical cyclone A major cyclonic weather system in the tropics, known locally as a hurricane or typhoon. (7)

tropical depression The first stage in a sequence of tropical weather systems that may lead to the formation of a hurricane; wind speeds averaged over one minute do not exceed 60 kmph. (7)

tropical grassland Grassland vegetation, also called savanna, that develops in the tropics where there is a long dry season. (21)

tropical rainforest Forest that develops in the constantly hot and wet climate around the equator, characterized by evergreen trees, a high primary production, and high species diversity. (21)

tropical seasonal forest Mixed deciduous and evergreen forest that marks the transition from tropical rainforest to the drier conditions of tropical grassland. (21)

tropical storm The second stage in a sequence of tropical weather systems that may lead to the formation of a hurricane; wind speeds averaged over one minute are between 60 and 115 kmph. (7)

tropics The Earth's surface lying between 23°30′ N and 23°30′ S—i.e., the area where the sun is overhead at some time during the year. (4)

tropopause The upper boundary of the troposphere, marked by a change from decreasing temperature with height to a zone where temperature changes little with height; this prevents weather disturbances in the troposphere from extending upwards. (3)

troposphere The lowest layer of the atmosphere in contact with the ground, and the layer in which movements take place that produce weather. (3)

tundra A biome found in the cold conditions near the poles and at high altitudes, with a typical vegetation of mosses, lichens, and some hardy herbs, shrubs and stunted trees. (21)

typhoon A tropical cyclone that occurs in the western Pacific Ocean. (7)

Ultisols One of the 11 soil orders defined under the Soil Taxonomy classification. Ultisols are the strongly weathered and leached soils of the middle and low latitudes. (20)

uniformitarianism The view that processes currently at work in Earth environments are those that have acted throughout geologic time. (18)

unloading A release of pressure that occurs when erosion removes overlying rocks and exposes a formerly buried rock at the surface. (13)

U-shaped valley A deep valley that has been eroded by a valley glacier. (15)

valley glacier A glacier that flows down a valley. (15)

valley train Deposits by meltwater in front of a glacier snout. (15)

ventifact A pebble or boulder that has been faceted and smoothed by wind abrasion in an arid region. (16)

Vertisols One of the 11 soil orders defined under the Soil Taxonomy classification. Vertisols have a high clay content and crack open when dry. Material falls into the cracks, leading to a gradual inversion of the soil. (20)

volcanic ash A deposit that forms as a result of explosive volcanic activity. The molten lava is broken into drops or clots that cool and solidify to form tephra. (10)

volcano A landform produced by the eruption of lava or ash through a vent. (11)

warm front A front at the leading edge of an advancing warm air mass. (7)

warm glacier A glacier in which the ice is close to melting point for nearly all the year. (5)

watershed The area drained by a river and its tributaries: the catchment or drainage basin. (14)

water table The upper level of groundwater saturation in an aquifer. (14)

water vapor The gaseous form of water that forms a variable gas in the atmosphere and absorbs long-wave radiation. (3)

wave A disturbance of a body or medium that moves from point to point. In the sea the water surface is deformed by an alternate rising and falling motion generated by wind action on the surface. (17)

wave refraction The process by which wave crests change direction on approaching the coast: the waves are slowed as they enter shallow water and this turns the crests toward a headland and away from each other in a bay. (17)

weather forecast An attempt to predict weather at a place for the next few hours or days. (7)

weathering The breakdown and decomposition of solid rock and rock fragments in response to atmospheric processes. (13)

weather system Distinct and repeated patterns of weather, such as a cyclone. (7)

westerly winds, currents Winds or currents that blow or flow from the west. (5)

whaleback forms Elongated, smoothed domes and basins formed by glacial abrasion. (15)

Wilson cycle The idea that plate tectonics is responsible for new oceans forming and continents splitting apart. (12)

wilting point Conditions of soil moisture under which the plants can obtain no water from the soil. Almost all the water in the soil at wilting point is hygroscopic water. (20)

wind A horizontal movement of air relative to Earth's surface. (5)

wind abrasion The process of erosion in which wind uses sand particles to wear away rock. (16)

xerophytic Drought-resistant plants that have adaptations to maximize their use of available water and minimize its loss. (21)

yardang A ridgelike landform elongated in the direction of the dominant wind which is formed by wind erosion in arid regions. (16)

young folded mountain A mountain system composed of relatively recent (up to 200 million years old) rocks that have been subject to compression and folding at a convergent plate margin. (12)

Topographic Map Symbols

Topographic maps are produced by the United States Geological Survey (USGS). Each map in a USGS series conforms to established specifications for size, scale, content, and symbolization. Except for maps that are formatted on a county or state basis, the USGS quadrangle series of topographic maps cover areas bounded by parallels for latitude and meridians of longitude.

Topographic maps are produced at various scales ranging from 1 : 24,000 to 1 : 1,000,000. These maps are created to show the Earth's surface. The surface is portrayed by contour lines. Map features are distinguished in the following manner:

- Brown—Contour lines
- Blue—Hydrographic features
- Black—Cultural features
- Green—Woodland cover, scrub, orchards, vineyards
- Red—Roads and lines of the public land survey system
- Purple—Features added from aerial photos during map revision.

The standard map symbols are as follows:

	Standard edition maps	*New or replacement standard edition maps*	*Provisional edition maps*
CONTROL DATA AND MONUMENTS			
Aerial photograph roll and frame number	Not Shown	Not Shown	3-20
Horizontal control:			
Third order or better, permanent mark	Neace △	Neace △	Neace
With third order or better elevation	BM △ 148	BM △ 45.1	Pike BM 45.1
Checked spot elevation	△64	△19.5	Not Shown
Coincident with section corner	Cactus	Cactus	Cactus
Unmonumented	Not Shown	Not Shown	
Vertical control:			
Third order or better, with tablet	BM × 53	BM × 16.3	BM × 53.4
Third order or better, recoverable mark	× 394	× 120.0	× 393.6
Bench mark at found section corner	BM 61	BM 18.6	BM 60.9
Spot elevation	× 17	× 5.3	× 17
Boundary monument:			
With tablet	BM 71	BM 21.6	BM 71
Without tablet	562	171.3	562
With number and elevation	67 988	67 301.1	67 988
U.S. mineral or location monument	▲	▲	USMM ▲
BOUNDARIES			
National			
State or territorial			
County or equivalent			
Civil township or equivalent			
Incorporated city or equivalent			
Park, reservation, or monument			
Small park			
LAND SURVEY SYSTEMS			
U.S. Public Land Survey System:			
Township or range line			
Location doubtful			
Section line			
Location doubtful			
Found section corner; found closing corner			
Witness corner; meander corner	WC MC	WC MC	WC MC

	Standard edition maps	*New or replacement standard edition maps*	*Provisional edition maps*
Other land surveys:			
Township or range line			
Section line			
Land grant or mining claim; monument			
Fence line			
ROADS AND RELATED FEATURES			
Primary highway			
Secondary highway			
Light duty road			
Unimproved road			
Trail			
Dual highway			
Dual highway with median strip			
Road under construction			U. C.
Underpass; overpass			
Bridge			
Drawbridge			
Tunnel			
BUILDINGS AND RELATED FEATURES			
Dwelling or place of employment: small; large			
School; church			
Barn, warehouse, etc.: small; large			
House omission tint			
Racetrack			
Airport			
Landing strip			
Well (other than water); windmill			
Water tank: small; large			
Other tank: small; large			
Covered reservoir			
Gaging station			
Landmark object			
Campground; picnic area			
Cemetery: small; large	Cem	Cem	Cem

Provisional edition maps
New or replacement standard edition maps
Standard edition maps

RAILROADS AND RELATED FEATURES

- Standard gauge single track; station
- Standard gauge multiple track
- Abandoned
- Under construction
- Narrow gauge single track
- Narrow gauge multiple track
- Railroad in street
- Juxtaposition
- Roundhouse and turntable

TRANSMISSION LINES AND PIPELINES

- Power transmission line: pole; tower — Trans Line
- Telephone or telegraph line — Telephone
- Aboveground oil or gas pipeline — Pipeline Aboveground
- Underground oil or gas pipeline — Pipeline

CONTOURS

- Topographic:
  - Intermediate
  - Index
  - Supplementary
  - Depression
  - Cut; fill
- Bathymetric:
  - Intermediate
  - Index
  - Primary
  - Index Primary
  - Supplementary

(Bathymetric contours on provisional edition maps: Not Shown)

MINES AND CAVES

- Quarry or open pit mine
- Gravel, sand, clay, or borrow pit
- Mine tunnel or cave entrance
- Prospect; mine shaft
- Mine dump — Mine dump
- Tailings — Tailings

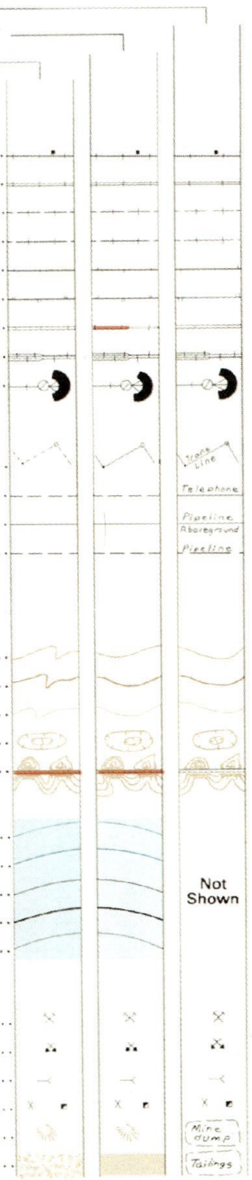

Provisional edition maps
New or replacement standard edition maps
Standard edition maps

SURFACE FEATURES

- Levee — Levee
- Sand or mud area, dunes, or shifting sand — Sand
- Intricate surface area — Strip mine
- Gravel beach or glacial moraine — Gravel
- Tailings pond — Tailings Pond

VEGETATION

- Woods
- Scrub
- Orchard
- Vineyard
- Mangrove — Mangrove

MARINE SHORELINE

- Topographic maps:
  - Approximate mean high water
  - Indefinite or unsurveyed
- Topographic-bathymetric maps:
  - Mean high water
  - Apparent (edge of vegetation)

(Topographic-bathymetric shoreline on provisional edition maps: Not Shown)

COASTAL FEATURES

- Foreshore flat — Mud
- Rock or coral reef — Reef
- Rock bare or awash
- Group of rocks bare or awash
- Exposed wreck
- Depth curve; sounding — Not Shown
- Breakwater, pier, jetty, or wharf
- Seawall

BATHYMETRIC FEATURES

- Area exposed at mean low tide; sounding datum
- Channel
- Offshore oil or gas: well; platform
- Sunken rock

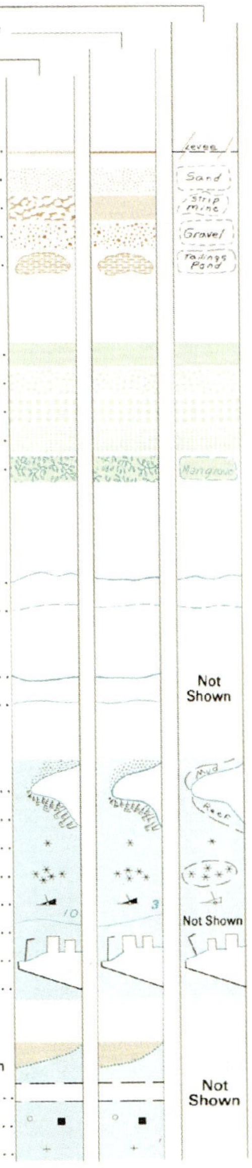

(Bathymetric features on provisional edition maps: Not Shown)

Provisional edition maps
New or replacement standard edition maps
Standard edition maps

RIVERS, LAKES, AND CANALS

- Intermittent stream
- Intermittent river
- Disappearing stream
- Perennial stream
- Perennial river
- Small falls; small rapids
- Large falls; large rapids
- Masonry dam
- Dam with lock
- Dam carrying road
- Intermittent lake or pond
- Dry lake
- Narrow wash — Wash
- Wide wash — Wash
- Canal, flume, or aqueduct with lock
- Elevated aqueduct, flume, or conduit
- Aqueduct tunnel
- Water well; spring or seep

GLACIERS AND PERMANENT SNOWFIELDS

- Contours and limits
- Form lines

SUBMERGED AREAS AND BOGS

- Marsh or swamp
- Submerged marsh or swamp
- Wooded marsh or swamp
- Submerged wooded marsh or swamp
- Rice field — Rice
- Land subject to inundation

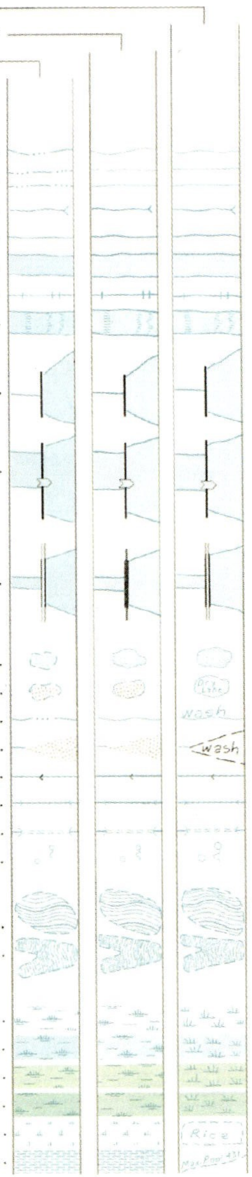

U.S. Weather Service Map Symbols

The following symbols have been developed by the National Weather Service, in Washington D.C. and they are used in the compilation of Daily Weather Maps. These standardized symbols are used to convey daily weather information in the United States and portions of Canada and Mexico.

Weather information is gathered at numerous stations throughout North America. The data from these stations is collected on a daily basis and is incorporated into the Daily Weather Map.

The Weather Map Symbols are provided as a courtesy to the instructor providing students with classroom weather activities.

Cloud Abbreviations

St—STRATUS

Fra—FRACTUS

Sc—STRATOCUMULUS

Cu—CUMULUS

Cb—CUMULONIMBUS

Ac—ALTOCUMULUS

Ns—NIMBOSTRATUS

As—ALTOSTRATUS

Ci—CIRRUS

Cs—CIRROSTRATUS

Cc—CIRROCUMULUS

Code No.	C_L	Description (Abridged From International Code)
1		Cu of fair weather, little vertical development and seemingly flattened
2		Cu of considerable development, generally towering, with or without other Cu or Sc bases all at same level
3		Cb with tops lacking clear-cut outlines, but distinctly not cirriform or anvil-shaped; with or without Cu, Sc, or St
4		Sc formed by spreading out of Cu; Cu often present also
5		Sc not formed by spreading out of Cu
6	—	St or StFra, but no StFra of bad weather
7	- - -	StFra and/or CuFra of bad weather (scud)
8		Cu and Sc (not formed by spreading out of Cu) with bases at different levels
9		Cb having a clearly fibrous (cirriform) top, often anvil-shaped, with or without Cu, Sc, St, or scud

No.	C_M	Description (Abridged From International Code)
1		Thin As (most of cloud layer semitransparent)
2		Thick As, greater part sufficiently dense to hide sun (or moon), or Ns
3		Thin Ac, mostly semitransparent; cloud elements not changing much and at a single level
4		Thin Ac in patches; cloud elements continually changing and/or occurring at more than one level
5		Thin Ac in bands or in a layer gradually spreading over sky and usually thickening as a whole
6		Ac formed by the spreading out of Cu or Cb
7		Double-layered Ac, or a thick layer of Ac, not increasing; or Ac with As and/or Ns
8		Ac in the form of Cu-shaped tufts or Ac with turrets
9		Ac of a chaotic sky, usually at different levels; patches of dense Ci are usually present also

Code No.	C_H	Description (Abridged From International Code) 3
1		Filaments of Ci, or "mares tails," scattered and not increasing
2		Dense Ci in patches or twisted sheaves, usually not increasing, sometimes like remains of Cb; or towers or tufts
3		Dense Ci, often anvil-shaped, derived from or associated with Cb
4		Ci, often hook-shaped, gradually spreading over the sky and usually thickening as a whole
5		Ci and Cs, often in converging bands, or Cs alone; generally overspreading and growing denser; the continuous layer not reaching 45° altitude
6		Ci and Cs, often in converging bands, or Cs alone; generally overspreading and growing denser; the continuous layer exceeding 45° altitude
7		Veil of Cs covering the entire sky
8		Cs not increasing and not covering entire sky
9		Cc alone or Cc with some Ci or Cs, but the Cc being the main cirriform cloud

PRECIPITATION AND WEATHER GROUP INDICATORS 4

i_R Precipitation Data

1 included
2 included but not in this section of the message
3 omitted—precipitation amount = 0
4 omitted—precipitaton amount not available

i_X Weather Data and Station Type

Code	Description	Station type
1	included	manned station
2	omitted—no significant weather	manned station
3	omitted—observed data not available	manned station
4	included	automatic station
5	omitted—no significant weather	automatic station
6	omitted—observed data not available	automatic station

SKY COVER 5

Code No.	N	SKY COVER
0	○	No clouds
1	⦶	One tenth or less, but not zero
2	◔	Two-tenths to three-tenths
3		Four-tenths
4	◑	Five-tenths
5		Six-tenths
6	◕	Seven-tenths to eight-tenths
7		Nine-tenths or overcast with openings
8	●	Completely overcast (ten-tenths)
9	⊗	Sky obscured

SKY COVER (Low And/Or Middle Clouds) 6

N_h	(Low And/Or Middle Clouds) SKY COVER
0	No clouds
1	One tenth or less, but not zero
2	Two-tenths to three-tenths
3	Four-tenths
4	Five-tenths
5	Six-tenths
6	Seven-tenths to eight-tenths
7	Nine-tenths or over cast with openings
8	Completely overcast
9	Sky obscured

HEIGHT IN FEET (Approximate) 7

h	(Approximate) HEIGHT IN FEET
0	0–100
1	100–300
2	300–600
3	600–900
4	900–1900
5	1900–3200
6	3200–4900
7	4900–6500
8	6500–8000
9	At or above 8,000, or no clouds

8

Code No.	a PRESSURE TENDENCY		Code No.	a	
0	Rising, then falling; same as or higher than 3 hours ago		4	Steady; same as 3 hours ago	
1	Rising, then steady; or rising, then rising more slowly	Barometric pressure now higher than 3 hours ago	5	Falling, then rising; same as or lower than 3 hours ago	
2	Rising steadily, or unsteadily		6	Falling, then steady; or falling, then falling more slowly	Barometric pressure now lower than 3 hours ago
3	Falling or steady, then rising; or rising, then rising more rapidly		7	Falling steadily, or unsteadily	
			8	Steady or rising, then falling; or falling, then falling more rapidly	

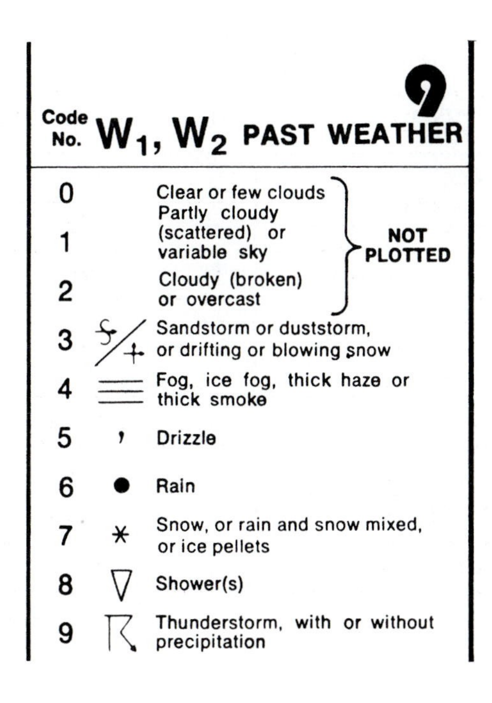

9

Code No.	W_1, W_2 PAST WEATHER	
0	Clear or few clouds	NOT PLOTTED
1	Partly cloudy (scattered) or variable sky	NOT PLOTTED
2	Cloudy (broken) or overcast	NOT PLOTTED
3	Sandstorm or duststorm, or drifting or blowing snow	
4	Fog, ice fog, thick haze or thick smoke	
5	Drizzle	
6	Rain	
7	Snow, or rain and snow mixed, or ice pellets	
8	Shower(s)	
9	Thunderstorm, with or without precipitation	

WW PRESENT WEATHER (Descriptions abridged from International Code)

	0	1	2	3	4
00	Cloud development NOT observed or NOT observable during past hour	Clouds generally dissolving or becoming less developed during past hour	State of sky on the whole unchanged during past hour	Clouds generally forming or developing during past hour	Visibility reduced by smoke
10	Light fog (mist)	Patches of shallow fog at station, NOT deeper than 6 feet on land	More or less continuous shallow fog at station, NOT deeper than 6 feet on land	Lightning visible, no thunder heard	Precipitation within sight, but NOT reaching the ground
20	Drizzle (NOT freezing) or snow grains (NOT falling as showers) during past hour, but NOT at time of observation	Rain (NOT freezing and NOT falling as showers) during past hour, but NOT at time of observation	Snow (NOT falling as showers) during past hour, but NOT at time of observation	Rain and snow or ice pellets (NOT falling as showers) during past hour, but NOT at time of observation	Freezing drizzle or freezing rain (NOT falling as showers) during past hour, but NOT at time of observation
30	Slight or moderate dust storm or sandstorm, has decreased during past hour	Slight or moderate dust storm or sandstorm, no appreciable change during past hour	Slight or moderate dust storm or sandstorm has begun or increased during past hour	Severe dust storm or sandstorm, has decreased during past hour	Severe dust storm or sandstorm, no appreciable change during past hour
40	Fog or ice fog at distance at time of observation, but NOT at station during past hour	Fog or ice fog in patches	Fog or ice fog, sky discernible, has become thinner during past hour	Fog or ice fog, sky NOT discernible, has become thinner during past hour	Fog or ice fog, sky discernible, no appreciable change during past hour
50	Intermittent drizzle (NOT freezing), slight at time of observation	Continuous drizzle (NOT freezing), slight at time of observation	Intermittent drizzle (NOT freezing), moderate at time of observation	Continuous drizzle (NOT freezing), moderate at time of observation	Intermittent drizzle (NOT freezing), heavy at time of observation
60	Intermittent rain (NOT freezing), slight at time of observation	Continuous rain (NOT freezing), slight at time of observation	Intermittent rain (NOT freezing), moderate at time of observation	Continuous rain (NOT freezing), moderate at time of observation	Intermittent rain (NOT freezing), heavy at time of observation
70	Intermittent fall of snowflakes, slight at time of observation	Continuous fall of snowflakes, slight at time of observation	Intermittent fall of snowflakes, moderate at time of observation.	Continuous fall of snowflakes, moderate at time of observation	Intermittent fall of snowflakes, heavy at time of observation
80	Slight rain shower(s)	Moderate or heavy rain shower(s)	Violent rain shower(s)	Slight shower(s) of rain and snow mixed	Moderate or heavy shower(s) of rain and snow mixed
90	Moderate or heavy shower(s) of hail, with or without rain, or rain and snow mixed, not associated with thunder	Slight rain at time of observation; thunderstorm during past hour, but NOT at time of observation	Moderate or heavy rain at time of observation; thunderstorm during past hour, but NOT at time of observation	Slight snow, or rain and snow mixed, or hail at time of observation; thunderstorm during past hour, but NOT at time of observation	Moderate or heavy snow, or rain and snow mixed, or hail at time of observation; thunderstorm during past hour, but NOT at time of observation

10

5	6	7	8	9
Haze	Widespread dust in suspension in the air, NOT raised by wind, at time of observation	Dust or sand raised by wind at time of observation	Well - developed dust whirl(s) within past hour	Dust storm or sandstorm within sight of or at station during past hour
Precipitation within sight, reaching the ground but distant from station	Precipitation within sight, reaching the ground, near to but NOT at station	Thunder storm, but no precipitation at the station	Squall(s) within sight during past hour or at time of observation	Funnel cloud(s) within sight of station at time of observation
Showers of rain during past hour, but NOT at time of observation	Showers of snow, or of rain and snow, during past hour, but NOT at time of observation	Showers of hail, or of hail and rain, during past hour, but NOT at time of observation	Fog during past hour, but NOT at time of observation	Thunderstorm (with or without precipitation) during past hour, but NOT at time of observation
Severe dust storm or sandstorm has begun or increased during past hour	Slight or moderate drifting snow, generally low (less than 6 ft)	Heavy drifting snow, generally low	Slight or moderate blowing snow, generally high (more than 6 ft)	Heavy blowing snow, generally high
Fog or ice fog, sky NOT discernible, no appreciable change during past hour	Fog or ice fog, sky discernible, has begun or become thicker during past hour	Fog or ice fog, sky NOT discernible, has begun or become thicker during past hour	Fog depositing rime, sky discernible	Fog depositing rime, sky NOT discernible
Continuous drizzle (NOT freezing), heavy at time of observation	Slight freezing drizzle	Moderate or heavy freezing drizzle	Drizzle and rain, slight	Drizzle and rain, moderate or heavy
Continuous rain (NOT freezing), heavy at time of observation	Slight freezing rain	Moderate or heavy freezing rain	Rain or drizzle and snow, slight	Rain or drizzle and snow, moderate or heavy
Continuous fall of snowflakes, heavy at time of observation	Ice prisms (with or without fog)	Snow grains (with or without fog)	Isolated starlike snow crystals (with or without fog)	Ice pellets or snow pellets
Slight snow shower(s)	Moderate or heavy snow shower(s)	Slight shower(s) of snow pellets, or ice pellets with or without rain, or rain and snow mixed	Moderate or heavy shower(s) of snow pellets, or ice pellets, or ice pellets with or without rain or rain and snow mixed	Slight shower(s) of hail, with or without rain or rain and snow mixed, not associated with thunder
Slight or moderate thunderstorm without hail, but with rain and/or snow at time of observation	Slight or moderate thunderstorm, with hail at time of observation	Heavy thunderstorm, without hail, but with rain and/or snow at time of observation	Thunderstorm combined with dust storm or sandstorm at time of observation	Heavy thunderstorm with hail at time of observation

ff WIND SPEED **11**

Symbol	Knots	miles per hour
	Calm	Calm
	1–2	1–2
	3–7	3–8
	8–12	9–14
	13–17	15–20
	18–22	21–25
	23–27	26–31
	28–32	32–37
	33–37	38–43
	38–42	44–49
	43–47	50–54
	48–52	55–60
	53–57	61–66
	58–62	67–71
	63–67	72–77
	68–72	78–83
	73–77	84–89
	103–107	119–123

Canadian Soil

THE CANADIAN SOIL CLASSIFICATION

The system of soil classification used in Canada is broadly similar in structure and purpose to the Soil Taxonomy system used in the United States. Both systems are hierarchical and are based on observable characteristics of the soil profile, particularly the nature and arrangement of the horizons. In contrast to Soil Taxonomy, however, the Canadian system deals only with soils found within Canada. It therefore gives greater emphasis to soils of cold climates and omits warm climate soils such as Oxisols and Ultisols. There are 9 soil orders in the Canadian system, each of which is described below. Each order is divided into 2 to 4 great groups, omitting the intervening level of suborder found in Soil Taxonomy.

Regosolic Soils

Regosols are young soils that show little or no evidence of pedogenic processes, and would be classified as Entisols in Soil Taxonomy. They are found in any place where the normal soil-forming processes do not have, or have not yet had, a chance to act. In Canada this includes extremely cold and dry climates, coarse-textured parent materials such as moraines, and recently deposited alluvial materials. They cover only about 1 percent of the land surface and are associated with a wide variety of climates and vegetation types. Until a revision of the classification in 1978 this order also included the present-day Cryosols, in which development is hindered by the presence of permafrost.

Cryosolic Soils

Cryosols are cold soils that contain permafrost within 1 meter of the soil surface (2 meters if the surface layers have been strongly disrupted by freeze-thaw action). Cryosols are the most extensive soils in Canada, covering about 40 percent of the land surface. They are dominant within the Arctic circle and extend, mixed with Brunisols, onto the northern part of the Canadian shield. The subdivision into great groups is on the basis of the degree of disruption by freeze-thaw action, or *cryoturbation*, and the nature of the soil material. All Cryosols have shallow rooting zones because the permafrost is close to the surface, and the low temperatures restrict the diversity and activity of soil organisms. The natural vegetation of Cryosols is arctic tundra in the north, grading into boreal forest on the Canadian Shield. The soils are generally unsuitable for agriculture. In the classification system of Soil Taxonomy, Cryosols are classified mainly as Entisols or Inceptisols that have mean annual temperatures below 0°C (32°F) in the root zone.

Organic Soils

Organic soils are composed mainly of decaying organic materials (over 30 percent organic matter by weight), and many of them are constantly waterlogged. Organic soils that have permafrost within 1 meter of the surface are classified separately as Organic Cryosols, a great group within the Cryosol order. Organic soils are subdivided on the basis of the thickness and properties—such as bulk density and porosity—of the organic material. In Canada, organic soils are mainly found within and close to forested areas and have their greatest extent in the shield area to the south of Hudson Bay. Organic soils are classified as Histosols in Soil Taxonomy.

Podzolic Soils

Podzolic soils are directly comparable to the Spodosols of Soil Taxonomy, are formed by the pedogenic regime of podsolization, and are the second most extensive soil order in Canada. They develop over the intensively leached, coarse-grained parent materials that cover the southern and south-eastern Canadian shield. A smaller area of podzolic soils is found in the mountains of the west coast, where volcanic ash is an important component of the topsoil. The dominant feature of all podzolic soils is the presence of an illuviated B horizon; the great groups are differentiated on the basis of this horizon's characteristics. The majority of podzolic soils in Canada are Humo-Ferric Podzols and contain relatively high amounts of organic carbon, and iron and aluminum. They support a natural vegetation of productive boreal forest. A small area of podzols is farmed for pasture, forage, and some grain crops in the milder climate of southeastern Canada, mainly in Quebec.

The Extent of Soil Orders and Rockland in Canada

SOIL ORDER	AREA		PERCENT OF TOTAL LAND AREA
	SQUARE KM	SQUARE MILES	
Brunisolic	789,780	304,855	8.6
Chernozemic	468,190	180,721	5.1
Cryosolic	3,672,080	1,417,431	40.1
Gleysolic	117,143	45,217	1.3
Luvisolic	809,046	312,293	8.8
Organic	373,804	144,286	4.1
Podzolic	1,429,111	551,633	15.6
Regosolic	73,442	28,349	0.8
Solonetzic	72,575	28,014	0.7
Rockland	1,375,031	3,543,548	15.0
Total	**9,180,202**	**6,556,347**	**100.0**

Luvisolic Soils

Luvisols have an illuvial B horizon in which clay has accumulated; they are equivalent to the Alfisol order in Soil Taxonomy. They are found primarily in the western Cordillera and in southern Ontario, and are mixed with other soil orders on the interior plains between the Canadian shield and the western mountains. In the subarctic areas Luvisols are unproductive, but in warmer climates they are well suited to productive forestry and, in the best areas, can be used for arable agriculture. Compared to the Chemozemic soils they are low in organic matter and require larger doses of fertilizer to maintain production. The short growing season also limits the choice of crop.

Brunisolic Soils

Brunisolic soils show more evidence of pedogenesis than Regosols but have not yet developed the diagnostic horizons that allow them to be firmly placed in a "mature" soil order. They are therefore at an intermediate stage of development and would be classified as Inceptisols in Soil Taxonomy. In Canada, they are found mainly but not exclusively under forest vegetation and have a typical brown color. This color indicates that some minerals present have already been altered by weathering, but that some weatherable minerals still remain. They are divided into great groups on the basis of the thickness of the A horizon and the pH. Brunisols are most often found in association with Podzolic and Luvisolic soils and in areas with large amounts of exposed rock.

Chernozemic Soils

Chernozemic soils have dark, well-structured A horizons that in the natural state support grassland or a mixture of grassland and trees. They are the approximate equivalent of cold Mollisols in Soil Taxonomy. Chernozemic soils are dominant in the southern part of the Interior Plains (between the Canadian Shield and the western Cordillera) that are the northern extension of the United States prairies. Chernozems cover only 5.1 percent of Canada's land surface, but support about half of the country's agricultural acreage. The soils are usually frozen for some period of the winter and are usually dry for some period of the summer. The great groups in this soil order are differentiated on the basis of the moisture regime (map). A natural vegetation of parkland—mixed trees and grass—is supported in the humid areas; in the drier areas trees are present only along watercourses. Very little natural vegetation cover remains on this soil order, however, as most of these soils are cropped or, in the drier areas, used for grazing.

Solonetzic Soils

Solonetz is a Russian word meaning salty. Solonetzic soils develop over saline parent materials after salinization has ceased. They cover only 0.7 percent of Canada's land surface and are found most frequently with Chernozems in the Interior Plains. Downward percolation of water carries dissolved salts and clay particles to the B horizon. The B horizon typically has a high pH (up to 10), tends to be hard when dry, and has a columnar or prismatic structure. It may also impede drainage so that the A horizon is wet as well as acid. The differentiation of A and B horizons is more or less distinct in different great groups; continued leaching leads to its disintegration. In their natural state solonetzic soils are unsuitable for cropping, and most of the 2.4 to 3.2 million hectares (6 to 8 million acres) they cover are used for grazing. Some experimental work has been carried out in which deep plowing is used to mix the alkali B horizon with the acidic A horizon.

Index

In this Index, **boldface** numbers indicate the place where a key term is defined; *italic* numbers indicate where an item occurs on an illustration. Combined ***italic boldface*** indicates both occur on one page.

U

V

W

X

Y

Z

ILLUSTRATION ACKNOWLEDGMENTS

Every attempt has been made to discover and contact the copyright holders of illustrations used in this book and to obtain their permission. If anyone holds a copyright that is not acknowledged in the following list, they should contact the publishers.

1-opener a National Aeronautics and Space Administration (NASA)
1-opener b National Wildlife Federation
1-opener c United States Geological Survey (USGS)
1-opener d USGS
1-3 Neil Rabinowitz
1-4 USGS
1-5 Michael Bradshaw
1-6 Pat Watson
1-7 NASA, Michael Helfert
1-8 NASA
1-9 USGS
Box 1 Photo: USGS
Diagram: Geographics, Portland OR

2-opener Colin Monteath/Mountain Camera
2-1 Neil Rabinowitz/Wayne Reckard
2-4 A Smith, University of Plymouth
2-5 A Smith, University of Plymouth
2-11 USGS
2-13 USGS
2-16 Natural Environment Research Council, United Kingdom
2-18 Natural Environment Research Council, United Kingdom
2-19 Natural Environment Research Council, United Kingdom
Inv 2 Satellite image: Mark Abbott, Oregon State University
Photo: Greg Vaughn/Tom Stack & Associates

3-opener Robert Frerck/Odyssey Production
3-1 NASA, Michael Helfert
3-4 V Ramanathan et al (1985), *Journal of Geophysical Research* 90:5557-5566; J Hansen et al (1988), *Journal of Geophysical Research* 93:9241-9364, Both copyright by the American Geophysical Union.
3-7 Utah Travel Council
Box 3-1 Photo: Pat Watson
Ozone: NASA, Mark Schoeberl
Box 3-2 Space Shuttle: NASA, Michael Helfert
Map: Witherby & Co Ltd

4-opener Robert Holmes
4-6 US Department of Agriculture (USDA)
4-7 Michael Bradshaw
4-8b R Barry and Chorley, *Weather and Climate*, Methuen & Co (1987)
4-11 NASA
4-17 NASA, Joel Susskind
4-18 NASA, Bill Rossow
4-19 NASA, Gene Feldman
Inv 4 Satellite images: Natural Environment Research Council, United Kingdom

5-opener John Bova/PhotoResearchers, Inc
5-1 Neil Rabinowitz
5-5 National Oceanic and Atmospheric Administration (NOAA)
5-8 Michael Bradshaw
5-10 Michael Bradshaw
5-11 NASA, Bill Rossow
5-18 John A Whitehead, Giant Ocean Cataracts, *Scientific American*, February 1989.
Copyright by Scientific American, Inc. All rights reserved.
5-19 NASA, Peter Cornillon

6-opener Bill Ross/Woodfin Camp & Associates
6-4 San Francisco Convention & Visitors Bureau
6-9a S D Burt
6-9b Ronald L Holle
6-9c Marion Foreman, Royal Meteorological Society
6-9d J F P Galvin
6-11 NASA, Michael Helfert
6-13 Michael Bradshaw
Box 6-2 Maps: NOAA data
Inv 6 Maps and images: Tom Carroll, NOAA

7-opener Cotton Coutson/Woodfin Camp & Associates
7-6 NOAA data
7-8 R K Pilsbury
7-9 P P K Verschure
7-10 John T Snow, The Tornado, *Scientific American*, April 1984. Copyright by Scientific American, Inc. All rights reserved.
7-11 Douglas Volz
7-12 NOAA data
7-14 NASA, Michael Helfert
7-16 NASA, Michael Helfert
7-17 NOAA
7-18 NOAA data
7-19 NASA, A Negri
7-20 University of Dundee
7-23 NOAA

8-opener Nicholas Foster/The Image Bank
8-2 United States Department of Defense
8-7 NASA
8-14 Source of information courtesy National Geographic Society
8-18 Paul Bradshaw
8-21 USDA
8-31a,b William P Lowry, The Climate of Cities, *Scientific American*, August 1967. Copyright by Scientific American, Inc. All rights reserved.
8-31c T J Chandler, *The Climate of London*, Hutchinson (1965).
Inv 8 Greg Vaughn/Tom Stack & Associates
Excerpt from Emilio Moran, *Developing the Amazon*, Indiana University Press (1981)

9-opener David Muench
9-1 USDA
9-3 *Earthquest*, Spring 1991, 5:1. Office for Interdisciplinary Earth Studies.
9-4 NOAA
9-6 H H Lamb; *Climate, History and the Modern World*, Methuen & Co (1982)
9-9 J M Barnola et al, Vostok Ice Core. Reprinted by permission from *Nature* 329:408-14; Copyright 1987 Macmillan Magazines Limited.
9-11 Michael Bradshaw

10-opener Robert E Ford/Terraphotographics/BPS
10-4a,b,d Hubbard Scientific Co
10-4c Science Graphics, Inc
10-6a-h Hubbard Scientific Co
10-7 Reprinted with the permission of Macmillan Publishing Company from *The Earth's Dynamic Systems*, Fifth Edition, by W. Kenneth Hamblin. Copyright 1989 by Macmillan Publishing Company.
10-08 John-Paul Lenney
10-11 John-Paul Lenney
10-13 Michael Bradshaw
10-14a Courtesy Macmillan Publishing

Co, Hamblin: Earth's Dynamic Systems, 5th Ed (1989)
10-14b USGS
10-15 Landform Slides
10-17 USGS
10-18 USGS
10-20 M J Selby, *Earth's Changing Surface*, Oxford University Press (1985)
10-23 B J Skinner, S C Porter, *The Dynamic Earth*, Wiley (1989)
10-24 B J Skinner, S C Porter, *The Dynamic Earth*, Wiley (1989)
Box 10-1 Map: G K Gilbert, USGS
Photos: USDA; Michael Bradshaw

11-opener Gregory G Dimuian/Photo Researchers
11-2 B Isacks et al, Seismicity and the new Global Tectonics, *Journal of Geophysical Research*, 73:5855-99 (1968), copyright by the American Geophysical Union.
11-3 Plate velocity data from NASA
11-6 USGS
11-8 USGS
11-14 NASA
Box 11-1 Map: NOAA/USGS
Photos: USGS

12-opener John Cleare
12-1 Bruce C Heezen, Marie Tharp (1977), copyright by Marie Tharp 1977. Reproduced by permission of Marie Tharp, 1 Washington Avenue, South Nyack, NY 10960
12-8 Michael Summerfield, *Global Geomorphology*, Longman (1991)
12-9 NASA, Michael Helfert
12-11 B J Skinner, S C Porter, *The Dynamic Earth*, Wiley (1989)
12-12 Adapted from Robert S Dietz, Geosynclines, Mountains and Continent-building, *Scientific American*, March 1972. Copyright 1972 by Scientific American, Inc. All rights reserved.
12-14 R Damian Nance, Thomas R Worsley, Judith B Moody, The Supercontinent Cycle, *Scientific American*, July 1988. Copyright 1988 by Scientific American, Inc. All rights reserved.
Box 12-1 Map and photo: USGS

13-opener Colin Monteath/Mountain Camera
13-1 NASA, Michael Helfert
13-2 H F Garner, *The Origin of Landscapes*, Oxford University Press (1974)
13-4 Reprinted with the permission of Macmillan Publishing Company from *The Earth's Dynamic Systems*, Fifth Edition, by W. Kenneth Hamblin. Copyright 1989 by Macmillan Publishing Company. Macmillan (1985)
13-7 Landform Slides
13-8 USDA
13-9 M Peter Gilday
13-10 N M Strakhov, *Principles of Lithogenesis*, Oliver & Boyd (1967)
13-11 M A Carson, M J Kirkby, *Hillslope Form and Process*, Cambridge University Press (1972)
13-12,
13-13 D J Varnes, *Slope Movement Types and Processes*, TRB Special Report 176, Transportation Research Board, National Research Council, Washington, D.C. (1978).
13-14 A Rapp, Recent Development of Mountain Slopes in Scandinavia, *Geografiska Annaler* 42:71-200 (1960), by permission of Scandinavian University Press, Oslo, Norway. Copyright is held by Scandinavian University Press.
13-16 Landform Slides
13-18 Timothy O'Keefe/Tom Stack & Associates
Box 13-1 USGS

14-opener Kenneth Murray/Photo Researchers, Inc
14-4 USGS
14-5 B J Skinner, S C Porter, *The Dynamic Earth*, Wiley (1989)
14-8 USDA
14-12 Michael Bradshaw
14-15 USGS
14-17 Landform Slides
14-18 USDA
14-20 Michael Bradshaw
14-21 USDA
14-22 Neil Rabinowitz
14-24 W B Bull, The Alluvial Fan Environment, *Progress in Physical Geography* 1:222-70 (1977). Edward Arnold.
14-25 Peter Kresan
14-30 Landform Slides
14-35 M G Wolman, A Cycle of Sedimentation and Erosion in Urban River Channels, *Geografiska Annaler* 49A:385-95 (1967), by permission of Scandinavian University Press, Oslo, Norway. Copyright is held by Scandinavian University Press.
Box 14-1 Photo: Joann Mossa
Box 14-2 K Richards, *Rivers*, Methuen & Co (1982)

15-opener Stephen J Krasemann/Nature Conservancy
15-1 Landform Slides
15-3 USGS
15-7 NASA, Michael Helfert
15-10 Landform Slides
15-13 M J Selby, *Earth's Changing Surface*, Oxford University Press (1985)
15-14 John-Paul Lenney
15-15 Michael Bradshaw
15-18 Landform Slides
15-19 R F Flint, *Glacial and Quaternary Geology*, Wiley (1971)
15-20a USGS
15-20b Landform Slides
15-22 Landform Slides
15-23 Landform Slides
15-24 Michael Bradshaw
Box 15-1 Landform Slides
Inv 15 Photos: David Crook; USGS; Neil Rabinowitz

16-opener OMIKRON/Photo Researchers, Inc; Excerpt from John Steinbeck, *Grapes of Wrath*, William Heinemann
16-1 USDA
16-3 NASA, Michael Helfert
16-8 J T Hack, Dunes of the Navajo Country, *Geographical Review* 31:240-63 (1941). Reproduction by permission of the American Geographical Society.
16-11 USGS
16-12 M J Selby, *Earth's Changing Surface*, Oxford University Press (1985)
16-15 Landform Slides
16-16 NASA, Michael Helfert
Box 16-1 USDA

17-opener David Muench
17-3 USGS
17-4 Scott Blackman/Tom Stack & Associates
17-6 J L Davies, *Geographical Variation in Coastal Development*, Oliver & Boyd (1980)
17-9 J S Pethick, *Introduction to Coastal Geomorphology*, Edward Arnold (1984)
17-10 D & R Bretzfelder
17-12 USDA
17-13 M Peter Gilday
17-14 R J Small, *The Study of Landforms*, Cambridge University Press (1970)
17-15 Joann Mossa
17-17, 17-18 Stephen P. Leatherman, *Barrier Island Handbook*, University of Maryland (1988)
17-19 NASA, Michael Helfert
17-20 W E Galloway, Process Framework for Describing the Morphologic and Stratigraphic Evolution of the Delatic Depositional Systems, in M L Broussard, ed, *Deltas, Models for Exploration*. Houston Geological Society (1968).
17-21 NASA, Michael Helfert
17-23 M J Selby, *Earth's Changing Surface*, Oxford University Press (1985)
17-25 NASA, Michael Helfert
17-27 Neil Rabinowitz
17-28 USGS
17-29 Michael Summerfield, *Global Geomorphology*, Longman (1991)
Box 17-1 Photos: NASA, Michael Helfert

18-opener Michael and Elvan Habicht/ Earth Scenes
18-1 Michael Bradshaw
18-2 R U Cooke, J C Doornkamp, *Geomorphology in Environmental Management*, Oxford University Press (1990)
18-3 A J Parsons, *Hillslope Form*, Routledge (1988)
18-5a,c,d Michael Bradshaw
18-8 Landform Slides
18-9 Julius Büdel, *Climatic Geomorphology*, University of Princeton Press (1982)
18-14 Virginia Braley

18-17 USGS
18-21 J Kelly Beatty

19-opener Mark Wright/Photo Researchers, Inc
19-2b John Shaw/Tom Stack & Associates
19-2c Mary Clay/Tom Stack & Associates
19-2c1 Michael Bradshaw
19-2c2 National Parks Service
19-2d Joe McDonald/Tom Stack & Associates
19-7 National Zoo
19-11 E R Degginger/Animals Animals
19-15 Doug Allen/Animals Animals
19-16 Blackwell Scientific Publications
19-18 Richard Gross
19-19a USDA
19-19b Neil Rabinowitz
Box 19-1 Photos: Clyde Walker

20-opener Robert Frerck/Odyssey Production
20-7 USDA
20-11 Hari Eswaran
20-12 Hari Eswaran
20-13 Hari Eswaran
20-14a Hari Eswaran
20-14b Kevin Magee/Tom Stack & Associates
20-15 Hari Eswaran
20-17 USDA
20-18a USDA
20-19 Ruth Weaver
20-20 Neil Rabinowitz
20-21 USDA
20-22a,b,d USDA
20-22c Neil Rabinowitz

21-opener David Muench
21-3a David A Furley, Walter W Newey, *Geography of the Biosphere,* Butterworth-Heinemann Ltd (1983)
21-3b Missouri Botanical Garden
21-4b Missouri Botanical Garden
21-5a Carol Gracie/New York Botanical Garden
21-5b Mella Panzella/Animals, Animals
21-5c Michael Dick/Animals, Animals
21-6 Carol Gracie/New York Botanical Garden
21-7, 21-8 USDA
21-10 Billy E Barnes/Stock Boston
21-11 David W Johnson
21-12 David L Brown/Tom Stack & Associates
21-14a David W Johnson
21-14b J Kelly Beatty
21-15 David A Furley, Walter W Newey, *Geography of the Biosphere,* Butterworth-Heinemann Ltd (1983)
21-16 David W Johnson
21-17a,b Stewart Halperin
21-18 Doug Sokell/Tom Stack & Associates
21-19 Richard Gross
21-20 Neil Rabinowitz
21-23 Neil Rabinowitz
Box 21-1 Photo: David W Johnson
Inv 12 Photo: Randall Hyman/Stock Boston